Proceedings of the
3rd Asia-Pacific Conference

FEW-BODY PROBLEMS IN PHYSICS

Proceedings of the
3rd Asia-Pacific Conference

FEW-BODY PROBLEMS IN PHYSICS

Editors

Yupeng Yan

C Kobdaj

P Suebka

Suranaree University of Technology
Thailand

World Scientific

NEW JERSEY • LONDON • SINGAPORE • BEIJING • SHANGHAI • HONG KONG • TAIPEI • CHENNAI

Published by

World Scientific Publishing Co. Pte. Ltd.
5 Toh Tuck Link, Singapore 596224
USA office: 27 Warren Street, Suite 401-402, Hackensack, NJ 07601
UK office: 57 Shelton Street, Covent Garden, London WC2H 9HE

British Library Cataloguing-in-Publication Data
A catalogue record for this book is available from the British Library.

FEW-BODY PROBLEMS IN PHYSICS
Proceedings of the Third Asia-Pacific Conference

ISBN 13 978-981-270-481-8
ISBN 10 981-270-481-7

Printed in Singapore by B & JO Enterprise

PREFACE

The Third Asia-Pacific Conference on Few-Body Problems in Physics (APFB05) was held from July 26 to 30 of 2005 in Nakhon Ratchasima, Thailand, with School of Physics, Suranaree University of Technology (SUT) as its host. It succeeded the first APFB meeting in Tokyo in 1999 and the second in Shanghai in 2002. The APFB05 was the first international conference in physics held in Thailand.

The APFB05 took place at a most appropriate time to let the participants to exchange very recent results from important laboratories and universities, to enhance contacts among the Asia-Pacific few-body-physics community and colleagues in other continents, to boost physics research and education in Thailand and countries in the region, and to introduce the culture and beauty of Thailand toscientists worldwide.

The scope of the conference covered research in the following fields: Few-nucleon systems (three-body forces and few-nucleon dynamics); Hadron structure and QCD; Exotic hadrons and atoms; Effective field theory in few-body physics; Electromagnetic and weak processes in few-body systems; Few-body dynamics in atoms, molecules, Bose-Einstein condensates and quantum dots; Few-body approaches to unstable nuclei, nuclear astrophysics and nuclear clustering aspects; Hypernucler physics (hadron-hyperon,hyperon-hyperon interactions); Relativity in few-body physics, and others.

Over 100 physicists from around the world participated in the scientific program of the APFB05. Many of the delegates were very senior and eminent scientists indeed (both from the theoretical and the experimental sectors), attracted to the conference variously by their scientific and personal relationship with the conference organizers, by the conference scientific program, or by a common interest in building science capacity in the region. Accordingly, the caliber of their contributions has ensured the high standard of the proceedings, and made its compilation a great pleasure.

Over 1000 students from 5 universities and 33 high schools in the northeastern region of Thailand attended the opening ceremony and other activities arranged by the conference. The atmosphere of the conference and especially the keynote lecture by Prof. Chen-Ning Yang (1957 Nobel laureate in Physics) on thematic melodies of 20-th century theoretical physics impressed them very much. Prof. Chen-Ning Yang is one of the few physicists whom the Thai pupils may know well. It is expected that more students would choose to study physics or related subjects for their university educations.

The APFB05 continued the convention set up in the APFB99 conference to offer an award with the aim of encouraging young scientists in this field. There appeared to be so many excellent young speakers that it was very hard to select only one to be awarded. After consultation among senior scientists, the award for the best presentation by a young scientist was given to Dr. Yu-Chiun Chen from the National Taiwan University.

While the essence of a conference and of the proceedings is the scientific material presented in talks and subsequent papers, there would not be a conference or proceedings without extensive sponsorship. The conference organizers greatly appreciate funding received from the various organizations displayed on the following page. Their contributions enabled the conference to support all of the participants.

We would like to thank all of the participants for attending the conference and contributing to the conference valuable talks. We would like to thank the International Advisory Committee, the Scientific Program Committee, and the members of the Local Organizing Committee for their contribution to the conference. We would like to thank SUT personnels and especially students from School of Physics for helping to organize the conference. We would like to thank Mr. Ayut Limphirat, Chakrit Nualchimplee, and Chalump Oonariya for getting all contributions in the correct format.

The Editors
Nakhon Ratchasima, November, 2005

ORGANIZING COMMITTEES

Local Organizing Committee

P. Suebka (SUT) – Chairman

B. Asavapibhop (Chula)
A. Chaiyasena (SUT)
S. Cheedket (Taksin)
N. Kheaomaingam (Burapha)
C. Kobdaj (SUT)
P. Manyum (SUT)
T. Ishii (SUT)
S. Limpijumnong (SUT)
E. B. Manoukian (SUT)
P. Pairor (SUT)
W. Pairsuwan (SUT)
S. Phatisena (SUT)
S. Rugmai (SUT)
S. Rujirawat (SUT)
P. Songsiriritthigul (SUT)
M. Thammachoti (SUT)
A. Tongraar(SUT)
W. Uchai (SUT)
J. Widjaja (SUT)
Y. Yan (SUT)

Scientific Program Committee

H. W. Fearing (TRIUMF)
E. Hiyama (Nara)
H. Kamada (Kitakyushu)
D. T. Khoa (Vietnam)
Y. Koike (Hosei)
P. Ko (KAIST)
C. Kobdaj (SUT)
H. K. Lee (SUT)
B. Q. Ma (Beijing)
T. Mart(Indonesia)
D. P. Min (Seoul)
K. S. Myint (Myanmar)
S. Oryu (Tokyo)
T.-S. Park (KIAS)
K. Sagara (Kyushu)
K. Sagarik (SUT)
P. Suebka (SUT)
F. G. Wang (Nanjing)
S. N. Yang (Taipei)
Z. Y. Zhang (IHEP)
B. S. Zou (IHEP)
Y. Yan (SUT)

International Advisory Committee

S. K. Adhikari (Sao Paulo)
C. G. Bao (Zhongshan)
V. V. Burov (JINR)
C. W. de Jager (Jlab)
H. W. Fearing (TRIUMF)
B. F. Gibson (Los Alamos)
R. M. Godbole (Bangalore)
P. Hoodboy (Quaid-E-Azam)
N. Kalantar-Nayestanaki (KVI)
K. Kar (Saha)
Y. E. Kim (Purdue)
Y. Koike (Hosei)
T.-S.H. Lee (Argonne)
W. G. Li (IHEP)
R. Machleidt (Idaho)
B. McKellar (Melbourne)
Ulf Meissner (Bonn)
D. P. Min (Seoul)
M. Oka (Tokyo)
E. Oset (Valencia)
J. M. Richard (Grenoble)
H. Sakai (Tokyo)
W. Saya-Kanit (Chulalongkorn)
W. Q. Shen (SINR)
H. Q. Song (SINR)
A. W. Thomas (JLab)
W. Tornow (TUNL)
T. Vilaithong (Chiang Mai)
F. G. Wang (Nanjing)
S. Yoksan (Srinakharinwirot)
B. L. G. Bakker (Vrije)
S. Boffi (Pavia)
K. T. Chao (Beijing)
A. Faessler (Tuebingen)
J. L. Friar (Los Alamos)
W. Gloeckle (Bochum)
F. Gross (TJNAF)
S. Ishikawa (Hosei)
M. Kamimura (Kyushu)
D. T. Khoa (Vietnam)
Y. S. Kim (Maryland)
H. K. Lee (Hanyang)
D. Leinweber (Adelaide)
B. Loiseau (Paris)
T. Mart (Indonesia)
R. D. McKeown (Caltech)
R. G. Milner (MIT)
S. Nagamiya (KEK)
S. Oryu (Tokyo)
W. Plessas (Graz)
K. Sagara (Kyushu)
P. U. Sauer (Hannover)
K. K. Seth (Northwestern)
S. A. Sofianos (South Africa)
P. Suebka (SUT)
L. Tomio (Sao Paulo)
W. T. H. van Oers (Manitoba)
R. Vinh Mau (Paris)
S. N. Yang (Taipei)
Z. Y. Zhang (IHEP)

PATRONAGES AND SPONSORSHIPS

Suranaree University of Technology

Commission on Higher Education

The Asia Pacific Center for Theoretical Physics

the Abdus Salam International centre of Theoretical Physics

PTT Public Company Limited

The Government Lottery Office

National Science and Technology Development Agency

Double A Paper Double Quality Paper

NanmeeBooks Publishing Company

National Synchrotron Research Center

GROUP PHOTO

CONFERENCE PROGRAM

July 26, 2005

Opening Session; 09:00 - 10:30

09:00 - 10:30 Opening Ceremony (Surapat II Building)

- Video presentation about Suranaree University of Technology (SUT)
- Welcome speech by The Honorable Mr. Pongpayom Wasaputi (Governor of Nakhon Ratchasima)
- Welcome speech by Prof. Dr. Tavee Lertpanyavit (SUT Rector)
- Reporting of the activities of the conference by Prof. Dr. Prasart Suebka (Dean, Institute of Science, SUT)
- Opening remarks by Prof. Dr. Wichit Srisa-an (President of the University Council)
- Key Note address on "Thematic melodies of 20-th century theoretical physics" by Prof. Dr. Chen Ning Yang (Nobel laureate in Physics)

10:30 - 11:00 **Tea Break** (Second Floor of Surasammanakhan)

Plenary Session I; 11:00 - 12:00

Chair: S. Oryu
11:00 - 11:30 Amand Faessler (Tübingen)
Description of hadrons in the Tübingen chiral quark model
11:30 - 12:00 A. W. Thomas (Jlab)
Strangeness Content of the Nucleon

12:30 - 13:50 **Lunch**

Plenary Session II; 13:50 - 15:50

Chair: B. F. Gibson
13:50 - 14:20 K. Sagara (Kyushu)
Anomalies in pd radiative capture and pd breakup reactions
14:20 - 14:50 Nasser Kalantar (KVI)
Experimental investigations of three-body systems at KVI
14:50 - 15:20 Masa Iwasaki (RIKEN)
Strange tribaryon and kaonic atom

15:20 - 15:50 B. S. Zou (IHEP)
"Exotic" hadron-hadron S-wave interactions

July 27, 2005

Plenary Session III; 08:00 - 10:00

Chair: W. Tornow
08:00 - 08:30 H. Stoecker (Frankfurt)
Jet correlations and quark matter
08:30 - 09:00 Shalev Gilad (MIT)
Studying $\gamma^* N \to \Delta \to \pi N$ with cross sections and polarization observables
09:00 - 09:30 Kamel Seth (Northwestern)
A review of the exciting developments in hadron spectroscopy-charmonium, bottomonium and the rest
09:30 - 10:00 L. Tomio (Sao Paulo)
Scaling in few-body nuclear physics

10:00 - 10:20 **Tea Break**

Plenary Session IV; 10:20 - 12:20

Chair: S. N. Yang
10:20 - 10:50 Aksel S. Jensen (Aarhus)
Three-body decay of nuclear resonances
10:50 - 11:20 T. S. Park (KIAS)
HEN process in effective field theory
11:20 - 11:50 B. Loiseau (Paris)
B decays into $\pi\pi K$ and $KK\overline{K}$: long-distance and final-state effects
11:50 - 12:20 Arun Saha (Jlab)
Study of the few nucleon systems with $(e, e'p)$ reactions

12:20 - 13:30 **Lunch**

Parallel Session A I; 13:30 - 15:30

Chair: V. B. Mandelzweig
13:30 - 14:00 S. N. Yang (Taipei)
Unitary model for the $\gamma p \to \gamma\pi^0 p$ reaction and the magnetic dipole moment of the $\Delta^+(1232)$
14:00 - 14:30 W. Tornow (TUNL)
Gamma-ray induced two- and three-body breakup of ^{3}He at low energies
14:30 - 15:00 Terry Mart (Indonesia)
Isobar model for $\gamma + p \to K^0 + \Sigma^+$ channel

15:00 - 15:30 Y. B. Dong (IHEP)
Photo- and electro-productions of the nucleon resonances in the point form relativistic quantum mechanics

Parallel Session B I; 13:30 - 15:30

Chair: P. Sauer
13:30 - 14:00 M. Kohno (Kyushu)
Isoscalar and isovector parts of the Σ single-particle potential from SCDW model analysis of (π^-, K^+) inclusive spectra
14:00 - 14:30 M. R. Robilotta (Sao Paulo)
Chiral symmetry and nuclear forces
14:30 - 15:00 Y. Fujiwara (Kyoto)
A practical method to solve the cut-off coulomb problem in the Lippmann-Schwinger RGM formalism
15:00 - 15:30 S. Ishikawa (Hosei)
Three-nucleon force effects in NN elastic and breakeup reactions

Parallel Session C I; 13:30 - 15:30

Chair: A. Dote
13:30 - 14:00 M. Kamimura (Kyushu)
Continuum-discretized coupled-channels method for four-body breakup reactions: application to ^{6}He+^{12}C and ^{6}He+^{209}Bi scattering
14:00 - 14:30 T. Fukuda (Osaka)
A new measurement of the ^{8}Li(α,n)^{11}B reaction for astrophysical interest
14:30 - 15:00 M. Bisset (Tsinghua)
Dalitz-esque treatment of new heavy particle pair production at the LHC
15:00 - 15:30 Marcus Bleicher (Frankfurt)
Transport model analysis of ultra-relativistic nucleus-nucleus collisions

15:30 - 15:50 **Tea Break**

Parallel Session A II; 15:50 - 17:30

Chair: T. Mart
15:50 - 16:10 Y. Shimizu (RCNP)
Spin correlation parameter C_{yy} of p + ^{3}He elastic backward scattering at intermediate energy
16:10 - 16:30 A. Deltuva (Lisboa)
Momentum-space treatment of Coulomb interaction in three-nucleon breakup reactions
16:30 - 16:50 E. Uzu (TUS)
Study of four-nucleon systems by four-body Faddeev-Yakubovsky equations
16:50 - 17:10 M. Hadizadeh (Tehran)

Four-body bound state calculations in 3-D approach (without angular momentum decomposition)
17:10 - 17:30 Tim Black (North Carolina)
Precision measurements of neutron scattering lengths in H, D, and ^{3}He

Parallel Session B II; 15:50 - 17:30

Chair: Z. B. Zou
15:50 - 16:10 Y. Yan (SUT)
Accurate evaluation of $\bar{p}D$ atoms
16:10 - 16:30 Akinobu Dote (KEK)
Systematic study of dense $\bar{K}$ nuclei with a revised $\bar{K}N$ potential
16:30 - 16:50 H. Kuboki (Tokyo)
Search for narrow dibaryon resonances via the $p + d$ scattering
16:50 - 17:10 T. Koike (RIKEN)
Cascade calculation of kaonic nitrogen atoms involving the electron refilling process
17:10 - 17:30 J. N. Hedditch (Adelaide)
1^{-+} exotic meson on the lattice using FLIC fermions

Parallel Session C II; 15:50 - 17:30

Chair: E. Hiyama
15:50 - 16:10 M. Yamaguchi (Hosei)
Faddeev three-cluster calculation of the helium isotopes including with core excitation
16:10 - 16:30 N. Furutachi (TUS)
Anti-symmetrized molecular dynamics simulation for O Isotopes
16:30 - 16:50 N. G. Kelkar (Los Andes)
Unstable bound states of η-mesic light nuclei
16:50 - 17:10 T. Watanabe (TUS)
Anti-symmetrized molecular dynamics by realistic NN potentials

18:00 - 20:30 **SUT 15th founding anniversary banquet**

July 28, 2005

Plenary Session V; 08:00 - 10:00

Chair: S. Ishikawa
08:00 - 08:30 S. Oryu (TUS)
Analyticity in two and three-body coulomb scattering in momentum space
08:30 - 09:00 Peter U. Sauer (Hannover)
The treatment of Coulomb in the description of three-nucleon reactions with two protons

09:00 - 09:30 Nobuhiro Yamanaka (Tokyo)
Time-dependent coupled channel studies of atomic coulomb three-body dynamics

09:30 - 10:00 Debades Bandyopadhyay (Saha)
Bose-Einstein condensation in neutron stars

10:00 - 10:20 **Tea Break**

Plenary Session VI; 10:20 - 12:20

Chair: A. W. Thomas

10:20 - 10:50 Emiko Hiyama (Nara)
Few-body aspects of strangness nuclear physics

10:50 - 11:20 K.S. Myint (Mandalay)
Structure of Kaonic nucleus $K^{-}pp$

11:20 - 11:50 Allena Opper (Ohio)
Charge symmetry breaking in few nucleon systems

11:50 - 12:20 Shung-ichi Ando (TRIUMF)
Neutron β-decay and electroweak processes of the deuteron in effective field theory

12:20 - 13:40 **Lunch**

Plenary Session VII; 13:40 - 15:40

Chair: K. S. Myint

13:40 - 14:10 Achim Czasch (Frankfurt)
Photon induced fragmentation of atoms and small molecules

14:10 - 14:40 Imam Fachruddin (Indonesia)
3N scattering at intermediate energies

14:40 - 15:10 H. Kamada (Kyushu)
Relativistic effects in neuteron-deuteron elastic scattering

15:10 - 15:40 H. Sakai (Tokyo)
The $n-d$ scattering compared to $p-d$ scattering at intermediate energies

15:40 - 16:00 **Tea Break**

Parallel Session A III; 16:00 - 18:00

Chair: Y. Fujiwara

16:00 - 16:20 J. Nagata (Hiroshima)
Energy-dependent phase-shift analyses for elastic pp scattering in COSY-data region

16:20 - 16:40 R. Alarcon (Arizona)
Polarization observables in deuteron electro-disintegration

16:40 - 17:00 C. H. Hyun (Sungkyunkwan)
Parity violation in *pp* scattering and vector-meson weak-coupling constants
17:00 - 17:20 Fu-Guang Cao (Massey)
The structure functions g_1 and g_2 of the nucleon in the meson cloud model
17:20 - 17:40 M. Nowakowski (Los Andes)
The electromagnetic potential of the neutron and its applications
17:40 - 18:00 S. C. Goyal (Agra)
Elastic properties of argon up to pressure 75 GPa

Parallel Session B III; 16:00 - 18:00

Chair: H. Bhang
16:00 - 16:20 A. Biegun (Silesia)
Three-nucleon force effects in observables for $\vec{d}p$ breakup at 130 MeV
16:20 - 16:40 Shigeyoshi Aoyama (Niigata)
Systematic analyses on cluster structures in light neutron-rich nuclei by using a new AMD approach
16:40 - 17:00 H. Nemura (RIKEN)
Full coupling dynamics of doubly strange hypernuclei
17:00 - 17:20 P. M. Milazzo (Italy)
Isotropic composition as a signature for different process leading to fragment production in midperipheral Ni+Al, Ni, Ag collisions at 30 MeV/nucleon
17:20 - 17:40 Mohammad Shoeb (India)
Alpha cluster model of ${}^{9}_{\Lambda}$Be and ${}^{10}_{\Lambda\Lambda}$Be hypernuclei and dispersive $\Lambda\alpha\alpha$ potential
17:40 - 18:00 M. Nasr (Libya)
Some interesting features of particles produced at high energy heavy ion collisions

Parallel Session C III; 16:00 - 18:00

Chair: H. Kamada
16:00 - 16:20 V. B. Mandelzweig (Reach)
Quazilinearization method and WKB
16:20 - 16:40 A. Sulaksono (Indonesia)
Regularity neutrino mean free path and relativistic mean field model
16:40 - 17:00 E. Ghanbari Adivi (Yazd)
Four-body Faddeev-Watson series description of proton-helium charge transfer process
17:00 - 17:20 Magnus Ogren (Lund)
Super-shell structure in two-component dilute fermionic gases
17:20 - 17:40 A. S. Kadyrov (Murdoch)
On scattering by arbitrary potentials

17:40 - 18:00 B. Desplanques (LPSC)
From factors in relativistic quantum mechanics approaches and Poincare space-time translation

July 29, 2005

Plenary Session VIII; 08:00 - 10:00

Chair: F. G. Wang

08:00 - 08:30 Serge Kox (LPSC-Grenoble)
Nucleon strange and axial form factors from PV experiments at Jlab

08:30 - 09:00 Frank Maas (Mainz)
Parity violating electron scattering at MAMI in Mainz

09:00 - 09:30 Iraj R. Afnan (Flinders)
The physics of the hypertriton

09:30 - 10:00 W. Mike Snow (Indiana)
Low energy neutrons on the weak interaction in few body systems

10:00 - 10:30 **Tea Break**

Plenary Session IX; 10:30 - 12:30

Chair: K. Seth

10:30 - 11:00 Nilanga Liyanage (Virginia)
Spin structure of the neutron and ^{3}He: new results from Jefferson Lab

11:00 - 11:30 Michael Kohl (MIT)
The charge form factor of the neutron at low Q^2

11:30 - 12:00 C. F. Perdrisat (William and Mary)
Proton form factor measurements at Jefferson Lab

12:00 - 12:30 Yu-Chun Chen (Teipei)
Two-photon exchange contribution to the elastic *ep* scattering at large momentum transfer with parton approach

12:30 - 14:00 **Lunch**

Plenary Session X; 14:00 - 16:00

Chair: R. Alarcon

14:00 - 14:30 Nurgali Takibayev (Kazakh)
The resonances of n, α, α system at astrophysical energy region

14:30 - 15:00 Zhongzhou Ren (Nanjing)
Few-body model of α decay and cluster decay

15:00 - 15:30 H. Bhang (Seoul)
Signatures of the three body process in the weak decay of Λ hypernuclei

15:30 - 16:00 L. Platter (Bonn)
Universal properties of four-body systems with large scattering length

16:00 - 16:20 **Tea Break**

July 30, 2005

Plenary Session XI; 08:00 - 10:00

Chair: M. Kamimura

08:00 - 08:30 L. Z. Bing (Zhongshan)
The method of fractional parentage coefficients applied to the spinor Bose-Einstein condensates

08:30 - 09:00 Yongle Yu (Lund)
Vortex structures in quantum dots at high magnetic fields

09:00 - 09:30 F. G. Wang (Nanjing)
Multi-quark study

09:30 - 10:00 Wen-Chen Chang (Teipei)
Recent results from LEPS/SPring-8 experiment

10:00 - 10:20 **Tea Break**

Plenary Session XII; 10:20 - 12:20

Chair: Amand Faessler

10:20 - 10:50 K. Hicks (Jlab)
Search for the Θ^+ in high statistics photoproduction experiments with CLAS

10:50 - 11:20 A. Hosaka (RCNP)
Production and decay of Θ^+

11:20 - 11:50 M. Oka (Tokyo)
Spectroscopy of pentaquark baryons

11:50 - 12:20 W. Eyrich (Erlangen)
The search for the pentaquark

12:20 - 14:00 **Lunch**

Plenary Session XIII; 14:00 - 15:30

Chair: P. Manyum

14:00 - 14:30 F. Huang (IHEP)
Baryon-meson interaction in the chiral SU(3) quark model

14:30 - 15:00 B. F. Gibson (Los Alamos)
Summary talk

15:00 - 15:30 P. Suebka
Closing the conference

Poster Session

(1) Y. Fujiwara, Y. Suzuki and M. Kohno
Almost Redundant Faddeev Components in the 3α Orthogonality Condition Model

(2) Y. Fujiwara, M. Kohno, K. Miyagawa and Y. Suzuki
Hypernuclear Physics: Hadron-Hyperon, Hyperon-Hyperon interactions

(3) B. Desplanques
A Method for Ensuring the Lorentz Invariance of Static Properties in the Instant Form of Relativistic Quantum Mechanics

(4) B. Desplanques
Relativistic Quantum Mechanics Approach to the Pion Charge Form Factor and its Asymptotic Behavior

(5) E. Z. Liverts, M. Ya. Amusia, E. G. Drukarev, R. Krivec and V. B. Mandelzweig
Wave Functions of Heliumlike Systems in Limiting Configurations I

(6) E. Z. Liverts, M. Ya. Amusia, E. G. Drukarev, R. Krivec and V. B. Mandelzweig
Wave Functions of Heliumlike Systems in Limiting Configurations II

(7) H. Nemura and C. Nakamoto
Stochastic Variational Calculations of Strange Systems

(8) Lauro Tomio, Sheila M. Holz, Victo S. Filho and Arnaldo Gammal
Dynamics of Nonlinear and Nonconservative Schrodinger-Type Equation Applied to Condensates in Nonharmonic Traps

(9) J. H. Esterline, A. S. Crowell, C. R. Howell, A. Hutcheson, R. A. Macri, S. Tajima, W. Tornow, B. J. Crowe, R. S. Pedroni and G. J. Weisel
Neutron-Helium-3 Analyzing Power at Low Energies

CONTENTS

HADRON STRUCTURE AND QCD

EXOTIC HADRONS AND ATOMS

EFFECTIVE FIELD THEORY IN FEW-BODY PHYSICS

ELECTROMAGNETIC AND WEAK PROCESSES IN FEW-BODY SYSTEMS

FEW-BODY DYNAMICS IN ATOMS, MOLECULES, BOSE-EINSTEIN CONDENSATES AND QUANTUM DOTS

FEW-BODY APPROACHES TO UNSTABLE NUCLEI, NUCLEAR ASTROPHYSICS AND NUCLEAR CLUSTERING ASPECTS

HYPERNUCLER PHYSICS: HADRON-HYPERON, HYPERON-HYPERON INTERACTIONS

OTHERS

Thematic Melodies of Twentieth Century Theoretical Physics: Quantization, Symmetry and Phase Factor

Chen Ning Yang

Chinese University, Hong Kong, and
Tsinghua University, Beijing 100084, China

I shall outline the 3 main melodies of 20-th century physics: quantization, symmetry and phases. Their interrelations and future influences will be emphasized.

It has been said that the 20th century was the century of physics. There are ample reasons to support this statement: It was in that century that man discovered, for the first time since our ancestors discovered fire, the second and the vastly stronger source of energy: nuclear power. It was in that century that man learned to manipulate electrons to create the transistor and the modern computer, transforming thereby human productivity and human lives. It was in that century that man learned how to probe into structures of atomic dimensions, discovering thereby the double helix, a key to the secrets of life. It was in that century that man ceased to be earth bound, taking first steps on the moon. In short, it was a century in which man made unprecedented progress on many fronts of human activities. And these progresses were largely ushered in by breathtaking advances in the science of physics.

It is hard not to be impressed by the decisive roles that the climatic developments in 20th century physics had played in human history. But decisive as they are in human history, they do not represent, in fact, the true glory of the development of the science of physics in the 20th century.

The true glory of physics in the 20th century lay instead in the deepened understandings of important primordial concepts which date from the beginning of human civilization: that of space, of time, of motion, of energy, and of force. In all of these primordial concepts, there have been profound revolutions in our understanding, revolutions that had brought forth a more beautiful, more subtle, more precise and more unified description of nature.

There have been in recent years studies of many aspects of the detailed history of twentieth century physics. It is not my purpose to delve into these subjects here. What I propose to do is instead to look into this history for the broad motifs of the developments, and to trace the three main strands that had persistently woven through its conceptual advances, appearing again and again in a variety of forms, singly or intertwined, like the thematic melodies of symphonic music. We shall see that these three melodies together define the tone and the flavour of the main developments of physics in the 20th century.

1. Quantization

The century opened with a paper by Planck (1858-1947) in which a constant, now called Planck's constant, designating the elementary quantum of action, was introduced. Like the opening bars in a prelude to symphonic music, it was to become the first thematic melody of the physics of the 20th century.

Planck was a very careful physicist. It was unusual for him to take the dramatically bold step of proposing a quantum of action. He did. But after more mature thinking about the proposal, he evidently got cold feet and started to backtrack in ensuing years. The torch then passed onto the younger generation: Einstein (1879-1955) first made his proposal about the photoelectric effect, and Bohr (1885-1962) followed by formulating his theory of quantized hydrogen atom. By 1918 when Planck gave his Nobel speech entitled "The Genesis and Present State of Development of the Quantum Theory", he said

> ..... If the various experiments and experiences .... from the different fields of physics provide impressive proof in favour of the existence of the quantum of action, the quantum hypothesis has, nevertheless, its greater support from the establishment and development of the atom theory by Niels Bohr. ...

He then outlined the triumphs and enormous frustrations of generalizing Bohr's theory, but ended his speech on a positive note:

> .... that which appears today so unsatisfactory will in fact eventually, seen from a higher vantage point, be distinguished by its special harmony and simplicity. ...

The eventual "higher vantage point" did arrive, through the development of quantum mechanics, but only after an agonizing period of great turmoil from 1913 to 1927. Pais, the distinguished historian of physics, has described this period as, borrowing from Dickens,

> It was the spring of hope, it was the winter of despair.

Perhaps the best characterization of the feelings of physicists during this period was Oppenheimer's (1904-1967) in his 1953 Reith lecture:

> "Our understanding of atomic physics, of what we call the quantum theory of atomic systems, had its origins at the turn of the century and its great synthesis and resolutions in the nineteen twenties. It was a heroic time. It was not the doing of any one man; it involved the collaboration of scores of scientists from many different lands, though from first to last the deeply creative and subtle and critical spirit of Niels Bohr guided, restrained, deepened, and finally transmuted the enterprise. It was a period of patient work in the laboratory, of crucial experiments and daring action, of many false starts and many untenable conjectures. It was a time of earnest correspondence and hurried conferences, of debate, criticism, and brilliant mathematical improvisation."
>
> "For those who participated, it was a time of creation; there was terror as well as exaltation in their new insight. It will probably not be recorded very completely as history. As

history, its recreation would call for an art as high as the story of Oedipus or the story of Cromwell, yet in a realm of action so remote from our common experience that it is unlikely to be known to any poet or any historian."

To illustrate the great up and down swings deeply felt by physicists during that period, let us look at a letter from Pauli (1900-1958) to Kronig (1904-1995) dated May 21, 1925,

> Physics is once again at a dead end at this time. For me, at any rate. It is much too difficult.

And another letter written by Pauli to Kronig five months later:

> Heisenberg's mechanics has restored my zest for life.

What happened that had excited Pauli in between the two letters was some vague perceptions by Heisenberg (1901-1976) through a historic groping "in the fog". In a remarkable passage in his old age, Heisenberg likened this groping in 1925 to mountain climbing:

> you sometimes want to climb some peak but there is fog everywhere you have your map or some other indication where you probably have to go and still you are completely lost in the fog. Then all of a sudden you see, quite vaguely in the fog, just a few minute things from which you say, "Oh, this is the rock I want." In the very moment that you have seen that, then the whole picture changes completely, because although you still don't know whether you will make the rock, nevertheless for a moment you say, " Now I know where I am; I have to go closer to that and then I will certainly find the way to go "

The vague perceptions of Heisenberg that day in 1925 was to lead to quantum mechanics, which was undoubtedly one of the greatest intellectual revolutions in human history.

2. Symmetry

A second thematic melody, symmetry, had its earliest origin in the 1905 paper of Einstein which introduced the theory of special relativity. This paper launched a great revolution that changed physicists' conceptions of space and time in a profound way. That this revolution was mathematically describable, in a very elegant way, by the four-dimensional symmetry between space and time was not known until the 1908 paper of Minkowski (1864-1909).

While Einstein himself was not originally impressed with the "superfluous learnedness" of Minkowski's work, he soon changed his mind and began to go further: He tried to greatly generalize the symmetry of special relativity, which in Minkowski's paper was mathematically formulated as the invariance of physical laws under Lorentz transformations. In his (Autobiographical Notes) written many years later, Einstein described how he had started on the idea of generalizing the invariance (i.e. symmetry):

> that the basic demand of the special theory of relativity (invariance of the laws under Lorentz-transformations) is too narrow, i.e. that an invariance of the laws must be postulated also relative to non-linear transformations of the coordinates in the four-dimensional continuum. This happened in 1908.

This larger invariance, after eight years of struggle by Einstein, finally gave rise to the theory of general relativity.

From today's vantage point, this Einstein-Minkowski-Einstein development was the first example of the theme of symmetry dictates interactions which we shall come back to later. At the time, this theme was not immediately picked up and further developed. It was to lie dormant for many years, while a variation of the theme was played out with the advent of quantum mechanics.

To follow this story we have to go back to melody one: quantization. With the great experimental developments of spectroscopy, and with Bohr's theory of 1913, quantum numbers entered into the very language of atomic physics. They were used to describe the states of a dynamic system. After the formulation of quantum mechanics it soon was realized that some of these quantum numbers are deeply related to the symmetry of dynamic systems, and that a beautiful and well developed branch of mathematics called group theory is the right approach to the concept of symmetry: E.g. the experimentally observed quantum numbers occur entirely naturally in group theory.

This new entry of symmetry into fundamental physics was, however, at first resisted by most physicists. The term "group pest" was coined to label this invasion by unfamiliar sophisticated mathematical concepts. It took half a dozen years for this resistance to dissolve, and with the developments of nuclear physics starting in the 1930s, and the later development of elementary particle physics starting in the 1950s, symmetry and group theory gradually became a dominant theme in physics: E.g. throughout much of the 1950s and 1960s, the efforts in elementary particle physics were directed mainly at finding the quantum numbers of the new particles and the symmetry behind them. A paragraph from a 1957 speech by the present author can serve to illustrate how symmetry was regarded in those years:

> It was, however, not until the development of quantum mechanics that the use of the symmetry principles began to permeate into the very language of physics. The quantum numbers that designate the states of a system are often identical with those that represent the symmetries of the system. It indeed is scarcely possible to overemphasize the role played by the symmetry principles in quantum mechanics. To quote two examples: The general structure of the periodic table is essentially a direct consequence of the isotropy of Coulomb's law. The existence of the antiparticles - namely the positron, the antiproton and the antineutron, were theoretically anticipated as consequences of the symmetry of physical laws with respect to Lorentz transformations. In both cases nature seems to take advantage of the simple mathematical representations of the symmetry laws. When one pauses to consider the elegance and the beautiful perfection of the mathematical reasoning involved and contrast it with the complex and far reaching physical consequences, a deep sense of respect for the power of the symmetry laws never fails to develop.

3. Phase Factor

Dirac (1902-1984) was one of the architects of quantum mechanics. He had erected the foundation of the mathematics of quantum mechanics on the noncommutativity of the dynamical variables p (momentum) and q (coordinate). So it was remarkable that in 1972 when he was 70 years old he said:

> So if one asks what is the main feature of quantum mechanics, I feel inclined now to say that it is not non-commutative algebra. It is the existence of probability amplitudes which underlie all atomic processes. Now a probability amplitude is related to experiment but only partially. The square of its modulus is something that we can observe. That is the probability which the experimental people get. But besides that there is a phase, a number of modulus unity which can modify without affecting the square of the modulus. And this phase is all important because it is the source of all interference phenomena but its physical significance is obscure. So the real genius of Heisenberg and Schrödinger, you might say, was to discover the existence of probability amplitudes containing this phase quantity which is very well hidden in nature and it is because it was so well hidden that people hadn't thought of quantum mechanics much earlier.

Phase was of course a concept well known in the theory of various kinds of waves in earlier centuries. Its entry into fundamental physics, however, had a tortuous history which we shall outline as follows:

A. In 1918 Weyl (1885-1955) tried to geometrize electromagnetism through the introduction of a linear differential form $d\varphi$, after Einstein's geometrization of gravity through the introduction of a quadratic differential form ds^2. He identified $d\varphi$ with $A_\mu dx^\mu$ times a constant, and considered a "stretch factor"

$$\exp\left[-\frac{e}{\gamma}\int A\mu dx^\mu\right] \tag{1}$$

which he applied to a particle transported along a four dimensional path, in analogy with the parallel displacement of a vector in general relativity. Weyl's ideas were criticized by Einstein who pointed out that if a meter stick is continually stretched along its world line, there would be no standardization of meter sticks possible, a devastating criticism. Weyl was not able to answer this criticism.

B. In 1922 Schrödinger (1887-1961) in a paper with the title "On a Remarkable Property of the Quantum-Orbits of a Single Electron" observed that the stretch factor of Weyl, when evaluated along a Bohr orbit, gives an exponent which is an integral multiple of a constant. He called this a remarkable property, saying

> It is difficult to believe that this result is merely an accidental mathematical consequence of the quantum conditions, and has no deeper physical meaning.

He then speculated on what value the constant γ should take. What is truly remarkable, in retrospect, was the fact that one of the possibilities Schrödinger mentioned was

$$\gamma = -i\hbar \tag{2}$$

> in which case the stretch factor would be equal to unity. ... I do not dare to judge whether this would make sense in the context of Weyl geometry.

All of this in 1922! Had Schrödinger pursued this possibility, he might have discovered wave mechanics in 1922! In the event, he apparently abandoned this whole line of thinking, only coming back to it after he read, in late 1925, de

Broglie's suggestion of wave lengths for particles. In a letter dated November 3, 1925, he wrote to Einstein as follows:

> The de Broglie interpretation of the quantum rules seems to me to be related in some ways to my note in the Zs. F. Phys. 12, 13, 1922, where a remarkable property of the Weyl 'gauge factor' exp $\left[-\int \phi dx\right]$ along each quasi-period is shown. The mathematical situation is, as far as I can see, the same, only from me much more formal, less elegant and not really shown generally. Naturally de Broglie's consideration in the framework of his large theory is altogether of far greater value than my single statement, which I did not know what to make of at first.

Two and a half months later, in early 1926, he submitted for publication his monumental paper creating wave mechanics.

Substituting (2) into (1) converts Weyl's stretch factor into a phase factor, and that is, in retrospect, the earliest entry of the thematic melody of phase into quantum mechanics, but only as a distant hesitant background refrain.

C. After 1926 Schrödinger did not return to this theme to further develop it. In fact, he strongly resisted to introduce $i = \sqrt{-1}$ into his wave equations. It was Fock and London who recognized the necessity of i in quantum mechanical electromagnetism. In particular, in a paper in 1927 by London with the title "Quantum-Mechanical Interpretation of Weyl's Theory", which quoted the Schrödinger 1922 paper, the stretch factor, which has become the phase factor, was the centre of the discussion.

D. Weyl in 1929 came back with an important paper that really launched what was called, and is still called, gauge theory of electromagnetism, a misnomer. (It should have been called phase theory of electromagnetism.) This amazing paper also discussed the 2 component neutrino theory which later in the 1950s became very important.

E. Weyl's papers were rambling and philosophical. Few physicists of my generation read them. We all learned about Weyl's gauge theory of electromagnetism from the review papers of Pauli in the Handbuch der Physik (1933) and Rev. Mod. Phys. (1941). Pauli, however, did not emphasize Weyl's concept of a stretch factor metamorphed into a phase factor for electromagnetism. Thus the importance of the phase factor had to lie dormant for several decades.

4. Development

All three thematic melodies were introduced in the first decades of the 20th century. Together they played roles in the rest of the century very much like the roles of thematic melodies in music: through developments, variations and entwining.

In the late 1940s, many new particles were discovered in cosmic ray experiments. They were unexpected, so they were called "strange particles". How these strange particles interact between themselves and with known particles became naturally a topic of discussion. For a few years these discussion followed the idea of writing down Lorentz-invariant couplings, such as scalar meson with vector couplings, vector meson with tensor couplings, etc., and deriving observable consequences of each possibility. Such ideas lacked a principle, *a general principle of interactions.* In

1954 Yang and Mills tried to formulate such a principle through generalizing Weyl's gauge theory of electromagnetism. They observed that electric charge conservation is related through Weyl's theory to an invariance: gauge invariance, or gauge symmetry. So perhaps another conservation law widely discussed at that time, conservation of isotopic spin, should be related to a generalized gauge invariance, or gauge symmetry. In their own words of 1954:

> The conservation of isotopic spin points to the existence of a fundamental invariance law similar to the conservation of electric charge. In the latter case, the electric charge serves as a source of electromagnetic field; an important concept in this case is gauge invariance which is closely connected with (1) the equation of motion of the electromagnetic field, (2) the existence of a current density, and (3) the possible interactions between a charged field and the electromagnetic field. We have tried to generalize this concept of gauge invariance to apply to isotopic spin conservation. It turns out that a very natural generalization is possible.

This natural generalization was later called Yang-Mills theory, or non-Abelian gauge theory.

Another way that also led to the same generalization was summarized in the beginning sentence of the abstract to another 1954 paper by these authors:

> It is pointed out that the usual principle of invariance under isotopic spin rotation is not consistent with the concept of localized fields.

This sentence is overstated. It should be amended to read

> It is pointed out that the usual principle of invariance under isotopic spin rotation is not consistent with the spirit of the concept of localized fields.

Both ways of approaching the generalization emphasized invariance, i.e. symmetry. The phase factor aspect was of course there, (the whole development being a quantum mechanical theory), but was not at that time central to the generalization. In particular, the stretch factor (changed into a phase factor), which was central to Weyl's thinking, was not correspondingly generalized. It was only in 1974 when it was finally generalized, and was called a nonintegrable non-Abelian phase factor.

With the nonintegrable phase factor the entwining of two of the thematic melodies, symmetry and phase factor, became intimate and intrinsic. Reviewing this history it is amazing that both melodies were close to the heart of Weyl throughout his life. Unfortunately, when he died in 1955, he did not have the opportunity to witness the great fruition of his ideas. He had this to say about himself in 1949:

> A lone wolf in Zurich, Hermann Weyl, also busied himself in this field; unfortunately he was all too prone to mix up his mathematics with physical and philosophical speculations.

Today, we would say that his speculations were amazingly insightful, and had served to change the course of the history of physics.

Non-Abelian gauge theory did not impress the physics community in the 1950s. The theory was patterned after electromagnetism in which the vector boson, i.e. the photon, has zero mass. It was therefore generally believed that in non-Abelian gauge theory the vector boson would also have zero mass. Since there were no such mesons found, the theory was not convincing. In 1954, Yang and Mills discussed this question in the last paragraph of their paper, and concluded

> In electrodynamics, by the requirement of electric charge conservation, it is argued that the mass of the photon vanishes. Corresponding arguments in the **b** field case do not exist even though the conservation of isotopic spin still holds. We have therefore not been able to conclude anything about the mass of the **b** quantum.

Then in the 1960s an important new idea about symmetry called *symmetry breaking* was formulated by several authors. The idea allows for complete symmetry in mathematical formalism, but at the same time allows for breaking of the symmetry in physical manifestations. The possibility then exists for the vector boson in a non-Abelian gauge theory to have nonzero mass. This idea was vigorously pursued both theoretically and experimentally, starting in the early 1970s, resulting in what is now called the standard model, which has been spectacularly successful in describing the multitude of experimental results in elementary particle physics.

In non-Abelian gauge theory, the interaction is determined by symmetry (i.e. invariance with respect to gauge transformations). Hence the principle: symmetry dictates interaction. As such it picked up the Einstein-Minkowski-Einstein melody that had earlier determined gravitational interaction through the requirement of invariance (i.e. symmetry) with respect to coordinate transformations.

Although the standard model has been spectacularly successful, it has a big defect: the way symmetry breaking is applied is ad hoc, and not unique. Few physicists believe therefore that the standard model is the final theory. This problem, together with the lack of success of putting general relativity and quantum theory together, remain basic problems of fundamental physics at the close of the 20th century.

The three melodies, quantization, symmetry and phase factor are beautifully and subtly entwined in the path integral formalism of Feynman, modified by requiring the integrand to be a time-ordered product,

$$propagator = \int exp[\frac{i}{\hbar}(action)]d(path) \tag{3}$$

for non-Abelian gauge theory: Quantization is an essential strand through the presence of the quantum of action, $\hbar$. Phase factor is an essential strand through the presence of $i = \sqrt{-1}$. And symmetry is an essential strand because the action for a non-Abelian gauge theory has a subtle multiplicative property under a gauge transformation.

It is amazing that all three thematic melodies were evolved from primordial concepts in the cognitive history of mankind: quantization through the recognition of units of measurement; symmetry through the recognition of the beauty of geometrical forms; phase through the observation of the phases of the moon. Twentieth century physics assigned precise mathematical meanings to each of these concepts and put them together in the path integral form (3), expressing, we might say, much that is the best in the tradition of physics since Newton and Huygens.

FEW-NUCLEON SYSTEMS: THREE-BODY FORCES AND FEW-NUCLEON DYNAMICS

Analyticity in Two- and Three-Body Coulomb Scattering in Momentum Space

S. Oryu and S. Nishinohara

Department of Physics, Faculty of Science and Technology, Tokyo University of Science,
Frontier Research Center of Computational Science of Tokyo University of Science,
2641-Yamazaki, Noda-city, 278-8510, Japan

Two- and three-charged particle systems are investigated in a mathematically rigorous approach in momentum space. A new method is proposed. The traditional screened Coulomb plus renormalization method is criticized in comparison with the new method. The new method is not only mathematically rigorous but also useful for numerical calculations as was the screened Coulomb potential method.

1. Introduction

The Rutherford scattering cross section has a singularity at forward angles. Therefore, the partial wave expansion converges only as a distribution.[1] On the other hand, the Fourier transformation of the Coulomb potential $V^C(r) = ZZ'e^2/r$ can only be obtained with the damping function $V^C(r) = \lim_{R\to\infty} V^R(r)$ with $V^R(r) = V^C(r)\xi(r,R)$ in which the Heaviside function $\xi(r,R) = \theta(r-R)$ or the exponential function $\xi(r,R) = \exp[-(r/R)^m]$ are popular choices for $\xi(r,R)$. By using such a potential, one can define the Lippmann-Schwinger equation,

$$T_l^{C}(p,p';\lambda;z) = V_l^{C}(p,p';\lambda) + \int_0^\infty V_l^{C}(p,p'';\lambda)G_0(p'';z)T_l^{C}(p'',p';\lambda;z)dp'', \quad (1)$$

where the limit of $\lambda \to 0$ leads to the so-called overlapping singularity in which the Green's function pole coincides with the logarithmic singularity of potential. We couldn't avoid such a singularity by any methods which were explored. Hence, our status quo is as follows: 1) The two-body Coulomb scattering problem can not be solved in momentum space. 2) Therefore, the three-body Faddeev equation can not be treated for the Coulomb problem in momentum space.[2] 3) However, nuclear reactions can not avoid Coulomb problems. 4) There exist historical techniques for treating Coulomb problems, but they suffer from a disease. Our final goal is to obtain a mathematically rigorous two-and three-body off-shell Coulomb amplitude and Coulomb-plus-nuclear scattering amplitude in momentum space.

2. Conventional Phase Shift Renormalization Methods

The traditional screening method is obtained by putting $V^{(R)} = V^S + V^R$, where V^S is the nuclear potential and V^R is the screened Coulomb potential. Therefore two-potential theory leads to,

$$T_l^{(R)} = T_l^{sR} + T_l^R = \overline{\omega}_l^R t_l^{sR} \omega_l^R + T_l^R, \tag{2}$$

$$t_l^{sR} = V_l^S + V_l^S G^R t_l^{sR}, \tag{3}$$

$$T_l^R = V_l^R + V_l^R G_0 T_l^R, \tag{4}$$

where the on-shell screened Coulomb amplitude is sandwiched by the Møller operators, using an approximation of $\overline{\omega}_l^R = 1 + T_l^R G_0 \approx e^{i\delta_l^R(k)}$, and $\omega_l^R = 1 + G_0 T_l^R \approx e^{i\delta_l^R(k)}$. Furthermore, they renormalize it by sandwiching it with the Gorshkov factor $e^{i\phi(k)}$.[3,4] Finally, the Coulomb scattering amplitude is defined by renormalizing $f_l^{(C)}(k) = -(\nu/2\pi)\, T_l^{(C)}(k)$, and

$$f_l^{(C)}(k) \rightarrow \lim_{R\rightarrow\infty} e^{i\phi(k,R)} f_l^{(R)}(k) e^{i\phi(k,R)} \equiv \lim_{R\rightarrow\infty} \tilde{f}^{(R)}(k) \tag{5}$$

with the renormalization phase $\phi(k,R) = \eta(k)[\ln 2kR - \gamma]$.[5] However, the r.h.s. of eq.(5) becomes

$$\tilde{f}^{(R)}(k) = e^{2i\phi(k,R)} \left(\frac{\exp[2i\{\delta_l^{sR}(k) + \delta_l^R(k)\}] - 1}{2ik} \right) \tag{6}$$

$$= e^{2i\{\phi(k)+\delta_l^R(k)\}} \left(\frac{e^{2i\delta_l^{sR}(k)} - 1}{2ik} \right) + e^{2i\phi(k,R)} \left(\frac{e^{2i\delta_l^R(k)} - 1}{2ik} \right) \tag{7}$$

$$\approx e^{2i\sigma_l(k)} \left(\frac{e^{2i\delta_l^{sR}(k)} - 1}{2ik} \right) + \frac{e^{2i\sigma_l(k)} - e^{2i\phi(k,R)}}{2ik}, \tag{8}$$

where the Coulomb phase shift is approximated by $\sigma_l(k) \approx \delta_l^R(k) + \phi(k,R)$. However, it is found that the second term of the numerator will oscillate rapidly due to $\phi(k,R) \rightarrow \infty$ when $R \rightarrow \infty$. Therefore, the r.h.s. of eq.(5) never converges to the Coulomb amplitude. The traditional renormalization amplitude doesn't converge to the Coulomb amplitude, but they replace it to conventionally converged function.[6–8]

3. New Method

3.1. *Two-body Coulomb amplitude*

Let us separate the Coulomb potential into a short range screened Coulomb potential (SCP) V^R and the auxiliary potential (AP) V^ϕ,

$$V^C = V^R + (V^C - V^R) = V^R + V^\phi. \tag{9}$$

We obtain the LS equation in terms of V^ϕ,

$$T^\phi = V^\phi + V^\phi G_0 T^\phi \equiv V^\phi \omega^\phi \equiv \overline{\omega}^\phi V^\phi, \tag{10}$$

with $\omega^{\phi} \equiv 1 + G_0 T^{\phi}$, and $\overline{\omega}^{\phi} \equiv 1 + T^{\phi} G_0$, respectively. If we could solve the auxiliary equation (10), then the screened Coulomb t-matrix $t^{R\phi}$ would be given by two-potential theory,

$$t^{R\phi} = V^R + V^R G^{\phi} t^{R\phi} \equiv V^R \omega^R \equiv \overline{\omega}^R V^R, \tag{11}$$

with $\omega^R \equiv 1 + G^{\phi} t^{R\phi}$, and $\overline{\omega}^R \equiv 1 + t^{R\phi} G^{\phi}$, and the resolvent

$$G^{\phi} = \frac{1}{z - H_0 - V^{\phi}} = G_0 + G_0 T^{\phi} G_0. \tag{12}$$

Therefore, the full off-shell Coulomb t-matrix would be defined by

$$T^C \Leftarrow T^{R\phi} + T^{\phi} = \overline{\omega}^{\phi} t^{R\phi} \omega^{\phi} + T^{\phi}. \tag{13}$$

We would like to solve eq.(10); however, the AP contains the Coulomb potential. Then, we have still the long range problem in the equation. The author (S.O.) has found a new theorem:
1) If $T^{\phi}(k,k;z) = 0$ exists, then $T^{\phi}(p,k;z) = T^{\phi}(k,p';z) = 0$ will be satisfied.
2) If theorem (1) is valid, then the off-shell $T^{\phi}(p,p';z)$ can be obtained by the K-matrix equation.
Consequently, it is found that the kernel of the K-matrix equation has no overlapping singularity.[9,10] Therefore, once we find the range $R = R_{cl}(k)$, we can obtain the off-shell $T^{\phi}(p,p';z)$. The AP on-shell t-matrix vanishes in two variables, k and R, for a fixed l. Therefore, one can solve the off-shell K-matrix for all energies for different ranges. The next problem is how to obtain the proper range $R_{cl}(k)$. One method is to obtain it from the differential equation, which was shown in Ref [9], and another method is to solve (10) directly. Finally, it is found that the range $R_{c0}(k)$ is approximately given by $120/\sqrt{z}$ for the proton-proton scattering.

3.2. *Two-body Nuclear amplitudes*

Let us consider a two-body nuclear charged particle system. The potential is given by

$$V^{(C)} = V^S + V^C = (V^S + V^R) + (V^C - V^R) = V^{(R)} + V^{\phi} \tag{14}$$

where V^S is a short range nuclear potential. By analogy with eq.(10) and eq.(11), the t-matrix is given as,

$$T^{(R)} = V^{(R)} + V^{(R)} G^{\phi} T^{(R)} = \overline{\omega}^R t^{sR} \omega^R + t^{R\phi}, \tag{15}$$

$$t^{sR} = V^S + V^S G^{R\phi} t^{sR}, \tag{16}$$

where $t^{R\phi}$ is given by (11). Then we easily deduce $T^{(C)}$ by using eqs.(10), (11), (13), and (15)

$$\begin{aligned} T^{(C)} &= \overline{\omega}^{\phi} T^{(R)} \omega^{\phi} + T^{\phi} = \overline{\omega}^{\phi} (\overline{\omega}^R t^{sR} \omega^R + t^{R\phi}) \omega^{\phi} + T^{\phi} \\ &= \overline{\omega}^{\phi} \overline{\omega}^R t^{sR} \omega^R \omega^{\phi} + \overline{\omega}^{\phi} t^{R\phi} \omega^{\phi} + T^{\phi} = \overline{\omega}^C t^{sR} \omega^C + T^C, \end{aligned} \tag{17}$$

where it is proved that the Coulomb Møller operators are given by $\omega^C = \omega^R\omega^\phi = (1+G^\phi t^{R\phi})(1+G_0T^\phi) = (1+G_0T^C)$ and $\overline{\omega}^C = \overline{\omega}^\phi\overline{\omega}^R = (1+T^\phi G_0)(1+t^{R\phi}G^\phi) = (1+T^C G_0)$. Furthermore, we can prove that $t^{sR} = t^{sC}$ by using the resolvent; that is,

$$G^{R\phi} = \frac{1}{z - H_0 - V^R - V^\phi} = \frac{1}{z - H_0 - V^C} = G_0 + G_0T^CG_0 \equiv G^C. \quad (18)$$

Here it should be stressed that eq.(18) contains an "important proof" of the equivalency of the screened Coulomb with the pure Coulomb Green's function; *i.e.* $G^{R\phi} = G^C$ at a "finite given range R". Thus, if we miss V^ϕ, then we have to take $R \to \infty$ to reach G^C. Therefore, all other calculations should be done with an infinite range. If the results converge at a finite range, such as in traditional methods, their results are inconsistent [6–8]. Then eq.(16) becomes,

$$\begin{aligned} t^{sR} &= V^S + V^SG^Ct^{sR} = V^S + V^S\{1 + G_0T^C\}G_0t^{sR} \\ &= V^S + \mathcal{K}^{SC}G_0t^{sR} \equiv t^{sC}. \end{aligned} \quad (19)$$

Therefore eq.(17) gives

$$T^{(C)} = \overline{\omega}^Ct^{sC}\omega^C + T^C. \quad (20)$$

3.3. *Three-body Nuclear amplitudes*

Our "regulation" for the two-body Coulomb t-matrix in eq.(13) and (17) is obviously written for the three-body transition t-matrix $T^{(C)}$ for a nuclear system in which the full potential is given by

$$V^{(C)} = V^S + W^0 + V^C = (V^S + W^0 + V^R) + (V^C - V^R) = V^{(R)} + V^\phi, \quad (21)$$

where V^S, W^0, V^C are a nuclear force, a short range three-body force, and the Coulomb force, respectively. Here we introduce the three-body Jacobi-coordinate channels: α, β, γ or $1, 2,$ and 3. The two-body potentials V are given by V_α, V_β, and V_γ, while the three-body force W^0 could be denoted by $W^0_{\alpha\beta}$. Therefore, eq.(5) indicates

$$\begin{aligned} V^{(C)}_{\alpha\beta} &= V^S_\alpha\delta_{\alpha\beta} + W^0_{\alpha\beta} + V^C_\alpha\delta_{\alpha\beta} = (V^S_\alpha\delta_{\alpha\beta} + W^0_{\alpha\beta} + V^R_\alpha\delta_{\alpha\beta}) + (V^C_\alpha - V^R_\alpha)\delta_{\alpha\beta} \\ &= V^{(R)}_{\alpha\beta} + V^\phi_\alpha\delta_{\alpha\beta}. \end{aligned} \quad (22)$$

Hereafter we suppress indices for simplicity except when necessary. Then the formal equation for such a three-body t-matrix can be represented by,

$$T^{(C)} = V^{(C)} + V^{(C)}G_0T^{(C)} = (V^{(R)} + V^\phi) + (V^{(R)} + V^\phi)G_0T^{(C)}. \quad (23)$$

Therefore, the three-body t-matrix can also be decomposed using two-potential theory,

$$\begin{aligned} T^{(C)} &= \overline{\omega}^\phi T^{(R)}\omega^\phi + T^\phi = \overline{\omega}^\phi(\overline{\omega}^R T^{sR}\omega^R + T^R)\omega^\phi + T^\phi \\ &= \overline{\omega}^\phi(\overline{\omega}^R(\overline{\Omega}^0 T\Omega^0 + T^0)\omega^R + T^R)\omega^\phi + T^\phi \\ &= \overline{\omega}^C\overline{\Omega}^0 T\Omega^0\omega^C + \overline{\omega}^C T^0\omega^C + T^C, \end{aligned} \quad (24)$$

where it has an onion-like structure. These t-matrices are labeled by the three-body Jacobi channels α, β, γ.

Finally, the three-body channel representation of eqs.(24) , (24) with α, β, γ, becomes as follows [11,12]

$$T^{(C)} = \sum_{\alpha,\beta=1}^{3} \sum_{\eta,\zeta=1}^{3} \sum_{\gamma,\delta=1}^{3} (\overline{\omega}^{C}_{\alpha\eta}(\overline{\Omega}^{0}_{\eta\gamma} T_{\gamma\delta} \Omega^{0}_{\delta\zeta} + T^{0}_{\eta\zeta})\omega^{C}_{\zeta\beta} + T^{C}_{\alpha\beta}). \tag{25}$$

4. Conclusion

We have presented a new method to solve nuclear charged particle systems in momentum space. One does not solve the two-body Lippmann-Schwinger equation but treats it by the mathematically rigorous two-potential theory. First of all, we introduced the Auxiliary potential which is defined by the difference between the pure Coulomb potential and a screened Coulomb potential. We can not treat the Lippmann-Schwinger equation for the Coulomb potential but we can calculate the K-matrix equation for the Auxiliary potential and impose the condition that the on-and half on-shell amplitudes will vanish. By using such a full off-shell amplitude, we can calculate the core amplitude with a screened Coulomb potential in which the kernel of the Lippmann-Schwinger equation for the screened Coulomb potential is modified with the off-shell auxiliary amplitude. The phase shift introduced by the solution will be the Coulomb phase shift.

References

1. J.R.Taylor, Nuovo Cimento **23** B, 313 (1974), and M. D. Semon and J. R. Taylor, Nuovo Cimento **26**B 48 (1975).
2. L.D.Faddeev, Zh. Eksp. Theor. Fiz. **39**, 1459 (1960) [Sov. Phys. JETP **12**, 1014 (1961).
3. V. G. Gorshkov, Zh. Eksp. Teor. Fiz. **40**, 1459 (1961) [Sov. Phys. JETP **12**, 1014 (1961)], *ibid.* ZhETF **40**, 1481 (1961) [Sov. Phys. JETP **13**, 1037 (1961)].
4. A. M. Veselova, Teor. Mat. Fiz. **3**, 326 (1970) [Theor. Math. Phys. **3**, 542 (1970)], *ibid*, **13**, 368 (1972) [**13**, 1200 (1972)], **35**, 180 (1978) [**35**, 395 (1978)].
5. E. O. Alt, W. Sandhas, H. Zankel, and H. Ziegelmann, Phys. Rev. Lett. **37**, 1537 (1976), *ibid.* Phys. Rev. **C17**, 1981 (1978), Nucl. Phys. **A445**, 429 (1985), *ibid.* E. O. Alt, W. Sandhas, Phys. Rev. **21** 1733 (1980)
6. G. H. Berthold, A. Stadler, and H. Zankel, Phy. Rev. **C41**, 1365 (1990).
7. A. Deltuva, A. C. Fonseca, P. U. Sauer, Phy. Rev. **C71**, 054005 (2005).
8. A. Deltuva, A. C. Fonseca, A. Kievsky, S.Rosati, P. U. Sauer, and M. Viviani, Phy. Rev. **C71**, 064003 (2005).
9. S. Oryu, S. Nishinohara, K. Sonoda, N. Shiiki, Y. Togawa and S. Chiba, *Proceedings of 19th European Conference of Few-Body Problems in Physics* (Groningen 2004).
10. S. Oryu, Phys. Rev. submitted.
11. S. Oryu, Few Body Systems Vol. **34**, 113-118 (2004).
12. S. Oryu and S. Gojuki, Prog. Theor. Phys. Suppl. Vol. **154**, 285-292 (2004).

Four-Body Bound State Formulation in Three-Dimensional Approach (Without Angular Momentum Decomposition)

M. R. Hadizadeh† and S. Bayegan‡

Department of Physics, University of Tehran,
P.O.Box 14395-547, Tehran, Iran
E-mail addresses: †*hadizade@khayam.ut.ac.ir* , ‡*bayegan@khayam.ut.ac.ir*

four-body bound state with two-body forces is formulated by the Three-Dimensional approach, which greatly simplifies the numerical calculations of few-body systems without performing the Partial Wave components. We have obtained the Yakubovsky equations directly as three dimensional integral equations.

1. Introduction

The four-body bound state calculations are traditionally carried out by solving coupled Yakubovsky equations in a partial wave basis. After truncation this leads to two coupled sets of finite number of coupled equations in three variables for the amplitudes. This is performed in configuration space[1] and in momentum space[2,3]. Though a few partial waves often provide qualitative insight, modern four-body calculations need 1572 or more different spin, isospin and angular momentum combinations[3]. It appears therefore natural to avoid a partial wave representation completely and work directly with vector variables. This is a common practice in four-body bound state calculations based on other techniques[4-9]. In recent years W. Glöckle and collaborators have introduced the three-dimensional approach which greatly simplifies the two- and three-body scattering and bound state calculations without using partial wave decomposition[10-12]. In this paper we extend this approach for four-body bound state with two-body interactions, we work directly with momentum vector variables in the Yakubovsky scheme. As a simplification we neglect spin and isospin degrees of freedom and treat four-boson bound state. Although the four-boson bound state has been studied with short-range forces and large scattering length at leading order in an effective quantum mechanics approach[13], but it is also based on partial wave approach.

2. Momentum Space Representation of Yakubovsky Equations in 3-D approach

The bound state of four identical bosons which interact via pairwise forces is given by coupled Yakubovsky equations[13]:

$$|\psi_1\rangle = G_0 t_{12} P[(1+P_{34})|\psi_1\rangle + |\psi_2\rangle]$$
$$|\psi_2\rangle = G_0 t_{12} \tilde{P}[(1+P_{34})|\psi_1\rangle + |\psi_2\rangle] \quad (1)$$

In order to solve coupled equations, *Eq.*(1), in momentum space we introduce the four-body basis states corresponding to each Yakubovsky component:

$$|\vec{u}_1\,\vec{u}_2\,\vec{u}_3\rangle$$
$$|\vec{v}_1\,\vec{v}_2\,\vec{v}_3\rangle \quad (2)$$

Let us now represent coupled equations, *Eq.*(1), with respect to the basis states have been introduced in Eq.(2):

$$\begin{aligned}
\langle \vec{u}_1\,\vec{u}_2\,\vec{u}_3|\psi_1\rangle = & \int D^3u' \int D^3u''\ \langle \vec{u}_1\,\vec{u}_2\,\vec{u}_3|G_0 t P|\vec{u}_1^{\,'}\,\vec{u}_2^{\,'}\,\vec{u}_3^{\,'}\rangle \\
& \langle \vec{u}_1^{\,'}\,\vec{u}_2^{\,'}\,\vec{u}_3^{\,'}|(1+P_{34})|\vec{u}_1^{\,''}\,\vec{u}_2^{\,''}\,\vec{u}_3^{\,''}\rangle\ \langle \vec{u}_1^{\,''}\,\vec{u}_2^{\,''}\,\vec{u}_3^{\,''}|\psi_1\rangle \\
+ & \int D^3u' \int D^3v'\ \langle \vec{u}_1\,\vec{u}_2\,\vec{u}_3|G_0 t P|\vec{u}_1^{\,'}\,\vec{u}_2^{\,'}\,\vec{u}_3^{\,'}\rangle \\
& \langle \vec{u}_1^{\,'}\,\vec{u}_2^{\,'}\,\vec{u}_3^{\,'}|\vec{v}_1^{\,'}\,\vec{v}_2^{\,'}\,\vec{v}_3^{\,'}\rangle\ \langle \vec{v}_1^{\,'}\,\vec{v}_2^{\,'}\,\vec{v}_3^{\,'}|\psi_2\rangle \\
\langle \vec{v}_1\,\vec{v}_2\,\vec{v}_3|\psi_2\rangle = & \int D^3v' \int D^3u'\ \langle \vec{v}_1\,\vec{v}_2\,\vec{v}_3|G_0 t \tilde{P}|\vec{v}_1^{\,'}\,\vec{v}_2^{\,'}\,\vec{v}_3^{\,'}\rangle \\
& \langle \vec{v}_1^{\,'}\,\vec{v}_2^{\,'}\,\vec{v}_3^{\,'}|(1+P_{34})|\vec{u}_1^{\,'}\,\vec{u}_2^{\,'}\,\vec{u}_3^{\,'}\rangle\langle \vec{u}_1^{\,'}\,\vec{u}_2^{\,'}\,\vec{u}_3^{\,'}|\psi_1\rangle \\
+ & \int D^3v'\ \langle \vec{v}_1\,\vec{v}_2\,\vec{v}_3|G_0 t \tilde{P}|\vec{v}_1^{\,'}\,\vec{v}_2^{\,'}\,\vec{v}_3^{\,'}\rangle \vec{v}_1^{\,'}\,\vec{v}_2^{\,'}\,\vec{v}_3^{\,'}\rangle\langle \vec{v}_1^{\,'}\,\vec{v}_2^{\,'}\,\vec{v}_3^{\,'}|\psi_2\rangle \quad (3)
\end{aligned}$$

Where $D^3A = d^3A_1\,d^3A_2\,d^3A_3$. After evaluating the following matrix elements:

$$\begin{aligned}
\langle \vec{u}_1\,\vec{u}_2\,\vec{u}_3|G_0 t P|\vec{u}_1^{\,'}\,\vec{u}_2^{\,'}\,\vec{u}_3^{\,'}\rangle = & \frac{\delta^3(\vec{u}_3-\vec{u}_3^{\,'})}{E-\frac{u_1^2}{m}-\frac{3u_2^2}{4m}-\frac{2u_3^2}{3m}} \\
& \{\ \delta^3(\vec{u}_2+\vec{u}_1^{\,'}+\frac{1}{2}\vec{u}_2^{\,'})\ \langle \vec{u}_1|t(\epsilon)|\frac{1}{2}\vec{u}_2+\vec{u}_2^{\,'}\rangle \\
& +\delta^3(\vec{u}_2-\vec{u}_1^{\,'}+\frac{1}{2}\vec{u}_2^{\,'})\ \langle \vec{u}_1|t(\epsilon)|\frac{-1}{2}\vec{u}_2-\vec{u}_2^{\,'}\rangle\} \\
\langle \vec{v}_1\,\vec{v}_2\,\vec{v}_3|G_0 t \tilde{P}|\vec{v}_1^{\,'}\,\vec{v}_2^{\,'}\,\vec{v}_3^{\,'}\rangle = & \frac{\delta^3(\vec{v}_2+\vec{v}_2^{\,'})\,\delta^3(\vec{v}_3-\vec{v}_1^{\,'})}{E-\frac{v_1^2}{m}-\frac{v_2^2}{2m}-\frac{v_3^2}{m}}\ \langle \vec{v}_1|t(\epsilon^*)|\vec{v}_3^{\,'}\rangle \\
\langle \vec{u}_1^{\,'}\,\vec{u}_2^{\,'}\,\vec{u}_3^{\,'}|(1+P_{34})|\vec{u}_1^{\,''}\,\vec{u}_2^{\,''}\,\vec{u}_3^{\,''}\rangle = & \ \delta^3(\vec{u}_1^{\,'}-\vec{u}_1^{\,''}) \\
& \times\{\ \delta^3(\vec{u}_2^{\,'}-\vec{u}_2^{\,''})\,\delta^3(\vec{u}_3^{\,'}-\vec{u}_3^{\,''}) \\
& +\delta^3(\vec{u}_2^{\,'}-\frac{1}{3}\vec{u}_2^{\,''}-\frac{8}{9}\vec{u}_3^{\,''})\,\delta^3(\vec{u}_3^{\,'}-\vec{u}_2^{\,''}+\frac{1}{3}\vec{u}_3^{\,''})\} \\
\langle \vec{v}_1^{\,'}\,\vec{v}_2^{\,'}\,\vec{v}_3^{\,'}|(1+P_{34})|\vec{u}_1^{\,'}\,\vec{u}_2^{\,'}\,\vec{u}_3^{\,'}\rangle = & \ \delta^3(\vec{u}_1^{\,'}-\vec{v}_1^{\,'}) \\
& \times\{\ \delta^3(\vec{u}_2^{\,'}+\frac{2}{3}\vec{v}_2^{\,'}-\frac{2}{3}\vec{v}_3^{\,'})\ \delta^3(\vec{u}_3^{\,'}+\frac{1}{2}\vec{v}_2^{\,'}+\vec{v}_3^{\,'}) \\
& +\delta^3(\vec{u}_2^{\,'}+\frac{2}{3}\vec{v}_2^{\,'}+\frac{2}{3}\vec{v}_3^{\,'})\ \delta^3(\vec{u}_3^{\,'}+\frac{1}{2}\vec{v}_2^{\,'}-\vec{v}_3^{\,'})\} \quad (4)
\end{aligned}$$

We can rewrite the coupled equations, $Eq.(3)$, as below coupled three-dimensional integral equations:

$$
\begin{aligned}
\langle \vec{u}_1\, \vec{u}_2\, \vec{u}_3 | \psi_1 \rangle &= \frac{1}{E - \frac{u_1^2}{m} - \frac{3u_2^2}{4m} - \frac{2u_3^2}{3m}} \int d^3u_2' \; \langle \vec{u}_1 | t_s(\epsilon) | \frac{1}{2}\vec{u}_2 + \vec{u}_2' \rangle \\
&\times \{ \; \langle \vec{u}_2 + \frac{1}{2}\vec{u}_2' \; \vec{u}_2' \; \vec{u}_3 | \psi_1 \rangle \\
&\quad + \langle \vec{u}_2 + \frac{1}{2}\vec{u}_2' \; \frac{1}{3}\vec{u}_2' + \frac{8}{9}\vec{u}_3 \; \vec{u}_2' - \frac{1}{3}\vec{u}_3 | \psi_1 \rangle \\
&\quad + \langle \vec{u}_2 + \frac{1}{2}\vec{u}_2' \; - \vec{u}_2' - \frac{2}{3}\vec{u}_3 \; \frac{1}{2}\vec{u}_2' - \frac{2}{3}\vec{u}_3 | \psi_2 \rangle \; \} \\
\langle \vec{v}_1\, \vec{v}_2\, \vec{v}_3 | \psi_2 \rangle &= \frac{\frac{1}{2}}{E - \frac{v_1^2}{m} - \frac{v_2^2}{2m} - \frac{v_3^2}{m}} \int d^3v_3' \; \langle \vec{v}_1 | t_s(\epsilon^*) | \vec{v}_3' \rangle \\
&\times \{ \; 2 \, \langle \vec{v}_3 \; \frac{2}{3}\vec{v}_2 + \frac{2}{3}\vec{v}_3' \; \frac{1}{2}\vec{v}_2 - \vec{v}_3' | \psi_1 \rangle \\
&\quad + \langle \vec{v}_3 \; - \vec{v}_2 \; \vec{v}_3' | \psi_2 \rangle \; \}
\end{aligned} \tag{5}
$$

Here ϵ and ϵ^* are two-body subsystem energies and $\langle \vec{a} | t_s(\varepsilon) | \vec{b} \rangle$ is the symmetrized two-body t-matrix[10]. The so obtained Y-amplitudes fulfill the below symmetry relations, as can be seen from $Eq.(5)$.

$$
\begin{aligned}
\langle \vec{u}_1\, \vec{u}_2\, \vec{u}_3 | \psi_1 \rangle &= \langle -\vec{u}_1\, \vec{u}_2\, \vec{u}_3 | \psi_1 \rangle \\
\langle \vec{v}_1\, \vec{v}_2\, \vec{v}_3 | \psi_2 \rangle &= \langle -\vec{v}_1\, \vec{v}_2\, \vec{v}_3 | \psi_2 \rangle \\
\langle \vec{v}_1\, \vec{v}_2\, \vec{v}_3 | \psi_2 \rangle &= \langle \vec{v}_1\, \vec{v}_2 \; - \vec{v}_3 | \psi_2 \rangle
\end{aligned} \tag{6}
$$

3. Choosing Coordinate Systems

The Y-components $|\psi_i(\vec{a}\; \vec{b}\; \vec{c})\rangle$ are given as function of Jacobi momenta vectors as solution of coupled three-dimensional integral equations, $Eq.(5)$. Since we ignore spin and isospin dependencies, for the ground state both Y-components $|\psi_i(\vec{a}\; \vec{b}\; \vec{c})\rangle$ are scalars and thus only depend on the magnitudes of Jacobi momenta and the angles between them. The first important step for an explicit calculation is the selection of independent variables, one needs six independent variables to uniquely specify the geometry of the three vectors[12]. Therefore in order to solve $Eq.(5)$ directly without introducing partial wave projection, we have to define suitable coordinate systems. For both Y-components we choose the third vector parallel to $Z-$axis, the second vector in the $X-Z$ plane and express the remaining vectors, the first as well as the integration vectors, with respect to them. With this choice of variables we can obtain the explicit representation for the Y-components $|\psi_1\rangle$ and $|\psi_2\rangle$ as [14]:

$$\psi_1(u_1\,u_2\,u_3\,x_1\,x_2\,x_{12}^3) = \frac{1}{E - \frac{u_1^2}{m} - \frac{3u_2^2}{4m} - \frac{2u_3^2}{3m}} \int d^3u_2^{'}\, t_s(u_1, \tilde{\pi}, \tilde{x}; \epsilon)$$
$$\times \{\ \psi_1(\pi_1\ u_2^{'}\ u_3\ x_{12}\ x_{13}\ x_{\pi_1 u_2'}^{u_3})$$
$$+\psi_1(\pi_1\ \pi_2\ \pi_3\ x_{22}\ x_{23}\ x_{\pi_1\pi_2}^{\pi_3})$$
$$+\psi_2(\pi_1\ \pi_4\ \pi_5\ x_{32}\ x_{33}\ x_{\pi_1\pi_4}^{\pi_5})\ \}$$
$$\psi_2(v_1\,v_2\,v_3\,X_1\,X_2\,X_{12}^3) = \frac{\frac{1}{2}}{E - \frac{v_1^2}{m} - \frac{v_2^2}{2m} - \frac{v_3^2}{m}} \int d^3v_3^{'}\, t_s(v_1, v_3^{'}, Y_{13'}; \epsilon^*)$$
$$\times \{\ 2\,\psi_1(v_3\ \Sigma_1\ \Sigma_2\ X_{12}\ X_{13}\ X_{v_3\Sigma_1}^{\Sigma_2})$$
$$\psi_2(v_3\ v_2\ v_3'\ X_{22}\ X_{23}\ X_{v_3 v_2}^{v_3'})\ \} \qquad (7)$$

The above coupled three-dimensional integral equations are the starting point for numerical calculations.

4. Summary

An alternative approach for four-body bound state calculations, which are based on solving the coupled Y-equations in a partial wave basis, is to work directly with momentum vector variables. We formulate the coupled Y-equations for identical particles as function of vector Jacobi momenta, specifically the magnitudes of the momenta and the angles between them. We expect that coupled three-dimensional Y-equations can be handled in a straightforward and numerically reliable fashion.

Acknowledgments

One of authors (M. R. H.)would like to thank H. Kamada and Ch. Elster for helpful discussions during EFB19 and APFB05 conferences.

References

1. N. W. Schellingerhout, J. J. Schut, and L. P. Kok, *Phys. Rev.* **C 46**, 1192 (1992).
2. W. Glöckle and H. Kamada, *Nucl. Phys.* **A 560**, 541 (1993).
3. A. Nogga, H. Kamada, W. Glöckle and B. R. Barrettl, *Phys. Rev.* **C 65**, 054003 (2002).
4. E. Hiyama et al., *Phys. Rev. Lett.* **85**, 270 (2000).
5. J. Usukura, K. Varga and Y. Suzuki, *Phys. Rev.* **B 59**, 5652 (1999).
6. M. Viviani, *Few Body Syst.* **25**, 177 (1998).
7. R. B. Viringa et al., *Phys. Rev.* **C 62**, 014001 (2000).
8. P. Navrátil, J. P. Vary and B. R. Barret, *Phys. Rev.* **C 62**, 054311 (2000).
9. N. Barnea, W. Leidemann and G. Orlandini, *Phys. Rev.* **C 61**, 054001 (2000).
10. Ch. Elster, W. Schadow, A. Nogga, W. Glöeckle, *Few Body Syst.* **27**, 83 (1999).
11. I. Fachruddin, W. Glöckle, Ch. Elster, A. Nogga, *Phys. Rev.* **C 69**, 064002 (2004).
12. H. Liu, Ch. Elster, W. Glöeckle, *arXiv:nucl-th/0410051*
13. L. Platter, H. W. Hammer, Ulf-G. Meissner, *Phys. Rev.* **A 70**, 052101 (2004).
14. M. R. Hadizadeh and S. Bayegan, *In Preparation*

Momentum-Space Treatment of Coulomb Interaction in Three-Nucleon Breakup Reactions*

A. Deltuva and A. C. Fonseca

Centro de Física Nuclear da Universidade de Lisboa, P-1649-003 Lisboa, Portugal

P. U. Sauer

Institut für Theoretische Physik, Universität Hannover, D-30167 Hannover, Germany

The Coulomb interaction between the two protons is included in the calculation of proton-deuteron breakup and of three-body electromagnetic disintegration of ^{3}He. The hadron dynamics is based on the purely nucleonic charge-dependent (CD) Bonn potential and its realistic extension CD Bonn + Δ to a coupled-channel two-baryon potential, allowing for single virtual Δ-isobar excitation. Calculations are done using integral equations in momentum space. The screening and renormalization approach is employed for including the Coulomb interaction. Convergence of the procedure is found at moderate screening radii. The reliability of the method is demonstrated. The Coulomb effect on breakup observables is seen at all energies in particular kinematic regimes.

The inclusion of the Coulomb interaction in the description of the three-nucleon continuum is one of the most challenging tasks in theoretical few-body nuclear physics. Whereas it has already been solved for elastic proton-deuteron (*pd*) scattering with realistic hadronic interactions using various procedures [12,2,5,4,8], there are only very few attempts [6–8] to calculate *pd* breakup and none of them uses realistic potentials allowing for a stringent comparison with the experimental data. Only very recently in Refs. [9,10] we included the Coulomb interaction in the description of hadronic and electromagnetic three-nucleon breakup reactions The description is based on the Alt-Grassberger-Sandhas (AGS) equations [3] in momentum space. The Coulomb potential is screened, standard scattering theory becomes applicable, and the renormalization procedure of Refs. [1,2] is applied to recover the unscreened limit. That approach is an extension of the Coulomb treatment for elastic *pd* scattering [8,14]. The treatment is applicable to any two-nucleon potential without separable expansion. Here we use the purely nucleonic charge-dependent (CD) Bonn potential [6] and its coupled-channel extension CD Bonn + Δ [7], allowing for a single virtual Δ-isobar excitation and fitted to the experimental data with the same degree of accuracy as CD Bonn itself. In the three-nucleon system the Δ isobar mediates an effective three-nucleon force and effective two- and three-nucleon currents, both consistent with the underlying two-nucleon force.

*Talk delivered by A. Deltuva

We emphasize that we work with the screened Coulomb potential w_R of particular form. It is screened around the separation $r = R$ between two charged baryons and in configuration space is given by

$$w_R(r) = w(r)\, e^{-(r/R)^n}, \tag{1}$$

with the true Coulomb potential $w(r) = \alpha_e/r$, α_e being the fine structure constant and n controlling the smoothness of the screening. We prefer to work with a sharper screening than the Yukawa screening ($n = 1$) of Ref. [6]. We want to ensure that the screened Coulomb potential w_R approximates well the true Coulomb one w for distances $r < R$ and simultaneously vanishes rapidly for $r > R$, providing a comparatively fast convergence of the partial-wave expansion, but avoids an unpleasant oscillatory behavior characteristic for the sharp cutoff ($n \to \infty$). We find the values $3 \leq n \leq 6$ to provide a sufficiently smooth, but at the same time a sufficiently rapid screening around $r = R$; like in Refs. [8–10] $n = 4$ is our choice for the results of this paper. The screening radius R is chosen much larger than the range of the strong interaction which is of the order of the pion wavelength $\hbar/m_\pi c \approx 1.4$ fm. Nevertheless, the screened Coulomb potential w_R is of short range in the sense of scattering theory. Standard scattering theory is therefore applicable. Although the choice of the screened potential improves the partial-wave convergence, the practical implementation of the solution of the AGS equation still places a technical difficulty, i.e., the calculation of the AGS operators for nuclear plus screened Coulomb potentials requires two-nucleon partial waves with pair orbital angular momentum considerably higher than required for the hadronic potential alone. In this context the perturbation theory for high two-nucleon partial waves developed in Ref. [10] is a very efficient and reliable technical tool for treating the screened Coulomb interaction in high partial waves.

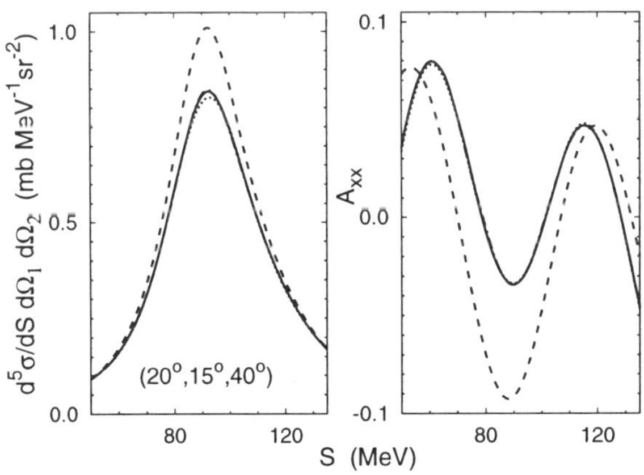

Fig. 1. Convergence of the pd breakup observables with screening radius R. The differential cross section and the deuteron analyzing power A_{xx} for pd breakup at 130 MeV deuteron lab energy are shown. Results for CD Bonn potential obtained with screening radius $R = 10$ fm (dotted curves), 20 fm (dash-dotted curves), and 30 fm (solid curves) are compared. Results without Coulomb (dashed curves) are given as reference for the size of the Coulomb effect.

Due to the finite-range nature of the breakup operators discussed in Refs. [2,10] the unscreened limit for observables of three-nucleon reactions is reached with sufficient accuracy at rather modest screening radii R. Convergence with screening radius R is the internal criterion for the reliability of our Coulomb treatment. Usually it is impressively fast as Fig. 1 demonstrates for a sample kinematical configuration and therefore strongly suggests the reliability of the present calculational scheme.

Selected physics results with significant Coulomb effects in *pd* breakup and three-body photodisintegration of ^{3}He are given in Figs. 2 and 3. The kinematical final-

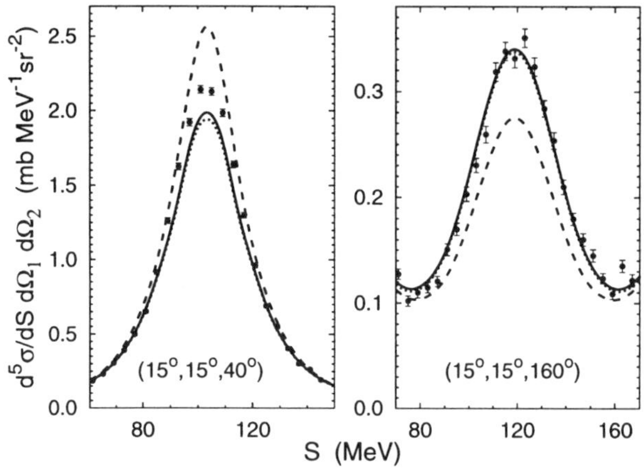

Fig. 2. Differential cross section for *pd* breakup at 130 MeV deuteron lab energy. Results including Δ-isobar excitation and the Coulomb interaction (solid curves) are compared to results without Coulomb (dashed curves). In order to appreciate the size of the Δ-isobar effect, the purely nucleonic results including Coulomb are also shown (dotted curves). The experimental data are from Ref. [5].

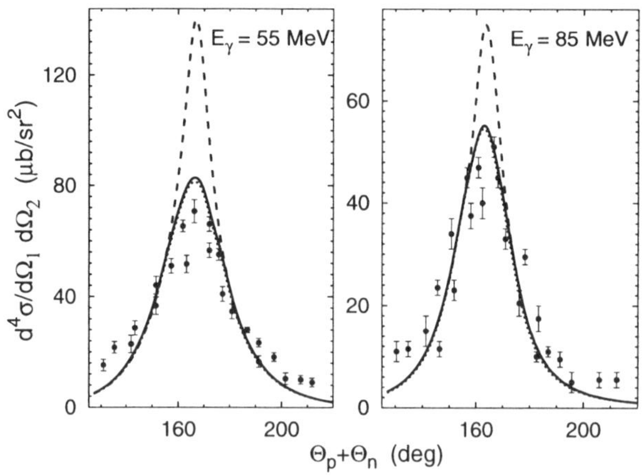

Fig. 3. The semi-inclusive fourfold differential cross section for $^3\mathrm{He}(\gamma, pn)p$ reaction at 55 MeV and 85 MeV photon lab energy as function of the np opening angle $\theta_p + \theta_n$ with $\theta_p = 81°$. CurvesRef. [19].

state configurations of *pd* breakup are characterized in a standard way by the polar angles of the two protons and by the azimuthal angle between them, $(\theta_1, \theta_2, \varphi_2 - \varphi_1)$, the fifth independent variable being the arclength S along the kinematical curve. The observed Coulomb effects are much more important than the effects of the three-nucleon force due to Δ-isobar excitation and are clearly supported by the experimental data.

The treatment of Coulomb in *pd* breakup is realistically achieved; the results are technically reliable and, physicswise, the inclusion of Coulomb interaction is often important for a successful comparison with experimental data.

Acknowledgments

A.D. is supported by the FCT grant SFRH/BPD/14801/2003, A.C.F. in part by the FCT grant POCTI/FNU/37280/2001, and P.U.S. in part by the DFG grant Sa 247/25.

References

1. A. Kievsky, M. Viviani, and S. Rosati, *Phys. Rev.* **C64**, 024002 (2001).
2. C. R. Chen, J. L. Friar, and G. L. Payne, *Few-Body Syst.* **31**, 13 (2001).
3. E. O. Alt, A. M. Mukhamedzhanov, M. M. Nishonov, and A. I. Sattarov, *Phys. Rev.* **C65**, 064613 (2002).
4. S. Ishikawa, *Few-Body Syst.* **32**, 229 (2003).
5. A. Deltuva, A. C. Fonseca, and P. U. Sauer, *Phys. Rev.* **C71**, 054005 (2005).
6. E. O. Alt and M. Rauh, *Few-Body Syst.* **17**, 121 (1994).
7. A. Kievsky, M. Viviani, and S. Rosati, *Phys. Rev.* **C56**, 2987 (1997).
8. V. M. Suslov and B. Vlahovic, *Phys. Rev.* **C69**, 044003 (2004).
9. A. Deltuva, A. C. Fonseca, and P. U. Sauer, *Phys. Rev. Lett.* **95**, 092301 (2005).
10. A. Deltuva, A. C. Fonseca, and P. U. Sauer, to be published.
11. E. O. Alt, P. Grassberger, and W. Sandhas, *Nucl. Phys.* **B2**, 167 (1967).
12. J. R. Taylor, *Nuovo Cimento* **B23**, 313 (1974); M. D. Semon and J. R. Taylor, *ibid.* **A26**, 48 (1975).
13. E. O. Alt, W. Sandhas, and H. Ziegelmann, *Phys. Rev.* **C17**, 1981 (1978); E. O. Alt and W. Sandhas, *ibid.* **C21**, 1733 (1980).
14. P. U. Sauer, talk at this conference.
15. R. Machleidt, *Phys. Rev.* **C63**, 024001 (2001)
16. A. Deltuva, R. Machleidt, and P. U. Sauer, *Phys. Rev.* **C68**, 024005 (2003).
17. A. Deltuva, K. Chmielewski, and P. U. Sauer, *Phys. Rev.* **C67**, 054004 (2003).
18. St. Kistryn *et al.*, nucl-ex/0508012.
19. N. R. Kolb *et al.*, *Phys. Rev.* **C44**, 37 (1991).

Quasilinear and WKB Solutions in Quantum Mechanics*

R. Krivec
J. Stefan Institute, P.O. Box 3000, 1001 Ljubljana, Slovenia

V. B. Mandelzaweig
Racah Institute of Physics, The Hebrew University, Jerusalem 91904, Israel

F. Tabakin
Department of Physics and Astronomy, University of Pittsburgh, Pittsburgh, PA 15260

Solutions of the Schrödinger equation by the quasilinearization method (QLM) and by the WKB method are compared. While the latter generates an expansion in powers of $\hbar$, QLM approaches the solution of the equivalent nonlinear Riccati equation by approximating nonlinear terms with a sequence of linear ones. QLM does not rely on the existence of a smallness parameter. It is shown that both energies and wave functions in the first QLM iteration are more accurate than in the WKB approximation. The first QLM iterate is represented by a closed expression, allowing analytical estimates of the effects of parameters on the properties of the quantum systems. Quadratic convergence assures extremely accurate energies and wave functions in a few QLM iterations.

The present work is devoted to comparison of QLM (see review[1] and references therein) and WKB. Indeed, the derivation of the WKB starts by casting the radial Schrödinger equation into the nonlinear Riccati form and solving by expansion in powers of $\hbar$. It is interesting instead to solve this nonlinear equation with the help of the quasilinearization technique and compare with the WKB results. Such a procedure was performed in works[2,3] where it was shown that the first QLM iteration reproduces the structure of the WKB series by generating an infinite series of WKB terms, but with different coefficients. Besides being a better approximation, the first QLM iteration is also expressible in a closed integral form. Similar conclusions are reached for higher QLM approximations and it can be shown[1] that the p-th QLM iteration yields the correct structure of the infinite WKB series and reproduces 2^p terms of the expansion of the solution in powers of $\hbar$ exactly, as well as a similar number of terms approximately.

That the first QLM iteration already provides a much better approximation to the exact solution than the usual WKB is obvious not only from comparison of terms of the QLM and WKB series[1-3] but also from the fact that the quantization

*Article based on the presentation by V. B. Mandelzweig at the Third Asia-Pacific Conference on Few-Body Problems in Physics.

condition in the first QLM iteration leads to exact energies for many potentials[1,9] such as Coulomb, harmonic oscillator, Pöschl-Teller, Hulthen, Hylleraas, Morse, Eckart and some other well known physical potentials, which have a simple analytic structure. The WKB method reproduces exact energies only for the first two.

The goal of this work is to point out that QLM iterates provide much better approximations than the usual WKB for other potentials with more complicated analytical structure too. Unlike WKB, if the initial QLM guess is properly chosen, the wave function in all QLM iterations is free of unphysical turning point singularities. Since the first QLM iteration is given by an analytic expression [1–6,8,9], it allows analytical estimates of the roles of parameters and the influences of their variations on the characteristics of a quantum system. The subsequent iterates display very fast quadratic convergence; accuracy of energies obtained after a few iterations is extremely high, reaching up to 18 significant figures for the sixth iterate, as we show here on the example of the physically important double well potential.

The usual WKB substitution $y(x) = \frac{\psi'(x)}{\psi(x)}$ converts the Schrödinger equation to the nonlinear Riccati form

$$\frac{dy(x)}{dx} + (k^2(x) + y^2(x)) = 0. \tag{1}$$

Here $k^2(x) = 2(E - V(x))$ where we use the units $m = 1, \hbar = 1$.

The proper bound state boundary condition for potentials falling off at $x \simeq x_0 \simeq \infty$ is $y(x) = \text{const}$ at $x \geq x_0$. This means that $y'(x_0) = 0$, so that Eq. (1) at $x \simeq x_0$ reduces to $k(x_0)^2 + y^2(x_0)) = 0$ or $y(x_0)) = \pm ik(x_0)$. We choose here to define the boundary condition with the plus sign, so that $y(x_0) = ik(x_0)$.

The quasilinearization[2,4,6] of Eq. (1) leads to the recurrence differential equation

$$\frac{dy_p(x)}{dx} + 2y_p(x)y_{p-1}(x) = y_{p-1}^2(x) - k^2(x) \tag{2}$$

with the boundary condition $y_p(x_0) = ik(x_0)$. The Green function $G_{p-1}(x,s)$ of this equation satisfies

$$\frac{dG_{p-1}(x,s)}{dx} + 2y_p(x)G_{p-1}(x,s) = \delta(x - s) \tag{3}$$

and is given by $G_{p-1}(x,s) = \Theta(x-s)exp\left[-2\int_s^x y_{p-1}(t)dt\right]$. Using this Green function with the boundary condition one can show[2,8] that the analytic solution[2] of Eq. (2) expressing the p-th iterate $y_p(x)$ in terms of the previous iterate is

$$y_p(x) = f_{p-1}(x) - \int_{x_0}^{x} ds \frac{d\, f_{p-1}(s)}{ds}\, e^{\left[-2\int_s^x y_{p-1}(t)dt\right]},\, f_{p-1}(x) = \frac{y_{p-1}^2(x) - k^2(x)}{2y_{p-1}(x)}. \tag{4}$$

Indeed, differentiation of both parts of Eq. (4) leads immediately to Eq. (2) which proves that $y_p(x)$ is a solution of this equation. The boundary condition is obviously satisfied automatically.

It is interesting to use Eqs. (2) to estimate ground state wave function and energy of the anharmonic oscillator $\frac{g^2x^2}{2} + \epsilon x^3$. Chosing the logarithmic derivative of the harmonic oscillator wave function $y_0(z) = -z$ as a zero guess, where $z = gx$ is a new variable, and looking for the first iteration $y_1(z)$ in the form $y_1(z) = -z + Q(z)$ one gets the following equation for $Q(z)$ and the first iteration energy E_1:

$$g\frac{dQ(z)}{dz} - 2zQ(z)) = \frac{2\epsilon z^3}{g^3} - 2E_1 + g \tag{5}$$

Solution of Eq. (5) gives $Q(z) = -\frac{\epsilon}{g^3}(z^2 + g)$ and $E_1 = \frac{g}{2}$ which corresponds to the wave function $\chi(x) = C\exp\left[-\frac{gx^2}{2} - \epsilon\left(\frac{x}{g^2} + \frac{x^3}{3g}\right)\right]$. The same calculation for the potential $\frac{g^2x^2}{2} + \epsilon x^4$ leads to the first QLM iteration wave function $\chi(x) = C\exp\left[-\frac{gx^2}{2} - \epsilon\left(\frac{3x^2}{4g^2} + \frac{x^4}{4g}\right)\right]$ and the first iteration energy $E_1 = \frac{g}{2} + \frac{3\epsilon}{4g^2}$. These QLM results coincide with the results obtained by Zhao[10] who used a recently developed iterative method of solution of the Schrödinger equation (see, for example,[11] and references therein).

The obtained wave functions obvously have incorrect asymptotics. To ensure proper wave function asymptotic behavior one has to make an adequate initial guess, choosing, for example, for the zeroth iterate $y_0(x)$ the zero WKB approximation, i.e., setting $y_0(x) = ik(x)$. This choice generates proper asymptotic behavior already in the zeroth approximation, and in addition automatically satisfies the boundary condition. However, one has to be aware that this choice has unphysical turning point singularities. To avoid this, we pick here[8,9] as the zeroth iteration the smooth Langer WKB wave function[8,9,12].

As an example we consider the two-power (double-well) $V(x) = \frac{1}{2}g^2(x^2 - a^2)^2$ potential, i.e., the quartic potential in one dimension with degenerate minima. This potential is typically studied in quantum field theory and in the framework of the tunneling problem in quantum mechanics. Its perturbation series does not converge and different alternative nonperturbative approaches are therefore being explored, since the description of tunneling between the two minima should be necessarily nonperturbative (see reference[11] and the references therein).

The exact energy of the ground state, for $g = \frac{1}{8}$ and $a = 16$, is 0.482 958 659 913 315 548 20. We obtained this result by the Runge-Kutta method in quadruple precision. The WKB energy is different by 0.23% and equals 0.484067, while the first-iteration QLM energy equals 0.483017 and differs from the exact energy only by 0.012%. The QLM energy coincides with the exact energy in eighteen digits after the sixth iteration.

Fig. 1 displays logarithmic differences between the exact and the WKB solution and the exact solution and the first QLM iteration. The difference between the exact solution and the first QLM iteration is two orders of magnitude smaller than the difference between the exact and the WKB solution, i.e., one QLM iteration increases the accuracy of the result by two orders of magnitude.

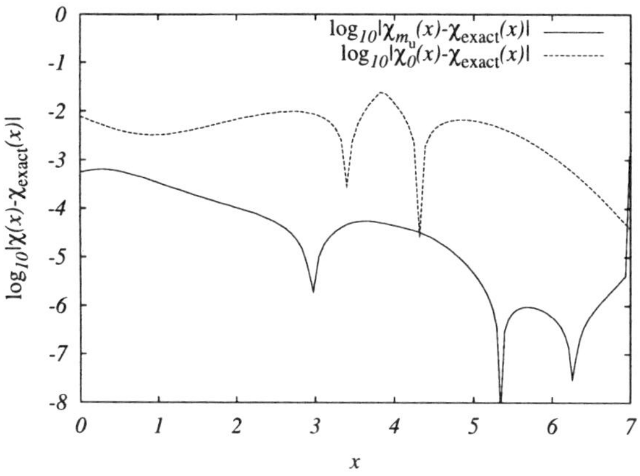

Fig. 1. Logarithm of the difference between the exact $\chi_{\rm exact}$ and WKB solutions χ_0 (dashed curve) and between the exact solution and the first QLM iterate χ_{m_u} (solid curve) for the ground state of the double well potential. The dips on the graphs are artifacts of the logarithmic scale, since the logarithm of the absolute value of the difference of two solutions goes to minus infinity at points where the difference changes sign. The overall accuracy of the solution can be inferred only at x values not too close to the dips.

In conclusion,it is shown on the example of the double well potential, that both energies and wave functions in the first QLM iteration are more accurate than in the WKB approximation. The first QLM iterate is represented by a closed expression, allowing analytical estimates of the effects of parameters on the properties of the quantum systems. Quadratic convergence assures extremely accurate energies and wave functions in a few QLM iterations.

Acknowledgments

This research was supported by Grant No. 2004106 from the United States-Israel Binational Science Foundation (BSF), Jerusalem, Israel.

References

1. V.B. Mandelzweig, Physics of Atomic Nuclei 68, 1227-1258 (2005); Yadernaya Fizika 68, 1277-1308 (2005).
2. K. Raghunathan and R. Vasudevan, J. Physics A: Math. Gen. 20, 839 (1987).
3. M. Jameel, J. Physics A: Math. Gen. 21, 1719 (1988).
4. V. B. Mandelzweig, J. Math. Phys. 40, 6266 (1999).
5. V. B. Mandelzweig, Few-Body Systems Suppl. 14, 185 (2003).
6. V. B. Mandelzweig and F. Tabakin, Computer Physics Comm. 141, 268 (2001).
7. R. Krivec and V. B. Mandelzweig, Computer Physics Comm. 152, 165 (2003).
8. R. Krivec, V. B. Mandelzweig and F. Tabakin, Few-Body Systems 34, 57 (2004).
9. R. Krivec and V. B. Mandelzweig, Phys. Lett. A337, 354-359 (2005).
10. Wei-Qin Zhao, arxiv:quant-ph/0310138 (2003).
11. R. Friedberg, T. D. Lee, Ann. Phys. **308**, 263 (2003).
12. R. E. Langer, Phys. Rev. 51, 669 (1937).

Scaling in Few-Body Nuclear Physics*

Lauro Tomio

Instituto de Física Teórica, UNESP, 01405-900, São Paulo, Brasil
E-mail: tomio@ift.unesp.br

T. Frederico[a], V. Timóteo[b], M. T. Yamashita[c], A. Delfino[d]

[a] *Departamento de Física, ITA, CTA, 12228-900, São José dos Campos, Brasil*
[b] *Centro Superior de Educação Tecnológica, Unicamp, 13484-370, Limeira, SP, Brasil*
[c] *Unidade Diferenciada de Itapeva, UNESP, 18409-010 Itapeva, SP, Brasil*
[d] *Instituto de Física, Universidade Federal Fluminense, 24210-900 Niterói, RJ, Brasil*

The nuclear matter calculations with realistic nucleon-nucleon potentials present a general scaling between the nucleon-nucleus binding energy, the corresponding saturation density, and the triton binding energy. The Thomas-Efimov three-body effect implies in correlations among low-energy few-body and many-body observables. It is also well known that, by varying the short-range repulsion, keeping the two-nucleon information (deuteron and scattering) fixed, the four-nucleon and three-nucleon binding energies lie on a very narrow band known as a Tjon line. By looking for a universal scaling function connecting the proper scales of the few-body system with those of the many-body system, we suggest that the general nucleus-nucleon scaling mechanism is a manifestation of a universal few-body effect.

When realistic two-nucleon interactions are used to calculate three-nucleon observables, the results exhibit some discrepancies [1], which are explained as from different strengths of the two-nucleon tensor force and short-range repulsions, provided that all realistic two-nucleon interactions have the correct one-pion exchange tail. In four-nucleon bound state (^{4}He) calculations the discrepancies still remain. But, as observed in the binding energies of ^{4}He (B_α) and triton (B_t), there is a correlation obtained when the short-range repulsion of the nucleon-nucleon interaction is varied while two-nucleon informations (deuteron and scattering) are kept fixed. This correlation is known as Tjon line [2]. B_α and B_t follows an almost straight line in the range of about 1-2 MeV of variation of the triton binding energy around the experimental value. As the long-range two-nucleon scales we have the deuteron binding energy (B_d) and the singlet virtual-state energy (B_v).

Two-body short-ranged interactions, supporting very low two-body binding energy and/or large scattering lengths, when used to calculate three-body systems approach the universal Thomas-Efimov limit [3]. By trying to find the range r_0 of the two-nucleon force, Thomas [4] showed that when $r_0 \rightarrow 0$, while the two-body

*This work is partially supported by FAPESP, CNPq, and FAPERJ

binding energy B_2 is kept fixed, the three-body binding energy goes to infinity (Thomas collapse). Much latter, Efimov [5] showed that, in the limit $B_2 = 0(r_0 \neq 0)$ the number of three-body bound states is infinite with an accumulation point at the common two- and three-body threshold. Note that both the Thomas and Efimov effects are claimed to be model independent, since they are due to a dynamically generated effective three-body potential acting at distances outside the range of the two-body potential. These apparently different effects are shown to be related to the same scaling mechanism [3]. One would expect that the Thomas-Efimov effect is manifested in weakly-bound quantum few-body systems which are much larger in size than the corresponding two-body effective range. The universal aspects of Efimov states are shown to generate a scaling limit in three-body systems [6], in applications to nuclear and atomic systems, using a renormalized zero-range model. This phenomenon, that was verified for low-energy three-body observables [6], was correspondingly observed when considering the three-body coupling constant [7] and recognized as a limit cycle phenomenon by Wilson [8].

Besides the fact that a two-body system with zero binding energy is not known in nature, in principle it is actually possible to produce it in laboratory. For trapped ultracold gases of certain atomic species it is expected that the Thomas-Efimov effect can be manifested, as it is possible to adjust the two-body scattering length at very large values, using Feshbach resonance techniques, by tuning the external magnetic field [9]. See Ref. [10] for a recent discussion on the scaling of observables with the three-body binding energy.

The Thomas-collapse of the three-body energy in systems of maximum wavefunction symmetry implies the existence of a three nucleon scale (identified with the triton binding energy), governing the short-range behavior of the wavefunction. Four nucleons can also form a state of maximum symmetry, allowing in principle the collapse of such configuration, independently of the three-nucleon collapse [11]. The Pauli principle allows only up to four nucleons at the same position. If more than four particles are allowed to overlap, it would imply that the asymptotic information from the interaction of the cluster would go beyond of those already fixed by the low-energy observables of two, three and four nucleons. If one wonders about the neutron matter within a non-relativistic quantum framework, in the limit of a zero-range force, we could say that the only scale in this case is the neutron-neutron scattering length. Therefore, the binding energy of neutron droplets will be strongly correlated to that quantity, which is the only physical scale in this situation allowed by the Pauli principle [12]. Moreover, stable tetraneutron droplets would imply in a major change in the neutron-neutron scattering length [13].

To study the possible correlations of the nuclear matter binding energy per nucleon with the two and three-body low-energy observables we consider the observables B_d, B_v, B_t and B_α as the scales determining the asymptotic properties of nuclei. In the limit of a zero-range interaction, we write the binding energy of a nuclei with mass number A and isospin projection I_z, considering isospin breaking effects, as

$$B_{(A,\ I_z)} = A\ B_t\ \mathcal{B}(\beta_v, \beta_d, \beta_\alpha, A, I_z)\,, \tag{1}$$

where $\beta_a = B_a/B_t$ with $a = v$, d and α.

According to the Tjon line, β_α remains approximately constant for a variety of two-nucleon potentials and the parametrization of the numerical results, given in MeV, for several two-nucleon potentials is $B_\alpha = 4.72\ (B_t - 2.48\text{MeV})$, which for $B_t^{exp} = 8.48$ MeV gives $B_\alpha^{exp} = 28.32$ MeV. Using this relation in (1), we obtain

$$R_{(A,\ I_z)} = B_{(A,\ I_z)}/A = \ B_t\ \mathcal{R}(B_t, A, I_z)\,, \tag{2}$$

where in the scaling function $\mathcal{R}_{(A,\ I_z)}$ the values of B_d and B_v are fixed to the experimental values. Equation (2) generalizes the concept of the Tjon line to nuclei. Indeed, there is some evidence from recent calculations using the AV18 nucleon-nucleon potential plus three-body forces [14] that there is a systematic improvement of the binding energy results for He, Li, Be and B isotopes simultaneously with the triton binding energy, when models are tuned to fit B_t.

Using a variety of two-nucleon potentials, in which the tensor strength was varied but the deuteron binding energy was kept fixed, it was shown that these interactions cannot quantitatively account for nuclear saturation [15]. In an energy versus density plot, the saturation points of nuclear matter obtained by employing different realistic potentials are located along a band (Coester band). Basically, nuclear matter saturates due to the composed repulsive and attractive short-range two-nucleon potential. It may also be seen as a typical low-energy problem if we compare the binding energy per nucleon, B_A/A, with the depth of the two-nucleon interaction. Then, it is natural to question whether any connection exists between the proper few-body scales, B_d, B_v and B_t with those of the many-body problem, like the B_A/A and the Fermi energy $E_F = \hbar^2 k_F^2/(2m_N)$. For light nuclei there is strong evidences of scaling between B_d, B_v and B_t as expressed by Eq. (2).

Here we are argue that the scales of nuclear matter, B_A/A and E_F, are determined by B_d, B_v and B_t. Therefore, we suppose that going towards the infinite isospin symmetrical nuclear matter, $A \to \infty$ and $I_z = 0$, the limit

$$\begin{aligned} \frac{B_A}{A} &\doteq \frac{B_t}{A} \lim_{A\to\infty} \mathcal{B}(\beta_v, \beta_d, \beta_\alpha, A, I_z = 0) \\ &= B_t\ \mathcal{G}(\beta_v, \beta_d, \beta_\alpha)\,, \end{aligned} \tag{3}$$

is well defined expressing the correlation between the binding energy of the nucleon in nuclear matter with the few-nucleon scales. The Fermi energy,

$$E_F = B_t\ \mathcal{E}_F(\beta_v, \beta_d, \beta_\alpha)\,, \tag{4}$$

will also be correlated to the few-nucleon binding energies.

In the present framework, the universal scaling functions connect the proper scales of the few-body system with those of the many-body system, as given by Eqs. (3) and (4). Different potentials, which describe the deuteron and the two-nucleon scattering properties give different values of B_t , B_α B_A/A and E_F.

As we have done in deriving Eq. (2) for a class of changes in the short-range part of the nuclear force that keeps the deuteron and low energy scattering properties unchanged, and taking into account that for these variations of the potential the ^{4}He and triton binding energies are strongly correlated as given by the Tjon line, one can rewrite Eqs. (3) and (4) in order to get a one parameter scaling:

$$\frac{B_A}{A} = B_t\ \mathcal{G}\left(B_t\right) \tag{5}$$

for fixed B_d and B_v, where the only true dependence in the class of potential variations is dominated by B_t. The analogous expression for the Fermi energy is

$$E_F\ =\ B_t\ \mathcal{E}_F\left(B_t\right), \tag{6}$$

where E_F scale with B_t.

In the perspective of the one parameter functions of Eqs. (5) and (6), it is clear that one could express E_t as a function of E_F and immediately get

$$\frac{B_A}{A\ E_F}\ = \mathcal{C}(E_F), \tag{7}$$

the correlation implied by the Coester band.

In order to enlighten our discussion we bring a variety of nuclear matter binding energies B_A/A at the corresponding saturation density, represented by the Fermi momenta k_F , calculated from different two-nucleon potentials. In Fig. 1, we present the well known Coester band in which the results for B_A/A and E_F are showed. The two distinct bands represent the nuclear matter calculations with and without the single-particle continuum contributions. The empirical values [16] are $B_A/A =$ 16 MeV and $E_F = 37.8$ MeV.

The one parameter dependence in Eqs. (5) and (6) suggests to plot the dimensionless quantity $B_A/(AB_t)$ as a function of the ratio E_F/B_t, which should look as an almost linear correlation. We display in the rhs of Fig. 1 the values for $B_A/(A\ B_t)$ versus E_F/B_t . As we could anticipate, the results show a clear linear correlation. We are tempted to say that, if the correlation is extrapolated and assuming that the binding and saturation densities somewhat decreases when three-body correlations are considered, it looks to be possible that the empirical values would be consistent with the correlation band.

In summary, we suggest a possible scaling of nuclei asymptotic properties with the triton binding energy, substantiated by some recent realistic calculations of light nuclei. We also found that the original correlation between the nuclear matter binding energy per nucleon with the Fermi momentum described by the Coester band can now be seen as robustly represented by the scaling of nuclear matter properties with the triton binding energy.

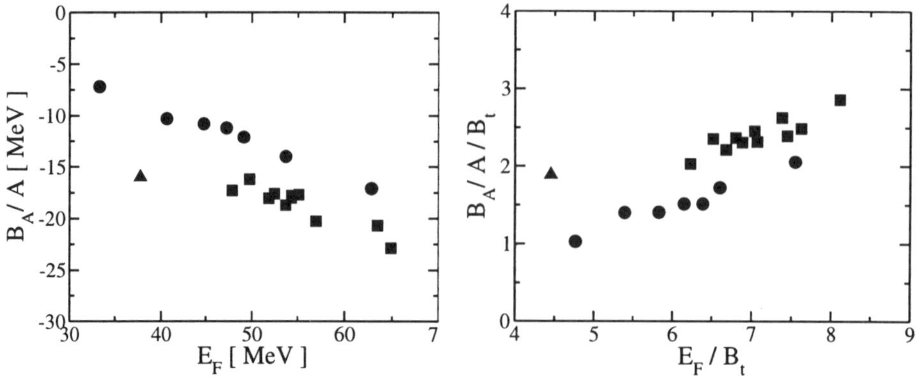

Fig. 1. Infinite nuclear matter binding energy as a function of E_F extracted from Ref. [17]. The squares includes the single particle contribution in continuum. The full triangle represents the empirical values. In the right-hand-side, all the quantities are given in units of the triton binding energy.

References

1. W. Glöckle, H. Witala, D. Huber, H. Kamada, J. Golak, Phys. Rep. **274** (1996) 107.
2. J.A. Tjon, Phys. Lett. B **56** (1975) 217; R.E. Perne and H. Kroeger, Phys. Rev. C **20** (1979) 340; J.A. Tjon, Nucl. Phys. A **353** (1981) 470.
3. S.K. Adhikari, A. Delfino, T. Frederico, I.D. Goldman and L. Tomio, Phys. Rev. A **37** (1988) 3666; S.K. Adhikari, A. Delfino, T. Frederico and L. Tomio, Phys. Rev. A **47** (1993) 1093.
4. L.H. Thomas Phys. Rev. **47** (1935) 903.
5. V. Efimov, Phys. Lett. B **33** (1970) 563; Nucl. Phys. A **362** (1981) 45.
6. A.E.A. Amorim, T. Frederico, L. Tomio **56** (1997) R2378; T. Frederico, L. Tomio, A. Delfino, and A.E.A. Amorim, **60** (1999) R9.
7. P.F. Bedaque, H.-W. Hammer, and U. van Kolck, Phys. Rev. Lett. **82** (1999) 463.
8. K.G. Wilson, hep-lat/0412043.
9. J.L. Roberts, N.R. Claussen, S.L. Cornish, and C.E. Wieman, Phys. Rev. Lett. **85** (2000) 728; S. Jonsell, H. Heiselberg and C.J. Pethick, Phys. Rev. Lett. **89** (2002) 250401.
10. A.S. Jensen, K. Riisager, D.V. Fedorov, E. Garrido, Rev. Mod. Phys. **76** (2004) 215.
11. S.K. Adhikari, T. Frederico, and I.D. Goldman, Phys. Rev. Lett. **74** (1995) 487; S.K. Adhikari and T. Frederico, Phys. Rev. Lett. **74** (1995) 4572; T. Frederico, A. Delfino, and L. Tomio, Phys. Lett. B **481** (2000) 143.
12. A. Delfino, and T. Frederico, Phys. Rev. C **53** (1996) 62.
13. S.C. Pieper, Phys. Rev. Lett. **90** (2003) 252501.
14. S.C. Pieper and R.B. Wiringa, Ann. Rev. Nucl. Part. Sci. **51** (2001) 53; R.B. Wiringa, S.C. Pieper, Phys. Rev. Lett. **89** (2002) 182501.
15. F. Coester, S. Cohen, B.D. Day, and C.M. Vincent, Phys. Rev. C **1** (1970) 769.
16. R. J.Furnstahl, nucl-th/0504043.
17. R. Machleidt, Adv. Nucl. Phys. **19** (1989) 189.

Spin Correlation Parameter C_{yy} of $p + {}^3$He Elastic Backward Scattering at Intermediate Energy

Y. Shimizu, K. Hatanaka, A.P. Kobushkin[a], T. Adachi[b], K. Fujita, K. Itoh[c], T. Kawabata[d], T. Kudoh[e], H. Matsubara, H. Ohira[e], H. Okamura[f], K. Sagara[e], Y. Sakemi, Y. Sasamoto[d], Y. Shimbara, H.P. Yoshida[e], K. Suda[d], Y. Tameshige, A. Tamii, M. Tomiyama[e], M. Uchida, T. Uesaka[d], T. Wakasa[e], and T. Wakui[d]

Research Center for Nuclear Physics (RCNP) Osaka University, Ibaraki, Osaka 567-0047, Japan
[a]*N.N. Bogulyubov Institute for Theoretical Physics, 252143 Kiev, Ukraine*
[b]*Department of Physics, Osaka University, Toyonaka, Osaka 560-0043, Japan*
[c]*Department of Physics, Saitama University, Urawa, Saitama 338-8570, Japan*
[d]*Center for Nuclear Study (CNS), University of Tokyo, Wako, Saitama 351-0198, Japan*
[e]*Department of Physics, Kyushu University, Hakozaki, Fukuoka 812-8581, Japan*
[f]*Cyclotron and Radioisotope Center (CYRIC), Tohoku University, Sendai, Miyagi 980-8578, Japan*

It is possible to use nucleon-nucleus scattering as a probe of the spin structure of the nuclei since target related observables are extremely sensitive to spin dependent parts of the target wave function. In addition, one can gain information about the nucleon-nucleus reaction mechanism, the spin dependent nucleon-nucleon interaction in the nuclear medium, and off-shell behavior of the nucleon-nucleon amplitudes. For ^{3}He(p,^{3}He)p elastic backward scattering, only small amount of data points exist for the differential cross section and no data exist for spin dependent observables. We developed a spin exchange type polarized ^{3}He target and measured the spin correlation parameter C_{yy} at 200, 300, and 400 MeV.

1. Introduction

For several decades considerable efforts have been done to investigate structure of the lightest nuclei (d, ^{3}He ^{4}He) at short distances between constituent nucleons. Significant progress was achieved both in theory and experiment, first of all because high quality data on spin dependent observables were obtained with both hadronic and electromagnetic probes. Large part of these investigations consists of studies of elastic backward (in the center of mass system) proton-nucleus scattering (EBS). This process involves large momentum transfer and therefore a belief exists that EBS can provide an access to high momentum components of the wave function of the lightest nuclei.

The p^3He EBS is studied in much less detail than the pd EBS. Today, however, high intensity beams of polarized protons in combination with polarized ^{3}He targets give an opportunity to perform detailed studies of p^3He EBS including spin dependent observables. This, in turn, demands careful theoretical studies of the reaction mechanism. There are several cross section data of the p^3He EBS, mainly at energies higher than 400 MeV. Analyzing powers were measured for the p^3He elastic scattering at TRIUMF at 200 ~ 500 MeV, but measurements were limited at relatively forward angles[1]. We measured the differential cross section and the spin correlation parameter C_{yy} of the p^3He EBS at 200, 300, and 400 MeV.

2. Experiment

The measurements were performed at the Research Center for Nuclear Physics (RCNP), Osaka University. We used vertically polarized protons at incident energies of 200, 300, and 400 MeV. The beam intensity was 10 to 40 nA, which was limited by counting rates. The proton polarization was about 70 %. Elastically scattered ^{3}He particles were measured by the Grand Raiden spectrometer[8] to be set at 0°. In order to stop the beam and integrate the current a Faraday cup was installed inside and near the exit of the first dipole magnet of the spectrometer.

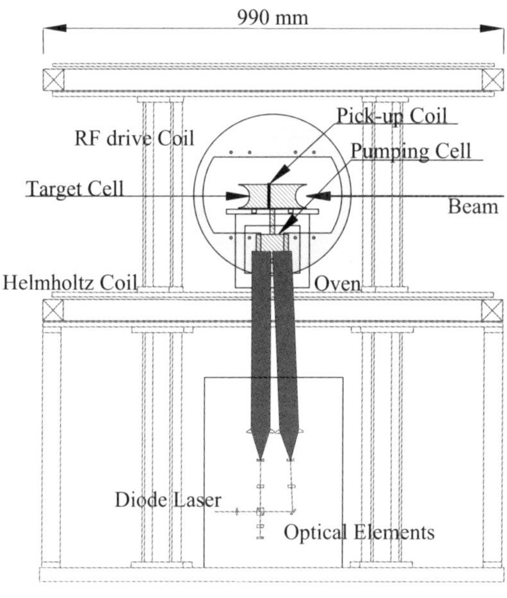

Fig. 1. The schematic setup of the ^{3}He target

A spin exchange type polarized ^{3}He target was developed at RCNP. The schematic setup is shown in Fig. 1. The cell which was made of borosilicate glass (Corning7056) consists of two parts, a target cell and a pumping cell, connected by a transfer tube. In order to reduce background, the cell windows were as thin as 150 μm. During operation the pumping cell was heated to about 460 K to get a sufficient Rb vapor density for optical pumping and spin exchange collisions. Polarized ^{3}He atoms are transferred to the target cell by diffusion. By using two cells we can avoid problems of large backgrounds from the Rb atoms and of depolarizing effects. A high power diode laser which is a Fiber Array Packaged Laser, COHERENT FAP79-30C-800LB, and optical elements were introduced to polarize the Rb atoms in the pumping cell. The ^{3}He polarization was measured by the Adiabatic Fast Passage (AFP) NMR method. However, this method gives only a relative value of the polarization. The target polarization was calibrated independently by using the $^3\mathrm{He}(p,\pi^+)^4\mathrm{He}$ reaction, where polarization is inferred from beam related asymmetries[3]. The spin signature of the reaction is $\frac{1}{2}^+ + \frac{1}{2}^+ \to 0^+ + 0^-$, and the spin correlation parameter C_{yy} takes the constant value of 1. Finally, the AFP NMR amplitude was calibrated by a following,

$$P_{^3\mathrm{He}}\ [\%] = (6.62 \pm 0.16) \times 10^2 \cdot V_{\mathrm{NMR}}\ [\mathrm{mV}]. \tag{1}$$

3. Result and Discussion

Figure 3 shows the differential cross section and the spin correlation parameter C_{yy} of the p^3He EBS at $E_p = 200$, 300, and 400 MeV. In this figure, the open circles show results of this work, and other symbols are values[4~7] extrapolated to $\theta_{cm} = 180°$ by us. For the differential cross section, present results are consistent with previous data. For the spin correlation parameter C_{yy}, these are measured for the first time.

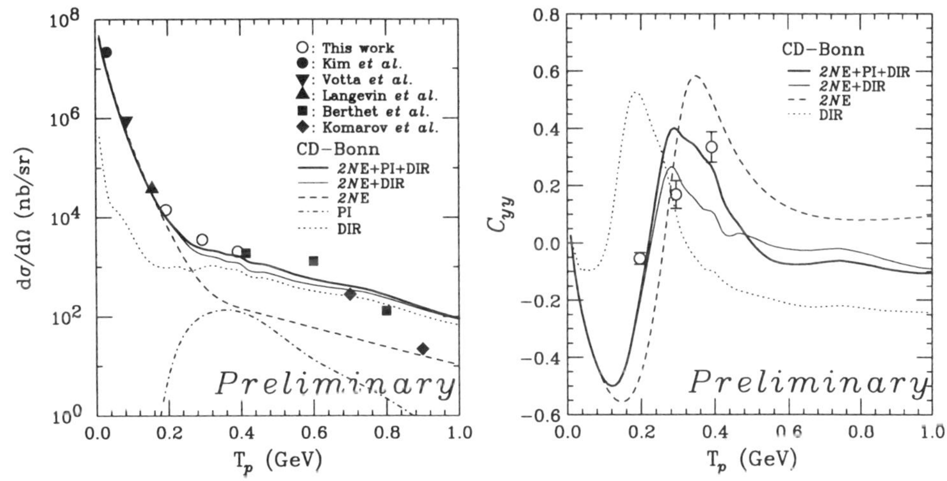

The left and right panel shows the differential cross section and the spin correlation parameter C_{yy} of the p^3He EBS, respectively. Open circles show results of this work, and other symboles are values[4~7] are extrapolated to $\theta_{cm} = 180°$ by us. The bold and thin solid curves represent *2N*E + PI + DIR and *2N*E + DIR mechanisms, respectively. The dashed, dot-dashed, and dotted curves represent *2N*E, PI, and DIR mechanism, respectively. All calculations were used ^{3}He wave function for CD-Bonn potential.

Recently, a careful theoretical study of the $p+^3$He EBS reaction mechanism appeared [8~11]. Kobushkin *et al.* [11] provide systematic analysis of the $p+^3$He EBS at intermediate energy. Their calculations were included three reaction mechanisms which are two nucleon exchange (*2N*E), pion mechanism (PI), and direct mechanism (DIR). Figure 3 shows the predictions for ^{3}He wave function with CD-Bonn potential. For the differential cross section, we can see that their calculation explains data well but at higher than 0.8 GeV they overestimate data. For the spin correlation parameter C_{yy}, there are discrepancies between our results and their calculation. The lack of agreement is likely due to the contribution of the non-nucleonic degrees of freedom. Present results of the spin correlation parameter C_{yy} will provide an impetus for more sophisticated theoretical models to be considered.

Acknowledgments

We thank the RCNP staff for their supports during the experiment. We also wish to thank Professor H. Toki for his encouragements throughout the work. This experiment was performed under Program No. E180 at the RCNP. This work was supported in part by the Grant-in-Aid for Scientific Research, Grant No. 14340074, of the Ministry of Education, Culture, Sports, Science and Technology of Japan.

References

1. D.K. Hasell *et al.*, Phys. Rev. Lett. **74**, 502 (1986); E.J. Brash *et al.*, Phys. Rev. C **52**, 807 (1995); R. Tacik *et al.*, Phys. Rev. Lett. **63**, 1784 (1989);
2. M. Fujiwara *et al.*, Nucl. Instrum. Meshodos Phys. Res. A **422**, 484 (1999).
3. G.G. Ohlsen, Rep. Prog. Phys. **35**, 760 (1972).
4. P. Berthet *et al.*, Phys. Lett. **106B**, 465 (1981); R. Frascaria *et al.*, Phys. Lett. **66B**, 329 (1977).
5. C.C. Kim *et al.*, Nucl. Phys. **58**, 32 (1964).
6. L.G. Votta *et al.*, Phys. Rev. C **10**, 520 (1974).
7. H. Langevin-Joliot *et al.*, Nucl. Phys. **A158**, 309 (1978).
8. A.V. Lado and Yu.N. Uzikov, Phys. Lett. **B279**, 16 (1992).
9. L.D. Blokhintsev *et al.*, Nucl. Phys. **A597**, 487 (1996).
10. Yu.N. Uzikov and J. Haidenbauer, Phys. Rev. C **68**, 014001 (2003); Yu.N. Uzikov, in *Proceedings of the Nuclear Many-Body and Medium Effects in Nuclear Interactions and Reactions, Fukuoka, Japan, 2002,* edited by K. Hatanaka, T. Noro, K. Sagara, H. Sakaguchi, and H. Sakai (World Scientific, Singapore, 2003), p. 137; Yu.N. Uzikov, Nucl. Phys. **A644**, 321 (1998).
11. A.P. Kobushkin *et al.*, nucl-th/0112078 (2003). Submitted to Phys. Lett. **B**.

Study of Four-Nucleon Systems by Four-Body Faddeev-Yakubovsky Equations

Eizo Uzu

Department of Physics, Faculty of Science and Technology, Tokyo University of Science,
2641 Yamazaki, Noda, Chiba, 278-8510, Japan
E-mail: j-uzu@ed.noda.tus.ac.jp

Yasuro Koike

Science Research Center, Hosei University,
2-17-1 Fujimi, Chiyoda-ku, Tokyo 102-8160, Japan,
Center for Nuclear Study, University of Tokyo,
2-1 Hirosawa, Wako, Saitama, 351-0198, Japan

Hiroyuki Kamada

Department of Physics, Faculty of Engineering, Kyushu Institute of Technology,
1-1 Sensuicho, Tobata, Kitakyushu 804-8550, Japan

Masahiro Yamaguchi

Science Research Center, Hosei University,
2-17-1 Fujimi, Chiyoda-ku, Tokyo 102-8160, Japan

Complex energy method is introduced to handle singularity of Green's function in four-body Faddeev-Yakubovsky equations. We find that it is applicable to finding resonance states. Finite range expansion method is introduced to reduce the size of matrices of the Faddeev-Yakubovsky equations. We think this is applicable even in the energy region above the four-body break-up threshold when we employ low-momentum NN interactions.

1. Introduction

Study of four-nucleon systems with the four-body Faddeev-Yakubovsky (FY) equations is one of the most challenging researches. For instance, we are interested in effects of three-nucleon forces (TNF) on four-nucleon systems. According to the studies of TNF in three-nucleon systems, these effects are getting larger with increasing energies. Thus it will be advantage if we study TNF in higher energy region, namely, above the four-body break-up threshold (4BBT). Another interest we have is to reproduce many resonance states. Four-nucleon systems are the smallest one which has many resonance states, and most of them are lower than 4BBT. Thus we need to solve the FY equations in all energy region.

When we solve the FY equations in the energy region above 4BBT, there is a hurdle to handle singularities of Green's functions. We found that complex energy method[1] (CEM) is available to handle the singularities even when we solve the FY equations above 4BBT[2]. By the way, the size of matrix is another hurdle. To reduce the size, we often utilize separable expansion of the three-body and two-two subamplitudes as well as two-body potentials. Recently we demonstrate[3] that finite range expansion (FRE)[4,5] method is applicable to clear this hurdle.

In this paper, we describe outlines of both of CEM and FRE and outlooks for applications to finding resonance states and solving the FY equations above 4BBT.

2. Complex Energy Method

2.1. *Outline of CEM*

Calculations for scattering systems in configuration space require boundary conditions which increase in complexity with growing particle numbers. These boundary conditions appear in the form of Green's functions in momentum space which carry singularities of increasing complexity. The Green's functions are expressed as $G_0 = 1/(E + i\varepsilon - H_0)$ where E and H_0 are the total and kinetic energy, respectively, and the limit $\varepsilon \to 0$ has to be taken.

In the two-body and three-body case, we can use the principal value prescription and (half) the residue theorem or contour deformation[6–9] technique. In the four-body case, however, behavior of singularities are much complicated due to H_0 containing three valuables. The singularities appear since we take the limit $\varepsilon \to 0$ before we solve the equations. Then if we solve the equations with finite ε's, we can solve them without any singularities. Obtained solutions are not physical, therefore, we perform analytic continuation to real energy, regarding ε as a sampling parameter. We leave details and example solutions in Ref.[2].

2.2. *Application CEM to Finding Resonance States*

Kukulin proposed analytical continuation in the coupling constant[10] for finding resonance states, and it is one of ideas gotten for CEM. Thus it is natural to extend CEM to find resonance states. Summary of CEM is as follows. First, we solve the equations (e.g. FY equations) on several complex energies and find solutions (e.g. scattering amplitudes). Next, we perform analytic continuation of the solutions to real energy, regarding the complex energies as a sampling parameter. To find resonance energy, we modify these terms as follows. First, we find eigenvalues on several real (or complex but its imaginary part is positive) energies. Next, we perform analytic continuation of the energies to the eigenvalue becoming one, regarding the calculated eigenvalues as a sampling parameter.

To check if this method works, we find a resonance energy and width for ^{9}He with regarding it as ^{8}He+n two-body system, and compared our solution with that in Ref.[12]. We employ set (I′) potential in this reference and obtained that the resonance energy is 1.162MeV and width is 1.618MeV which agree well with the values in the reference. We would like to apply this method to four-nucleon system and find many resonances using FY equations.

3. Finite Range Expansion Method

3.1. *Outline of FRE*

Two-body Lippmann-Schwinger equation is expressed in operator form as

$$T = V + VG_0T, \tag{1}$$

where T is a T-matrix, V is a potential, and G_0 is Green's function. When we study nuclear physics, the potential V is of a finite range. Any state vector $|\psi\rangle$ in the Hilbert space can be expanded with a set of basis functions $|B_i\rangle$ which are complete in the range, or

$$V|\psi\rangle = \sum_i V|B_i\rangle \xi_i, \tag{2}$$

with expansion coefficients ξ_i. This is a fundamental idea of FRE. The basis functions simply can be polynomial.

Defining form factors $|g_i\rangle$ and $\langle g_i|$ as

$$|g_i\rangle = V|B_i\rangle \quad \text{and} \quad \langle g_i| = \langle B_i|V, \tag{3}$$

we expand the potential and T-matrix in Eq. (1) as

$$V = \sum_{ij} |g_i\rangle \lambda_{ij} \langle g_j|, \quad \text{and} \quad T = \sum_{ij} |g_i\rangle \tau_{ij} \langle g_j|, \tag{4}$$

respectively, where

$$\left[\lambda^{-1}\right]_{ij} = \langle B_i|V|B_j\rangle, \quad \text{and} \quad \left[\tau^{-1}\right]_{ij} = \left[\lambda^{-1}\right]_{ij} - \langle g_i|G_0|g_j\rangle. \tag{5}$$

Eq. (1) is reduced as

$$\tau_{ij} = \lambda_{ij} + \sum_{km} \lambda_{ik} \langle g_k|G_0|g_m\rangle \tau_{mj}, \tag{6}$$

which are $N\times N$ matrix equations with a truncation of the sum in Eq. (2) with N.

Equations for three-body and two-two subamplitudes are

$$X = Z + Z\tau X \quad \text{and} \quad Y = W + W\tau Y, \tag{7}$$

where X and Y are three-body and two-two subamplitudes, Z and W are Born terms which are described in Ref.[3], and τ is a two-body propagator which is equivalent to that in eq. (6). Eqs. (7) are identical to eq. (1), thus following procedure for separable expansion is the same as that of eqs. (2)-(6), on one condition. If energy is higher than 4BBT, Z and W reaches to infinity in configulation space and FRE is not applicable. In other words, the energy of the system must be lower than 4BBT when we employ FRE. We leave details and example solutions in Ref.[3].

3.2. *Outlook of FRE to Apply Above 4BBT*

We are planning to employ Low-Momentum NN interaction (LMNN)[13] for applying FRE in the energy region above 4BBT. LMNN is a potential in momentum representation which is divided into two subspaces, the model space and its complement, with a unitary transformation. The model space is within some cutoff momentum and the potential is zero in the outside of this region.

When we employ LMNN to solve the FY equations, Z and W in eqs. (7) are restricted in some finite range in momentum space. Thus we can apply FRE even when the energy is above 4BBT.

4. Summary

CEM is introduced to handle singularity of Green's function in the energy region above 4BBT. We find that the basic idea of CEM is applicable to finding resonance states that there are many states in the four-nucleon systems and most of them are not well studied yet. FRE is introduced to reduce the size of matrices of the FY equations. We find that this method gives well converged solutions below 4BBT. And we think FRE is applicable even in the energy region above 4BBT when we employ LMNN. Thus we get good tools to solve the four-body FY equations and this is the start of the study of four-nucleon systems.

References

1. H. Kamada, Y. Koike and W. Glöckle, Prog. Theor. Phys. **109**, 869L (2003).
2. E. Uzu, H. Kamada, and Y. Koike, Phys. Rev. C **68**, 061001 (2003).
3. E. Uzu and Y. Koike, Prog. Theor. Phys. submitting.
4. Y. Koike, Prog. Theor. Phys. **87**, 775 (1992).
5. Y. Koike, W. C. Parke, L. C. Maximon, and D. R. Lehman, Few-Body Sys., **23**, 53 (1997).
6. C. Lovelace, Phys. Rev. **C135**, 1125 (1964).
7. R. T. Cahill and I. H. Sloan, Nucl. Phys. **A165**, 161 (1971).
8. W. Ebenhöh, Nucl. Phys. **A191**, 97 (1972).
9. Y. Koike, Nucl. Phys. **A301**, 411 (1978).
10. V. I. Kukulin, V.M. Krasnopolsky, and M. Miselkhi, Yad Fiz. **29**, 818 (1979) [Sov. J. Nucl. Phys. **92**, 421 (1979)].
11. S. Nakaichi, T.K. Lim, Y. Akaishi, and H. Tanaka, Phys. Rev. A **26**, 32 (1982).
12. S. Aoyama, K. Katō, and K. Ikeda, Phys. Rev. C **55**, 2379 (1997).
13. S. Fujii, E. Epelbaum, H. Kamada, R. Okamoto, K. Suzuki, and W. Glöckle, Phys. Rev. C **70**, 024003 (2004).

The Treatment of Coulomb Interaction in the Description of Three-Nucleon Reactions with Two Protons*

A. Deltuva and A. C. Fonseca

Centro de Física Nuclear da Universidade de Lisboa, P-1649-003 Lisboa, Portugal

P. U. Sauer

Institut für Theoretische Physik, Universität Hannover, D-30167 Hannover, Germany

The Coulomb interaction between the two protons is included in the calculation of three-nucleon hadronic and electromagnetic reactions. The hadron dynamics is based on the purely nucleonic charge-dependent (CD) Bonn potential and its realistic extension CD Bonn + Δ to a coupled-channel two-baryon potential, allowing for single virtual Δ-isobar excitation. Calculations are done using integral equations in momentum space. The screening and renormalization approach is employed for including the Coulomb interaction. Convergence of the procedure is found at moderate screening radii. The reliability of the method is demonstrated. The Coulomb effect on observables is seen at low energies for the whole kinematic regime. In proton-deuteron elastic scattering at higher energies the Coulomb effect is confined to forward scattering angles; the Δ-isobar effect found previously remains unchanged by Coulomb.

The inclusion of the Coulomb interaction in the description of the three-nucleon continuum is one of the most challenging tasks in theoretical few-body nuclear physics. The Coulomb interaction is well known, in contrast to the strong two-nucleon and three-nucleon potentials mainly studied in three-nucleon scattering. However, due to its $1/r$ behavior, the Coulomb interaction does not satisfy the mathematical properties required for the formulation of standard scattering theory. When the theoretical description of three-particle scattering is attempted in integral form, the Coulomb interaction renders the standard equations ill-defined; the kernel of the equations is noncompact. When the theoretical description is based on differential equations, the asymptotic boundary conditions for the wave function have to be numerically imposed on the trial solutions and, in the presence of the Coulomb interaction, those boundary conditions are nonstandard.

Our treatment of the Coulomb interaction is based on the ideas proposed in Ref. [1] for two charged particle scattering and extended in Ref. [2] for three-particle scattering. The Coulomb potential is screened, standard scattering theory for short-range potentials is used in the Alt-Grassberger-Sandhas formulation [3], and the obtained results are corrected by the renormalization technique for the unscreened

*Talk delivered by P. U. Sauer

limit. In contrast to previous works [4,5] based on the same idea of screening and renormalization which were limited to the use of low-rank separable potentials, we use modern two-nucleon potentials and three-nucleon forces in full without separable expansion. In particular, the results of this paper are based on the purely nucleonic charge-dependent (CD) Bonn potential [6] and on its coupled-channel extension CD Bonn + Δ [7], allowing for a single virtual Δ-isobar excitation and fitted to the experimental data with the same degree of accuracy as CD Bonn itself. In the three-nucleon system the Δ isobar mediates an effective three-nucleon force and effective two- and three-nucleon currents, both consistent with the underlying effective two-nucleon force. The screening and renormalization technique used by us is described in detail in Refs. [8,9]. The specific features of our method are recalled below.

a) We work with a new type of screened Coulomb potential

$$w_R(r) = w(r)\, e^{-(r/R)^n}, \tag{1}$$

where $w(r) = \alpha_e/r$ is the true Coulomb potential, α_e being the fine structure constant and n controlling the smoothness of the screening. We prefer to work with a sharper screening than the Yukawa screening ($n = 1$) of Refs. [4,5]. We want to ensure that the screened Coulomb potential w_R approximates well the true Coulomb one w for distances $r < R$ and simultaneously vanishes rapidly for $r > R$, providing a comparatively fast convergence of the partial-wave expansion, but avoids an unpleasant oscillatory behavior characteristic for the sharp cutoff ($n \to \infty$). Like in Refs. [8,9] $n = 4$ is our choice for the results of this paper.

b) Although the choice of the screened potential improves the partial-wave convergence, the practical implementation of the solution of the AGS equation still places a technical difficulty, i.e., the calculation of the AGS operators for nuclear plus screened Coulomb potentials requires two-nucleon partial waves with pair orbital angular momentum considerably higher than required for the hadronic potential alone. In this context the perturbation theory for high two-nucleon partial waves developed in Ref. [10] is a very efficient and reliable technical tool for treating the screened Coulomb interaction in high partial waves.

After the renormalization procedure the predictions for observables of three-nucleon reactions have to show independence from the choice of the screening radius R, provided it is chosen sufficiently large. That convergence is our internal criterion for the reliability of our Coulomb treatment. Figure 1 studies the convergence of our method with increasing screening radius R for elastic proton-deuteron (pd) scattering. The convergence is impressive and strongly suggests the reliability of the chosen Coulomb treatment. Furthermore, Ref. [11] makes a detailed comparison between the results obtained by the present technique and those of Ref. [12] obtained from the variational solution of the three-nucleon Schrödinger equation in configuration space with the inclusion of an *unscreened* Coulomb potential between the protons and imposing the proper Coulomb boundary conditions explicitly. The agreement, across the board, between the results derived from two entirely different methods, clearly indicates that both techniques for including the Coulomb interaction are reliable.

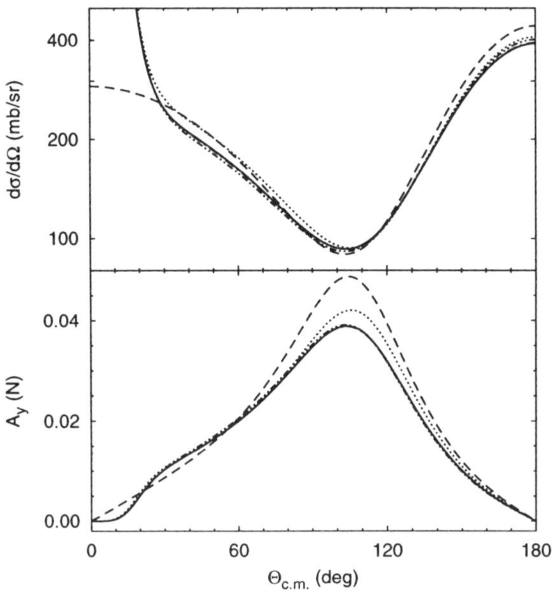

Fig. 1. Convergence of the differential cross section and of the proton analyzing power $A_y(N)$ for *pd* elastic scattering at 3 MeV proton lab energy with screening radius R. The observables are shown as functions of the c.m. scattering angle. The hadronic potential is CD Bonn + Δ. Results obtained with screening radius $R = 5$ fm (dotted curves), 10 fm (dashed-double-dotted curves), 15 fm (dashed-dotted curves), 20 fm (double-dashed-dotted curves), 25 fm (solid curves) are compared. Results without Coulomb (dashed curves) are given as reference for the size of the Coulomb effect.

Figure 2 gives characteristic low-energy results, just above deuteron breakup threshold. The Coulomb effect is quite appreciable at all scattering angles. In contrast, on the scale of the observed Coulomb effect, the Δ-isobar effect is minute at those low energies. The inclusion of Coulomb is essential for a successful account of data for the spin-averaged differential cross section and for the deuteron tensor analyzing powers. However, the inclusion of Coulomb increases the discrepancy between theoretical predictions and experimental data in the peak region of proton and deuteron vector analyzing powers, the so-called A_y puzzle. Our findings are consistent with the results of Refs. [5,12].

Figure 3 shows selected results at 135 MeV proton lab energy. The Coulomb effect is confined to the forward direction, i.e., to c.m. scattering angles smaller than 30 degrees where the Δ-isobar effect is not visible. The Δ-isobar effect shows up rather strongly in the region of the diffraction minimum, where its effect is beneficial and the Coulomb effect is gone. Δ-isobar and Coulomb effects are nicely separated. Thus, the Δ-isobar effect found previously [7] on the Sagara discrepancy and on spin observables remains essentially unchanged by the inclusion of the Coulomb interaction. The predictions of Fig. 3 are characteristic for all observables at higher energies.

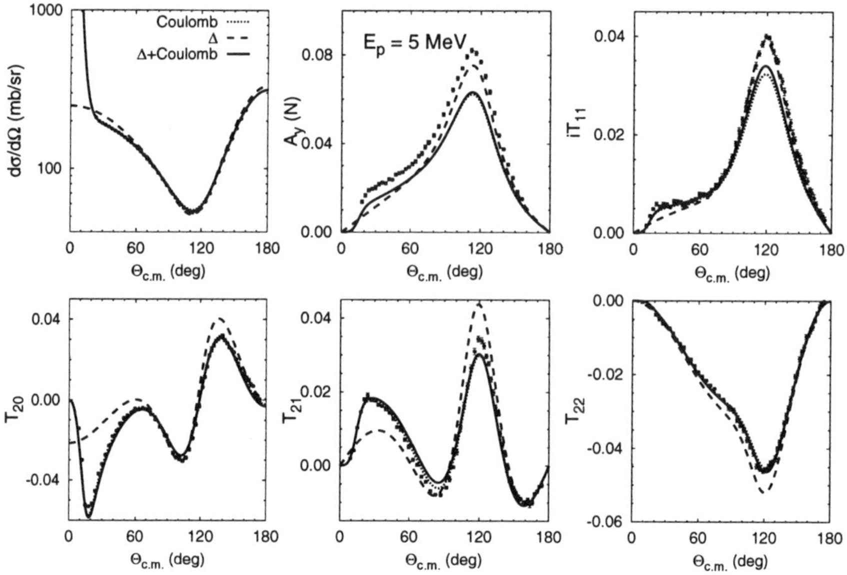

Fig. 2. Differential cross section and analyzing powers for *pd* elastic scattering at 5 MeV proton lab energy as functions of the c.m. scattering angle. Results including Δ-isobar excitation and the Coulomb interaction (solid curves) are compared to results without Coulomb (dashed curves). In order to appreciate the size of the Δ-isobar effect the purely nucleonic results including Coulomb are also shown (dotted curves). The experimental data are from Ref.[13].

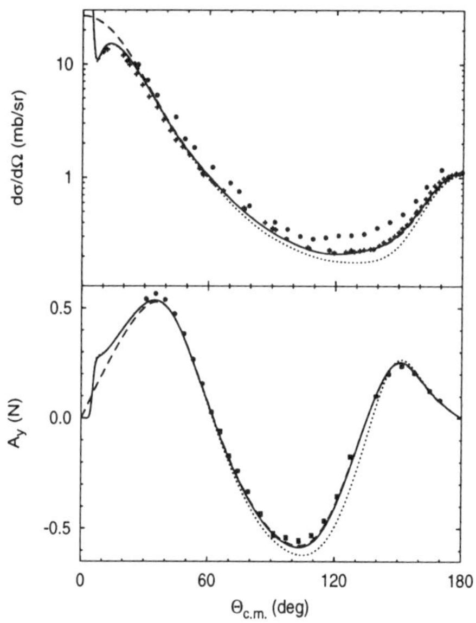

Fig. 3. Differential cross section and proton analyzing power $A_y(N)$ for *pd* elastic scattering at 135 MeV proton lab energy as functions of the c.m. scattering angle. The curves are explained in the caption of Fig. 2. The experimental data are from Ref.[14] (crosses) and from Ref.[15] (full circles) for the differential cross section, and from Ref.[16] for the analyzing power.

The treatment of Coulomb in *pd* elastic scattering is realistically achieved; the results are technically reliable and, physicswise, the inclusion of Coulomb interaction is often important for a successful comparison with experimental data. The extension of that Coulomb treatment to *pd* breakup is given in another talk at this conference [17]. The extension of the Coulomb treatment to the corresponding electromagnetic reactions is carried out in Refs. [8,9].

Acknowledgments

A.D. is supported by the FCT grant SFRH/BPD/14801/2003, A.C.F. in part by the FCT grant POCTI/FNU/37280/2001, and P.U.S. in part by the DFG grant Sa 247/25.

References

1. J. R. Taylor, *Nuovo Cimento* **B23**, 313 (1974); M. D. Semon and J. R. Taylor, *ibid.* **A26**, 48 (1975).
2. E. O. Alt, W. Sandhas, and H. Ziegelmann, *Phys. Rev.* **C17**, 1981 (1978); E. O. Alt and W. Sandhas, *ibid.* **C21**, 1733 (1980).
3. E. O. Alt, P. Grassberger, and W. Sandhas, *Nucl. Phys.* **B2**, 167 (1967).
4. G. H. Berthold, A. Stadler, and H. Zankel, *Phys. Rev.* **C41**, 1365 (1990).
5. E. O. Alt, A. M. Mukhamedzhanov, M. M. Nishonov, and A. I. Sattarov, *Phys. Rev.* **C65**, 064613 (2002).
6. R. Machleidt, *Phys. Rev.* **C63**, 024001 (2001).
7. A. Deltuva, R. Machleidt, and P. U. Sauer, *Phys. Rev.* **C68**, 024005 (2003).
8. A. Deltuva, A. C. Fonseca, and P. U. Sauer, *Phys. Rev.* **C71**, 054005 (2005).
9. A. Deltuva, A. C. Fonseca, and P. U. Sauer, *Phys. Rev. Lett.* **95**, 092301 (2005).
10. A. Deltuva, K. Chmielewski, and P. U. Sauer, *Phys. Rev.* **C67**, 054004 (2003).
11. A. Deltuva, A. C. Fonseca, A. Kievsky, S. Rosati, P. U. Sauer, and M. Viviani, *Phys. Rev.* **C71**, 064003 (2005).
12. A. Kievsky, M. Viviani, and S. Rosati, *Phys. Rev.* **C64**, 024002 (2001).
13. K. Sagara *et al.*, *Phys. Rev.* **C50**, 576 (1994); K. Sagara, private communication.
14. K. Sekiguchi *et al.*, *Phys. Rev.* **C65**, 034003 (2002).
15. K. Ermisch *et al.*, *Phys. Rev.* **C68**, 051001(R) (2003).
16. K. Ermisch *et al.*, *Phys. Rev. Lett.* **86**, 5862 (2001).
17. A. Deltuva, talk at this conference.

Three-Nucleon Force Effects in Observables for $\vec{d}p$ Breakup at 130 MeV*

A. Biegun, B. KŁos, A. Micherdzińska, E. Stephan, W. Zipper
Institute of Physics, University of Silesia, PL-40007 Katowice, Poland

K. Bodek, J. Golak, St. Kistryn, J. Kuroś-ŻoŁnierczuk, R. Skibiński, R. Sworst, H. WitaŁa, J. Zejma
Institute of Physics, Jagiellonian University, PL-30059 Kraków, Poland

A. Kozela
Institute of Nuclear Physics PAN, PL-31342 Kraków, Poland

K. Ermisch, N. Kalantar-Nayestanaki, M. Kiš, M. Mahjour-Shafiei
Kernfysisch Versneller Instituut, NL-9747 AA Groningen, The Netherlands

W. Glöckle
Institut für Theoretische Physik II, Ruhr-Universität, D-44780 Bochum, Germany

H. Kamada
Department of Physics, Kyushu Institute of Technology, Kitakyushu 804-8550, Japan

E. Epelbaum
Jefferson Laboratory, Theory Division, Newport News, VA 23606, USA

A. Nogga
Forschungszentrum Jülich, Institut für Kernphysik (Theorie), D-52425 Jülich, Germany

P. Sauer
Institut für Theoretische Physik, Universität Hannover, D-30167 Hannover, Germany

A. Deltuva
Centro di Fisica Nuclear da Universidade de Lisboa, P-1649-003 Lisboa, Portugal

Measurement of the cross section and spin observables for the kinematically complete $^1\mathrm{H}(\vec{\mathrm{d}},\mathrm{pp})\mathrm{n}$ breakup process was carried out at 130 MeV beam energy. The obtained cross section results, spanning the large part of the phase space, are compared with the the-

*This work is supported by the Polish Committee for Scientific Research under Grant No 1P03B02627 oretical predictions based on realistic potentials and ChPT. Confronting the calculated cross sections with the experimental data shows a clear advantage of the predictions in which the 3NF contributions are included. Non-negligible Coulomb effects have been also observed in some kinematical regions.

1. Introduction

Three-nucleon (3N) system is the simplest testing ground for probing the basic nucleon-nucleon (NN) interaction in a non-trivial environment. It allows also to search for subtle effects of the additional dynamics, so-called three-nucleon force (3NF). Among 3N systems the deuteron-proton breakup process, with continuum of final states, provides the richest basis for such studies. Theoretical calculations for this process based on the so-called realistic potentials (CD-Bonn, Argonne, Nijmegen I and II) and the Tucson-Melbourne (TM99) model of 3NF show that relative influence of 3NF effects increases with beam energy. At deuteron beam energy of 130 MeV significant contributions of the genuine 3NF are predicted for cross sections, vector and tensor analyzing powers in numerous regions of the phase space [1]. An alternative mechanism of generating a 3NF is based on the so-called explicit Δ-isobar excitation. These calculations are performed in a coupled channel approach and the predicted effects of 3NF in cross section are generally smaller then in the case of realistic potentials combined with TM99. Recently, the new calculations based on chiral perturbation theory (ChPT) became also available. In order to provide appropriate testing basis for various theoretical approaches and to study details of the 3N dynamics, the breakup reaction has been measured with a polarized deuteron beam, with the use of a detection system covering a large part of the phase space.

2. Experimental Setup and Data Analysis

The experiment was performed at the Kernfysisch Versneller Instituut (KVI), Groningen, The Netherlands. The beam of vector and tensor polarized deuterons with energy of 130 MeV was focused to a spot of approximately 2 mm diameter on a liquid hydrogen target of 4 mm thickness. The experimental setup consisted of a three-plane multiwire proportional chamber MWPC and two layers of segmented scintillator hodoscope: transmission ΔE and stopping E detectors. MWPC was used for precise reconstruction of the particle emission angles, while the hodoscope allowed to identify the particles, to determine their energies and to define trigger conditions. The ΔE-E wall covered a large fraction of the phase-space: for polar angles θ the range from 10° to 35° and a full (2π) range of the azimuthal angles ϕ. We registered coincidences of charged reaction products: two outgoing protons from the breakup reaction or proton and deuteron from the elastic scattering.

From the collected data the cross sections for 72 kinematically complete configurations have been extracted. The selected configurations span a grid of 9 different θ_1, θ_2 pairs and 8 values of ϕ_{12}. Examples of the cross-section distributions are presented in Fig. 1. The data were compared with the results of the realistic potential

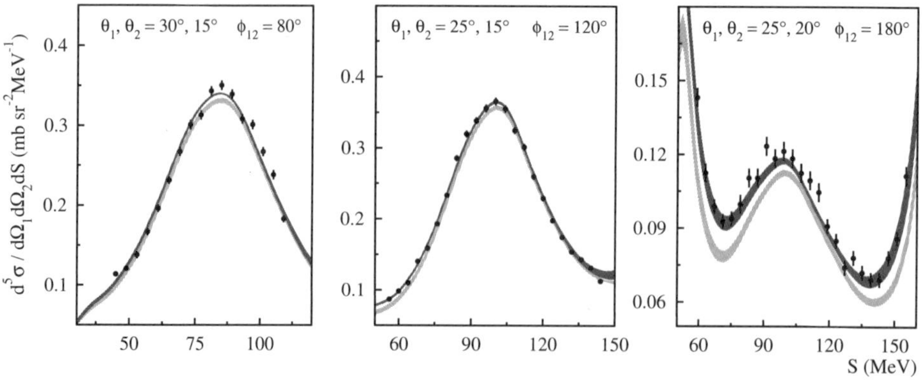

Fig. 1. Breakup cross-section data (dots) for three kinematical configurations. The light-grey bands represent predictions of the realistic NN potentials, the dark-grey bands show predictions with the same potentials combined with the TM99 3NF model.

plus model 3NF approach, with the ChPT predictions and with the calculations employing the coupled-channel potential with the explicit Δ-isobar treatment. Detail discussion of the comparison is given in Refs [2,3]. It has been observed that for a majority of the investigated configurations the predictions reproduce well the measured cross-section distributions. The agreement is further improved when the 3NF's are included in the treatment of the 3N system dynamics.

However, there are some configurations in which significant discrepancies have been observed between the measured cross sections and their theoretically predicted values (see Figure 2). One possible explanation is the influence of the Coulomb interaction. The first results of calculations for the breakup reaction with included Coulomb effects have just became available [4]. Examples of our cross-section data compared with such calculations are presented in Figure 2. Clearly, inclusion of the

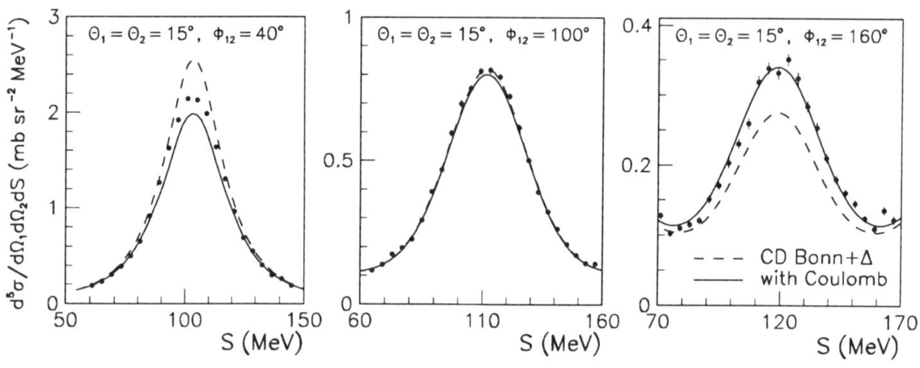

Fig. 2. Cross-section results (dots) in three kinematical configurations characterized by the same polar angles, compared to the results of the coupled-channel calculations. The dashed lines represent the predictions obtained with the inclusion of the virtual Δ-isobar excitation effects. The solid lines show the results of the calculations taking into account also the Coulomb interaction effects.

Coulomb interaction reduces significantly the observed discrepancies, both overestimations and underestimations of the data.

Since the $^{1}H(\vec{d},pp)n$ experiment was performed for 7 various polarization states of the deuteron beam, the analysis is currently extended to extract polarization observables. A few examples of preliminary results are shown in Figure 3.

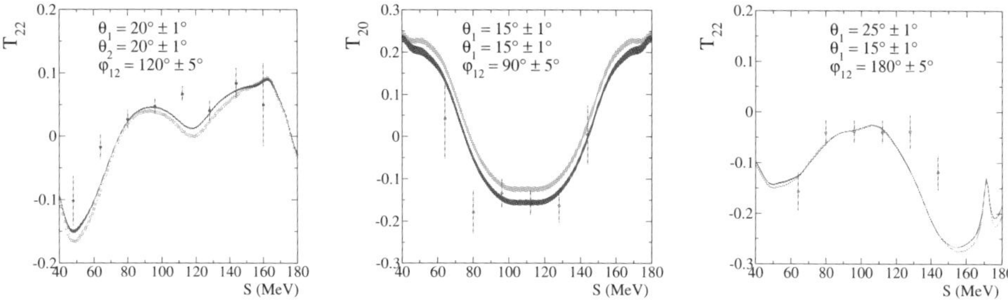

Fig. 3. Preliminary results for tensor analyzing powers T_{22} and T_{20} in selected kinematical configurations. The meaning of the theoretical bands is the same as in Figure 1.

References

1. J. Kuroś-Żołnierczuk, H. Witała, J. Golak, H. Kamada, A. Nogga, R. Skibiński, and W. Glöckle, Phys. Rev. C **66**, 024004 (2002).
2. St. Kistryn *et al.*, Phys. Rev. C **68**, 054004 (2003).
3. St. Kistryn *et al.*, accepted for publication in Phys. Rev. C, arXiv:nucl-ex/0508012.
4. A. Deltuva, A. C. Fonseca and P. Sauer, to be published in Phys. Rev. C, arXiv:nucl-th/0509034.

Three-Nucleon Force Effects in Nucleon-Deuteron Elastic and Breakup Reactions

S. Ishikawa

Department of Physics, Science Research Center, Hosei University,
2-17-1 Fujimi, Chiyoda, Tokyo 102-8160, Japan
E-mail: ishikawa@i.hosei.ac.jp

We constructed a model that simulates the long-range part of meson theoretical three-nucleon forces in a simple mathematical form with central and tensor operators. We studied thereby possible effects on polarization observables in nucleon-deuteron elastic and breakup reactions when the effect of the tensor component is inversed, which was successful in explaining tensor analyzing powers at low energy proton-deuteron scattering.

1. Introduction

A three-nucleon force (3NF) arising from the exchange of two pions among three nucleons (2πE-3NF) is known to get rid of the discrepancies between experimental data and theoretical calculations with realistic two-nucleon forces (2NFs) for the three-nucleon (3N) binding energies and nucleon-deuteron (ND) differential cross sections around the minima of angular distributions.

On the other hand, the 2πE-3NF is not successful in explaining some of spin observables for the ND scattering. Well known examples are the vector analyzing powers of nucleon and deuteron at low energies, for which 2NF calculations and experimental data differ considerably and the 2πE-3NF gives only a minor effect as compared to the differences. Besides these, we have pointed out that tensor components of the 2πE-3NF give undesirable contributions to tensor analyzing powers (TAP's) in proton-deuteron (*pd*) elastic scattering at 3.0 MeV[1,2]. In Ref. 2, we reported that a 3NF having an opposite effect to the tensor contribution in the 2πE-3NF improves the fit to experimental data of the TAP's.

In this paper, we will report the effects of such tensor-inversed 3NF on ND scattering observables for higher energies up to 30 MeV.

2. Model 3NFs

Since the mathematical structure of the 2πE-3NF is so complicated, first, we introduce a model 3NF that simulates the 2πE-3NF in a simple way[2]. It consists of two-body central (spin-independent) and tensor operators accompanied by a three-

Table 1. Strength parameters of the Gaussian 3NFs[2].

Model	V_0 (MeV)	V_T (MeV)
C+T	-38	+20
C-T	-45	-20

body Gaussian form factor:

$$V = \sum_{i \neq j \neq k} \exp\{-(\frac{r_{ik}}{r_G})^2 - (\frac{r_{jk}}{r_G})^2\} \left\{V_0 + V_T S_T(ij)\hat{P}_{11}\right\}, \qquad (1)$$

where $\hat{P}_{11}$ is the projection operator to the spin- and isospin triplet state of the pair (i, j). We fixed the range parameter r_G to be 1.0 fm for simplicity and determined two strength parameters to simulate the calculated 3N binding energy and the tensor analyzing power $T_{21}(\theta)$ of *pd* scattering at $E_p = 3.0$ MeV with the Argonne V_{18} model of 2NF (AV18)[3] and the Brazil model of the 2πE-3NF (BR)[4]. The values are shown in Table 1 as "C+T".

Next, we change the sign of the tensor strength V_T in the C+T 3NF and readjust the spin-independent strength V_0 to fit the 3N binding energy (tensor-inversion). The values determined are shown in Table 1 as "C-T". This C-T 3NF gives an effect with opposite tensor contribution to that of the 2πE-3NF and explains the low energy $T_{21}(\theta)$ experimental data quite well[2].

Since it is known that the 2πE-3NF gives only a minor effect on the vector analyzing powers, we use a spin-orbit 3NF introduced by Kievsky[5], which has a form of

$$V = V_{LS} \exp\left(-\alpha\sqrt{\frac{2}{3}\left(r_{12}^2 + r_{23}^2 + r_{31}^2\right)}\right) \sum_{i>j} \boldsymbol{L}_{ij} \cdot (\boldsymbol{S}_i + \boldsymbol{S}_j). \qquad (2)$$

We used the smallest range version of the model with $\alpha = 1.5\ \mathrm{fm}^{-1}$ and the strength of $V_{LS} = -16$ MeV, which is slightly modified from the original value. The potential will be denoted as "SO" below.

3. Numerical Results

3.1. *Notes on the numerical calculations*

Our numerical method to solve the Faddeev equation is based on coordinate space integral equation approach[6,7]. In Table 2, we compare our results of phase shift parameters for the doublet and quartet scattering at 4.0, 14.1, and 42.0 MeV for the Malfliet-Tjon I-III potential with the results by the Los-Alamos/Iowa group that are cited in the benchmark calculations[8].

In the present calculation, 3N partial-wave states for which 2NF acts are restricted to those with total two-nucleon angular momenta up to 3 and the total 3N angular momentum is truncated at 19/2. We do not take into account for the Coulomb force, and thus, all calculations below are for the neutron-deuteron (*nd*) scattering. We use the AV18 model for the 2NF.

Table 2. Phase shifts and inelasticities for *nd* elastic scattering with the Malfliet-Tjon I-III potential.

E_n (MeV)		Doublet LA/Iowa[8]	Doublet This work	Quartet LA/Iowa[8]	Quartet This work
4.0	$\Re(\delta)$	143.7	143.7	101.5	101.6
	η	0.964	0.964	1.000	1.000
14.1	$\Re(\delta)$	105.4	105.4	68.9	69.0
	η	0.463	0.465	0.978	0.980
42.0	$\Re(\delta)$	41.3	41.6	37.8	37.9
	η	0.501	0.500	0.906	0.894

3.2. *Elastic observables*

In all figures below, the results with the AV18 are denoted by solid curves; those with the AV18+BR+SO, which is essentially equivalent to the AV18+C+T+SO, by dashed curves; those with the AV18+C-T+SO by dotted curves.

Examples of numerical results are shown in Fig. 1: (a) the vector analyzing power of neutron $A_y(\theta)$ at $E_n = 14.1$ MeV and (b) the tensor analyzing power of deuteron $T_{21}(\theta)$ at $E_d = 56$ MeV ($E_n = 28$ MeV). In Fig. 2, we plot the energy dependence of the observables at scattering angles of 90° and 125°.

In Fig. 1 (a), a noticeable improvement in $A_y(\theta)$ at $E_n = 14.1$ MeV is obtained around $\theta = 125°$ mainly due to the SO-3NF. This improvement is observed up to $E_n \sim 20$ MeV in Fig. 2 (a).

On the other hand, in Fig. 1 (b), an unpleasant change occurs for $T_{21}(\theta \sim 90°)$ at $E_n = 28$ MeV by the introduction of the 2πE-3NF as in the low energy case[1]. However, it looks like that the tensor-inversion effect cancels the change caused by the 2πE-3NF around $\theta = 90°$. As a result, $T_{21}(\theta \sim 90°)$ by the AV18 and those by the AV18+C-T+SO almost coincide each other throughout the energy range up to $E_n = 30$ MeV as shown in Fig. 2 (b). This effect may arise from both the change in the spin-independent strength and the sign change of the tensor component in the C-T 3NF, see Table 1, and the fact that a scattering observable is given as a binary form of the scattering amplitudes, *e.g.*, the product of scalar amplitudes and tensor amplitudes as dominant terms for the tensor analyzing powers.

Results of such complexity are observed also for $T_{21}(\theta \sim 125°)$ and $A_y(\theta \sim 90°)$.

3.3. *Breakup observables*

In Fig. 3, examples of breakup observables are presented. The figure shows the results of the differential cross sections and the neutron vector analyzing power for so-called collinear configuration in *nd* breakup reaction at $E_n = 13$ MeV. Although the effects of the 3NFs are small for the breakup cross sections, we see small but visible effects in the vector analyzing power showing the difference between the 3NF models.

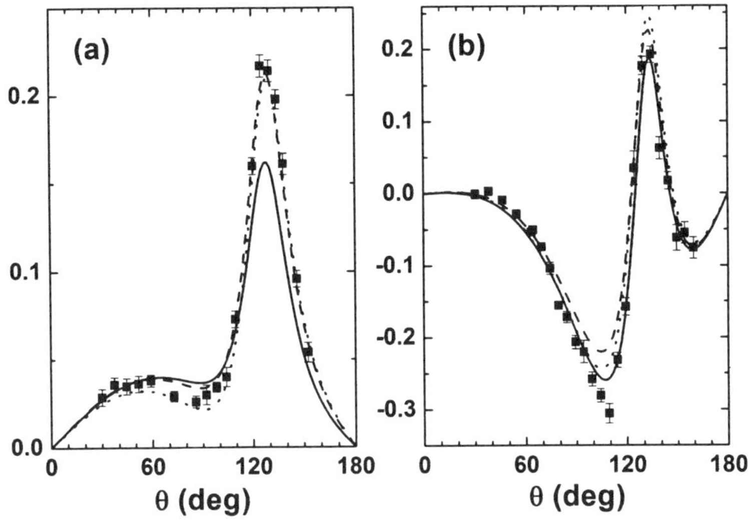

Fig. 1. (a) Vector analyzing power of neutron $A_y(\theta)$ for nd elastic scattering at $E_n = 14.1$ MeV and (b) tensor analyzing power of deuteron $T_{21}(\theta)$ for dp elastic scattering at $E_d = 56$ MeV. Data are from Ref. 9 for (a) and from Ref. 10 for (b). See the text for the meaning of the curves.

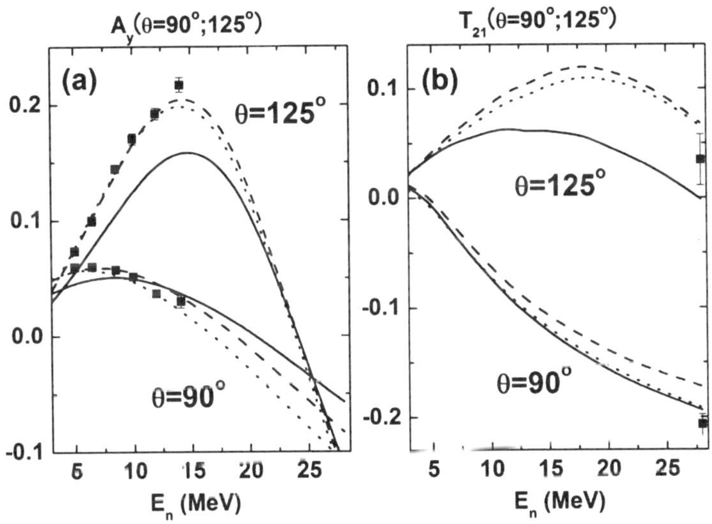

Fig. 2. Energy dependence of (a) $A_y(\theta)$ for nd elastic scattering and (b) $T_{21}(\theta)$ for dp elastic scattering at $\theta = 90°$ and $\theta = 125°$. Data are from Refs. 9, 11, and 12 for (a) and from Ref. 10 for (b). See the text for the meaning of the curves.

4. Summary

We have studied effects of tensor and spin-orbit contributions in 3NF on ND scattering observables by introducing model 3NFs that are easy to handle.

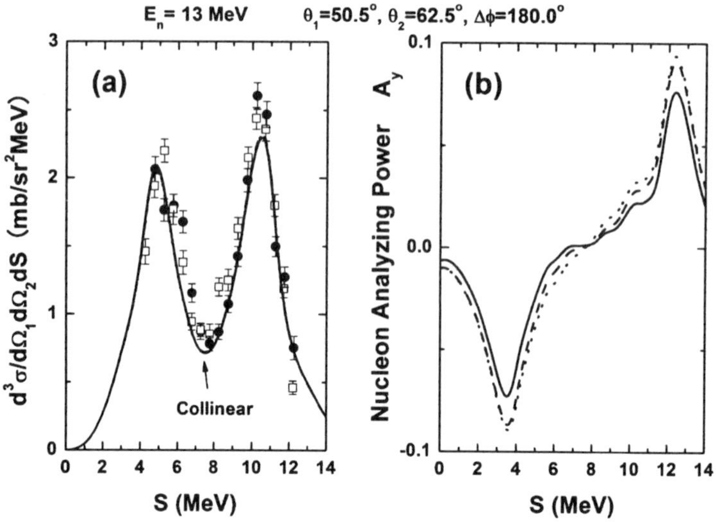

Fig. 3. Differential cross section and neutron analyzing power for *nd* breakup at 13 MeV as a function of the arc length along the kinematical curve for configuration: $\theta_{n1} = 50.5°$, $\theta_{n2} = 62.5°$, and $\phi_{12} = 180°$. The data are from Ref. 13 for the solid circles and Ref. 14 for open squares. See the text for the meaning of the curves.

The tensor-inversion effect of 2πE-3NF, which was successful in 3.0 MeV, also provides sizable effects in $A_y(\theta)$ and $T_{21}(\theta)$ for higher energies.

It is an interesting problem whether the mechanism of the exchange of mesons other than the pion can produce such tensor effects that gives the opposite contribution to that of the 2πE-3NF.

References

1. S. Ishikawa, M. Tanifuji, and Y. Iseri, *Phys. Rev.* **C67**, 061001-R (2003).
2. S. Ishikawa, M. Tanifuji, and Y. Iseri, *Proceedings of the Seventeenth International IUPAP Conference on Few-Body Problems in Physics*, Durham, North Carolina, USA, June 5-10, 2003, edited by W. Glöckle and W. Tornow (Elsevier, 2004) p.S61.
3. R. B. Wiringa, V. G. J. Stoks, and R. Schiavilla, *Phys. Rev.* **C51**, 38 (1995).
4. H. T. Coelho, T. K. Das, and M. R. Robilotta, *Phys. Rev.* **C28**, 1812 (1983).
5. A. Kievsky, *Phys. Rev.* **C60**, 034001 (1999).
6. S. Ishikawa, *Nucl. Phys.* **A463**, 145c (1987).
7. S. Ishikawa, *Few-Body Syst.* **32**, 229 (2003).
8. J. L. Friar, *et al.*, *Phys. Rev.* **C42**, 1838 (1990).
9. W. Tornow, *et al.*, *Phys. Rev.* **C27**, 2439 (1983).
10. K. Hatanaka, *et al.*, *Nucl. Phys.* **A426**, 77 (1984).
11. W. Tornow, *et al.*, *Phys. Lett.* **B257**, 273 (1991).
12. C. R. Howell, *et al.*, Few-Body Syst. **2**, 19 (1987).
13. J. Strate, *et al.*, *Nucl. Phys.* **A501**, 51 (1989).
14. H. R. Setze, *et al.*, *Phys. Rev.* **C71**, 034006 (2005).

Use of the AMD Method Employing Realistic NN Potentials for Few Nucleon Systems

T. Watanabe and S. Oryu

Department of Physics, Faculty of Science and Technology, Frontier Research Center for Computational Science, Tokyo University of Science, Noda, Chiba 278-8510, Japan.
E-mail: oryu@ph.noda.tus.ac.jp

A new method for calculating the ground states of light nuclei is proposed in which an improved trial function for the Antisymmetrized Molecular Dynamics (AMD) method is stipulated. Bound states for 2N, 3N, and 4N systems are calculated by using a Volkov potential, and the results are compared with those obtained from the two-body Lippmann-Schwinger equation, the three-body Faddeev equations, and the four-body Faddeev-Yakubovsky equations, respectively. Furthermore, we could successfully take into account the LS and tensor interactions of the realistic Paris potential for the first time within the AMD framework. Using our new method, the deuteron binding energy, the root mean square radius, the magnetic moment, and the quadrupole moment were calculated. The results reproduce those for physical observables obtained using the Lippmann-Schwinger equation.

1. Introduction

The antisymmetrized molecular dynamics (AMD) method [1,2] is a powerful technique for investigating light and medium-light nuclei from "first principles", or by the "only input" with a given NN potential. It is known that the AMD method can not only describe the "coexistence of the cluster picture and the shell-model picture" but also the neutron halo property in light and medium-light nuclei. Earlier calculations [3–5] were performed using very simple effective local potentials like the Volkov potential. So called "realistic NN potentials" were not previously employed in the analysis, because the bound state was not obtained when the potentials contains strong repulsive core [6–8]. Furthermore, including tensor and LS parts of NN interactions in the naive AMD framework is tedious. Therefore, the AMD method is not optimal for the "few-body problem" because the deuteron, triton, and α particle are strongly affected by the tensor component of the NN force. One of the main purposes of this paper is to modify the AMD method to solve the few-body problem using realistic NN interactions.

2. Standard AMD Method

In the standard AMD method [1,2], a single particle wave function is expressed as a Gaussian wave packet in which a complex parameter set $\{\boldsymbol{Z}\}$ =

$\{Z_{1x}, Z_{1y}, Z_{1z}; Z_{2x}, Z_{2y}, Z_{2z}; \cdots, Z_{Nz}\}$ is introduced for the N-nucleon system. The wave function can be written as follows:

$$\begin{aligned}\varphi_i(j) &= \phi_{\boldsymbol{Z}_i}(\boldsymbol{r}_j)|\sigma_i\rangle|\tau_i\rangle \\ &= (\frac{2\nu}{\pi})^{\frac{3}{4}} \exp[-\nu(\boldsymbol{r}_j - \frac{\boldsymbol{Z}_i}{\nu})^2 + \frac{1}{2}\boldsymbol{Z}_i^2]|\sigma_i\rangle|\tau_i\rangle,\end{aligned} \tag{1}$$

where $\boldsymbol{r}_j$ is a j-th nucleon position vector from the origin, $\boldsymbol{Z}_i$ denotes a central coordinate of the wave packet, and ν is width parameter, $|\sigma_i\rangle$ and $|\tau_i\rangle$ denote the spin and isospin functions, respectively. The wave function from an N-nucleon system is given by the Slater determinant,

$$\Phi = \frac{1}{N!}\det[\varphi_i(j)]. \tag{2}$$

In order to obtain the ground state of the system, we minimize the expectation value of the Hamiltonian by a variational technique with respect to the parameter $\{\boldsymbol{Z}\}$. Numerically, the frictional cooling method [9] is used.

3. A New AMD Trial Wave Function

First, we introduce the basis wave function written in terms of Jacobi coordinates. We write the Jacobi coordinates for an N-particle system as $\boldsymbol{\rho} = (\boldsymbol{\rho}_1, \boldsymbol{\rho}_2, \boldsymbol{\rho}_3, \cdots, \boldsymbol{\rho}_N)$. Here we can denote the relation between usual cartesian coordinates $\boldsymbol{r}$ and the Jacobi coordinates $\boldsymbol{\rho}$ by the transformation matrix K_{ij}

$$(\boldsymbol{\rho}_i) = (K_{ij})(\boldsymbol{r}_j) \qquad (i,j) = 1, 2, \cdots, N. \tag{3}$$

Therefore, $\{\boldsymbol{Z}\}$ is also transformed by K_{ij}

$$(\boldsymbol{A}_i) = (K_{ij})(\boldsymbol{Z}_j) \qquad (i,j) = 1, 2, \cdots, N. \tag{4}$$

We define our wave function by a Gaussian-type function in the Jacobi coordinate basis

$$\Psi = \sum_{k=1}^{M_c} C_k \phi_k(\{\boldsymbol{\rho}, \boldsymbol{\nu}_k, \boldsymbol{A}_k, \boldsymbol{\xi}_k\}), \tag{5}$$

where

$$\begin{aligned}&\phi_k(\{\boldsymbol{\rho}, \boldsymbol{\nu}_k, \boldsymbol{A}_k, \boldsymbol{\xi}_k\}) \\ &= \mathcal{A}\Big[\prod_{i=1}^{N} \exp[-\nu_{ki}(\boldsymbol{\rho}_i - \boldsymbol{A}_{ki})^2]\Big]\Big[\prod_{i=1}^{N} |\sigma_{ki}\rangle|\tau_i\rangle\Big],\end{aligned} \tag{6}$$

and $\{\boldsymbol{C}\} = \{C_1, C_2, C_3, \cdots, C_{M_c}\}$ are linear coefficients. The $|\sigma_{ki}\rangle$ is the spin wave function using variational parameters $\{\boldsymbol{\xi}\} = \{\boldsymbol{\xi}_1, \boldsymbol{\xi}_2, \boldsymbol{\xi}_3, \cdots, \boldsymbol{\xi}_N\}$ as follows:

$$|\sigma_{ki}\rangle = (\frac{1}{2} + \xi_{ki})|\uparrow\rangle + (\frac{1}{2} - \xi_{ki})|\downarrow\rangle, \tag{7}$$

where the $|\uparrow\rangle$, $|\downarrow\rangle$ are the spin-up state and the spin-down state respectively. The $|\tau_i\rangle$ is the isospin wave function. $\mathcal{A}$ is the antisymmetrization operator. We use this wave function instead of Eq. (2), and perform a variational minimization. [a]

4. Simulation Results

Results for the ground state of ^{2}H, ^{3}H, ^{3}He and ^{4}He using a Volkov potential are presented in Table 1. It is found that those are consistent with numerical results of Faddeev calculations.

Table 1. Binding energies of 2N, 3N, and 4N systems calculated by AMD with a Volkov potential. The results are compared with those of the Lippmann-Schwinger eq., Faddeev eq., and Faddeev-Yakubovsky eq. The three kinds of results in the 4N system correspond to three different Jacobi coordinates, the so-called K-type, H-type, and K+H-type.

	2N	3N	4N
AMD(Ours)	-0.545	-8.45	-30.14(K) -29.81(H) -30.27(K+H)
Numerical Integration	-0.545[13]	-8.43 [14]	-30.27 [14]

Table 2. Various deuteron properties obtained in our model with the Paris potential, the binding energy, the root mean square radius (R.M.S.), the magnetic moment, and the quadrupole moment are compared with the original quantities from the Paris potential.

	E_B(MeV)	R.M.S.(fm)	$\mu_D(\mu_0)$	Q_D(fm^2)
Paris Original [11]	-2.22	1.972	0.853	0.279
Ours	-2.20	1.983	0.874	0.280

Next, we show the simulation results for the deuteron using the Paris potential (Tables 2,3), which contains a strong repulsive core. The results obtained from the modified AMD method agree with those from the Lippmann-Schwinger (LS) equation precisely. It should be noted that the expectation value of the tensor force is reproduced quite well.

Table 3. Expectation values of the kinetic energy, LS force, tensor force, and spin-orbit force in MeV.

	$\langle T\rangle$	$\langle V_C\rangle$	$\langle V_{LS}\rangle$	$\langle V_T\rangle$	$\langle V_{SO_2}\rangle$
Original[11,15]	18.91	4.81	-1.04	-16.01	-1.54
Ours	18.78	4.78	-1.03	-15.90	-1.52

[a]We call this the AMD$^+$ model, which is discussed in Ref. 12.

5. Summary and Discussion

In this paper, we investigated various light nuclei such as ^{2}H, ^{3}H, ^{3}He, and ^{4}He within the AMD method.

Firstly, we calculated the bound states with two effective potentials, the Volkov and the Paris potentials. We have investigated ground state energies for nuclei ranging from the two-nucleon system to an eight-nucleon system using the Volkov potential [6–8]. The realistic Paris potential was only used for the central part of NN interaction to calculate the binding energies in which the short range part was modified without any change in the phase shift. Because the Paris potential, in particular, has a strong repulsive core, it destabilizes the computation in the naive formulation of the AMD approach..

Secondly, in order to investigate fully the short range correlation between nucleons, we employed Jacobi coordinates for the numerical bases. For that purpose, we proposed an improved AMD method as a hybrid model which consists of AMD plus the eigenvalue problem. Here we call it the "AMD$^+$ model". As a result, we succeeded in obtaining the bound states with the Paris potential. With this method, we could also deal with the tensor correlation which is important for the deuteron properties. The new method is a powerful way to treat realistic potentials which gives not only the ground state but also resonance states, even though it is based on the variational method. We confirmed that our results are consistent with those of Faddeev calculations when Jacobi coordinates of the 3N and 4N systems are utilized.

References

1. A.Ono, H.Horiuchi, T.Maruyama and A.Ohnishi, Prog. Theor. Phys. **87**, 1185 (1992).
2. A.Ono, H.Horiuchi, T.Maruyama and A.Ohnishi, Phys. Rev. Lett. **68**, 2898 (1992).
3. Y.Kanada-En'yo, A.Ono, H.Horiuchi, Phys.Rev. **C52**, 628 (1995).
4. Y.Kanada-En'yo, H.Horiuchi, Phys.Rev. **C52**, 648 (1995).
5. Y.Kanada-En'yo, H.Horiuchi, Phys.Rev. **C54**, 468 (1995).
6. T. Watanabe, Y.Taniguchi, N. Sawado, and S. Oryu, Modern Physics Letters **A18** , 182-185 (2003).
7. Y. Taniguchi, T. Watanabe, N. Sawado, and S. Oryu, Few-Body Systems Suppl. **15**, 247-252 (2003).
8. Y. Taniguchi, T. Watanabe, N. Sawado, and S. Oryu, Proceedings of the Seventeenth International IUPAP Conference on Few-Body Problems in Physics (Few-Body 17), S198- S200 (2004 Elsevier B.V.).
9. L.Wilet, E.M.Henley, M.Kraft and A.D.MacKellar, Nucl. Phys. **A282**,314 (1977).
10. A.Volkov, Nucl.Phys. **75**, 128 (1958).
11. M.Lacombe, B.Loiseau, J.M.Richard and R.Vinh Mau, Phys. Rev. **C21**, 861 (1980)
12. T. Watanabe, S. Oryu, Phys. Rev. submitted. (2005).
13. K.Varga,Y.Suzuki and I.Tanihata, Phys. Rev. **C52**, 3013 (1995).
14. H.Kamada and W.Glöckle, Nucl. Phys. **A458**, 205 (1992).
15. M.Lacombe,B.Loiseau,J.M.Richard, R.Vinh Mau, J.Cote, P.Pires, and R.de Tourreil, Phys.Lett **101B**,139 (1981).

Anomalies in *pd* Radiative Capture and *pd* Breakup Reactions

K. Sagara, T. Kudoh, M. Tomiyama, H. Ohira, S. Shimomoto, T. Yagita,
Dept. of Physics, Kyushu University, Fukuoka, 812-8581 Japan

K. Hatanaka, Y. Tameshige, A. Tamii, Y. Shimizu, J. Kamiya
RCNP, Osaka University, Osaka, 567-0047 Japan

H. Kamada
Faculty of Engineering, Kyushu Inst. of Technology, Kitakyushu, 804-8550 Japan

H. Witala
Dept. of Physics, Yagellonian University, Cracow, Poland

In a $p+d$ radiative capture experiment at E_d = 200MeV, a large discrepancy between the experiment and calculations was found in A_{xx} value together with a curious relation of $A_{xx} \approx A_{yy}$. A confirming experiment at E_d = 140 MeV is in progress. Recent experiment at KVI confirmed $A_{xx} \approx A_{yy}$ relation, however, no large discrepancy was fond in A_{xx}. We measured also D(p,p)pn cross section at E_p = 250 MeV, and found large enhancement of cross section compared to calculations. Coulomb effects in *pd* breakup was investigated at E_p = 13 MeV by making experiments and by inventing an approximate Coulomb calculation. Coulomb force suppresses the breakup cross section at *pp* FSI, and the disagreement at 250 MeV can not be explained by Coulomb force.

1. Anomalies in *pd* radiative capture

Introduction of 2π-exchange 3-nucleon force (2π3NF) is very effective to reproduce the nuclear binding energies and Nd scattaring cross section. Introduction of 2π3NF alone is, however, not enough to reproduce spin observables of Nd scattaring. The disagreement in spin observables seems to indicate the necessity of 3NF other than 2π3NF, such as $\pi\rho$3NF and $\rho\rho$3NF. As they include heavy-meson exchanges, they are short-range forces.

To investigate short-range 3NF, we made experiments of *pd* radiative capture at E_d = 200 MeV and 140 MeV. In this reaction, p and d in scattering states fuse to make a compact ^{3}He. Momentum transfer is high, and short-range forces may play important roles.

Cross section of *pd* capture is very small, less than 1μbarn, hence we need special experimental instruments and methods to obtain accurate data.

A 200-MeV polarized d-beam from RCNP ring cyclotron was incident on a liquid hydrogen target to induce d+p → ^{3}He+γ reaction. The beam polarization axis was in the vertical direction and the beam polarization was about 60% which was measured by a beam-line polarimeter using d+p scattering. We used a liquid hydrogen target of 1.5 mm in thickness (11mg/cm^2) having very thin window foils made of 4.4 μm thick aramide(0.6 mg/cm^2). The energy loss in the target of ^{3}He induced by pd capture was about 2 MeV and the angular spread of ^{3}He in the target was typically 0.08°.

In the laboratory system, ^{3}He particles from pd capture are recoiled out in forward angles within 4.7°. To detect ^{3}He recoils we used a large acceptance spectrometer (LAS) which has acceptance of ±5.7° and ±3.4° in the vertical and horizontal planes, respectively. With LAS at 0°, whole the ^{3}He recoils in the vertical plane were simultaneously detected. Next LAS was placed at 2.5°, and ^{3}He recoils from 1.5° to 4.7° in the horizontal plane were simultaneously detected. Since the d-beam polarization axis was in the vertical direction, A_{xx} of pd capture was measured in the vertical plane, and A_y and A_{yy} together with the cross section of pd capture were measured in the horizontal plane. The backgrounds (BG) were measured with an empty target, that is, the liquid hydrogen was evaporated and the gas hydrogen was evacuated.

Experimental results are shown in Fig. 1. Curves are recent meson-exchange-current caluculations and calculations in Siegert approximation both with and without 2π3NF. Remarkable two features are seen in the tensor analyzing powers; (1) relation of $A_{xx} \approx A_{yy}$ holds in the experimental data, although calculated A_{xx} and A_{yy} are much different, and (2) measured A_{xx} largely disagrees with calculated A_{xx}. Recently A_{xx} data at 90° was analyzed again by one on us, and essentially the same result was obtained.

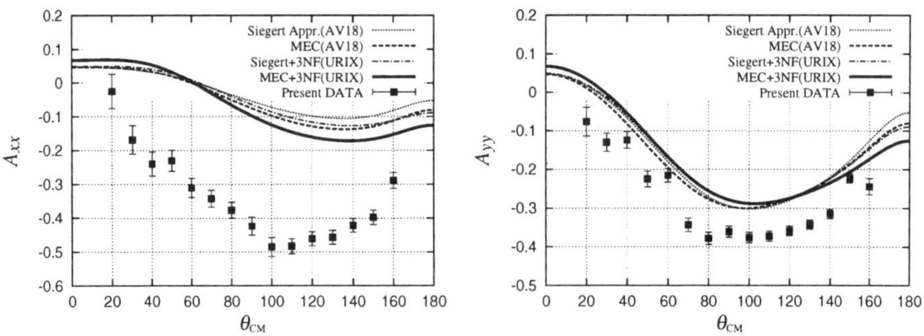

Fig. 1. Analyzing powers of d+p → ^{3}He+γ reaction at E_d = 200 MeV. Solid and dotted curves are recent meson-exchange-current calculations with and without 2pi3NF, respectively.

At E_d = 17.5 MeV, A_{xx}, A_{yy}, and A_{zz} were measured precisely [1]. At this energy, the $A_{xx} \approx A_{yy}$ relation holds, and both A_{xx} and A_{yy} (also A_{zz}) are well reproduced by calculations, that is, there is no A_{xx} anomaly at low energy.

In order to see energy dependence of A_{xx} anomaly, we made another *pd* capture experiment at E_d = 140 MeV at RCNP. The experimental procedure was the same as that at E_d = 200 MeV except that the liquid target was a little thinner as 1.2 mm in thickness and that analyzing powers of the *d*-beam polarimeter were taken from ref. [2]. The background level was higher at 140MeV than before and the data analysis has not been finished yet. The preliminary data at 90° indicate that A_{xx} $\approx$ A_{yy} relation holds and A_{xx} anomaly appears also at 140 MeV.

Recently A_{yy} and A_{zz} of *pd* capture at E_d = 133 MeV and 180 MeV were measured at KVI [3]. Their A_{yy} and A_{xx} (= - A_{yy} - A_{zz}) at 90° are shown in Fig. 2 together with our data. KVI data are about a factor of 2 smaller than ours in the absolute value, and A_{xx} anomaly does not appear in KVI data although slight disagreement remains.

We are now analyzing our data at 140 MeV very carefully. A new experiment at 270 MeV has been made at RIKEN and the forthcoming data will bring some judgment.

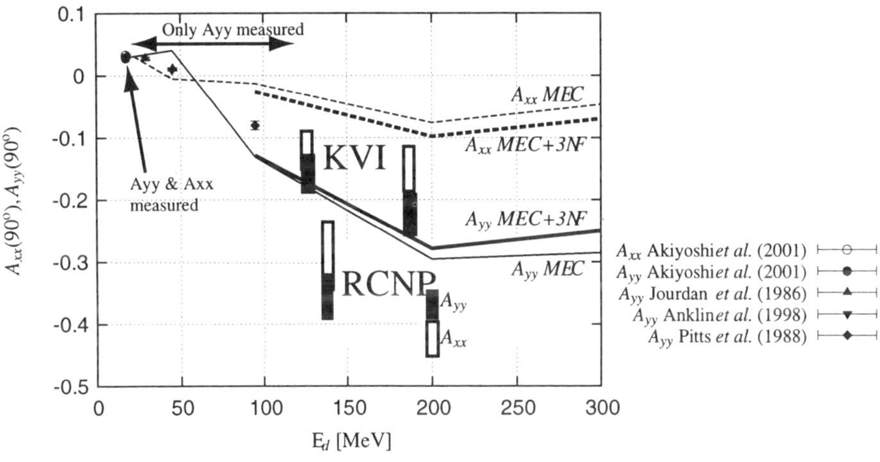

Fig. 2. Energy dependence of A_{xx} and A_{yy} of $d+p \rightarrow {}^3He+\gamma$ reaction at θ_{cm} =90°. Thick and thin curves are recent meson-exchange-current calculations with and without 2π3NF, respectively.

The $A_{xx} \approx A_{yy}$ relation has been confirmed in all the *pd* capture experiments below 200 MeV. This relation seems to be special, because so far as we know, $A_{xx} \approx$ - A_{yy} relation roughly holds in other *d*-induced reactions such as elastic and inelastic scatterings, nucleon transfer reactions and breakup reactions.

If $A_{xx} \approx$ - A_{yy} relation holds in a reaction, the reaction is enhanced (or suppressed) by a vertically polarized deuteron and is suppressed (or enhanced) by a horizontally polarized d euteron. A deuteron has a plorate shape, and a *d*-induced

reaction depends on the direction of the axis of the plorate shape. The $A_{xx} \approx$ - A_{yy} relation means that the reaction mainly takes place in the peripheral region of the nucleus.
On the contrary, the $A_{xx} \approx A_{yy}$ relation in *pd* capture shows symmetry with respect to the beam axis (z-axis), and it means that the reaction proceeds not in the peripheral region but in the central region. We consider that short-range forces may play important roles in *pd* capture and that A_{xx} and A_{yy} of *pd* capture may be suitable observables to investigate short-range 3NF.

As seen in Fig. 2, 2π3NF has a power to decrease the difference between A_{xx} and A_{yy} of *pd* capture but the power is insufficient to equate A_{xx} and A_{yy}. A new 3NF or new reaction mechanism that forcefully equates A_{xx} and A_{yy} of *pd* capture may be necessary.

2. Anomalies in *pd* breakup reaction

We also made an experimental search for short-range 3NF in *pd* breakup at intermediate energy. A polarized proton beam of 250 MeV from RCNP ring cyclotron was incident on a liquid deuterium target of 150 mg/cm^2 in thickness, and one of the protons from *pd* breakup was detected by a broad-range magnetic spectrometer LAS.

Preliminary results for the cross section at $\theta_p = 10°$, $15°$ are shown in Fig. 3. The *pd* breakup experimental results are considerably larger than *nd* calculations. Effects of 2π3NF decrease the disagreement, but are insufficient to reproduce the data. Candidates for the origin of the disagreement may be short-range 3NF, relativity, and Coulomb force.

So we investigated Coulomb effects at a low energy of $E_p = 13$ MeV. An exclusive measurement of D(p,pp)n cross section was made around configuration of *pp* final state interaction (FSI). The results at $\theta_1 = \theta_2 = 30°$, ϕ_{12}=11.2° are shown in Fig. 4. Suppression of *pp* FSI peak by Coulomb force was very well described by Watson-Migdal formula.

In Faddeev formula, 3N reaction amplitude consists of 3 components according to FSI. We multiply the *nn* FSI component of the *nd* amplitude by a ratio of *pp* amplitude to *nn* amplitude of Watson-Migdal formula, and we obtain a "*pd*" amplitude. This WM + Faddeev calculation works well for *pd* breakup at 13 MeV around *pp* FSI, as seen in Fig. 4. We made also an inclusive measurement of *pd* breakup cross section at 13 MeV at 20°-60°. One of the protons from *pd* breakup was particle-identified and counted by a counter telescope. The experimental results are considerably well described by the WM + Faddeev calculation, as seen in Fig. 4.

The WM + Faddeev calculation is applied for the breakup at 250 MeV. At 250 MeV, Coulomb correction by WM modification slightly decreases the cross section as seen in Fig. 3. Disagreement becomes even larger by the Coulomb correction.

Since 2π3NF increase the cross section at 250 MeV, short-range 3NF is indeed a possible origin at 250 MeV. Also relativity is a the possible origin. Investigations of these candidates are very interesting works to be made.

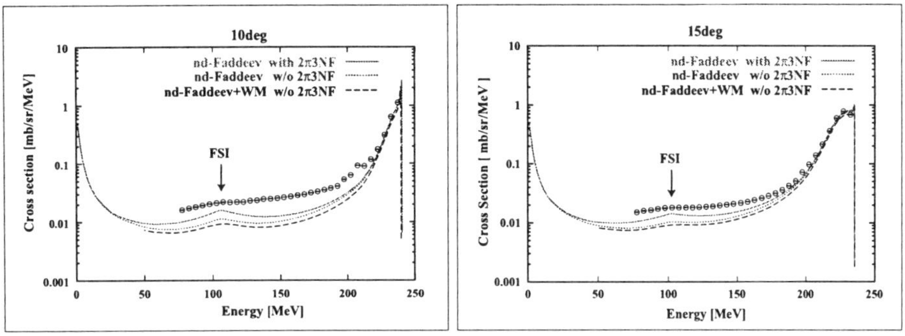

Fig. 3. Inclusive cross section of D(p,p)pn at E_p = 250 MeV. Dashed curve represents WM-Faddeev calculation without 2π3NF.

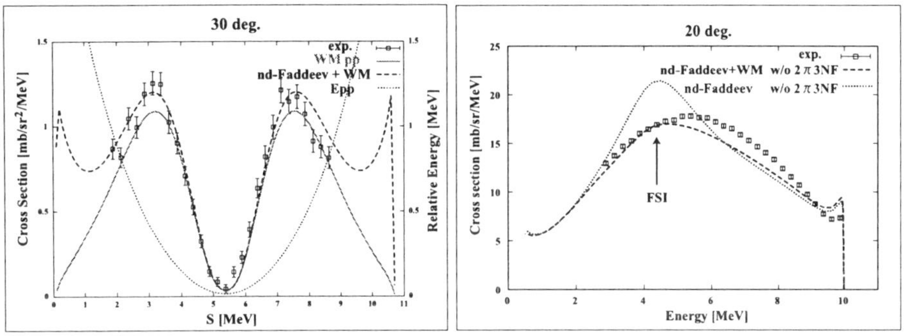

Fig. 4. (Left) Exclusive cross section of D(p,pp)n reaction at θ_1=θ_2=30°, ϕ_{12}= at 11.2, E_p = 13 MeV. Solid and Dashed curves are WM calculation and WM-Faddev calculations, respectively. (Right) Inclusive cross section of D(p,p)pn reaction at Ep = 13 MeV. Dotted and dashed curves are nd Faddeev calculation and WM-Faddev calculations, respectively.

References

1. H. Akiyoshi et al. ,*Phys. Rev.* **C64**, 034001 (2001).
2. K. Sekiguchi et al. , *Phys. Rev.* **C65**, 034003 (2002).
3. A. A. Mehandoost-Khajeh-Dad et al. , *Phys. Lett.* **B617** 18 (2005).

Experimental Investigations of Three-Body Systems at KVI

N. Kalantar-Nayestanaki

Kernfysisch Versneller Instituut (KVI),
Zernikelaan 25, 9747 AA Groningen, The Netherlands
E-mail: nasser@kvi.nl

Three-body systems have been studied in detail at KVI in the past few years. Two categories of reactions have been chosen to investigate these systems, namely elastic and break-up reactions in proton-deuteron scattering in which only hadrons are involved, and proton-deuteron capture reaction involving real and virtual photons in the final state. Some results from both types of reactions will be presented along with theoretical calculations showing that despite a relatively good understanding of these systems, there are still discrepancies between the experimental results and the theoretical calculations.

The three-nucleon system has attracted considerable attention over the past two decades as it is the simplest possible system where one can include effects beyond the well-studied *NN* force. Even though the *NN* forces are not obtained from first principles, it is possible to use the wealth of the nucleon-nucleon scattering data and obtain realistic *NN* potentials such as Nijmegen I, Nijmegen II, Reid93, CD Bonn and Argonne v_{18} (AV18)[1]. These potentials can be used, in turn, to make predictions for observables in three-body systems. One extra ingredient that must in principle be included in the calculations for these systems is the three-body force (TBF). A well-known evidence for the existence of three-body forces is the fact that two-body potentials alone fail to predict the binding energies of three- and four-body systems. This is even seen for the binding energy and the excitation energy levels of heavier systems for which exact calculations can be done[2]. How large the effect of including these forces is and which interactions should be included in three-body forces have been a matter of concern for many years. A number of three-body forces exist in the literature, albeit the level of sophistication is far below what has been achieved for the two-nucleon sector[3,4]. These forces can be added to *NN* forces in Faddeev-type calculations from which precise results can be expected in proton-deuteron elastic and break-up reactions[5]. If one is interested in studying the large momentum component of the overlap wave functions of the proton-deuteron system, one should investigate the proton-deuteron capture reaction in which a photon is also involved. However, due to the presence of electromagnetic currents, one needs to properly include the effects of the Meson Exchange Currents (MEC) as well. Both types of reactions mentioned above have been studied at KVI the results of which will be partly presented here.

The experiments reported here were performed at KVI with the superconducting cyclotron AGOR. The detection system was the Big-Bite Spectrometer (BBS)[6] in conjunction with the Euro-Supernova focal-plane detection system (ESN)[7] for the detection of protons, deuterons and ^{3}He, and the Plastic Ball[8] for the detection of the photons. For the study of the elastic scattering, (polarized) protons of energies of 108, 120, 135, 150, 170 and 190 MeV, and For the study of the radiative capture, polarized deuterons with energies of 110, 133 and 180 MeV were facilitated. Solid CH_2, CD_2 and liquid hydrogen targets were used for various reactions. For the capture reaction, the BBS was placed at around 0° to detect the outgoing ^{3}He particles while for the study of the elastic scattering, it was placed between 5° and 50° to detect protons and deuterons in order to cover a large range of c.m. scattering angles. For the details of the two experiments reported here, refer to Refs.[9,10].

Elastic Deuteron-Proton Scattering

Measured differential cross sections are shown in Fig. 1. For the analyzing-power results the reader is referred to Ref.[9]. For The statistical uncertainty, which is in general of the order of 2%, is smaller than the size of the symbol. The total systematic uncertainty, which is the quadratic sum of the uncertainty in the normalization factor, the uncertainty in the polarization where polarized protons were used, and the uncertainty in the detection efficiency, is in general less than 7%, and should be considered as a point-to-point systematic uncertainty. The systematic error bands are shown in the right panel of both figures. With these high precision data, one can now discriminate between different calculations with and without three-body force. Here, the calculations do not include Coulomb force the effect of which is shown to be small for the kinematics shown (see the contribution of P.U. Sauer elsewhere in these proceedings.).

In the figure, cross-section data from other laboratories[11–15] are also included where the incident energies were near each other. It is clearly seen that there is a general agreement between the data from KVI and other laboratories with the exception of the cross-section data at 135 MeV from RIKEN[11]. It is also clear that the present experiment presents the data for a wide range of intermediate energies and scattering angles.

As can be seen from the figure and on a global scale, the overall agreement with the calculations is reasonable given the large range of variables (energy and angle). However, closer inspection (see the right panel in Fig. 1) immediately reveals that the calculations based solely on two-nucleon potentials have shortcomings. The inclusion of the three-body forces, whether it is the Tucson-Melbourne (TM$'$)[3] or Urbana-IX[4] improves the results to a large extent. This improvement is also observed when one uses the approach in Refs.[16,17]. At small angles, there are clear disagreements between the data and the predictions. This is most probably due to Coulomb effects not taken into account in the present calculations. The results of the calculation based on χPT[18] also agree reasonably well with the data lending confidence to this approach. Efforts to increase the energy range of applicability of χPT are much required.

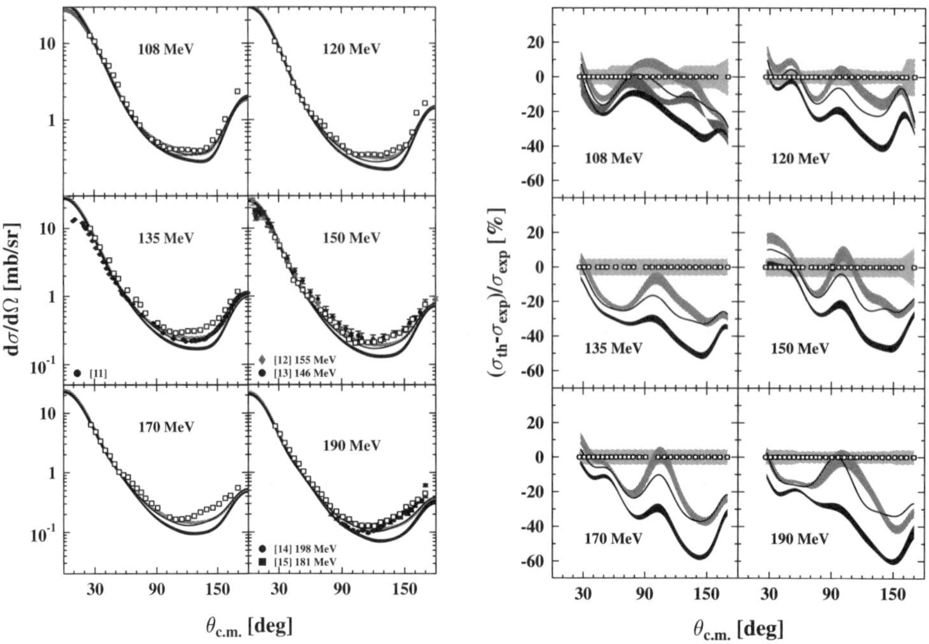

Fig. 1. Differential cross sections as a function of $\theta_{\text{c.m.}}$ (left) and the difference between measured and calculated cross sections (right). The data measured in this work are plotted as open squares. The curves shown are calculations based solely on *NN* potentials (black band), calculations from AV18+Urbana-IX (solid line) and *NN*+TM′ (gray band). Data at 135 MeV (circles)[11], 155 MeV (diamond)[12], 146 MeV (circle)[13], 198 MeV (circle)[14], and 181 MeV (solid square)[15] are shown as well. At 108 MeV, the calculation based on χPT is shown as a dark-gray band.

Inspecting the figure further, one sees that the region of the minimum of the cross section is rather well explained after the inclusion of three-body forces. This was one of the regions where disagreement was expected to be largest if one ignores the effects of TBF. The deviations after the inclusion of TBF are largest at the backward angles. What is remarkable is that this disagreement increases with increasing the incident beam energy[9]. The source of this discrepancy is not clear at the moment. A plausible explanation for this discrepancy can be higher-order effects, such as π-ρ or ρ-ρ exchange, which have not been included in the calculations. Part of the disagreement can also be due to relativistic effects[19] which have also not been properly taken into account in the calculations. The same conclusions hold, more-or-less, for the results of the analyzing powers not shown here[9].

For the elastic channel, also spin-transfer coefficients have been measured and the final results are emerging. In addition, the break-up reaction has been studied at KVI by the Polish-Dutch collaboration[20]. For a sample of these data, see the contribution of Biegun et al. in these proceedings. New measurements for this reaction are planned with the newly installed detection system, BINA at KVI.

Deuteron-Proton Capture

Figure 2 compares the results of the deuteron-proton radiative capture experiment with the predictions of the Hanover group[21], based on the purely nucleonic CD-Bonn potential and its coupled-channel extension, CD Bonn + Δ, allowing for a single excitation of a nucleon to a Δ isobar. The Δ mediates TBF and generates effective two- and three-nucleon currents in addition to irreducible one- and two-baryon contributions as described in detail in Ref.[21]. Note that the Hanover calculation agrees reasonably well with our data for all polarization observables and all energies. The effect of the Δ isobar is small at these energies and for these observables. The effect of TBF is also shown to be small in the Faddeev calculations done by Bochum-Cracow group[22,23] not shown here.

The predictions for tensor-analyzing powers A_{zz} and A_{yy} by the Hanover and Bochum-Cracow groups are very similar. Since the tensor-analyzing power A_{xx} is related to A_{yy} and A_{zz}, via $A_{xx}+A_{yy}+A_{zz}=0$, we conclude that also this observable is rather well predicted by all models including MECs. Surprisingly, a recent experiment conducted at RCNP with a 100 MeV/nucleon incident deuteron beam showed large deviations for A_{xx} in comparison with similar model predictions (see the contribution by K. Sagara elsewhere in these proceedings). Our data taken at an energy of 90 MeV/nucleon clearly do not show such large discrepancies, and therefore contradict the preliminary data of RCNP. Also at lower energies, no anomaly is observed for the tensor-analyzing powers in the deuteron-proton radiative capture process.

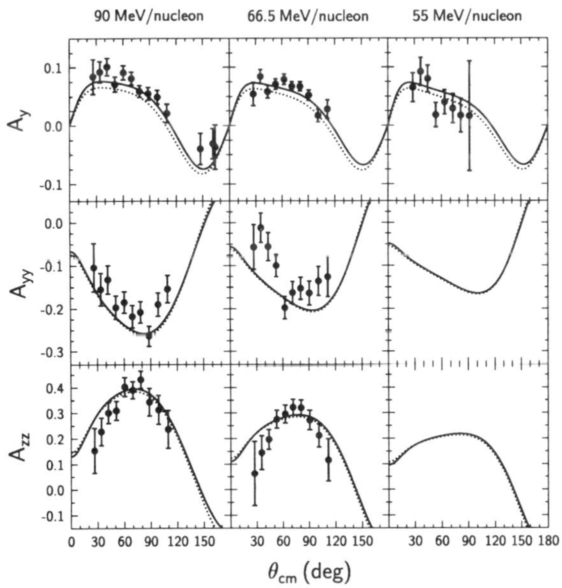

Fig. 2. Polarization data for the deuteron-proton radiative capture reaction are compared with the predictions of the Hanover group. The dashed line represents the calculation based on the CD-Bonn potential, whereas the solid line includes, in addition, contributions from the Δ isobar.

In conclusion, a series of measurements at KVI have produced very accurate data for cross sections and analyzing powers in the elastic proton-deuteron scattering. The results have been compared to the state-of-the-art three-body Faddeev calculations using modern two and three nucleon forces. The results show, that even though the bulk of the data is well explained by the calculations, there remain some discrepancies at the backward angles which are still not resolved. The measured spin observables in the deuteron-proton radiative capture reaction agree reasonably well with the theoretical calculations including MEC and TBF. The present results at 180 MeV deuteron energy, however, disagree with the preliminary results of the data taken at RCNP. The KVI activities are continuing with the measurements of the spin-transfer coefficients of the elastic proton-deuteron scattering and extensive measurements of the break-up channel in the three-body system with the recently commissioned BINA detector at KVI.

Acknowledgments

The author acknowledges the assistance of AGOR cyclotron group for providing all the beams for this experiment. He also thanks the help of those involved in the experiment. In particular, the work of Karsten Ermisch and Ali Mehmandoost-Khajeh-Dad which led to these results should be mentioned.

References

1. V.G.J. Stoks et al., Phys. Rev. C49, 2950 (1994); R. Machleidt et al., Phys. Rev. C53, R1483 (1996); R.B. Wiringa et al., Phys. Rev. C51, 38 (1995).
2. S.C. Pieper et al., Phys. Rev. C64, 014001 (2001).
3. S.A. Coon and H.K. Han, Few Body Syst. 30, 131 (2001).
4. B. Pudliner et al., Phys. Rev. Lett. 74, 4396 (1995).
5. W. Glöckle et al., Phys. Rep. 274, 107 (1996).
6. A.M. van den Berg, Nucl. Inst. and Meth. in Phys. Res. B99, 637 (1995).
7. H.J. Wörtche, Nucl. Phys. A687, 321c (2001).
8. A. Baden et al., Nucl. Instr. and Meth. 203, 189 (1982).
9. K. Ermisch et al., Phys. Rev. C71, 064004 (2005).
10. A.A. Mehmandoost-Khajeh-Dad et al. Phys. Lett. B617, 18 (2005).
11. H. Sakai et al., Phys. Rev. Lett. 84, 5288 (2000).
12. K. Kuroda, A. Michalowicz and M. Poulet, Nucl. Phys. 88 (1966) 33.
13. H. Postma and R. Wilson, Phys. Rev. 121, 1129 (1961).
14. R.E. Adelberger and C.N. Brown, Phys. Rev. D5, 2139 (1972).
15. G. Igo et al., Nucl. Phys. A195, 33 (1972).
16. S. Nemoto et al., Phys. Rev. C58, 2599 (1998).
17. A. Deltuva et al., Phys. Rev. C68, 024005 (2003).
18. E. Epelbaum et al., Phys. Rev. C66, 064001 (2002) and references therein.
19. H. Kamada et al., Phys. Rev. C66, 044010 (2002).
20. St. Kistryn et al., Phys. Rev. C68, 054006 (2003).
21. A. Deltuva et al., Phys. Rev. C69, 034004 (2004).
22. R. Skibinski et al., Phys. Rev. C67, 054001 (2003).
23. J. Golak et al., Phys. Rev. C62, 054005 (2000).

Relativistic Effects in Neutron-Deuteron Elastic Scattering

H. Witała and J. Golak

M. Smoluchowski Institute of Physics, Jagiellonian University, PL-30059 Kraków, Poland
E-mail:witala@if.uj.edu.pl and ufgolak@cyf-kr.edu.pl

W. Glöckle

Institut für theoretische Physik II, Ruhr-Universität Bochum, D-44780 Bochum, Germany
E-mail: walter.gloeckle@tp2.ruhr-uni-bochum.de

H. Kamada*

Department of Physics, Faculty of Engineering, Kyushu Institute of Technology, Kitakyushu 804-8550, Japan
E-mail: kamada@mns.kyutech.ac.jp

We solved the three-nucleon (3N) Faddeev equation including relativistic features at incoming neutron lab energies E_n^{lab} = 28, 65, 135 and 250 MeV. Those features are relativistic kinematics, boost effects and Wigner spin rotations. As dynamical input a relativistic nucleon-nucleon (NN) interaction exactly on-shell equivalent to the AV18 NN potential has been used. The boost effects are significant at higher energies. The effects of Wigner rotations for elastic scattering observables were found to be small. There are rather small effects in the cross section, which are mostly restricted to the backward angles.

1. Introduction

In the nucleon-deuteron(Nd) scattering the studied discrepancies between a theory based on NN potentials only and experiment become larger with increasing energy of the 3N system. Adding now a 3NF to the pairwise interactions leads in some cases to a better description of the data. The elastic nucleon-deuteron angular distribution in the region of its minimum and at backward angles is the best studied example [1–3]. At energies higher than ≈ 100 MeV current 3NFs only partially improve the description of cross section data and the remaining discrepancies, which increase with energy, indicate the possibility of relativistic effects. The need for a relativistic description of 3N scattering was also raised when precise measurements of the total cross section for neutron-deuteron (nd) scattering [4] were analyzed within the framework of nonrelativistic Faddeev calculations [5].

The estimation of relativistic effects on the binding energy of three nucleons has

*oral presentation in the conference

been the focus of a lot of work. Basically two different aproaches have been followed: one is a manifestly covariant scheme linked to a field theoretical approach [6–8], the other one is based on relativistic quantum mechanics formulated on spacelike hypersurfaces in Minkowski space [9–14].

An analytical scale transformation of momenta which relates NN potentials in the nonrelativistic and relativistic Schrödinger equations in such a way, that exactly the same NN phase shifts are obtained by both equations, was employed [15]. Though this transformation is not a substitute for a NN potential with proper relativistic features it can serve as a first step to illustrate the effects of Lorentz boosts on NN potentials. Such an approach was applied in [13] and we also will follow it in the present study to get the first estimation of relativistic effects in the 3N continuum. In this first study we would like to find out what are the changes of elastic nd scattering observables when the nonrelativistic form of the kinetic energy is replaced by the relativistic one and a proper treatment of **boost effects** and effects due to **Wigner rotations** [12,16] of spin states is performed.

The next section explains our formulation. We refer to ref.17 for details, especially regarding the treatment of moving sigularities[18] and Wigner rotations.

2. Formulation

The Nd scattering with neutron and protons interacting through a NN potential V alone is described in terms of a breakup operator T satisfying the Faddeev-type integral equation [19–21]

$$T|\phi> = tP|\phi> + tPG_0T|\phi> . \tag{1}$$

The two-nucleon (2N) t-matrix t results from the interaction V through the Lippmann-Schwinger (LS) equation. The permutation operator $P = P_{12}P_{23} + P_{13}P_{23}$ is given in terms of the transposition P_{ij} which interchanges nucleons i and j. The incoming state $|\phi >= |\vec{q}_0 > |\phi_d >$ describes the free nucleon-deuteron motion with relative momentum $\vec{q}_0$ and the deuteron wave function $|\phi_d >$. Finally G_0 is the free 3N propagator. The standard nonrelativistic 2N LS equation turns now into a relativistic one, which in a general frame reads

$$t(\vec{k},\vec{k}\,';\vec{q}\,) = V(\vec{k},\vec{k}\,';\vec{q}\,) + \int d^3k'' \frac{V(\vec{k},\vec{k}\,'';\vec{q}\,)t(\vec{k}\,'',\vec{k}\,';\vec{q}\,)}{\sqrt{(2\omega(\vec{k}\,')\,^2+\vec{q}\,^2} - \sqrt{(2\omega(\vec{k}\,'')\,^2+\vec{q}\,^2} + i\epsilon} \tag{2}$$

We refer to $t(\vec{k},\vec{k}\,';\vec{q}\,)$ as the boosted 2N t-matrix like we talk of the boosted 2N potential [22] as

$$V(\vec{q}) \equiv \sqrt{(2\omega(\vec{k})+v)^2+\vec{q}\,^2} - \sqrt{(2\omega(\vec{k}))^2+\vec{q}\,^2}. \tag{3}$$

In this first study we do not treat the boosted potential matrix element in all its complexity as given in ref. [13] but restrict ourselves to the leading order term in a q/ω and v/ω expansion

$$V(\vec{k},\vec{k}\,';\vec{q}\,) = v(\vec{k},\vec{k}\,')$$

$$\times \left[1 - \frac{\vec{q}^{\,2}}{8\sqrt{m^2 + \vec{k}^{\,2}}\sqrt{m^2 + (\vec{k}\,')^2}} \right]. \tag{4}$$

In addition, the results for two more drastic approximations are given. In the first one the boost effects are neglected completely

$$V(\vec{k}, \vec{k}\,'; \vec{q}\,) = v(\vec{k}, \vec{k}\,'), \tag{5}$$

and in the second one the k-dependence of the first order relativistic correction term is omitted

$$V(\vec{k}, \vec{k}\,'; \vec{q}\,) = v(\vec{k}, \vec{k}\,') \left(1 - \frac{\vec{q}^{\,2}}{8m^2} \right). \tag{6}$$

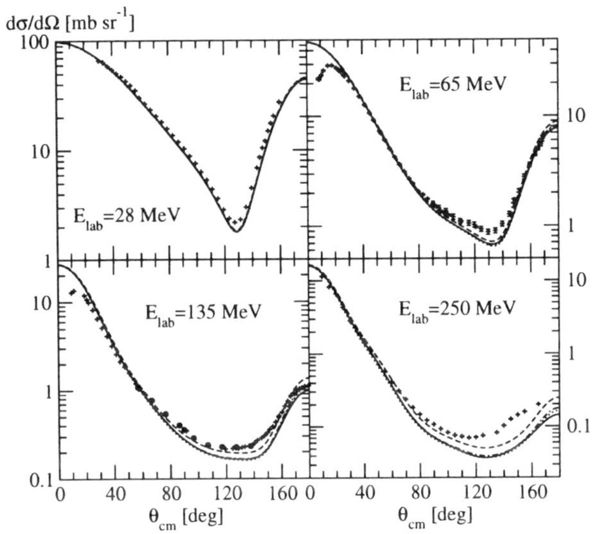

Fig. 1. The differential cross sections for Nd elastic scattering at various energies. The solid line is the result of the nonrelativistic Faddeev calculation with the AV18 potential [23]. The relativistic predictions based on the approximations (5), (6), and (4) for the boosted potential keeping all other relativistic features unchanged are shown by the dashed, dotted, and dashed-dotted lines. The pd data at 28, 65, 135 and 250 MeV are from ref. [24,25,2] and [26], respectively.

3. Results

In Fig.1 we show our results for the nd elastic scattering cross sections at four energies together with experimental pd data. In addition to the nonrelativistic prediction, based on the solution of the 3N Faddeev equation with the nonrelativistic form of the free propagator G_0 and partial wave states constructed with standard Jacobi momenta, also our relativistic results using the approximations according to the Eqs.(5), (6), and (4) for the boosted potential matrix elements are presented. It is only the total neglection of the boost effect in $V(\vec{q})$ which leads to a clearly visible deviation from the nonrelativistic results. Taking the boost effect into account according to the approximations (4) and (5) reduces the effect drastically and only at the backward angles deviations from the nonrelativistic results are discernible.

In Fig.2 we compare relativistic and nonrelativistic predictions for the deuteron vector analyzing power iT_{11}. It turns out that the relativistic effects are relatively small and they stay below $\approx 5\%$ in the angular regions outside of zero crossings.

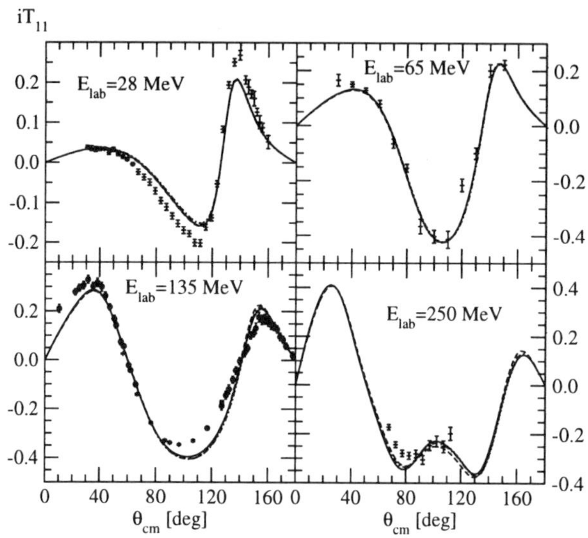

Fig. 2. The deuteron vector analyzing power iT_{11} in Nd elastic scattering. For description of lines see Fig.1. The pd data at 28, 65, 135 and 250 MeV are from ref. [24,27,2] and [28], respectively. The pd data from [28] were taken at 190 MeV.

4. Outlook

In view of presented results the discrepancies between pure 2N force predictions and Nd cross section data must be due to 3NFs. At lower energies (up to about ≈ 135 MeV) this discrepancy can be removed when current three-nucleon forces, mostly of 2π-exchange character [29,30], are included in the nuclear Hamiltonian. At the higher energies, however, a significant part of the discrepancy remains and increases further with increasing energy. This indicates that additional 3N forces should be added to the 2π-exchange type forces. Natural candidates in the traditional meson-exchange picture are exchanges like $\pi - \rho$ and $\rho - \rho$. This has to be expected since in χPT [31] in the order in which nonvanishing 3NF's appear the first time there are three topologies of forces, the 2π-exchange, a one-pion exchange between one nucleon and a two-nucleon contact interaction and a pure 3N contact interaction. They are of the same order and have to be kept together. Therefore it appears very worthwhile to persue a strategy adding in the traditional meson exchange picture further 3N forces. Our results here showing that relativistic effects based on relativistic kinematics and boost effects of the NN force are small support the usefulness of high energy elastic Nd scattering to study 3N force properties.

This work has been supported by the Polish Committee for Scientific Research under Grant no. 2P03B00825, by the NATO grant no. PST.CLG.978943, and by the Japan Society for the Promotion of Science. The numerical calculations have been performed on the Cray SV1 of the NIC in Jülich, Germany.

References

1. H. Witała et al., Phys. Rev. Lett. **81**, 1183 (1998).
2. K. Sekiguchi et al., Phys. Rev. C**65**, 034003 (2002).
3. H. Witała et al., Phys. Rev. C**63**, 024007 (2001).
4. W.P. Abfalterer et al., Phys. Rev. Lett. **81**, 57 (1998).
5. H. Witała et al., Phys. Rev. C**59**, 3035 (1999).
6. G. Rupp and J.A. Tjon, Phys. Rev. C**45**, 2133 (1992).
7. F. Sammarruca and R. Machleidt, Few-Body Syst. **24**, 87 (1998).
8. A. Stadler, F. Gross, and M.Frank, Phys. Rev. C**56**, 2396 (1997).
9. B. Bakamjian and L.H. Thomas, Phys. Rev. **92**, 1300 (1952).
10. R.A. Krajcik et al., Phys. Rev. D**10**, 1777 (1974).
11. B.D. Keister, in Few-Body Problems in Physics, edite by Paul Schoessow, AIP Conf. Proc. No. 334, (AIP, Woodbury, NY, 1995), p.164.
12. B.D. Keister and W.N. Polyzou, Adv. Nucl. Phys. **20**, 225 (1991).
13. H. Kamada, W. Glöckle, J. Golak, and Ch. Elster, Phys. Rev. C**66**, 044010 (2002).
14. W. Glöckle, T-S. H. Lee, and F. Coester, Phys. Rev. C**33**, 709 (1986).
15. H. Kamada, W. Glöckle, Phys. Rev. Lett. **80**, 2547 (1998).
16. S. Weinberg, *The quantum theory of fields*, vol. I, Cambridge University Press, 1995.
17. H. Witała, J. Golak, W. Glöckle, H. Kamada, Phys. Rev. C 71, 054001 (2005).
18. H. Kamada, Few-Body Syst. Suppl.**12**, 433 (2000).
19. H.Witała, T.Cornelius and W.Glöckle, Few-Body Syst. **3**, 123 (1988).
20. W. Glöckle, H.Witała, D.Hüber, H Kamada, J.Golak, Phys.Rep. **274**, 107 (1996).
21. W.Glöckle, The Quantum Mechanical Few-Body Problem, Springer-Verlag 1983.
22. F. Coester, Helv. Phys. Acta **38**, 7 (1965).
23. R.B. Wiringa, V.G.J. Stoks, R. Schiavilla, Phys. Rev. C**51**, 38 (1995).
24. K. Hatanaka et al., Nucl. Phys. **A 426**, 77 (1984).
25. S. Shimizu et al., Phys. Rev. C**52**, 1193 (1995).
26. K. Hatanaka et al., Phys. Rev. C**66**, 044002 (2002).
27. H.Witała et al., Few-Body Syst. **15**, 67 (1993).
28. R.V. Cadman et al., Phys. Rev. Lett. **86**, 967 (2001).
29. S.A. Coon et al., Nucl. Phys. **A317**, 242 (1979); S.A. Coon and W. Glöckle, Phys. Rev. C**23**, 1790 (1981).
30. B.S. Pudliner et al., Phys. Rev. C**56**, 1720 (1997).
31. E. Epelbaum et al., Phys. Rev. C **66**, 064001 (2002).

Three-Nucleon Scattering at Intermediate Energies

I. Fachruddin

Departemen Fisika, Universitas Indonesia, Depok 16424, Indonesia

Ch. Elster

Department of Physics and Astronomy, Ohio University, Athens, OH 34701, USA

W. Glöckle

Institut für Theoretische Physik II, Ruhr-Universität Bochum, D-44780 Bochum, Germany

By means of a technique, which does not employ partial wave (PW) decompositions, the nucleon-deuteron break-up process is evaluated in the Faddeev scheme, where only the leading order term of the amplitude is considered. This technique is then applied to calculate the semi-exclusive proton-deuteron break-up reaction $d(p,n)pp$ for proton laboratory energies E_{lab} of a few hundred MeV. A comparison with PW calculations is performed at 197 MeV projectile energy. At the same energy rescattering processes, which are not included in the 3D calculations yet, are shown to be still important in the full Faddeev PW calculations, especially for the cross section and the analyzing power A_y. Next kinematical relativistic effects are investigated for projectile energies up to about 500 MeV. At the higher energies, those relativistic effects start not to be negligible, especially in the peak of the cross section.

1. Introduction

In order to investigate the short distance behavior of three nucleon (3N) as well as nucleon-nucleon (NN) forces it is necessary to consider 3N scattering at higher energies of a few hundred MeV. The partial wave (PW) technique, which takes as basis states a specific number of angular momentum eigenstates, has a long history of being employed to solve Faddeev equations for 3N scattering [1]. However as the energy increases the number of contributing angular momenta proliferates, leading to increased algorithmic difficulties as well as more tedious algebraic work. Therefore, a new and alternative approach is needed, which is not based on PW basis.

In momentum space the choice is to work directly with momentum vector states. For three boson scattering this approach has been pioneered successfully carried out in Refs. [2]. We develop our approach [3], referred here as the 3D technique, for three nucleon scattering, allowing the use of realistic nuclear forces, since spin and isospin are also taken into account. Since the NN system serves as input for our 3N calculations, we successfully applied the 3D technique to both NN scattering [4] and the deuteron [5].

We evaluate the nucleon-deuteron (Nd) break-up process using the Faddeev scheme. However, we do not solve the full Faddeev equations but rather concentrate on the first order term in the multiple scattering series given by them. Thus, we assume that at intermediate energies of a few hundred MeV the leading term may sufficiently describe the scattering process. We include relativistic effects in the kinematics to investigate their size as function of energy. For our application we concentrate on the semi-exclusive $d(p,n)pp$ reaction and calculate the spin averaged differential cross section, the neutron polarization, the proton analyzing power and the polarization transfer coefficients. There are in fact experimental data in the energy range up to about 500 MeV [6]-[8]. The 3D technique works with any potentials given in operator form, not the PW projected ones. We employ the NN potentials AV18 [9] and Bonn-B [10].

2. The Nd Break-Up Amplitude

The break-up process is given by the Nd break-up operator U_0 given by

$$U_0 = (1+P)TP, \tag{1}$$

where $P \equiv P_{12}P_{23} + P_{13}P_{23}$ is a permutation operator, T stands for NN t-matrix. The term TP is the leading term of the 3N transition operator T_F obeying Faddeev equation $T_F = TP + TG_0PT_F$, with G_0 being the free three-nucleon propagator. The operator U_0 is fully anti-symmetrized and can be written as a sum of three terms in TP, the first of which is $U_0^{(1)} = TP$. The other terms $U_0^{(2)} = P_{12}P_{23}TP$ and $U_0^{(3)} = P_{12}P_{23}TP$ are related to $U_0^{(1)}$ by means of permutations.

Without employing PW decompositions, except for the deuteron, the Nd break-up amplitude $U_0(\mathbf{p},\mathbf{q})$ is given in Eq. (2) (see Ref. [3] for derivation), where m_{si}, τ_i $(i=1,2,3)$ are the spins and isospins of the three nucleons in final state, m_{s1}^0, τ_1^0 the spin and isospin of nucleon 1 acting as the projectile, $\Psi_d^{M_d}$ the deuteron state, M_d the projection of total angular momentum of the deuteron along an arbitrary z-axis, $\psi_l(\pi')$ the deuteron wave function components (s and d waves), $T^{\pi St}_{\Lambda\Lambda'}(p,\pi,\cos\theta';E_p)$ the NN t-matrix elements in a momentum-helicity basis [4] for given parity π, total spin S and isospin t, final and initial helicities Λ and Λ', and a center of mass energy E_p of the 23-subsystem, $d^S_{\Lambda'\Lambda}(\theta)$'s being the rotation matrices [11]. The Jacobi momenta $\mathbf{p}$ and $\mathbf{q}$ describe the 3N kinematics in the final state, where $\mathbf{p}$ is the relative momentum between nucleon 2 and 3, $\mathbf{q}$ the relative momentum of nucleon 1 to the 23-subsystem, and $\mathbf{q}_0$ the relative momentum of the projectile to the deuteron.

The operator $U_0(\mathbf{p},\mathbf{q})$ given in Eq. (2) is derived within the framework of the nonrelativistic Faddeev scheme. By adopting the formulation given in Ref. [12], relativistic kinematics is introduced. Thus, we re-evaluate the Jacobi momenta, carry out corresponding Lorentz transformations to the two- and three-particle c.m. subsystems, and employ relativistic energy-momentum relations. To calculate the observables, a relativistic description of the cross section is employed.

$$
\begin{aligned}
U_0(\mathbf{p},\mathbf{q}) &\equiv \left\langle \mathbf{p}\mathbf{q}m_{s1}m_{s2}m_{s3}\tau_1\tau_2\tau_3 \middle| U_0 \middle| \mathbf{q}_0 m_{s1}^0 \tau_1^0 \Psi_d^{M_d} \right\rangle \\
&= \frac{(-)^{\frac{1}{2}+\tau_1}}{4\sqrt{2}} \delta_{\tau_2+\tau_3,\tau_1^0-\tau_1} \sum_{m_s'} e^{-i(\Lambda_0\phi_p-\Lambda_0'\phi_\pi)} C\left(\frac{1}{2}\frac{1}{2}1; m_s' m_{s1}\right) \\
&\times \sum_l C\left(l11; M_d - m_s' - m_{s1}, m_s' + m_{s1}\right) Y_{l,M_d-m_s'-m_{s1}}(\hat{\boldsymbol{\pi}}')\psi_l(\pi') \\
&\times \sum_{S\pi t} \left(1-\eta_\pi(-)^{S+t}\right) C\left(\frac{1}{2}\frac{1}{2}t;\tau_2\tau_3\right) C\left(\frac{1}{2}\frac{1}{2}t;\tau_1^0,-\tau_1\right) \\
&\times C\left(\frac{1}{2}\frac{1}{2}S; m_{s2}m_{s3}\Lambda_0\right) C\left(\frac{1}{2}\frac{1}{2}S; m_{s1}^0 m_s'\Lambda_0'\right) \\
&\times \sum_{\Lambda\Lambda'} d^S_{\Lambda_0\Lambda}(\theta_p) d^S_{\Lambda_0'\Lambda'}(\theta_\pi) e^{i(\Lambda'\phi'-\Lambda\Omega)} T^{\pi St}_{\Lambda\Lambda'}(p,\pi,\cos\theta';E_p) \\
&+ U_0^{(2)}(\mathbf{p},\mathbf{q}) + U_0^{(3)}(\mathbf{p},\mathbf{q}),
\end{aligned} \tag{2}
$$

with

$$
\boldsymbol{\pi} \equiv \frac{1}{2}\mathbf{q} + \mathbf{q}_0 \qquad \boldsymbol{\pi}' \equiv -\mathbf{q} - \frac{1}{2}\mathbf{q}_0 \tag{3}
$$

$$
\cos\theta' = \cos\theta_p \cos\theta_\pi + \sin\theta_p \sin\theta_\pi \cos(\phi_p - \phi_\pi) \tag{4}
$$

$$
e^{i(\Lambda'\phi'-\Lambda\Omega)} = \frac{\sum_{N=-S}^{S} e^{iN(\phi_p-\phi_\pi)} d^S_{N\Lambda}(\theta_p) d^S_{N\Lambda'}(\theta_\pi)}{d^S_{\Lambda'\Lambda}(\theta')}. \tag{5}
$$

3. The Semi-exclusive Proton-Deuteron Break-Up Reaction

Our formulation is applied to the semi-exclusive proton-deuteron (pd) break-up reaction $d(p,n)pp$. We calculate the spin averaged differential cross section, the neutron polarization, the proton analyzing power and the polarization transfer coefficients for proton laboratory energies E_{lab} up to about 500 MeV.

First, we perform comparisons with calculations based on the well established PW technique, which also include only the leading term of the Faddeev amplitude. Both schemes agree for projectile energies $E_{lab} < 200$ MeV. However, at about 200 MeV deviations occur in the cross section peak, as shown in Fig. 1 for the cross section at $E_{lab} = 197$ MeV, where the PW calculation has not sufficiently converged to the 3D calculation. The PW calculation shown in Fig. 1 takes into account NN angular momenta up to $j = 5$ and 7, and 3N states for total 3N angular momenta up to $J = 31/2$, which is a technical maximum at present.

Since we do not take rescattering terms into account, we nevertheless have to estimate their effects. At $E_{lab} = 197$ MeV we do this by comparing the 3D calculation with a full Faddeev PW calculation, and display the results in Fig. 2. We see that the higher order multiple scattering contributions lower the cross section and are essential for the description of the analyzing power especially for the small energies of the outgoing neutron.

Next, we investigate effects of relativistic kinematics on semi-inclusive observables as function of the projectile energy. In Fig. 3 we compare our nonrelativistic 3D calculations with the corresponding relativistic ones, and see that the position as well as the height of the quasi-free peak are influenced. Since the position of the quasi-free peak is solely determined by kinematics, it is satisfying to see that the use of relativistic kinematics put the calculated peak at the right position with respect to the data. As expected the effect of relativistic kinematics increases with increasing projectile energy.

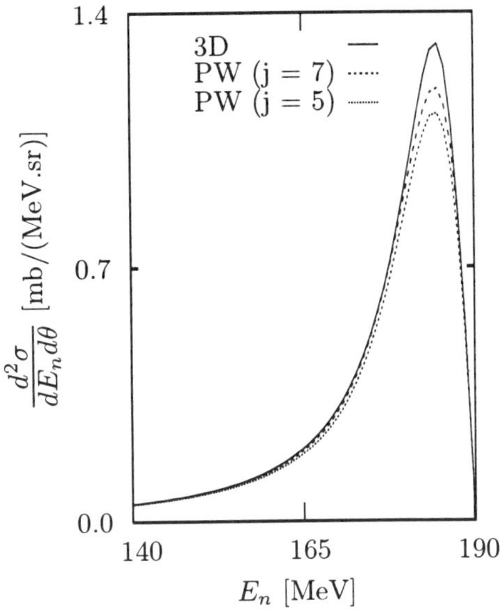

Fig. 1. The spin averaged differential cross section at $E_{lab} = 197$ MeV, $\theta = 13^0$, based on the Bonn-B potential.

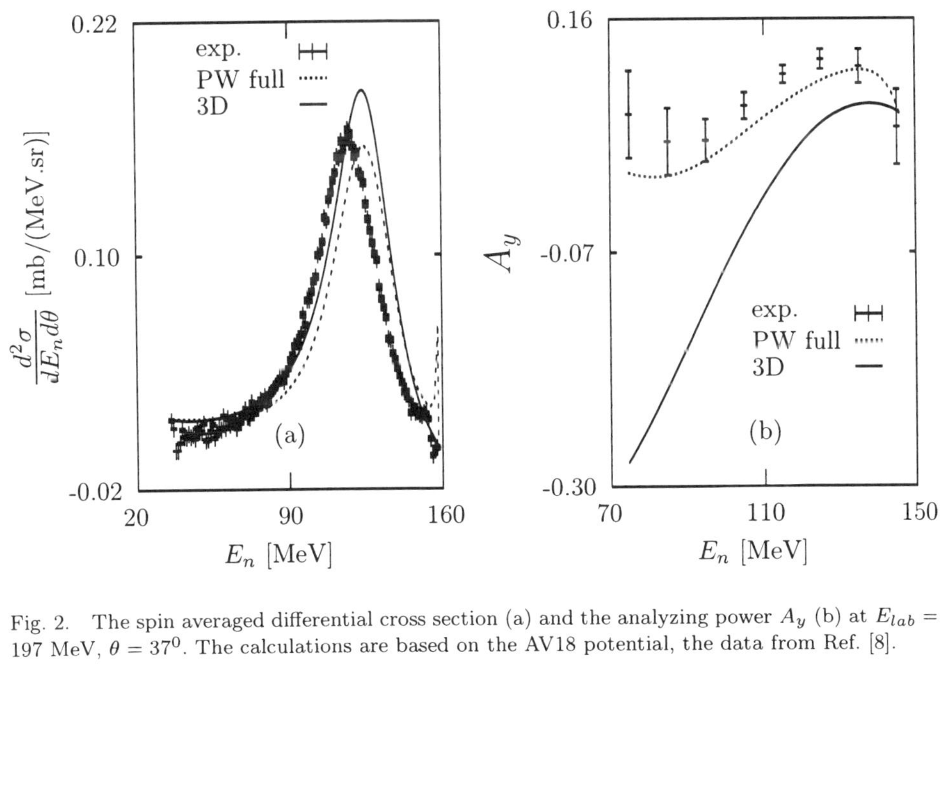

Fig. 2. The spin averaged differential cross section (a) and the analyzing power A_y (b) at $E_{lab} =$ 197 MeV, $\theta = 37^0$. The calculations are based on the AV18 potential, the data from Ref. [8].

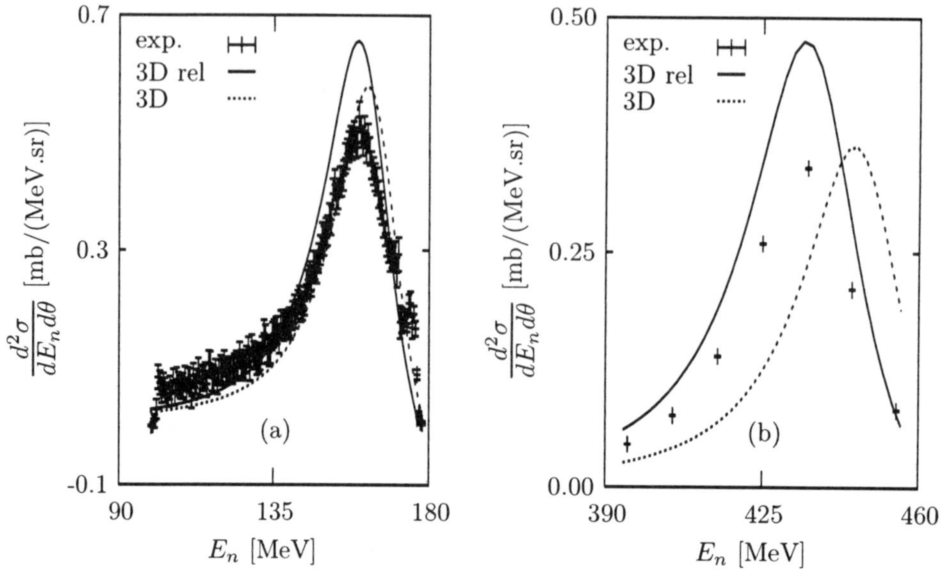

Fig. 3. The spin averaged differential cross section at (a) E_{lab} = 197 MeV, $\theta = 24^0$ and (b) E_{lab} = 495 MeV, $\theta = 18^0$. The calculations are based on the Bonn-B potential, the data from Ref. [8] (a) and Ref. [6] (b).

4. Summary and Conclusions

We calculated the semi-exclusive pd break-up process within the Faddeev scheme in the leading order in a multiple scattering expansion in a 3D formulation based directly on momentum vectors. This technique has proved to be a viable alternative to traditional PW decompositions. For energies smaller than 200 MeV our 3D calculations show perfect agreement with PW calculations. At energies larger than 200 MeV the inclusion of relativistic kinematics proves essential to obtain the correct position of the quasi-free peak in the (p,n) charge exchange reaction.

References

1. L. D. Faddeev, Sov. Phys. JETP **12**, 1014 (1961).
2. H. Liu, Ch. Elster, W. Glöckle, nucl-th/0410051 and to appear in Phys. Rev. C; Ch. Elster, et. al., Few-Body Systems 27, 83 (1999).
3. I. Fachruddin, Ch. Elster, W. Glöckle, Phys. Rev. **C68**, 054003 (2003).
4. I. Fachruddin, Ch. Elster, W. Glöckle, Phys. Rev. **C62**, 044002 (2000).
5. I. Fachruddin, Ch. Elster, W. Glöckle, Phys. Rev. **C 63**, 054003 (2001).
6. X. Y. Chen, et. al., Phys. Rev. **C47**, 2159 (1993).
7. T. Wakasa, et. al., Phys. Rev. **C59**, 3177 (1999).
8. D. L. Prout, et. al., Phys. Rev. **C65**, 034611 (2002).
9. R. B. Wiringa, V. G. J. Stoks, R. Schiavilla, Phys. Rev. **C51**, 38 (1995).
10. R. Machleidt, Adv. Nucl. Phys. **19**, 180 (1989).
11. M. E. Rose, *Elementary Theory of Angular Momentum* (Wiley, New York, 1957).
12. R. Fong and J. Sucher, J. Math. Phys. **5**, 456 (1964).

Three-Body Decay of Nuclear Resonances

A. S. Jensen, D. V. Fedorov and H. O. U. Fynbo

Dep. of Physics and Astronomy, Univ. of Aarhus, DK-8000 Aarhus C, Denmark
E-mail: asj@phys.au.dk

E. Garrido

Instituto de Estructura de la Materia, CSIC, E-28006 Madrid, Spain
E-mail: e.garrido@iem.cfmac.csic.es

The concept known from emission of an α-particle is generalized to apply for decays of a nuclear resonance into three particles. We use the hyperspherical adiabatic expansion method. Structure and decay are discussed for two ^{17}Ne-states. The energy distributions of the final state fragments carry the signatures of the various decay mechanisms. Characteristic features of the energy distributions are discussed for ^{6}He and ^{11}Li-states especially those related to decay via two-body virtual s-states.

1. Introduction

We assume that a nuclear resonance is populated for example in a reaction or by beta decay. If this resonance partially decays into three fragments in the final state we have strictly a three-body problem at large distance. At small distance this is not necessarily true since the cluster assumption may be inappropriate. This is precisely the same problem encountered in the classical treatment of α-emission[1]. The generalization to three-body decay is needed in connection with the kinematically complete experiments in both nuclear and molecular physics [2–4]. The basic observables are the energy distributions of the final state fragments. They carry the information about the resonance structure and the decay mechanism. These distributions are basically determined by the momentum space wavefunction where the divergence at large distance has to be properly treated to extract the physically significant part[5]. It is a challenge to understand these decays even though they seem to be very simple extensions of well-known two-body decays. Furthermore these decays should be accessible through our three-body methods[6] which previously were applied to structure and reactions of quantum halos[7].

2. Resonance structure

We use the hyperspherical adiabatic expansion method[6]. The masses, charges, coordinates and momenta are denoted m_i, eZ_i, $\mathbf{r}_i$, and $\mathbf{p}_i$, $i = 1, 2, 3$, where e is the (positive) unit charge. The hyperspherical coordinates consist of the hyperradius ρ

defined by $\rho^2 = \sum_{i<j} m_i m_j \mathbf{r}_{ij}^2/(mM)$ ($\mathbf{r}_{ij} = \mathbf{r}_i - \mathbf{r}_j$, $M = m_i + m_j + m_k$, m is an arbitrary mass), and five dimensionless angles collectively denoted Ω. The particles interact pairwise through two-body potentials $V_{ij}(r_{ij})$. If necessary a short-range three-body potential depending on ρ is added.

Energies and wave functions of the resonances are computed by use of the complex scaling method[8] where $\mathbf{r}_i \to \mathbf{r}_i \exp(i\theta)$, i.e. the coordinates are rotated in the complex plane by an angle θ. We fix ρ and solve the Faddeev equations in angular space which provides the adiabatic potentials. The coupled radial set of equations are finally solved with the appropriate boundary conditions resulting in complex resonance energies and wavefunctions.

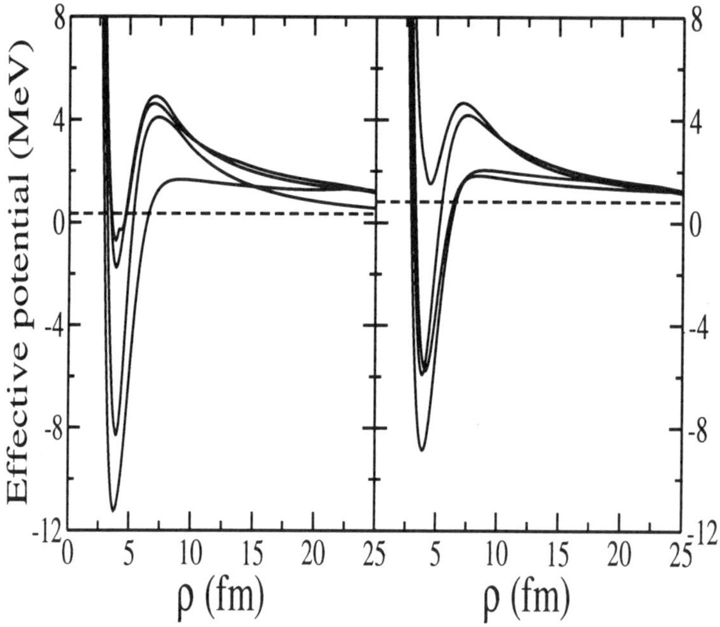

Fig. 1. The lowest adiabatic potentials as function of ρ for the states of $\frac{3}{2}^-$ (left) and $\frac{5}{2}^-$ (right) in ^{17}Ne(^{15}O+p+p). The resonance energies at 0.34 MeV and 0.82 MeV are indicated by the horizontal lines.

In fig.1 we show an example of the lowest adiabatic potentials as function of hyperradius. The total wavefunction is a linear combination of terms corresponding to each of these potentials. The almost horizontal stretch on the lowest potential for $3/2^-$ is consistent with a (in fact two) ^{16}F-resonances and one proton moving away. This is sequential decay, except that the energy is too low to allow this intermediate structure, and consequently the process is only virtual. However, this mechanism is crucial for the size of the width.

In fig.2 we show the structure of the wavefunctions related to the lowest poten-

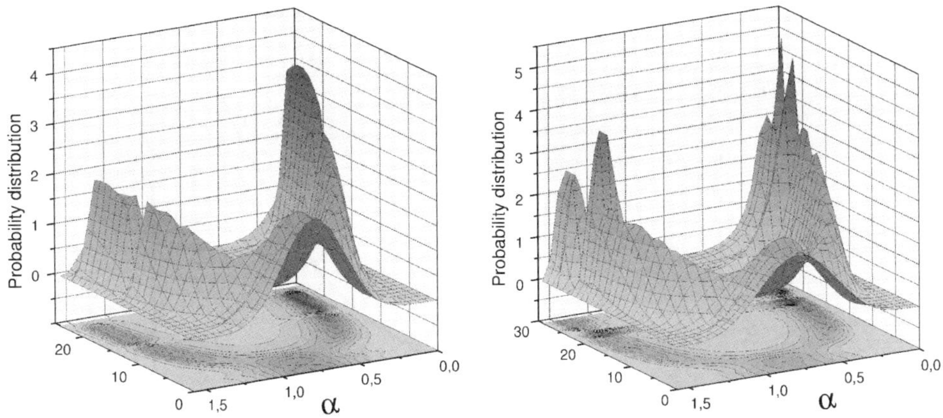

Fig. 2. The probability distribution for the dominating adiabatic potentials of ^{17}Ne($\frac{3}{2}^-$) (left) and ^{17}Ne($\frac{5}{2}^-$) (right) as function of hyperradius ρ and α related to the distance (r) between the proton and the ^{15}O-core by $r \propto \rho \sin\alpha$.

tial. As ρ increases the probabilities peak at very small distances between proton and oxygen-core, i.e. (virtual) sequential decay. Direct decays correspond to a structure where all distances simultaneously become large.

3. Energy distributions

The energy distribution of the final state fragments are essentially obtained from the probability distributions for large values of ρ. Thus from fig.2 we see that the emitted proton has a very large kinetic energy (small α) corresponding to one proton moving opposite to ^{16}F in the center of mass system. The proton emitted from the subsequently decaying ^{16}F would then have a very low energy while ^{15}O would appear with intermediate energy. In the same way distributions with peaks at intermediate energy result from direct decay whereas a peak at large ^{15}O-energy appear if the two protons are emitted in the same direction. All these processes can be described by the present method.

To understand the details without Coulomb complications we turn to two-neutron halos like ^{6}He(2^+) and ^{11}Li(1^-). In fig.3 we show the result when many adiabatic potentials and partial waves are included. The lowest potential leads to a peak at large α-particle energy which remotely has some resemblance with the measurements, but in fact is very far from a good description. Including more potentials in the calculation change the distributions but only the interference between two adiabatic potentials produce a distribution resembling the measurements. The explanation is then a special combination of basically three potentials, i.e. one for ^{5}He-emission and two for two-neutron emission corresponding to ground and first excited state relative motion.

The dominating decay mode with a high-energy peak for the α-distribution is the

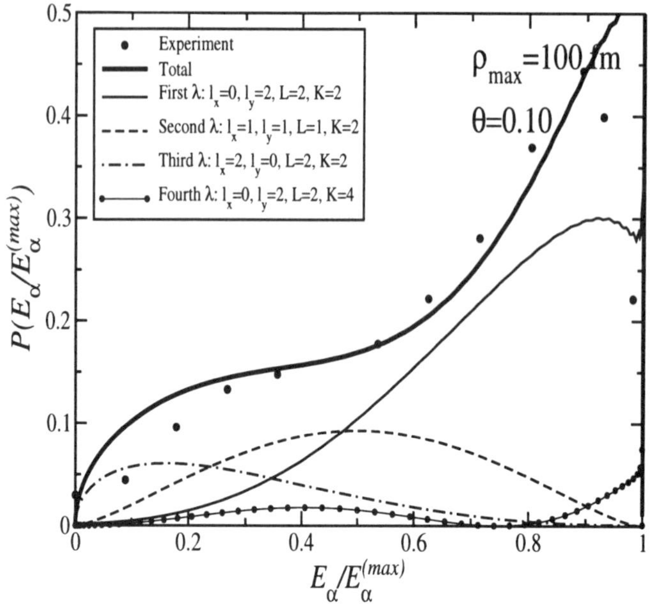

Fig. 3. The energy distribution of the α-particle after decay of the 2^+-resonance in ^{6}He. The points are extracted from measurements[9]. Contributions from different adiabatic potentials are shown.

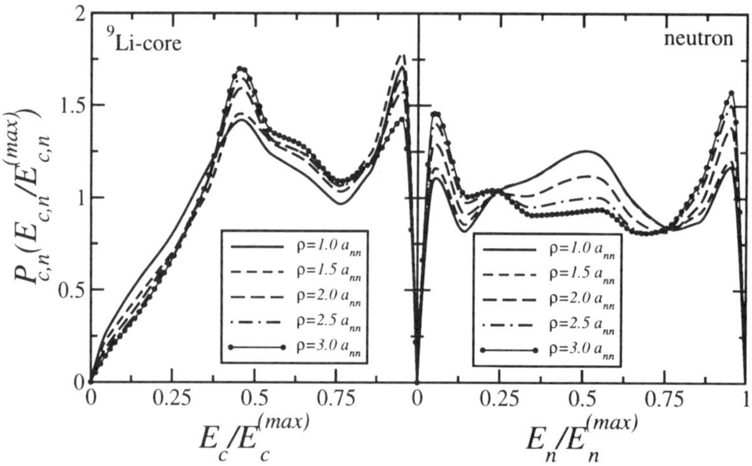

Fig. 4. The energy distribution of the ^{9}Li-core (left) and neutrons (right) after decay of ^{11}Li(1^-) for realistic interactions[10], i.e. scattering lengths $a_{nn} = a_{nc} \approx 20$ fm, and effective range $R_{eff} \approx 5$ fm. All contributing partial waves and adiabatic potentials are included.

signature of two-neutron emission. The intermediate structure is then two neutrons in relative s-states which is an unstable (virtual s-state) configuration without a lifetime. The neutron-neutron scattering length determines the attraction. If more than one scattering length is large we approach the Efimov regime[6] as exemplified in fig.4. The neutrons appear in three peaks corresponding to one-neutron emission (low and high energy) and two-neutron (or core) emission (intermediate energy). The ^{9}Li-core appears at high (two-neutron emission) and intermediate (one-neutron emission) energies. These decays proceed via virtual states where only s-wave two-body attractions are exploited. The mechanism arises from the same origin as the Efimov effect, i.e. interference between spatially separate configurations[6]. This effect is responsible for the inverse square effective potential which dominates the lowest adiabatic potential up to relatively large distances. In turn the characteristic energy distribution emerges with a dominating component from the same potential.

4. Conclusion

The hyperspherical adiabatic expansion method is used to study three-body decays. Resonance structures, decays and energy distributions of the fragments after the decay are computed. Different decay mechanisms are discussed especially sequential decay through an intermediate configuration like two-body resonances or virtual s-states. The Efimov effect is shown to leave fingerprints on the energy distributions corresponding to peaks at specific energies.

References

1. E. Garrido et al., *Nucl. Phys. A* **748**, 27 and 39, (2005).
2. H.O.U. Fynbo et al., *Phys.Rev.Lett.* **91**, 082502 (2003).
3. B. Blank et al., *C.R. Physiques* **4**, (2003).
4. U. Galster et al. *Phys. Rev. Lett.* **92**, 073002 (2004)
5. D.V. Fedorov et al., *Few-body systems* **34**, 33 (2004).
6. E. Nielsen et al., *Phys. Rep.* **347**, 373 (2001).
7. A.S. Jensen et al., *Rev. Mod. Phys.* **76**, 215 (2004).
8. D.V. Fedorov, E. Garrido, and A.S. Jensen, Few-body systems, **33**, 153 (2003).
9. B.V. Danilin, et al., *Sov. J. Nucl. Phys.* **46**, 225 (1987).
10. E. Garrido, D.V. Fedorov and A.S. Jensen, *Nucl. Phys. A* **708**, 277 (2002).

HADRON STRUCTURE AND QCD

Baryon Spectroscopy and the Constituent Quark Model

A. W. Thomas and R. D. Young

Thomas Jefferson National Accelerator Facility
12000 Jefferson Ave., Newport News VA 23185 USA

We explore further the idea that the lattice QCD data for hadron properties in the region $m_\pi^2 > 0.2\text{GeV}^2$ can be described by the constituent quark model. This leads to a natural explanation of the fact that nucleon excited states are generally stable for pion masses greater than their physical excitation energies. Finally we apply these same ideas to the problem of how pentaquarks might behave in lattice QCD, with interesting conclusions.

1. Introduction

Studies of lattice QCD have revealed a transition in the behavior of hadron properties in the region of quark mass $m_\pi \sim 0.4 - 0.5$ GeV. Beyond this point hadron properties exhibit smooth, slowly varying behavior as a function of quark mass. The constituent quark mass is expected to depend linearly on the current quark mass, with $M = M_0 + cm_q$ and $c \sim 1$. In addition, in a constituent quark model (CQM) hadron masses are roughly linear in the number of constituent quarks ($\approx n_H M$, with n_H the number of constituent quarks), magnetic moments are proportional to $1/M$ and so on. Since this is consistent with what is observed it is clear that the lattice simulations are qualitatively consistent with the constituent quark picture in the region $m_\pi > 0.5$ GeV.

On the other hand, in the region where the pion mass approaches zero, we know on model-independent grounds that *all* hadron properties exhibit rapidly varying, non-analytic behavior as a function of m_q – as a consequence of the pion loops resulting from spontaneous chiral symmetry breaking [1]. These loops can change physical properties by up to 50% from the predictions based on naive chiral extrapolation [2]. Since precise lattice computations at sufficiently low quark mass to see this curvature unambiguously are not yet possible [a], the accurate determination of physical hadron properties is not possible unless we know the corresponding chiral coefficients. Thus, while nucleon properties such as mass, magnetic moments, charge radii and moments of parton distribution functions can be extracted by chiral extrapolation, hadron spectroscopy presents a serious challenge. In the case of pentaquarks, since if they exist their structure is completely unknown, nothing is

[a]One possible exception, which relies on the non-unitary nature of quenched QCD, is the comparison between N and Δ properties in quenched QCD [3,4].

known about their chiral behavior. Hence, even if one finds a signal at large m_q, the ambiguity in its physical mass is hundreds of MeV.

In this report we shall concentrate on the mass region in which constituent quark behavior dominates. We aim to understand the lattice data which has been obtained so far in terms of a constituent quark picture – with remarkable success.

2. Access Quark Model – AccessQM

In spite of the obvious, qualitative success of the CQM in describing the quark mass dependence of hadron properties calculated in lattice QCD in the currently accessible mass region, there has so far been only one quantitative application of the CQM to it! This approach, labelled AccessQM by the authors [5], involved the magnetic moments of the baryon octet. The procedure was to use the simplest CQM to give the octet moments in the mass region where $m_u \sim m_d \sim m_s$ and then to extrapolate to the region of physical, light quark masses using the correct chiral coefficients. The result was in excellent agreement with the experimental data. Further improvements in the CQM should be investigated in order to see just how good the agreement can be made.

On a more general level, this success suggests a novel approach to the CQM, where it would be developed to match lattice QCD data at relatively large quark mass, rather than trying to match it to experimental data where one knows that pion corrections can be very large. In addition, this work suggests that a comparison with lattice data at larger light quark mass would serve as a useful test of *any* model of hadron structure [6]. So far only the chiral quark soliton model and the cloudy bag model have taken up this challenge [7–9].

2.1. *N and Δ masses*

As the next check on the physical consistency of the CQM at large quark mass we compare the data for N and Δ masses from full QCD with expectations in AccessQM. In Ref. [5] the constituent quark mass was written as

$$M = 0.421 + 0.301 m_\pi^2 \tag{1}$$

with m_π in GeV. In a careful comparison of the N and Δ masses in full and quenched QCD, Young *et al.* [3] found that the masses in full QCD (from the MILC Collaboration [10]) were well fit by:

$$\begin{aligned} m_\Delta &= 1.43(3) + 0.75(8) m_\pi^2 + \Sigma_\Delta(\text{chiral loops}) \\ m_\mathrm{N} &= 1.24(2) + 0.92(5) m_\pi^2 + \Sigma_N(\text{chiral loops}), \end{aligned} \tag{2}$$

where the chiral loops were calculated using a dipole form factor at the baryon-pion vertices with mass parameter 0.8 GeV. In the region where the pion mass is large these chiral loops are small and vary slowly. Thus it makes sense to compare the slope of the average of the N and Δ masses in that region, namely 0.83 ± 0.09 GeV^{-1}, with three times the coefficient of m_π^2 in Eq.(1), that is 0.90 GeV^{-1}. In terms of the absolute values, when the light quark mass is comparable with the strange quark mass, that is when $m_\pi^2 \sim 0.48\,\mathrm{GeV}^2$, $3M$ is roughly 1.68 GeV, while the average of the N and Δ masses (from the MILC data) is of order 1.60 GeV. Clearly the CQM does a very reasonable job of describing the data quantitatively in the large mass region. A more sophisticated treatment would also model the hyperfine interaction associated with (say) gluon exchange with a term proportional to $1/M$.

2.2. *Unstable Hadrons* – $\Delta(1232)$

Now we compare the behavior of the mass of the $\Delta(1232)$ resonance with the mass of the corresponding decay channel, namely a pion and a nucleon. Provided that the self-energies of the N and Δ associated with chiral loops are either nearly equal or small, we can write the difference between the Δ mass and that of the threshold for the open N π channel as:

$$m_\Delta - m_N - m_\pi = 0.19 - 0.17 m_\pi^2 - m_\pi\,. \tag{3}$$

This would vanish and hence the Δ would become bound at $m_\pi \sim 0.18$ GeV. However, at such a low mass the self-energies associated with chiral loops cannot be neglected and it is not hard to see that to a good approximation we would expect the lhs of Eq.(3) to vanish when m_π is approximately equal to the physical Δ-N mass difference. This is indeed consistent with the lattice results.

2.3. *Other Excited States*

In terms of the application of lattice QCD to hadron spectroscopy, the result of the previous subsection, namely that the Δ resonance actually becomes a stable state when the pion mass exceeds the physical Δ-N mass difference, is an important result. It is far easier to use lattice techniques to find a stable particle, which is a true eigenstate of the QCD Hamiltonian, than to extract information about a resonance which is not the lowest energy state with a particular set of quantum numbers. Of course, in the case of the Δ the situation is even better because the π-N channel must have angular momentum one ($L = 1$) for which the minimum, non-zero momentum on a lattice of size Na is $2\pi/Na$. Hence the Δ is stable on the lattice unless $m_\Delta < \sqrt{m_N^2 + (2\pi/Na)^2} + \sqrt{m_\pi^2 + (2\pi/Na)^2}$, which is typically several hundred MeV higher than the naive threshold, $m_N + m_\pi$.

Another case of considerable interest involving $L = 1$ is the ρ meson which has a p-wave decay into two pions. Of course, for the physical ρ this decay involves momenta that are too high for the comfortable application of effective field theory. Nevertheless, it must be accounted for if one hopes to obtain a meaningful value of the physical mass. It helps that the width of the ρ is well known and this provides a powerful constraint on the phenomenological treatment of this channel [11,12]. In any case, for the present discussion what matters is that the ρ is actually stable on the lattice even when m_ρ is below $2m_\pi$ and a data point at such a low mass provides a

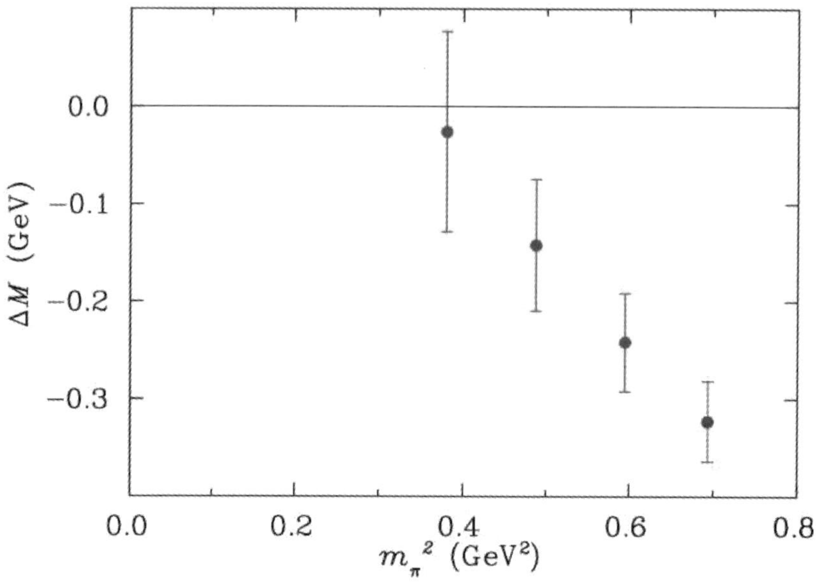

Fig. 1. Mass difference between the lowest-lying negative parity excited nucleon bound state, the $I(J^P) = \frac{1}{2}(\frac{1}{2}^-)$ $N^*(1535)$, and the S-wave $N + \pi$ two-particle scattering state [13].

powerful constraint on the chiral extrapolation needed to obtain the physical mass of the ρ.

However, the challenge of greatest interest to us here is baryon spectroscopy and it is interesting to see whether it is more generally true that nucleon excited states become stable as the quark mass increases. Figure 1, from Lasscock *et al.* [13], shows the difference between the mass of the $1/2^-$ $N^*(1535)$ and its s-wave decay threshold, $m_N + m_\pi$, as a function of pion mass. It is clearly stable for $m_\pi^2 > 0.3$–$0.4\,\mathrm{GeV}^2$. As in the case of the Δ this is very close to the point where the pion mass is equal to the difference between the physical resonance and nucleon masses ($m_\pi \sim 0.6\,\mathrm{GeV}$).

A similar result seems to hold for all nucleon excited states [14–17] which look as though they might match the corresponding experimental state after a reasonable chiral extrapolation. As we have seen earlier, this is what we expect as a first approximation in the case of any nucleon excited state, given that lattice QCD shows that the CQM seems to describe the behavior of non-perturbative QCD in the region where the light quark masses are large ($m_\pi > 0.5$ GeV). The key to this behavior is the rapid variation of the pion mass and hence the corresponding threshold energy, while the energy difference between the mass of the nucleon and the excited state varies relatively slowly.

3. Possible Pentaquarks

It is difficult to imagine a more important discovery in modern strong interaction physics should the pentaquark story have a positive conclusion. At the present time the experimental situation is certainly confusing, with almost as many negative findings as positive and some earlier announcements now contradicted with much higher statistics [18,19]. Theoretically the situation is just as inconclusive. It is difficult to take predictions based on various quark models seriously for a state where we have so little experience. As a result there has been a particular interest in studies based on lattice QCD, with a considerable amount of effort already applied to the problem.

Apart from the challenge of finding an appropriate source which has a large amplitude for producing a pentaquark, one has the bigger challenge of deciding exactly what signal one is expecting to find. We saw in the previous section that it has been possible to study nucleon excited states on the lattice because they become stable as the light quark mass increases. Without this property it would have been a much harder job. The question then is what one might expect with respect to the stability of the pentaquark.

To investigate this question we begin with an expression for the mass of a pentaquark as a function of light quark mass motivated by a CQM which should be valid at sufficiently large m_q:

$$m_5 = M_s + 4M + H_5^{\rm hyp} \,. \qquad (4)$$

Here M_s is the constituent mass of the strange quark, which is fixed (at 0.565 GeV in the AccessQM) in this study of light quark mass variation. Using Eq.(1) we find

$$m_5 = 2.25 + 1.20 m_\pi^2 + H_5^{\rm hyp} \,. \qquad (5)$$

In order to compare with the corresponding threshold, in this case KN, we need an expression for the kaon mass for fixed strange quark mass. Using the Gell-Mann–Oakes–Renner relation one can write $m_K^2 = (m_K^{(0)})^2 + m_\pi^2/2$, where $m_K^{(0)}$ is the kaon mass in the SU(2) chiral limit. To keep the discussion simple, we expand the kaon mass to leading order in m_π^2

$$m_K = m_K^{(0)} + m_\pi^2/(4 m_K^{(0)}) = 0.485 + 0.515 m_\pi^2 + \delta_K \qquad (6)$$

All higher order terms are collected into the term δ_K, where to indicate the scale at $m_\pi^2 = 0.4\,{\rm GeV}^2$, $\delta_K = 30\,{\rm MeV}$ and at $m_\pi^2 = 0.8\,{\rm GeV}^2$, $\delta_K = 100\,{\rm MeV}$.

Finally, we can now compute the difference between the pentaquark mass and the KN threshold (ignoring issues of finite lattice volume for the moment) at sufficiently large m_q that we can ignore chiral loops:

$$m_5 - m_N - m_K = 0.53 - 0.24 m_\pi^2 + H_5^{\rm hyp} - \delta_K \,. \qquad (7)$$

The clear difference between the situation for nucleon excited states and pentaquark states is that in the region of large quark mass, where the chiral loops for the nucleon and (presumably) the hyperfine interaction (including chiral loops) for the pentaquark are smooth and slowly varying, there is not expected to be any rapid variation with m_π. In particular, there is no term linear in m_π and the coefficient of m_π^2 is relatively small.

A plot of $m_5 - m_N - m_K$ would be expected to show only a very smooth variation with m_π in the region above $m_\pi^2 \sim 0.2\mathrm{GeV}^2$, with typically rapid chiral variation only below this region *if* this region is accessible.

The only way that the pentaquark could be stable (i.e., the lhs of Eq.(7) negative) in the large mass region explored by lattice QCD is if the hyperfine interaction in the pentaquark were very large and attractive in that region – an order of magnitude larger than the estimate reported recently in Ref. [20]. If this were the case, then it suggests that the only way that this state could become unbound at the physical point would be through the increase at low m_π in the known, large chiral corrections to the nucleon mass. These would need to cancel a good part of the large pentaquark hyperfine interaction at the physical point. In this case the interaction responsible for producing the pentaquark is unlikely to be chiral in origin. Rather it is more likely to have as its origin something like gluon exchange, which is expected to exhibit a relatively smooth variation as in the CQM (typically like $1/M$).

The other alternative, should the pentaquark exist, is that the hyperfine interaction is relatively weak. In this case, Eq. (7) indicates that the pentaquark is likely to remain unstable at larger quark masses. This would force one into the situation of needing to provide a careful analysis to identify one- and two-particle states in a finite-volume lattice simulation. To make a connection with the physical limit, the chiral corrections must necessarily be important, and one must consider

$$m_5 - m_N - m_K = 0.53 - 0.24 m_\pi^2 + \Sigma_5(\text{chiral loops}) - \Sigma_N(\text{chiral loops})\,, \qquad (8)$$

where H_5^{hyp} and δ_K are assumed negligible. At the physical point, $\Sigma_N \sim -300\,\text{MeV}$ and thus for the pentaquark to exist at $\sim 100\,\text{MeV}$ above threshold, the chiral corrections must be of order $\sim -700\,\text{MeV}$. The dynamical origin of the pentaquark in this case would therefore be chiral in nature.

4. Conclusion

We have explored the degree to which the constituent quark model can quantitatively describe the behavior of baryon properties in the region of relatively large quark mass currently accessible to lattice QCD. In fact it works very well, providing further support to calls to develop modern constituent quark models in this mass regime and to make the connection with experimental data through chiral extrapolation. Of particular interest with respect to baryon spectroscopy is a very natural explanation of the reason why nucleon excited states are stable for m_π greater than roughly the difference of the physical mass of the excited state and the mass of the nucleon.

When the same ideas are applied to the possible existence of a pentaquark, the conclusions are somewhat different. For example, a plot of $m_5 - m_N - m_K$ would

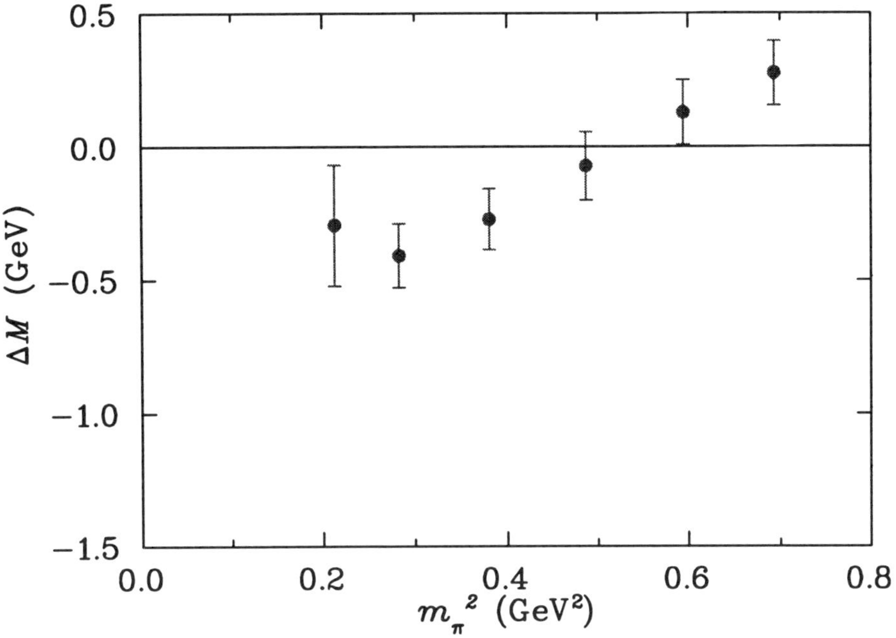

Fig. 2. Mass difference between the lowest-lying 5-quark $I(J^P) = 0(\frac{3}{2}^+)$ and the $N+K$ 2-particle state [21].

be expected to show only a very smooth variation with m_π in the region above $m_\pi^2 \sim 0.2\text{GeV}^2$, with typically rapid chiral variation only below this region – *if* it is accessible. It is interesting that this is precisely the type of behavior seen in the recent investigation of a possible spin-3/2, positive parity pentaquark state by the CSSM-JLab collaboration – see Fig. 2. Clearly this needs further investigation.

Furthermore, these ideas lead us to the conclusion that, if the signal observed in Fig. 2 is genuine, the interaction responsible for producing pentaquarks is unlikely to be chiral in origin. Rather it is more likely to have as its origin something like gluon exchange, which is expected to exhibit a relatively smooth variation as in the CQM (typically like $1/M$). Finally, we note that while the slope of ΔM at large m_π in Fig. 2 differs from that in Eq. (7), this may be consistent with a $1/M$ dependence of a large, attractive hyperfine interaction.

Acknowledgments

It is a pleasure to acknowledge many helpful conversations with D. Leinweber concerning the AccessQM, M. Paris regarding quark-model studies and B. Lasscock and collaborators regarding the lattice simulations of pentaquark states. This work was supported by DOE contract DE-AC05-84ER40150, under which SURA operates Jefferson Laboratory.

References

1. L. F. Li and H. Pagels, Phys. Rev. Lett. **26**, 1204 (1971).
2. W. Detmold, W. Melnitchouk, J. W. Negele, D. B. Renner and A. W. Thomas, Phys. Rev. Lett. **87**, 172001 (2001) [arXiv:hep-lat/0103006].
3. R. D. Young, D. B. Leinweber, A. W. Thomas and S. V. Wright, Phys. Rev. D **66**, 094507 (2002) [arXiv:hep-lat/0205017].
4. R. D. Young, D. B. Leinweber and A. W. Thomas, Nucl. Phys. Proc. Suppl. **128**, 227 (2004) [arXiv:hep-lat/0311038].
5. I. C. Cloet, D. B. Leinweber and A. W. Thomas, Phys. Rev. C **65**, 062201 (2002) [arXiv:hep-ph/0203023].
6. W. Detmold, D. B. Leinweber, W. Melnitchouk, A. W. Thomas and S. V. Wright, Pramana **57**, 251 (2001) [arXiv:nucl-th/0104043].
7. D. B. Leinweber, D. H. Lu and A. W. Thomas, Phys. Rev. D **60**, 034014 (1999) [arXiv:hep-lat/9810005].
8. D. B. Leinweber, A. W. Thomas, K. Tsushima and S. V. Wright, Phys. Rev. D **61**, 074502 (2000) [arXiv:hep-lat/9906027].
9. K. Goeke, J. Ossmann, P. Schweitzer and A. Silva, arXiv:hep-lat/0505010.
10. C. W. Bernard *et al.*, Phys. Rev. D **64**, 054506 (2001) [arXiv:hep-lat/0104002].
11. C. R. Allton, W. Armour, D. B. Leinweber, A. W. Thomas and R. D. Young, Phys. Lett. B, *in press* [arXiv:hep-lat/0504022].
12. D. B. Leinweber, A. W. Thomas, K. Tsushima and S. V. Wright, Phys. Rev. D **64**, 094502 (2001) [arXiv:hep-lat/0104013].
13. B. G. Lasscock *et al.*, Phys. Rev. D **72**, 014502 (2005) [arXiv:hep-lat/0503008].
14. S. Sasaki, T. Blum and S. Ohta, Phys. Rev. D **65**, 074503 (2002) [arXiv:hep-lat/0102010].
15. M. Gockeler, R. Horsley, D. Pleiter, P. E. L. Rakow, G. Schierholz, C. M. Maynard and D. G. Richards [QCDSF Collaboration], Phys. Lett. B **532**, 63 (2002) [arXiv:hep-lat/0106022].
16. W. Melnitchouk *et al.*, Phys. Rev. D **67**, 114506 (2003) [arXiv:hep-lat/0202022].
17. D. B. Leinweber, W. Melnitchouk, D. G. Richards, A. G. Williams and J. M. Zanotti, Lect. Notes Phys. **663**, 71 (2005) [arXiv:nucl-th/0406032].
18. A. R. Dzierba, C. A. Meyer and A. P. Szczepaniak, J. Phys. Conf. Ser. **9**, 192 (2005) [arXiv:hep-ex/0412077].
19. K. H. Hicks, Prog. Part. Nucl. Phys. **55**, 647 (2005) [arXiv:hep-ex/0504027].
20. M. W. Paris, Phys. Rev. Lett., *in press* [arXiv:nucl-th/0507061].
21. B. G. Lasscock *et al.*, arXiv:hep-lat/0504015.

A Theoretical Study on the Spin Dependent Structure Functions of the Nucleon*

F. Bissey, F.-G. Cao and A. I. Signal

Institute of Fundamental Sciences PN461, Massey University,
Private Bag 11 222, Palmerston North, New Zealand

We calculate the spin dependent structure functions $g_1(x)$ and $g_2(x)$ of the nucleon in the meson cloud model of nucleon structure. The contributions from kinematic terms which mix transverse and longitudinal spin components are included. The structure functions of the "bare" hadrons are calculated by using the MIT bag model and the method of the Adelaide group. The contributions from polarized gluon distributions are taken into account in a phenomenological way. The calculations agree with the experimental measurements for $g_1(x)$ and $g_2(x)$ of the proton and the neutron in a large range of x $(0.01 < x < 0.6)$.

1. Introduction

The spin dependent structure functions of the nucleon, $g_1^N(x)$ and $g_2^N(x)$ are the subject of much theoretical and experimental interest[1-5]. The main reason for this interest has been the large amount of evidence, starting with the EMC experiment[6] that strongly suggests that constituent quark models are inadequate to describe the spin structure of nucleons. This has led to considerable activity in order to determine how the spin of nucleons is built up from the intrinsic spin and orbital angular momentum of their constituent quarks and gluons. The $g_2(x)$ structure functions are of interest because they do not have a simple interpretation in the quark-parton model, but are related to transverse momentum of quarks and higher twist operators which measure correlations between quarks and gluons.

We report a theoretical study[7] on the spin dependent structure functions of the nucleon using the meson cloud model (MCM)[8,9] which has been applied successfully in spin independent DIS. In Section 2 we present briefly the framework to investigate the spin dependent structure functions in the MCM, including kinematic terms which lead to g_2 (g_1) of cloud components contributing to g_1 (g_2) of the nucleon. The numerical results are shown and discussed in Section 3. In the last section we summarise our findings.

*This work is supported by the Marsden Fund of the Royal Society of New Zealand.

2. Spin Dependent Structure Functions in the Meson Cloud Model

In the meson cloud model (MCM) the nucleon can be viewed as a "bare" nucleon plus some baryon-meson Fock states which result from the fluctuation of nucleon to baryon plus meson. The model assumes that the lifetime of a virtual baryon-meson Fock state is much longer than the interaction time in the deep inelastic or Drell-Yan process, thus scattering from the virtual baryon-meson Fock states can contribute to the observed structure functions of the nucleon.

Baryon or meson components of the cloud with spin $\geq 1/2$ will contribute directly to the $g_1^N(x)$ and $g_2^N(x)$. Three cases of interest are spin $\frac{1}{2}$ baryons, spin $\frac{3}{2}$ baryons, and spin 1 mesons. The contributions from a spin $\frac{1}{2}$ baryon and a spin 1 meson can be written as[a]

$$\delta g_1^N(x,Q^2) = \frac{1}{1+\gamma^2}\int_x^1 \frac{dy}{y}\sum_{i=1,2}[\Delta f_{iL}(y)+\Delta f_{iT}(y)]g_i^{B(M)}(x/y,Q^2), \tag{1}$$

$$\delta g_2^N(x,Q^2) = -\frac{1}{1+\gamma^2}\int_x^1 \frac{dy}{y}\sum_{i=1,2}\left[\Delta f_{iL}(y)+\frac{\Delta f_{iT}(y)}{\gamma^2}\right]g_i^{B(M)}(x/y,Q^2), \tag{2}$$

where $g_i^{B(M)}$ are the spin dependent structure functions of the baryons (mesons) in the Fock states. The functions $\Delta f_{1(2)\,L(T)}$ are the longitudinally and transversely polarized fluctuation functions. The calculation details and the expressions for the contributions from spin $\frac{3}{2}$ baryons can be found in Refs. [7,10].

From equations (1,2) we see that the nucleon structure functions pick up contributions from both g_1 and g_2 of the baryon and meson in the cloud. This occurs because the spin vectors of the cloud components, s_B^μ and s_M^μ are not parallel to the initial nucleon spin vector s_N^μ. So if the initial nucleon is longitudinally polarized, the hadrons in the cloud will have both longitudinal and transverse spin components, and hence g_2^B and g_2^M will give a finite contribution to g_1^N. Similarly g_2^N will get a contribution from g_1^B and g_1^M. As the "bare" g_2^N structure functions are expected to be small, this kinematic contribution from the baryon-meson cloud could be a major portion of these structure functions.

3. Numerical Results and Discussions

The Fock states we consider are $|N\pi\rangle$, $|N\rho\rangle$, $|\Delta\pi\rangle$ and $|\Delta\rho\rangle$. The contributions from Fock states involving higher invariant mass squared are very small (a few per cents). We estimate all the structure functions of the baryons and mesons in the Fock states, including those of the "bare" nucleons, using the MIT bag model and the methods developed by the Adelaide group[11] and ourselves[12,13]. This approach gives a reasonable description of the unpolarized structure functions of the nucleons when compared to experimental data. We take into account the contributions from polarized gluon distributions by introducing a phenomenological $\Delta g(x)$ for

[a]The convention for the sign in equations (1,2) is different from Ref. [10].

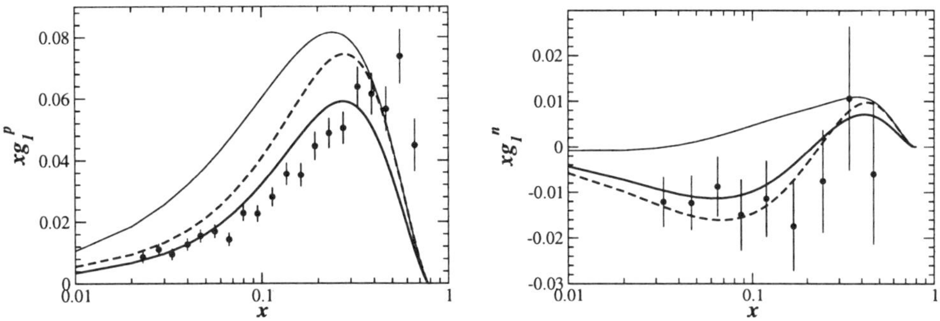

Fig. 1. Spin dependent structure functions xg_1^p and xg_1^n at $Q^2 = 2.5$ GeV^2. The thin solid curves are the bag model calculations. The thick dashed curves are the bag model calculations plus contributions from the polarized gluon. The thick solid curves are the total results in the MCM calculations.

the polarized parton distributions of each flavour calculated in the bag model. The structure functions g_2 for the bare hadrons are estimated via the Wandzura-Wilczek relation[14].

In figure 1 we compare theoretical calculations for xg_1^p and xg_1^n with experimental measurements from HERMES Collaboration[1,2]. Our calculations describe the data very well in a large range of x $(0.01 < x < 0.6)$. This overall good agreement is achieved by considering the effects associated with the meson cloud and polarized gluon at the same time. The meson cloud can lower the bag model calculation for g_1 in the entire range of x since the meson cloud may carry (as angular momentum) some of the spin of the nucleon[15]. However the calculations without polarized gluon contributions fail to describe the experimental data in the small x region. The importance of the polarized gluon contributions is more obvious in the calculation of xg_1^n. Without these contributions, theoretical calculations will not be able to reproduce the shape suggested by the experimental data.

The values for the Ellis-Jaffe sum rule for the proton and neutron, $\Gamma^{p(n)} = \int_0^1 g_1^{p(n)}(x)dx$ are found to be 0.120 and -0.027 respectively, which are very close to the experimental values of 0.122 ± 0.003(stat.) $\pm$ 0.010(syst.)[1] and -0.037 ± 0.013(stat.) $\pm$ 0.005(syst.)[2]. The Bjorken sum rule, $\Gamma^p - \Gamma^n$ is found to be 0.147 which is 10% smaller than the experimental measurement. In this work, the structure functions have been calculated by considering only the contributions from the valence partons. It has been pointed out[16] that the flavour asymmetry in the polarized nucleon sea may contribute $5\% \sim 10\%$ to the Bjorken sum rule.

The results for xg_2^p and xg_2^n are presented in figure 2 along with experimental data from SLAC-E155[3] and Jlab[4,5]. The data are given in a range of Q^2, $0.8\,\text{GeV}^2 < Q^2 < 8.2\,\text{GeV}^2$ for xg_2^p and $0.57\,\text{GeV}^2 < Q^2 < 4.83\,\text{GeV}^2$ for xg_2^n. There is a good agreement between the theoretical calculations and experimental measurements in the range of $0.05 < x < 0.7$. The calculations for xg_2^n are about 10% smaller than the most recent precision measurement by the JLab[5] at $x \simeq 0.2$. The experimental information on the shape of xg_2^n is not conclusive due to large error bars. Once again

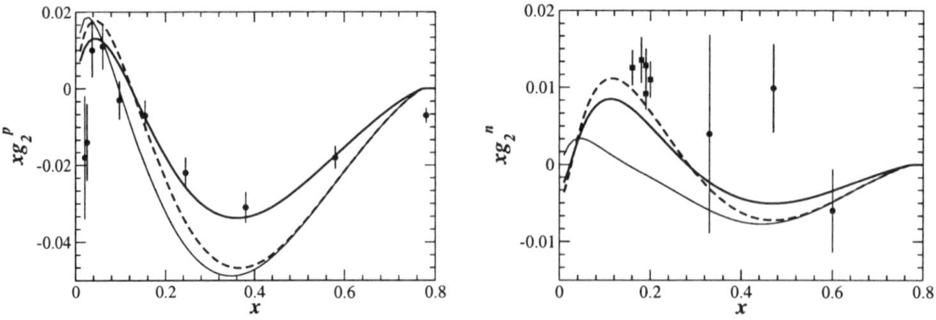

Fig. 2. As in figure 1 but for the spin dependent structure functions xg_2^p and xg_2^n.

we found that the polarized gluon contributions are crucial to the calculations in the $x < 0.2$ region, especially for xg_2^n. The cloud contributions are more important for xg_2^p and xg_2^n than for xg_1^p and xg_1^n.

4. Summary

We calculated the spin dependent structure functions of the nucleon xg_1^p, xg_1^n, xg_2^p and xg_2^n in the meson cloud model of nucleon structure, including the contributions from kinematic terms which mix transverse and longitudinal spin components, and the contributions from the polarized gluon distributions. Our results are consistent with the experimental measurements for those structure functions in a large range of x ($0.01 < x < 0.6$).

References

1. A. Airapetian *et al.*, HERMES Collaboration, *Phys. Lett.* **B442**, 484 (1998)
2. K. Ackerstaff *et al.*, HERMES Collaboration, *Phys. Lett.* **B404**, 383 (1997).
3. P. L. Anthony *et al.*, E155 Collaboration, *Phys. Lett.* **B553**, 18 (2003).
4. X. Zheng *et al.*, *Phys. Rev.* **C70**, 065207 (2004).
5. K. Kramer *et al.*, *Phys. Rev. Lett.* **95**, 142002 (2005).
6. J. Ashman *et al.*, EMC Collaboration, *Phys. Lett.* **B206**, 364 (1988);
7. F. Bissey, F.-G. Cao and A. I. Signal, in preparation.
8. J. D. Sullivan, *Phys. Rev.* **D5**, 1732 (1972).
9. A. W. Thomas, *Phys. Lett.* **B126**, 97 (1983).
10. S. Kumano and M. Miyama, *Phys. Rev.* **D65**, 034012 (2002).
11. see *e.g.*, A. I. Signal and A. W. Thomas, *Phys. Rev.* **D40**, 2832 (1989); C. Boros and A. W. Thomas, *Phys. Rev.* **D60**, 074017 (1999).
12. A. I. Signal, *Nucl. Phys.* **B497**, 415 (1997).
13. F.-G. Cao and A. I. Signal, *Phys. Rev.* **D68**, 074002 (2003).
14. S. Wandzura and F. Wilczek, *Phys. Lett.* **B72**, 195 (1977).
15. F. M. Steffens, H. Holtmann, and A. W. Thomas, *Phys. Lett.* **B358**, 139 (1995).
16. F.-G. Cao and A. I. Signal, *Eur. Phys. J.* **C21**, 105 (2001).

Proton Form Factor Measurements at Jefferson Lab

C. F. Perdrisat[1], E. J. Brash[2,3], M. K. Jones[3], L. Pentchev[1], V. Punjabi[4] and F. R. Wesselmann[4]

[1]*College of William and Mary, Williamsburg, VA 23187*
[2]*Christopher Newport University, Newport News, VA 23606*
[3]*Jefferson Lab, Newport News, VA 23606*
[4]*Norfolk State University, Norfolk, VA 23504*
E-mail: perdrisa@jlab.org

We report the results of a series of measurements of the ratio of the electric- and magnetic form factors of elastic ep scattering, G_{Ep}/G_{Mp} up to $Q^2 - 5.6$ GeV^2, using the recoil polarization method. A new experiment to measure G_{Ep}/G_{Mp} to Q^2=9 GeV^2 is currently being prepared. The understanding of the structure of the nucleon is of fundamental importance in nuclear and particle physics. The internal structure of the nucleon is revealed by its root mean square radius, anomalous magnetic moment, charge and current distributions, polarizabilities and transition amplitudes to excited baryonic states.

The quantities of interest here are the elastic electric and magnetic Sachs form factors, G_{E_p} and G_{M_p}, related to the Dirac and Pauli form factors F_{1_p} and F_{2_p} by:

$$G_{E_p} = F_{1_p} - \tau\kappa F_{2_p} \text{ and } G_{M_p} = F_{1_p} + \kappa F_{2_p}, \tag{1}$$

where $\tau = \frac{Q^2}{4M_p^2}$, Q^2 is the four-momentum transfer squared and M_p the mass of the proton; κ_p is the anomalous part of the magnetic moment of the proton $\mu_p = \kappa_p + 1$ in units of the nuclear magneton.

A separation of the electric and magnetic terms is evident in the cross section formula when the Sachs form factors are used. It is then possible to obtain both G_{Ep}^2 and G_{Mp}^2 separately by the Rosenbluth method. In the Born (or one-photon exchange) approximation, the cross section is:

$$\frac{d\sigma}{d\Omega} = \frac{\alpha^2\, E_e\, \cos^2\frac{\theta_e}{2}}{4E_{beam}^3\, \sin^4\frac{\theta_e}{2}} \left[G_{Ep}^2 + \frac{\tau}{\epsilon}G_{Mp}^2\right]\left(\frac{1}{1+\tau}\right), \tag{2}$$

where $\epsilon = \left[1 + 2(1+\tau)\tan^2(\frac{\theta_e}{2})\right]^{-1}$ is the longitudinal polarization of the virtual photon, with values between 0< ϵ <1, E_{beam} and E_e are the energies of the incident and scattered electron, respectively, and θ_e is the electron scattering angle in the laboratory frame. Figure 1 and 2 show results of G_{Ep} and G_{Mp} obtained by Rosenbluth separation [1-9], plotted as the ratios G_{Ep}/G_D and $G_{Mp}/\mu_p G_D$ versus Q^2, up to 6 GeV^2 and 35 GeV^2, respectively. Here $G_D = (1 + Q^2/0.71)^{-2}$ is the dipole form factor. For $Q^2 < 1$ GeV^2, the uncertainties for both G_{Ep} and G_{Mp} are only a few percent, and $G_{Ep}/G_D \simeq G_{Mp}/\mu_p G_D \simeq 1$. For G_{Ep} above $Q^2 = 1$ GeV^2, the large uncertainties result in a scatter of the results from different experiments.

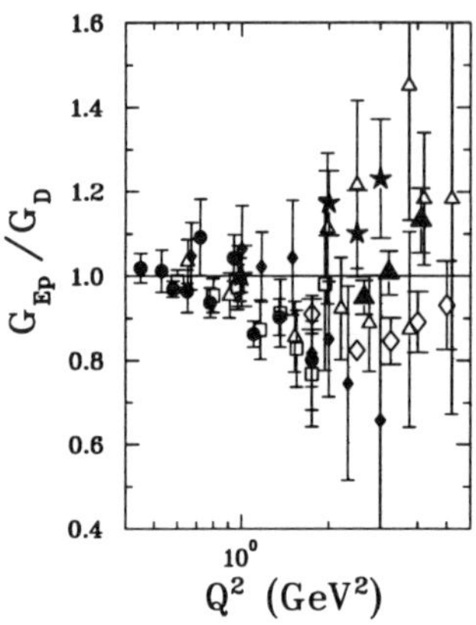

Fig. 1. World data for G_{Ep}/G_D.

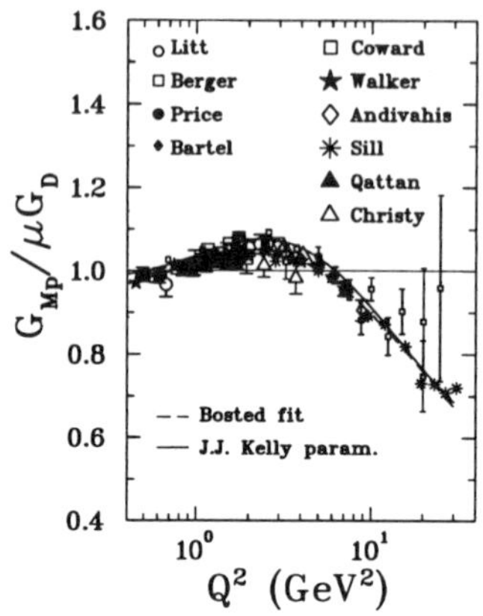

Fig. 2. World data for $G_{Mp}/\mu_p G_D$.

The proton form factor ratio G_{Ep}/G_{Mp} can be obtained from polarization observables of the $\vec{e}p \rightarrow e\vec{p}$ or $\vec{e}\vec{p} \rightarrow ep$ reaction, the recoil proton polarization transfer or the beam-target polarization asymmetry, respectively. Double spin experiments [10,11] measure the product $G_{Ep}G_{Mp}$ as well as $(G_{Mp})^2$, hence are more sensitive to small values of G_{Ep} than cross section measurements.

In Born approximation for $\vec{e}p \rightarrow e\vec{p}$, the scattering of longitudinally polarized electrons results in a transfer of polarization to the recoil proton with components, P_t perpendicular to, and P_ℓ parallel to the proton momentum in the scattering plane; for 100 % longitudinally polarized electrons, these are:

$$I_0 P_t = -2\sqrt{\tau(1+\tau)}G_{Ep}G_{Mp}\tan\frac{\theta_e}{2} \tag{3}$$

$$I_0 P_\ell = \frac{1}{M_p}(E_{beam}+E_e)\sqrt{\tau(1+\tau)}G_{Mp}^2\tan^2\frac{\theta_e}{2} \tag{4}$$

$$I_0 = G_{Ep}^2 + \frac{\tau}{\epsilon}G_{Mp}^2 \tag{5}$$

Together these equations give [12]:

$$\frac{G_{Ep}}{G_{Mp}} = -\frac{P_t}{P_\ell}\frac{(E_{beam}+E_e)}{2M_p}\tan\frac{\theta_e}{2} \tag{6}$$

This method was used in 1998 to measure G_{Ep}/G_{Mp} up to Q^2 of 3.5 GeV^2 [13,14]. Protons and electrons were detected in coincidence in the two high-resolution spectrometers (HRS) of Hall A. The azimuthal distribution of the protons rescattered in a graphite analyzer was measured in a focal plane polarimeter (FPP). In 2000 new measurements were made up to Q^2 of 5.6 GeV^2 [21]. To extend the measurement to these higher Q^2 required an increase of the FPP figure-of-merit using CH_2 as analyzer, and detection of the electron in a calorimeter for complete solid angle matching.

The proton spin precesses in the magnetic elements of the HRS. The HRS's in Hall A consist of 3 quadrupoles and one dipole with shaped entrance and exit edges, as well as a radial field gradient. The polarization vector at the polarimeter, $\vec{P}^{fpp}$, can be related to that at the target, $h\tilde{\mathbf{P}}$, through a spin rotation matrix, $\mathbf{S}$, $P^{fpp} = h(\mathbf{S})\mathbf{P}$. The spin matrix elements can be calculated using a model of the HRS event-by-event with the differential-algebra-based transport code COSY [22].

The combined results from both polarization transfer experiments are plotted in Fig. 3 as the ratio $\mu_p G_{Ep}/G_{Mp}$; the error bars shown are statistical only; the systematic uncertainties are shown by the polygons. The observed sharp decrease of

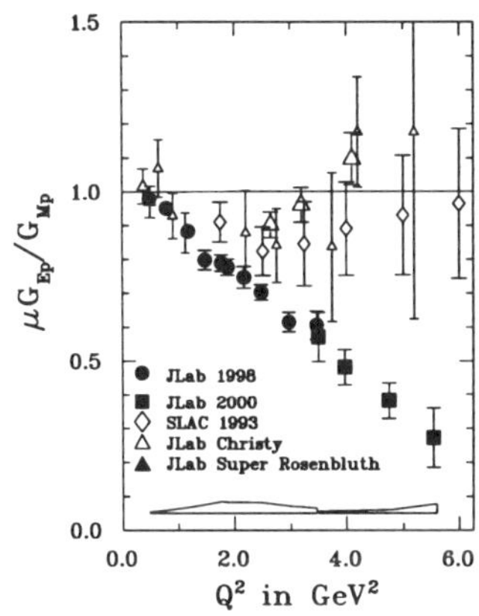

Fig. 3. The ratios $\mu_p G_{Ep}/G_{Mp}$ from two JLab experiments, compared to the Rosenbluth separation data.

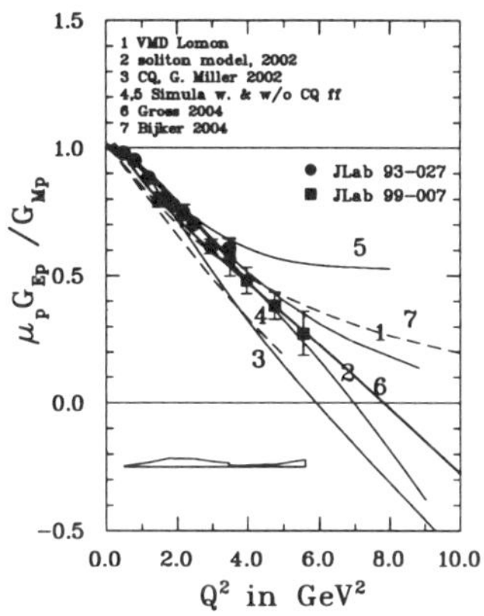

Fig. 4. The ratios $\mu_p G_{Ep}/G_{Mp}$ from the two JLab recoil polarization experiments, compared with a selection of calculations.

the ratio $\mu G_{Ep}/G_{Mp}$ from 1, starting at Q^2= 0.5 GeV^2 to a value of ~ 0.28 at Q^2= 5.6 GeV^2 indicates that G_{Ep} falls faster than G_{Mp}. Should the $\mu_p G_{Ep}/G_{Mp}$-ratio continues this linear decrease, it would cross zero at $Q^2 \approx 7.5\ GeV^2$.

We compare the data with a selection of calculations in Fig. 4. The recent vector meson dominance (VMD) calculation of Lomon [23] is shown as curve 1. In 2003 Iachello and collaborators reconsidered their 1973 work [24]; their model consists of a small structure besides the pion cloud described by VMD; they retain the original slope for G_{Ep}/G_{Mp} in the Q^2 range of this experiment [19,20] (curve 7).

In the soliton model Holzwarth[25] used the skyrmion as an extended object with one vector meson propagator and relativistic boost to the Breit frame (curve 2).

The constituent quark model reproduces the data presented here provided relativistic effects are included (RCQM). Frank *et al.* [26] have calculated G_{Ep} and G_{Mp} in the light-front constituent quark model and predicted a change of sign for G_{Ep}

near 5.6 GeV^2 (curve 3). Cardarelli *et al.* [27,28] calculated the ratio with light-front dynamics and investigated the effects of SU(6) symmetry breaking. In Ref. [29], they pointed out that an effective one-body electromagnetic current, using a constituent quark form factors, gives a reasonable description of pion and nucleon form factors. Their results, with two different quark form factors, are curves 4 and 5. Recently Gross and Agbakpe [30] revisited the RCQM using a covariant spectator model which allows exact handling of all Poincaré transformations (curve 6).

In pQCD Brodsky and collaborators[31] argued that up to leading order in $1/Q^2$, the magnetic form factor is proportional to α_s/Q^4 times slowly varying logarithmic terms. The absorbed virtual photon cannot induce a quark helicity flip, hence F_{2p} must decrease faster than F_{1p}, by a factor Q^{-2}; hence Q^2F_{2p}/F_{1p} should be constant at very high Q^2. In Fig. 5 the JLab data together with data of Andivahis *et al.*[6], are shown as Q^2F_{2p}/F_{1p}; the JLab results do not show yet the specific pQCD Q^2 dependence. Recent revisions of the pQCD prediction by Brodsky [32] and by

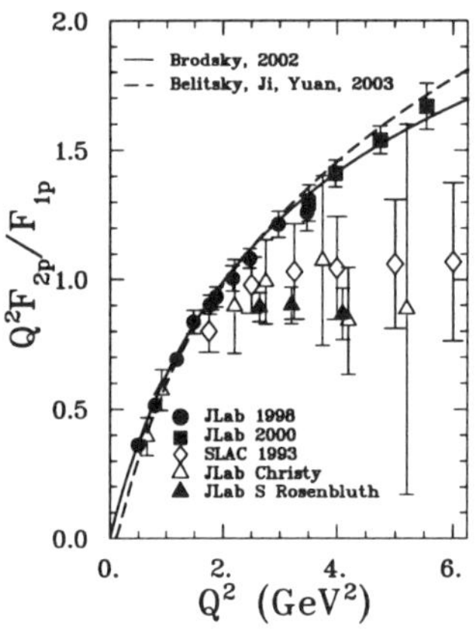

Fig. 5. Data from two JLab experiments are shown as Q^2F_{2p}/F_{1p} vs. Q^2.

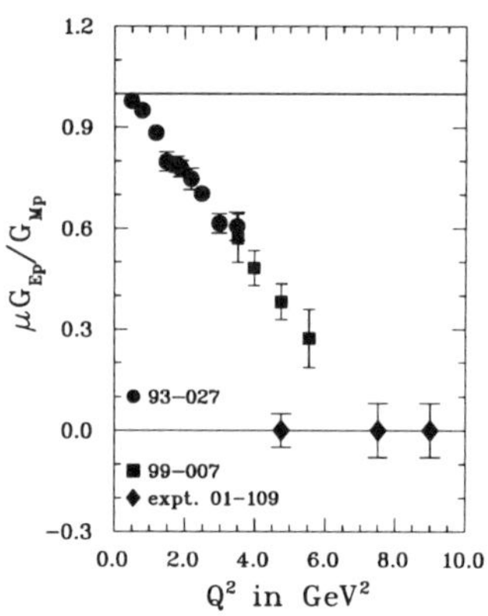

Fig. 6. Predicted statistical error bars for the 2 new data points proposed, and a control point at 4.2 GeV^2.

Belitsky *et al.* [33] fit the data nicely (see Fig. 5;) in the latter reference the quark orbital angular momentum is explicitly taken into account and enhances F_2.

In a different approach, Miller and Frank [34] have shown that imposing Poincaré invariance leads to violation of helicity conservation rule, resulting in the behavior of F_{2p}/F_{1p} observed in the JLab data.

Experiment G_{Ep}-III will extend the Q^2-range to 9 GeV^2 [35]6; it will use the High Momentum Spectrometer (HMS) in Hall C to detect the recoiling proton, and a new calorimeter with a large solid angle to detect the scattered electron. The measurements to Q^2 of 9 GeV^2 also requires a better polarimeter than used in the two previous experiments. We have chosen a configuration for the new polarimeter with two identical FPPs in series. The continuation of the recoil polarization measurements of the G_{Ep}/G_{Mp} ratio to larger Q^2 is possible only if the electron detector solid angle matches the spectrometer solid angle matches the spectrometer acceptance.

This work was supported by DOE contract DE-AC05-84ER40150 Modification No. M175, under which the Southeastern Universities Research Association (SURA) operates the Thomas Jefferson National Accelerator Facility. This work was supported in part by grant PHY-0140100 from the National Science Foundation.

References

1. J. Litt *et al.* Phys. Lett. B **31** 40 (1970).
2. Ch. Berger *et al.* Phys. Lett. B **35** 87(1971).
3. L.E. Price *et al.* Phys. Rev. D **4** 45(1971).
4. W. Bartel *et al.* Nuc. Phys. B **58** 429 (1973).
5. R.C. Walker *et al.* Phys. Rev. D **49** 5671 (1994).
6. L. Andivahis *et al.* Phys. Rev. D **50** 5491 (1994).
7. A.F. Sill *et al.* Phys. Rev. D **48** 29 (1993).
8. M.E. Christy *et al.* Phys.Rev. C **70** 015206 (2004).
9. I. A. Qattan *et al.* Phys. Rev. Lett. **94** 142301 (2005).
10. A.I. Akhiezer and M.P. Rekalo, Sov. Phys. - Doklady **13** 572 (1968).
11. R.G. Arnold, C.E. Carlson, and F. Gross Phys. Rev. C **23** 363 (1981).
12. C.F. Perdrisat and V. Punjabi, Jefferson Lab experiment E89-014 (1989).
13. M.K. Jones *et al.* Phys. Rev. Lett. **84** 1398 (2000).
14. V. Punjabi *et al.* Phys.Rev. C **71** 055202 (2005); Err. Phys.Rev. C **71** 069902 (2005).
15. O. Gayou *et al.* Phys. Rev. Lett. **88** 092301 (2002).
16. M. Bertz COSY INFINITY v. 7, NSCL Tech. Rep., MSUCL-977, (1995) Michigan State U.
17. E.L. Lomon Phys.Rev. C **64** 035204 (2001).
18. F. Iachello, A.D. Jackson, and A. Landé Phys. Lett. B **43** 191 (1973).
19. R. Bijker and F. Iachello Phys, Rev. C **69** 068201 (2004).
20. F. Iachello and Q. Wan Phys. Rev. C **69** 055204
21. O. Gayou *et al.* Phys. Rev. Lett. **88** 092301 (2002).
22. M. Bertz COSY INFINITY version 7, Tech. Rep., MSUCL-977, (1995) Michigan State University.
23. E.L. Lomon Phys.Rev. C **64** 035204 (2001).
24. F. Iachello, A.D. Jackson, and A. Landé Phys. Lett. B **43**(1973)
25. G. Holzwarth Z. Phys. A **356** 339 (1996); G. Holzwarth hep-ph/0201138 v1 (2002).
26. M.R. Frank, B.K. Jennings, and G.A. Miller Phys. Rev. C **54** 920 (1996).
27. F. Cardarelli and S. Simula Phys. Rev. C **62** 65201 (2000).
28. M. De Sanctis et al. Phys. Rev. C **62** 25208 (2000).
29. S. Boffi *et al.* Eur. Phys. J. A **14** 17 (2002).
30. Franz Gross and Peter Agbakpe nucl-th/0411090 v1 (2004).
31. S.J. Brodsky and G.R. Farrar Phys. Rev. D **11** 1309 (1975).
32. S.J. Brodsky hep-ph/0208158 v1 (2002).
33. A.V. Belitsky, Xiandong Ji, and Feng Yuan Phys.Rev.Lett. **91** 092003 (2003).
34. G.A. Miller and M.R. Frank, Phys. Rev. C **65**, 065205 (2002).
35. C.F. Perdrisat, V. Punjabi, M.K. Jones, and E. Brash, JLab expt 04-108(2004).

Photoproduction of ϕ Mesons with Linearly Polarized Photons at LEPS/SPring-8 Experiment

W.C. Chang for the LEPS collaboration*

Institute of Physics, Academia Sinica, Taipei 11529, Taiwan
E-mail: changwc@phys.sinica.edu.tw

Diffractive photoproduction of ϕ-meson near threshold involves Pomeron exchange, pseudo-scalar exchanges and possible exotic components such as glueball and scalar-meson exchange. Simultaneous measurements of cross section and decay asymmetry with linearly polarized photons make possible disentangling these mechanisms with different exchange parity. We reported such a study by LEPS/SPring-8 experiment.

1. Introduction

Photoproduction reactions of vector mesons $(\rho, \omega and \phi)$ are useful in understanding the dynamics of the hadron interactions.[1] Photoproduction of ϕ-meson is particularly interesting since the conventional meson exchanges (π, η) in the t-channel are strongly suppressed due to OZI-rule. This reaction provides a unique opportunity to study Pomeron exchange and other non-conventional production mechanisms, e.g. scalar-glueball($J^{PC} = 0^{++}$) exchange[2–4], and $s\bar{s}$ knockout [5], which are usually masked off by meson exchange at low energies. On the other hand, the measurement of decay angular distributions of ϕ-meson with the use of linearly-polarized-photon beam is a powerful tool to decompose the scattering amplitude into a natural-parity-exchange (Pomeron and glueball) part and an unnatural-parity-exchange (π, η) part.

In this paper, we report the results from LEPS experiment with a linearly-polarized-photon beam at very forward production angles[6] where only measurements with unpolarized-photon beams were reported previously.[7]

2. Experiment

Linearly polarized photons were produced by means of the backward-Compton scattering of laser photons off the 8 GeV electron at the SPring-8 BL33LEP beamline

*LEPS collaboration: D.S. Ahn, J.K. Ahn, H. Akimune, Y. Asano, W.C. Chang, S. Datè, H. Ejiri, H. Fujimura, M. Fujiwara, K. Hicks, T. Hotta, K. Imai, T. Ishikawa, T. Iwata, H. Kawai, Z.Y. Kim, K. Kino, H. Kohri, N. Kumagai, S. Makino, T. Matsumura, N. Matsuoka, T. Mibe, K. Miwa, M. Miyabe, Y. Miyachi, M. Morita, N. Muramatsu, T. Nakano, M. Niiyama, M. Nomachi, Y. Ohashi, T. Oba, H. Ohkuma, D.S. Oshuev, C. Rangacharyulu, A. Sakaguchi, T. Sasaki, P.M. Shagin, Y. Shiino, H. Shimizu, Y. Sugaya, M. Sumihama, A.I. Titov, H. Toyokawa, A. Wakai, C.W. Wang, S.C. Wang, K. Yonehara, T. Yorita, M. Yoshimura, M. Yosoi and R.G.T. Zegers

(LEPS: Laser Electron Photons at SPring-8 facility).[8] The photon energy was determined by measuring recoil electrons using a tagging counter with a resolution (σ) of 15 MeV and maximum energy of the photon beam was 2.4 GeV. The typical photon flux was about 10^6 s^{-1}, which was monitored by counting scattered electrons with the tagging system. The degree of linear polarization varied with photon energy; it was 95 % at the maximum energy, 60 % at the threshold. A liquid hydrogen target with a length of 50 mm was used in the experiment.

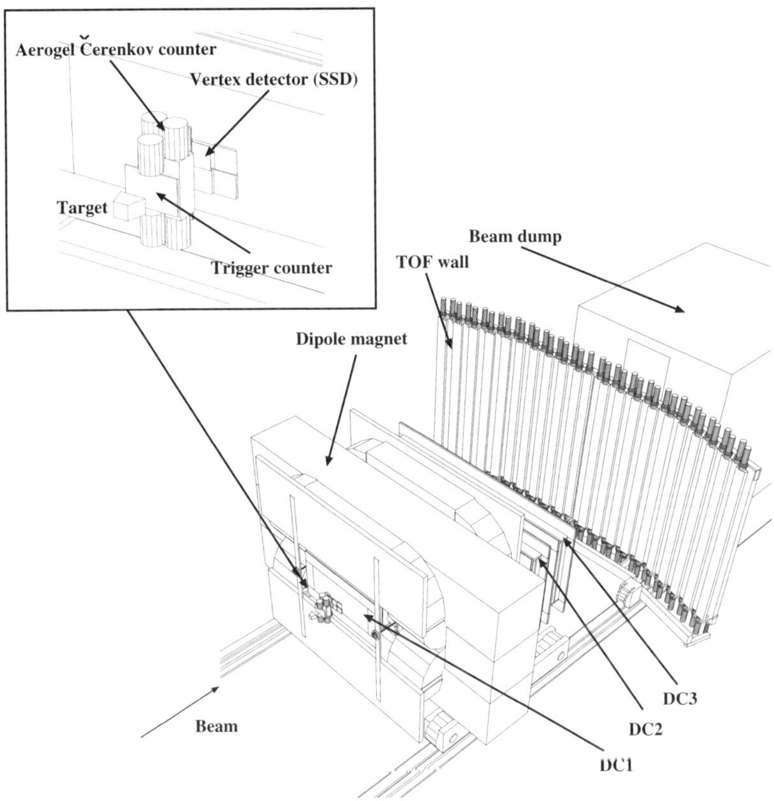

Fig. 1. The LEPS charged particle spectrometer

The momenta and the time-of-flight (TOF) of charged particles were measured with a magnetic spectrometer shown in Fig. 1. The momentum resolution (σ) for 1 GeV/c particles was 6 MeV/c. The TOF resolution (σ) was 150 psec for a typical flight path length of 4 m. The mass resolution (σ) was 30 MeV/c^2 for a 1 GeV/c^2 kaon. The incident photon energy and the momenta of K^+K^- tracks or $K^\pm p$ tracks were measured to identify the reaction $\gamma p \rightarrow K^+K^-p$ followed by the $\phi \rightarrow K^+K^-$ decay. Based on the detected particles, we define two types of event topology: K^+K^--reconstructed events (KK mode), and $K^\pm p$-reconstructed events

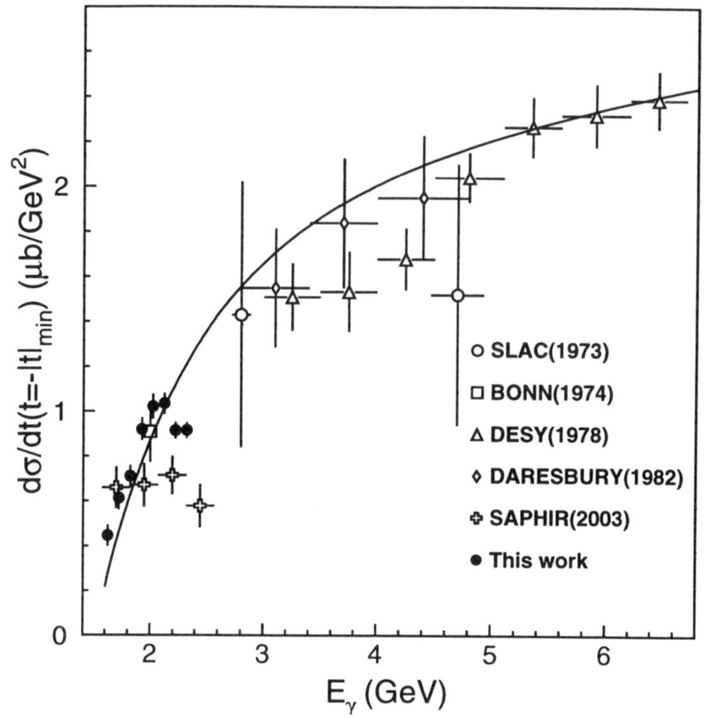

Fig. 2. Energy dependence of $(d\sigma/dt)_{t=-|t|_{min}}$. The closed circles are the results of the present work. Other data points are old experimental data.The error bars represent statistical errors. The solid curve represents the prediction of a model including the Pomeron trajectory, π and η exchange processes.

(Kp mode). In KK mode, the K^+K^- pairs from the ϕ-meson decay was selected by applying an invariant mass cut of K^+K^- on ϕ meson and a missing mass cut on proton.

3. Results

The differential cross sections were measured in terms of $t + |t|_{min}$ where $|t|_{min}$ is the minimum 4-momentum transfer from the incident photon to the ϕ-meson. A fit to the $d\sigma/dt$ was performed with an exponential function; i.e. $(d\sigma/dt)_{t=-|t|_{min}} e^{b(t+|t|_{min})}$ with $(d\sigma/dt)_{t=-|t|_{min}}$ and b as free parameters. No strong energy dependence of the slope b was found beyond the statistical errors.

Fig. 2 shows the energy dependence of $(d\sigma/dt)_{t=-|t|_{min}}$ when b is set to the average slope. The energy dependence of the $(d\sigma/dt)_{t=-|t|_{min}}$ shows a non-monotonic behavior with a local maximum at $E_\gamma \sim 2$ GeV. Data were compared with the prediction of a model including the Pomeron exchange , π and η exchange processes.[4]

The decay angular distributions in the Gottfried-Jackson frame were obtained at forward angles ($-0.2 < t + |t|_{min} \leq 0$ GeV^2) in the two different energy-regions: (1) around the local maximum of the cross section (ΔE_1:1.97 $< E_\gamma <$ 2.17 GeV) and (2) above the local maximum (ΔE_2:2.17 $< E_\gamma <$ 2.37 GeV) where there is enough statistics and the acceptance is fairly flat over all angular variables, as shown in Fig. 3. In both energy regions, $W(\cos\theta)$ behaves as $\sim (3/4)\sin^2\theta$, indicating the dominance of the helicity-conserving processes. The distribution $W(\phi - \Phi)$ shows a positive angular modulation value which means that the contributions from natural parity exchanges are stronger than those for unnatural parity exchanges (π, η-meson exchange). Same magnitudes of modulation within errors implies that the relative contribution of the natural parity exchange and the unnatural parity exchange remains constant in the two energy regions. Therefore, it is difficult to attribute the origin of the local maximum in the cross section to different strengths of the unnatural-parity exchange processes in the two energy regions. Considering a monotonic decrease of Pomeron exchange toward threshold, additional natural-parity exchange beyond the Pomeron exchange process is needed to account for the observations of local structure in cross section and stable relative contribution of the natural parity exchange and the unnatural parity exchange.

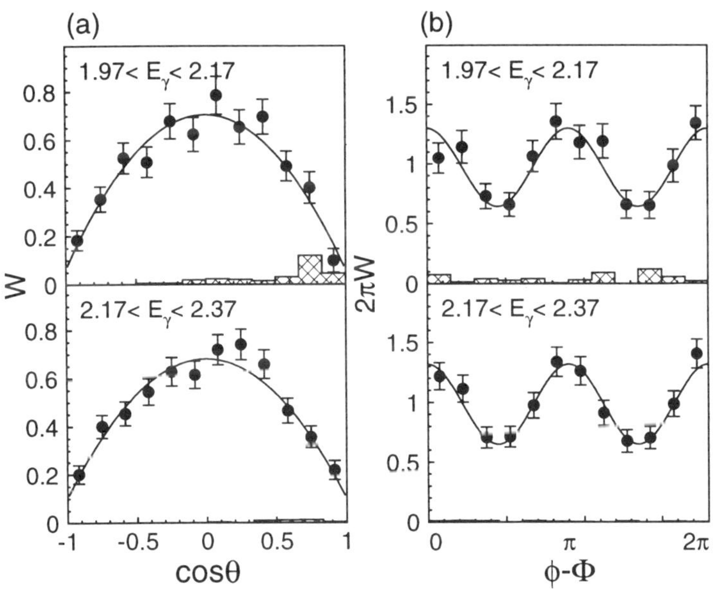

Fig. 3. Decay angular distributions of (a) $W(\cos\theta)$ and (b) $W(\phi - \Phi)$ for -0.2$< t + |t|_{min}$ at two E_γ bins in the Gottfried-Jackson frame. The solid curves are the fit to the data. The hatched histograms are systematic errors.

4. Summary

Photoproduction of the ϕ-meson was studied for the first time by means of linearly polarized photons at forward angles in the low energy region from the threshold energy of E_γ=1.57 GeV to 2.37 GeV. The differential cross-sections at $t = -|t|_{min}$ increase non-monotonically as a function of E_γ, and show a local maximum at around 2.0 GeV. The polar angle distributions demonstrate dominance of the helicity non-conserving processes. The azimuthal angle distributions over the local maximum suggest that the local maximum is not only due to unnatural-parity processes, but also due to a new dynamics of natural-parity exchange beyond the Pomeron exchange process.

References

1. T.H. Bauer, R.D. Spital, D.E. Yennie and F.M. Pipkin *Rev. Mod. Phys.* **50**, 261 (1978).
2. T. Nakano and H. Toki, in *Proceedings of the International Workshop on Exciting Physics and New Accelerator Facilities*, SPring-8, Hyogo, 1997 (World Scientific Singapore, 1998), p.48
3. A.I. Titov, T.-S.H. Lee, H. Toki, and O. Streltsova, *Phys. Rev. C* **60**, 035205 (1999).
4. A.I. Titov, T.-S.H. Lee, *Phys. Rev. C* **67**, 065205 (2003).
5. A.I. Titov, Y. Oh, and S.N. Yang, *Phys. Rev. Lett.* **79**, 1634 (1997); A.I. Titov, Y. Oh, and S.N. Yang, *Phys. Rev. C* **58**, 2429 (1998).
6. T. Mibe *et al.*, nucl-ex/0506015, to be published in Phys. Rev. Lett. (2005).
7. J. Ballam *et al. Phys. Rev. D* **7**, 3150 (1973); H.J. Besch *et al. Nucl. Phys.* **B70**, 257 (1974); Behrend *et al. Nucl. Phys.* **B 144**, 22 (1978); Barber *et al. Zeit. Phys. C* **12**, 1 (1982); J. Barth *et al., Eur. Phys. J. A* **17**, 269 (2003).
8. T. Nakano *et al. Nucl. Phys.* **A 684**, 71(2001).

Search for Narrow Dibaryon Resonances via the $p + d$ Scattering

H. Kuboki [a*], A. Tamii [b], H. Sakai [a], K. Yako [a], M. Hatano [a], T. Saito [a], M. Sasano [a], K. Hatanaka [b], Y. Sakemi [b], Y. Shimizu [b], K. Fujita [b], Y. Tameshige [b], J. Kamiya [c], T. Uesaka [d], Y. Maeda [d], K. Sekiguchi [e], K. Sagara [f], T. Wakasa [f], T. Kudoh [f], M. Shiota [f], and S. Shimomoto [f]

[a] *Department of Physics, University of Tokyo, Bunkyo, Tokyo 113-0033, Japan*
[b] *Research Center for Nuclear Physics (RCNP), Osaka University, Ibaraki, Osaka 567-0047, Japan*
[c] *Accelerator Group, Japan Atomic Energy Research Institute, Tokai, Ibaraki 319-1195, Japan*
[d] *Center for Nuclear Study (CNS), University of Tokyo, Bunkyo, Tokyo 113-0033, Japan*
[e] *The Institute of Physical and Chemical Research, Wako, Saitama 351-0198, Japan*
[f] *Department of Physics, Kyushu University, Fukuoka, Fukuoka 812-8581, Japan*

Nucleon-nucleon decoupled dibaryons, called Super Narrow Dibaryons (SNDs), were searched for by the *pd* scattering at E_p =295 MeV over the mass range of 1898 to 1953 MeV. The experiment was carried out at the Research Center for Nuclear Physics with a two-arm magnetic spectrometer and a liquid deuterium target. A mass resolution of 1 MeV and a low background condition were obtained. No resonance structure was observed in the missing mass spectra and the upper limits of the cross section were determined from the data.

1. Introduction

One of interesting predictions of the QCD is possible existence of multi-quark states with more than three quarks. Among the predicted multi-quark states, so called dibaryon has been studied most intensively through experimental search as well as theoretical approach. However, no evidence of dibaryons has been found.

In this work, we discuss Super Narrow Dibaryons (SNDs) which have the following two characteristics. Firstly, the SNDs have a symmetric wave function in terms of nucleons. Their decay into the two-nucleon (NN) system is forbidden because of Pauli's exclusion principle[1]. Secondly, the SNDs have a mass of less than two nucleons plus one π meson. This condition forbids SNDs from decaying into the two-nucleon and a π meson ($NN\pi$) system. As a result, the SNDs decay through only electromagnetic process emitting γ rays with a narrow decay width of less than 1 keV. The allowed decay channels of the SNDs are $D \to NN\gamma$ and $D \to d\gamma$.

Recently, candidates of the SND resonances were reported by the group at the Moscow Meson Factory (MMF)[2,3]. Narrow peaks in missing mass spectra have

*E-mail: kuboki@nucl.phys.s.u-tokyo.ac.jp

been observed at 1904±2, 1926±2, and 1942±2 MeV in the $pd \rightarrow ppX$ reaction by using a proton beam at an energy of 305 MeV. Since the observed widths of the resonances were equivalent to the experimental resolution of 4 MeV, the resonances were attributed to SNDs. The production cross section of the resonance at 1904 MeV was estimated to be 3±2 μb/sr on the assumption that the dibaryon mainly decays into the $pn\gamma$ decay channel. However the observed spectrum was accompanied by a large number of unknown background events most probably due to carbon in the CD_2 target together with low statistics, so that the presence of the resonances is marginal.

The purposes of this work are to study the existence of SND candidates and to search for SNDs over the mass range of 1898 to 1953 MeV in the same kinematical conditions of the MMF experiment but with higher sensitivity.

2. Experiment

The experiment was performed at the Research Center for Nuclear Physics (RCNP). The proton beam at an energy of 295 MeV was transported to the experimental hall, where it bombarded a target in the scattering chamber. A liquid deuterium target (LDT) system[4] was used as a deuteron target. The thickness of the LDT was 50 mg/cm^2. Since aramid foils of the target cell swelled in vacuum and bubbles might be produced in the LDT, the thickness of the LDT changed by 4% during the measurement. Therefore we newly installed a luminosity monitor[5] for monitoring the thickness of the LDT by measuring the pd elastic scattering. A deuterated polyethylene (CD_2) sheet[6] with a thickness of 44 mg/cm^2 was employed[7] as a deuterium target to normalize the luminosity. Two magnetic spectrometers shown in Fig. 1 were used to measure the particle momenta with high accuracy. The left arm was called Grand Raiden (GR)[8], which was placed at 70° and detected the scattered protons. The angle of GR was the same as that of the MMF measurement. The right arm was the Large Acceptance Spectrometer (LAS)[9], which detected the SND decay protons for the $pn\gamma$ decay channel or deuterons for the $d\gamma$ decay channel. The scattered protons and the decay particles were detected in coincidence.

The mass resolution was checked by the pd elastic scattering measurement (see Fig. 2). A mass resolution of 0.96 MeV (FWHM) was obtained. Assuming the decay width of an SND is less than 1 keV, an SND resonance should be observed as a peak with a width equal to the experimental resolution. The expected histogram on the assumption that the peak width is 1 MeV and the differential cross section is 3 μb/sr is shown in Fig. 3 together with the obtained missing mass spectrum over the mass range of 1898–1911 MeV. Since the signal to noise ratio is 3 to 20 over the whole mass range, the peak structure is not affected by a continuum background mainly due to the target cell. The obtained spectra were fitted by a linear function and the background was subtracted from the value of each bin.

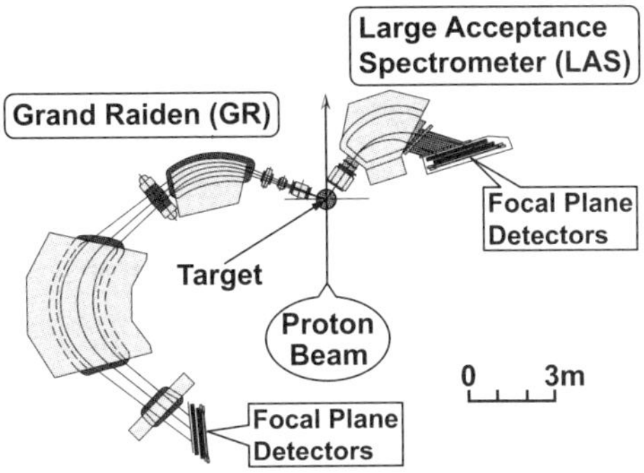

Fig. 1. Overview of the two-arm spectrometer system, consisting of Grand Raiden (GR) and the Large Acceptance Spectrometer (LAS).

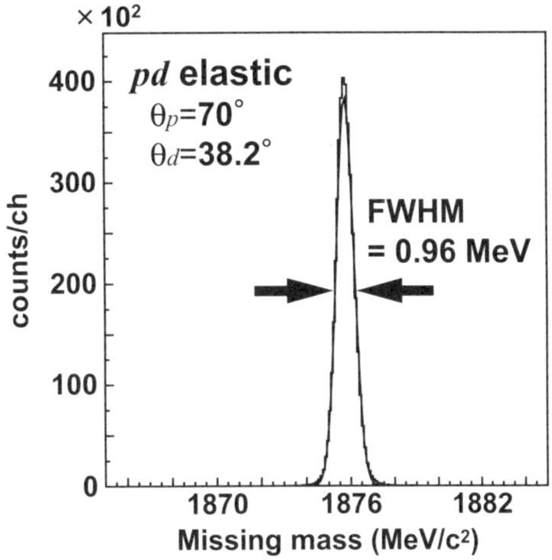

Fig. 2. The missing mass spectrum of the *pd* elastic scattering. A mass resolution of 0.96 MeV was obtained.

3. Results and Discussion

The missing mass spectra after background subtraction are shown in Fig. 4 for the (a) $pn\gamma$ and (b) $d\gamma$ decay channels. The vertical axis is the double differential cross section of the dibaryon production multiplied by the branching ratio for the corresponding channel. The experimental data are plotted by solid circles with statistical error bars. The solid curves in Fig. 4 (a) represent the expected histogram calculated from the differential cross section of 3 μb/sr and isotropic decay of the dibaryon in its rest frame. No significant peak is observed. We determined the upper limits of the dibaryon production cross section at the 90% confidence level. The obtained upper limits are plotted as a function of the missing mass in Fig. 4 (c) for the $pn\gamma$ and $d\gamma$ decay channels. The daggers shown in Fig. 4 (c) represent the values of the differential cross section of 3±2 μb/sr. The upper limits of the cross section for the $pn\gamma$ decay channel are less than 1 μb/sr over the mass range of 1898–1930 MeV, and less than 2.5 μb/sr over the mass range of 1930–1953 MeV. Those for the $d\gamma$ decay channel are less than 0.03 μb/sr over the mass range of 1898–1953 MeV.

4. Summary

We searched for SND resonances by the $pd \to ppX$ and $pd \to pdX$ reactions at E_p =295 MeV over the mass range of 1898–1953 MeV, where candidates of SND were reported at the MMF. We obtained a mass resolution of 0.96 MeV and a low background condition by using a liquid deuterium target system. The results of the dibaryon production cross section as a function of the missing mass are flat with no

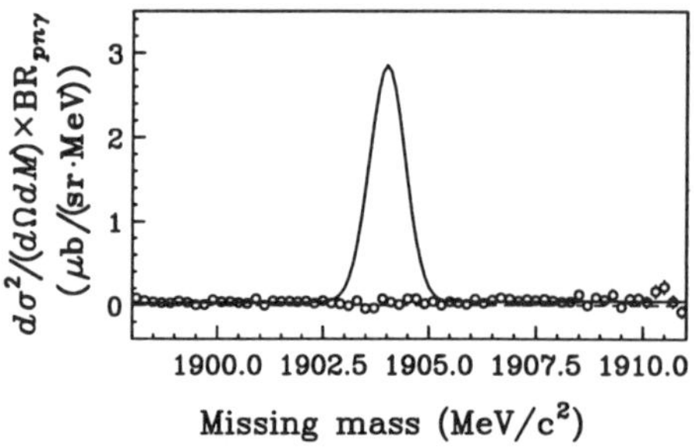

Fig. 3. The missing mass spectra over the mass range of 1898–1911 MeV for the $pn\gamma$ decay channel. The vertical axis is the SND production cross section multiplied by the branching ratio for the $pn\gamma$ decay channel. The solid curve represents the expected histogram. A signal to noise ratio of 20 is obtained.

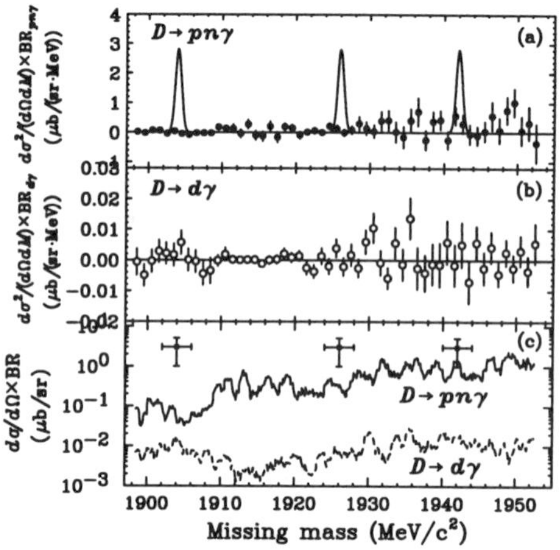

Fig. 4. The double differential cross section of the dibaryon production is shown for the (a) $pn\gamma$ and (b) $d\gamma$ decay channels. The upper limits of the dibaryon production cross section at the 90% confidence level are plotted in Fig. (c). Details are described in the text.

resonance structure within the statistical accuracy for each channel. The obtained upper limits at the 90% confidence level are less than 1 μb/sr and 2.5 μb/sr over the mass range of 1898–1930 MeV and 1930–1953 MeV, respectively, for the $pn\gamma$ decay channel. The upper limits for the $d\gamma$ decay channel are less than 0.03 μb/sr.

Acknowledgments

We would like to thank the RCNP cyclotron crew for the excellent proton beam.

References

1. P.J. Mulders *et al.*, *Phys. Rev. D* **21**, 2653 (1980).
2. L.V. Fil'kov *et al.*, *Phys. Rev. C* **61**, 044004 (2000)
3. L.V. Fil'kov *et al.*, *Euro. Phys. J.* A**12**, 369 (2001)
4. K. Sagara *et al.*, *RCNP Annual Report*, p. 158 (1995).
5. H. Kuboki *et al.*, *RCNP Annual Report*, p. 131 (2002).
6. Y. Maeda *et al.*, *Nucl. Inst. Meth.* **A490**, 518 (2002).
7. K. Hatanaka *et al.*, *Phys. Rev. C* **66**, 044002 (2002).
8. M. Fujiwara *et al.*, *Nucl. Inst. Meth.* **A422**, 484 (1999).
9. N. Matsuoka *et al.*, *RCNP Annual Report*, p. 186 (1991)

Spectroscopy of Pentaquark Baryons*

Makoto Oka
Department of Physics, H27, Tokyo Institute of Technology
Meguro, Tokyo 152-8551, JAPAN
E-mail: oka@th.phys.titech.ac.jp

A review is given to pentaquark mass predictions in quark models and QCD. It is pointed out that no successful quark model prediction is available for low-lying pentaquark states. Some new results of direct application of QCD, QCD sum rules and lattice QCD, are also presented.

1. Introduction

In 2003, the LEPS group at SPring-8 observed a sharp Θ^+ peak in $\gamma n \to K^-\Theta^+$ reaction.[1] This is an evidence of an exotic baryon resonance with strangeness +1, and in the following year, it was confirmed by DIANA, CLAS, SAPHIR, HERMES, ZEUS, and COSY groups.[2] It is, however, challenged by multiple negative results by now, mostly from high energy experiments, by the CDF, HyperCP, E690, BES, BELLE, BaBar, HERA-B, PHENIX groups.[2] This year new high statistics data by the CLAS collaboration at JLab have revealed that a preceding observation of Θ^+ by the same group was no more valid.[3] The negative results have high statistics and are quite convincing, but they may not completely wash away the "evidence" yet. A new result from LEPS shows further evidence of forward photoproduction $\gamma + d \to \Lambda(1520) + \Theta^+$.[4] The LEPS group claims that the result is not inconsistent with the other experimental data. As it certainly requires confirmation, it will take some more time before the final conclusion. Meanwhile, many theoretical works have been published.[5] A purpose of this article is to review the results in quark models and also to present some new results from direct application of QCD.

Θ^+ is considered to be a baryon with $S = +1$, $B = 1$, $Q = +1$ and $Y = B + S = 2$. Required "valence quark" combination is $uudd\bar{s}$ and thus it is called pentaquark. Except for a preliminary indication by the STAR group at RHIC,[6] no pK^+ ($I_3 = +1$) state is found, which suggests that the isospin of Θ^+ is 0. The simplest SU(3) irreducible representation is $F = \overline{10}$, which is unique within the flavor representations produced by five valence quarks. The other members of the antidecuplet include genuine pentaquark states, $\Xi_{I=3/2}$.

*in collaboration with T. Doi, H. Iida, N. Ishii, Y. Nemoto, F. Okiharu, H. Suganuma and J. Sugiyama.

An often asked question is what we shall learn if such states exist. In fact, QCD allows not only quark-anti-quark mesons and three-quark baryons, but also various exotic states, *i.e.*, four-quark mesons, $qq\bar{q}\bar{q}$, pentaquark baryons, $qqqq\bar{q}$, hybrid mesons, $\bar{q}gq$, dibaryons, $qqqqqq$, and so on. Such exotic states will reveal multi-quark dynamics that cannot be accessed in non-exotic hadrons. In particular, some important questions could be answered, such as (1) whether the color confinement is simply bag-like, or string-like, or more complicated in multi-quark systems, (2) whether valence quark model is valid in hadronic interactions, (3) what is the mechanism of dissociation of multi-quark states into two (or more) color-singlet hadrons, and so on. So far, the pentaquarks do not give definite answers to these questions, but it is clear that if Θ^+ is a pentaquark state, then the standard constituent quark model should be modified so that a five quark state is as light as the standard three quark baryons.

In this article, we first summarize current status of quark model calculations and then present some recent results from QCD sum rules and lattice QCD.

2. Theory: I. Quark Models

It is fair to say that there has been no successful calculation that fully explains $\Theta^+(1540)$. First of all, most mass predictions are much higher than the observed mass, 1540 MeV. Most calculations predict several low-lying excited states, but none is found. For instance, many models predict low-lying $1/2^-$, $1/2^+$, $3/2^-$ and $3/2^+$ states ($I = 0$ and/or 1).

Several recent calculations try to find exact ground states of 5-quark (5Q) systems. Such attempts include a molecular dynamics calculation, and variational calculations with and without coupling to continuum states. En'yo et al. applied an antisymmetrized molecular dynamics to five quarks and predicted several narrow states as the lowest energy states.[7] They showed that the intrinsic 5Q configuration prefers diquark clustering as was conjectured by Jaffe and Wilczek.[8] The absolute masses of the pentaquark states, however, had to be adjusted to the observed one and thus only the excitation energies can be predicted.

A variational calculation was performed by Takeuchi and Shimizu using one-gluon exchange plus Nambu-Goldstone boson exchanges between quarks.[9] They predicted several low-lying states within 200 MeV above the ground state $1/2^-$, which they regard as a nonresonant state. But the lowest resonance state comes at above 2 GeV, that is about 500 MeV heavier than the desired pentaquark. Hiyama et al. carried out a variational calculation of $1/2^-$ and $1/2^+$ pentaquarks with full couplings to KN continuum states.[10] They found that most of the low-lying states obtained in the bound-state approach (which constrains the wave function of five quarks to be localized) are dissolved into KN continuum. Instead, they found a very narrow resonance state at about 500 MeV above the KN threshold. Another highly-excited broad resonance was found in the spin $1/2^+$. Origins of these highly-excited resonances are not known yet.

It is clear now that the lowest $1/2^-$ state, which appears naturally in any quark model calculations, will not make a sharp resonance, but is almost completely mixed with KN scattering states. Thus the next question is whether a $J = 1/2^+$ state is a possible candidate of narrow Θ^+. It was shown by Hosaka et al. that it can be as narrow as a few MeV according to the quark structure of the p-wave pentaquark,[11] but is always associated with a $J = 3/2^+$ state close-by. Their splitting is given by the LS force and will be about a few tens MeV. Because the partner is to have a similar narrow width, it should also be detected in the same production channel.

The next possibility is a $J = 3/2^-$ state, which is a viable candidate of Θ^+, although mass predictions are still too high. Because it decays only into d-wave KN scattering states, it is expected to be quite narrow. The spectroscopic factor is also highly suppressed, if we assume that the main component of the quark wave function consists of s-waves only. In all, $1/2^+$, $3/2^+$ and $3/2^-$ may be candidates of the lowest resonance state of five quarks, but the predicted masses are too high to be identified with the observed Θ^+.

3. Theory: II. QCD Sum Rules

It is highly desirable to apply QCD directly to the pentaquark problem under the situation. We report here the results of two approaches in QCD. The first one is to apply QCD sum rule to the pentaquarks of spin 1/2 and 3/2 with masses about 1.5–2GeV. The second approach is lattice QCD (LQCD) in quenched approximation.

In both the approaches, one calculates a two-point correlation function of a local product of quark field operators. This local operator is called interpolating field operator (IFO). We choose IFO for $J = 1/2$ $(I = 0)$ pentaquark systems as

$$J(x) \equiv \epsilon_{abc}\epsilon_{ade}\epsilon_{bfg}[u_d^T(x)C\gamma^5 d_e(x)][u_f^T(x)Cd_g(x)]C\bar{s}_c^T(x), \tag{1}$$

and Rarita-Schwinger forms for $J = 3/2$ $(I = 0)$ pentaquark,

$$J_\mu^{(1)}(x) \equiv \epsilon_{abc}\epsilon_{def}\epsilon_{cfg}[u_a^T(x)Cd_b(x)][u_d^T(x)C\gamma_5\gamma_\mu d_e(x)]\gamma_5 C\bar{s}_g^T(x), \tag{2}$$

$$J_\mu^{(2)}(x) \equiv \epsilon_{abc}\epsilon_{def}\epsilon_{cfg}[u_a^T(x)C\gamma^5 d_b(x)][u_d^T(x)C\gamma_5\gamma_\mu d_e(x)]C\bar{s}_g^T(x). \tag{3}$$

These operators are constructed from scalar, pseudoscalar and vector diquark combinations. We also employ IFO's which are products of N and K or K* hadrons. Although they are all local operators and are related by Fierz transform among them, each may have different coupling strengths to pentaquark resonance states, if they exist.

In the QCD sum rule,[12] we parametrize the imaginary part of the correlation function in terms of a supposed resonance peak and continuum background and determine the resonance parameters by comparing the correlation function with that calculated perturbatively at a large Euclid momentum.

We study whether the QCD spectral function for the relevant quantum numbers satisfies the positivity condition. We have found that the correlation function of $J = 1/2^+$ $(I = 0)$ pentaquark becomes almost zero or negative in the region of

$M \sim 1 - 2$ GeV. Thus it is concluded that there should be no resonance state in the same mass region. In contrast, $J = 1/2^-$ ($I = 0$) pentaquark is not rejected in the same energy region. Using the Borel sum rule technique, we determine its mass, although the result is sensitive to the chosen continuum threshold. We find that the mass is around $1.3 - 1.7$ GeV.

Similarly, $J = 3/2$ possibility is studied in the sum rules.[13,14] It is found that the sum rule depends sensitively on the choice of IFO, which indicates that the operators have different coupling strengths to a resonance state. The second operator, $J_\mu^{(2)}$, that consists of a scalar and vector diquarks, happens to have a stronger strength at low energy. On the other hand, $J_\mu^{(1)}$ does not give a positive spectral function in the $M \sim 1 - 2$ GeV region.

We have found that low lying $3/2^-$ and $3/2^+$ pentaquarks may exist from the positivity analysis and mass extraction from the $J_\mu^{(2)}$ sum rules. In particular, the $3/2^-$ state is lower in mass, ~ 1.5 GeV, which is almost degenerate to $1/2^-$ state obtained in the sum rule. Because the $3/2^-$ state does not couple to s-wave KN scattering states, its spectral function is expected to be suppressed near the threshold. It is reasonable to conclude that the $3/2^-$ pentaquark is a physical state at around 1.5 GeV.

4. Theory: III. Lattice QCD

Several lattice QCD calculations have been performed for the pentaquarks with $J = 1/2$ and $J = 3/2$.[15] They all take quenched approximation, while fermion formulations vary. The conclusions are somewhat scattered at this moment in the sense that a few results indicate a pentaquark resonance state, while the majority denies existence of a low-lying sharp resonance.

We here present a new result in studies using an anisotropic lattice so that the precision of the mass estimate is high enough.[16] We employ a lattice size $12^3 \times 96$, and $\beta = 5.75$, which correspond to $(2.2)^3 \times 4.4$ fm^4 in physical unit. 504 gauge configurations are generated on an anisotropic lattice ($a_s/a_t = 4$), and the clover Wilson quark is employed. Plots of effective masses show clear plateaus in most of the cases, which indicate a state with a definite energy is reached. We have found that the diquark operators show less clear plateau structures than the operator constructed as a product of two hadrons, in particular for $J = 3/2$.

It is important to distinguish resonance states from hadronic scattering states, in the present case, mostly NK scatterings. In order to judge whether a plateau state in an effective mass plot is a compact resonance state or not, we have invented a simple method in which the boundary conditions for some of the quarks are twisted. For the Θ^+ state, we perform a LQCD calculation where we impose antiperiodic boundary condition in the spatial directions on u and d quarks, keeping the s quark in the periodic boundary condition (hybrid boundary condition). Under such boundary conditions, pentaquark states, which have four compact $u + d$ quark contents, may not be affected much, while the nucleon and kaon, both of which contain odd number of $u + d$ quarks, should not be allowed to occupy the node-less ground state configuration. Thus it is expected that the NK threshold energy will rise when the pentaquarks stay at the same masses.

Using the hybrid boundary condition method, we have tested whether the obtained five quark plateaus are compact pentaquarks or not. The results are unfortunately negative, i.e., all the plateaus are shifted according to the threshold so that they are all consistent with NK scattering states. The results can be summarized as follows. (1) A $J = 1/2^-$ state observed at 1.75 GeV is likely to be a $L = 0$ NK scattering state. The hybrid boundary condition method shows no compact 5Q states. (2) A $1/2^+$ appears at 2.25 GeV, which is too heavy to be identified with $\Theta^+(1540)$. (3) A $J = 3/2^-$ state observed at 2.11 GeV on the lattice is consistent with NK* $(L = 0)$ scattering state. (4) Similarly, $3/2^+$ states observed at 2.42 GeV and 2.64 GeV are consistent with NK* $(L = 1)$ and N*K* $(L = 0)$ scattering states, respectively.

Thus no evidence of compact 5Q resonance state is obtained in the present LQCD calculation. It should however be noted that the current calculation is under the quenched approximation and also contains ambiguity due to extrapolation from the results for rather large quark masses.

5. Conclusion

In conclusion, we find no definite evidence of pentaquarks either in the quark model approaches or direct QCD calculations. The quark model calculations of the 5Q mass have a difficulty that the predicted masses are still too high and also that the predictions are associated with some missing low-lying states.

Both the LQCD and QCD sum rule results reject existence of a low-lying $J = 1/2^+$ 5Q state. Although a low-lying $J = 1/2^-$ state is predicted, it is found that the state is not a compact resonance, but is likely to be a NK S-wave scattering state. We have found that the hybrid boundary condition method is useful in determining whether a low-lying state is a sharp compact resonance or a scattering state of two hadrons. The QCD sum rule allows $J = 3/2^-$, and $3/2^+$ 5Q states, but the same states seen in LQCD are all consistent with two-hadron (NK or N*K scattering states.

This discrepancy between the results of the sum rules and lattice QCD on $J = 3/2$ pentaquarks requires further study. Both of the approaches have sone defects. In the sum rule, convergence of the operator product expansion has to be checked. Because the pentaquark operators have higher dimensions than three-quark baryon operators, the pentaquark sum rules may suffer slower convergence. The lattice QCD so far is under the quenched approximation. It is known that the extrapolation in quark mass may cause large ambiguity in the quenched approximation. It is highly desirable to perform unquenched lattice calculation with smaller quark masses.

Acknowledgments

I acknowledge all the collaborators listed in the footnote of the headline. I also acknowledge Drs. S. Takeuchi, T. Shinozaki, T. Nishikawa, A. Hosaka and E. Hiyama for discussions and information.

References

1. T. Nakano *et al.*, Phys. Rev. Lett. **91** (2003) 012002.
2. For a review of experiments, see K. Hicks, Prog. Part. Nucl. Phys. **55** (2005) 647.
3. R. De Vita, presented at the APS April Meeting at Tampa, 2005.
4. T. Nakano, presented at the Int. Conf. QCD and Hadronic Physics, Beijing (2005).
5. Many theoretical papers on this subject have been published. See the papers cited in M. Oka, Prog. Theor. Phys. **112** (2004) 1; and also the papers in Proc. Int. Workshop on PENTAQUARK 04, ed. by A. Hosaka and T. Hotta (World Scientific, 2005).
6. J. Ma, presented at the APS April meeting at Tampa, 2005.
7. Y. Kanada-En'yo, O. Morimatsu and T. Nishikawa, Phys. Rev. **C71** (2005) 045202.
8. R.L. Jaffe and F. Wilczek, Phys. Rev. Lett. **91** (2003) 232003.
9. S. Takeuchi and K. Shimizu, Phys. Rev. **C71** (2005) 062202; hep-ph/0411016.
10. E. Hiyama, M. Kamimura, A. Hosaka, H. Toki and M. Yahiro, hep-ph/0507105; Proc. PENTAQUARK04, 274 (World Scientific, 2005).
11. A. Hosaka, M. Oka and T. Shinozaki, Phys. Rev. **D71** (2005) 074021.
12. J. Sugiyama, T. Doi and M. Oka, Phys. Lett. **B581** (2004) 167; Nucl. Phys. **A755** (2005) 391.
13. T. Nishikawa, Y. Kanada-En'yo, Y. Kondo and O. Morimatsu, Phys. Rev. **D71** (2005) 076004.
14. J. Sugiyama, T. Doi and M. Oka, to be published.
15. F. Csikor, et al., JHEP **0311** (2003) 070; S. Sasaki, Phys. Rev. Lett. **93** (2004) 152001; N. Mathur et al., Phys. Rev. **D70** (2004) 074508; T.T. Takahashi et al., Phys. Rev. **D71** (2005) 114509; T.W. Chiu et al., Phys. Rev. **D72** (2005) 034505; B.G. Lasscock et al., Phys. Rev. **D72** (2005) 014502.
16. N. Ishii, T. Doi, H. Iida, Y. Nemoto, M. Oka, F. Okiharu and H. Suganuma, Phys. Rev. **D71** (2005) 034001; hep-lat/0506022.

Two-Photon Exchange Contribution to the Elastic e-p Scattering at Large Momentum Transfer within a Partonic Approach

Y.C. Chen*

Department of Physics. National Taiwan University,
Taipei 10617, Taiwan
E-mail: snyang1@phys.ntu.edu.tw

M. Vanderhaeghen†

College of William & Mary / Jefferson Laboratory,
Williamsburg, VA, USA
E-mail: marcvdh@jlab.org

It was found that the elastic electron-nucleon scattering experiments using the Rosenbluth method and the polarization extractions give different results for G_E/G_M at large Q^2. This apparent discrepancy between the Rosenbluth and the polarization transfer methods can be explained by a two-photon exchange correction. We estimate the two-photon exchange contribution to elastic electron-proton scattering at large momentum transfer by using a quark-parton representation of virtual Compton scattering. We thus can relate the two-photon exchange amplitude to the generalized parton distributions. We find that the interference of one- and two-photon exchange contributions is able to substantially resolve the difference between electric form factor measurements from Rosenbluth and polarization transfer experiments. Two-photon exchange has additional consequences which could be experimentally observed, including nonzero polarization effects and a positron-proton/electron-proton scattering asymmetry.

The electric and magnetic form factors(G_E and G_M) are basic quantities in our knowledge of the nucleon structure. In upolarized experiments, using the Rosenbluth method, the ratio $R = G_E^p/G_M^p$ is extracted by measuring the differential cross section $d\sigma/d\Omega_{Lab} \propto G_M^2(Q^2) + \frac{\varepsilon}{\tau}G_E^2(Q^2)$, with the proportionality factor being well known, and isolating the ε dependent term, where $1/\varepsilon \equiv 1 + 2(1+\tau)\tan^2\frac{\theta_{Lab}}{2}$, θ_{Lab} is the laboratory scattering angle, and $0 \leq \varepsilon \leq 1$.

Recently, polarized lepton beam give another way to extract the ratio R by measuring the polarizations of the recoiling proton, which gives: $P_s/P_l = -\sqrt{\frac{2\varepsilon}{\tau(1+\varepsilon)}}\frac{G_E}{G_M}$, where P_l is the polarization parallel to its momentum and P_s is the polarization perpendicular to its momentum(still in the scattering plane). Recent unpolarized experiments at the Thomas Jefferson Laboratory[1,2] have confirmed the earlier Rosenbluth measurements from SLAC[3]. However, at large Q^2 region, there's a large discrepancy between Rosenbluth and polarization experiments[4]. One possible explanation[5] for

*Work partially supported by the Taiwanese NSC under contract 92-2112-M002-049
†Work partially supported by the U.S. DOEunder contract DE-FG02-04ER41302

the discrepancy is that there are radiative corrections, which are important in the Rosenbluth experiments at large Q^2.

For elastic e-N scattering: $l(k) + N(p) \to l(k') + N(p')$, we adopt the usual definitions: $P = (p + p')2, K = (k + k')/2, s = (p + k)^2, u = (k - p')^2$, and choose $q = k - k' = p' - p, Q^2 = -q^2, \nu = K \cdot P$. For a theory which respects Lorentz, parity, and charge conjugation invariance, the T matrix can be expanded in terms of six independent Lorentz structures[6]. To simplify, we do not consider here the lepton helicity flip amplitudes, which are of the order of the electron mass. Then, the T-matrix can expressed as:

$$T = \frac{e^2}{Q^2}\bar{u}(k',h)\gamma_\mu u(k,h) \cdot \bar{u}(p') \left[\tilde{G}_M \gamma^\mu - \tilde{F}_2 \frac{P^\mu}{M} + \tilde{F}_3 \frac{\gamma \cdot K P^\mu}{M^2}\right] u(p), \qquad (1)$$

where $\tilde{G}_M$, $\tilde{F}_2$ and $\tilde{F}_3$ are complex functions of ν and Q^2. In 1γ approximation(Born approximation), $\tilde{F}_3$ and the phase of $\tilde{G}_M$ and $\tilde{F}_2$ vanish. Similar to the 1γ form factors, we can define $\tilde{G}_E \equiv \tilde{G}_M - (1+\tau)\tilde{F}_2$, $Y_{2\gamma}(\nu, Q^2) \equiv R\left(\frac{\nu \tilde{F}_3}{M^2|\tilde{G}_M|}\right)$, and assume that $\tilde{G}_M \sim \tilde{G}_M^{1\gamma}$ and $\tilde{G}_E \sim \tilde{G}_E^{1\gamma}$, then we have:

$$(R^{exp}_{Rosenbluth})^2 \simeq \frac{|\tilde{G}_E|^2}{|\tilde{G}_M|^2} + 2\left(\tau + \frac{|\tilde{G}_E|}{|\tilde{G}_M|}\right) Y_{2\gamma}, \qquad (2)$$

$$(R^{exp}_{Polarization}) \simeq \frac{|\tilde{G}_E|}{|\tilde{G}_M|} + \left(1 - \frac{2\varepsilon}{1+\varepsilon}\frac{|\tilde{G}_E|}{|\tilde{G}_M|}\right) Y_{2\gamma}. \qquad (3)$$

By solving Eqs. (2,3) for $|\tilde{G}_E|/|\tilde{G}_M|$ and $Y_{2\gamma}$ for each Q^2, we can see that the result for ε indicates that corrections to the Born approximation are actually small in absolute value and display an ε dependence. And $R^{exp}_{1\gamma+2\gamma} = |\tilde{G}_E|/|\tilde{G}_M|$ is close to $R^{exp}_{Polarization}$ [3].

To separate the 1γ and 2γ exchange contributions, we introduce the decompositions : $\tilde{G}_M = G_M + \delta\tilde{G}_M$, and $\tilde{G}_E = G_E + \delta\tilde{G}_E$, where G_M,G_E are the form factors originate from 1γ exchange, which are functions of Q^2 only. To estimate the 2γ contribution to $\tilde{F}_3$, $\delta\tilde{G}_M$, and $\delta\tilde{G}_E$, we consider a partonic calculation based on a calculation of the elastic e-q scattering amplitude for massless quarks.

In elastic lepton-quark scattering, $l(k) + q(p_q) \to l(k') + q(p'_q)$, the Mandelstam invariants are given by $\hat{s} = (k + p_q)^2$, Q^2, and $\hat{u} = (k - p'_q)^2$, satisfying $\hat{s} + \hat{u} = Q^2$. The T-matrix for the 2γ exchange e-q scattering can be written as:

$$H_{h,\lambda} = \frac{(ee_q)^2}{Q^2}\bar{u}(k',h)\gamma_\mu u(k,h) \cdot \bar{u}(p'_q,\lambda)\left(\tilde{f}_1\gamma^\mu + \tilde{f}_3 \gamma \cdot K P_q^\mu\right) u(p_q,\lambda), \qquad (4)$$

with $P_q \equiv (p_q + p'_q)/2$, where e_q is the fractional quark charge (for a flavor q), and the quark helicity $\lambda = \pm 1/2$ is conserved in the hard scattering process. For massless quarks, there is no analog of the F_2 term.

To calculate the hard amplitudes $H_{h,\lambda}$ of Eq. (4) at order $O(e^4)$, we consider the 2γ exchange direct and crossed box diagrams. For further use, we separate the amplitude $\tilde{f}_1$ for the scattering of massless electrons and quarks into a soft and hard part, i.e. $\tilde{f}_1 = \tilde{f}_1^{soft} + \tilde{f}_1^{hard}$. The soft part corresponds with the situation where one of the exchange photons carries zero four-momentum, and is obtained by replacing the other photon's four-momentum by q in the numerator and in its propagator in the loop integral. The amplitude of $\tilde{f}_3$ is IR finite, and the real part of these amplitude at the quark level is given by: $\left(\bar{f}_1^{soft}\right) = \frac{e^2}{4\pi^2}\left\{\ln\left(\frac{\lambda^2}{\sqrt{-\hat{s}\hat{u}}}\right)\ln\left|\frac{\hat{s}}{\hat{u}}\right| + \frac{\pi^2}{2}\right\}$, $R\left(\bar{f}_1^{hard}\right) = \frac{e^2}{4\pi^2}\left\{\frac{1}{2}\ln\left|\frac{\hat{s}}{\hat{u}}\right| + \frac{Q^2}{4}\left[\frac{1}{\hat{u}}\ln^2\left|\frac{\hat{s}}{Q^2}\right| - \frac{1}{\hat{s}}\ln^2\left|\frac{\hat{u}}{Q^2}\right| - \frac{1}{\hat{s}}\pi^2\right]\right\}$, and $R\left(\bar{f}_3\right) = \frac{e^2}{4\pi^2}\left\{\hat{s}\ln\left|\frac{\hat{s}}{Q^2}\right| + \hat{u}\ln\left|\frac{\hat{u}}{Q^2}\right| + \frac{\hat{s}-\hat{u}}{2}\left[\frac{\hat{s}}{\hat{u}}\ln^2\left|\frac{\hat{s}}{Q^2}\right| - \frac{\hat{u}}{\hat{s}}\ln^2\left|\frac{\hat{u}}{Q^2}\right| - \frac{\hat{u}}{\hat{s}}\pi^2\right]\right\}$. The imaginary parts of $\tilde{f}1$ and $\tilde{f}3$ originate solely from the direct exchange box diagram and can be given by: $I\left(\bar{f}_1^{soft}\right) = -\frac{e^2}{4\pi^2}\ln\left(\frac{\lambda^2}{\hat{s}}\right)$, $I\left(\bar{f}_1^{hard}\right) = -\frac{e^2}{4\pi^2}\left\{\frac{Q^2}{2\hat{u}}\ln\left(\frac{\hat{s}}{Q^2}\right) + \frac{1}{2}\right\}$, and $I\left(\bar{f}_3\right) = -\frac{e^2}{4\pi^2}\frac{1}{\hat{u}}\left\{\frac{\hat{s}-\hat{u}}{\hat{u}}\ln\left(\frac{\hat{s}}{Q^2}\right) + 1\right\}$.

Having calculated the hard subprocess, we next discuss how to embed the quarks in the nucleon. We begin by discussing the soft contributions.One can show that the IR contributions from these processes, which are proportional to the products of the charges of the interacting quarks, added to the soft contributions from the handbag diagrams give the same result as the soft contributions calculated with just a nucleon intermediate state. The hard parts when the photons couple to different quarks are subleading in Q^2 because of momentum mismatches in the wavefunctions.

For the real parts, the IR divergence arising from the direct and crossed box diagrams, at the nucleon level, is cancelled when adding the bremsstrahlung contribution from the interference of diagrams where a soft photon is emitted from the electron and from the proton, which was calculated in Ref. [7]. This provides a radiative correction term from the soft part of the boxes plus e-p bremsstrahlung which added to the lowest order term may be written as: $\sigma_{R,soft} = \sigma_{1\gamma}\left(1 + \delta_{2\gamma}^{soft} + \delta_{brems}^{ep}\right)$, where $\sigma_{1\gamma}$ is the 1γ cross section. The sum of these soft effect is IR finite and gives a small correction factor to the 1γ cross section[7–9].

We next discuss the hard 2γ exchange contribution in the kinematical regime of $s, u, Q^2 \gg M^2$. The hard scattering amplitude is calculated as a convolution between an e-q hard scattering and a soft nucleon matrix element. It is convenient to choose a frame where $q^+ = 0$, where the light-cone variables $a^\pm$ proportional to $(a_0 \pm a_3)$ and we choose $P_3 > 0$. In this frame, the + momentum fractions of electrons and partons are defined as $\eta = K^+/P^+$ and $x = P_q^+/P^+$ respectively. We extend the handbag formalism, used in wide angle Compton scattering[10,11], to the 2γ exchange process in elastic e-p scattering, and keep the x dependence in the hard scattering amplitude. This yields the T -matrix which can be expressed as:

$$T^{hard}_{h,\lambda'_N,\lambda} = \int_{-1}^{1} \frac{dx}{x} \sum_q \frac{1}{2} \left[H^{hard}_{h,+\frac{1}{2}} + H^{hard}_{h,-\frac{1}{2}} \right]$$
$$\times \left[H^q(x,0,q^2) \bar{u}(p',\lambda'_N) \gamma \cdot n u(p',\lambda'_N) + E^q(x,0,q^2) \bar{u}(p',\lambda'_N) \frac{i\sigma^{\mu\nu} n_\mu q_\nu}{2M} u(p',\lambda'_N) \right]$$
$$+ \int^{1} \frac{dx}{} \sum \frac{1}{} \left[H^{hard}_{h,+\frac{1}{}} - H^{hard}_{h,-\frac{1}{}} \right] sign(x) \times \tilde{H}^q(x,0,q^2) \bar{u}(p',\lambda'_N) \gamma \cdot n \gamma_5 u(p',\lambda'_N), \quad (5)$$

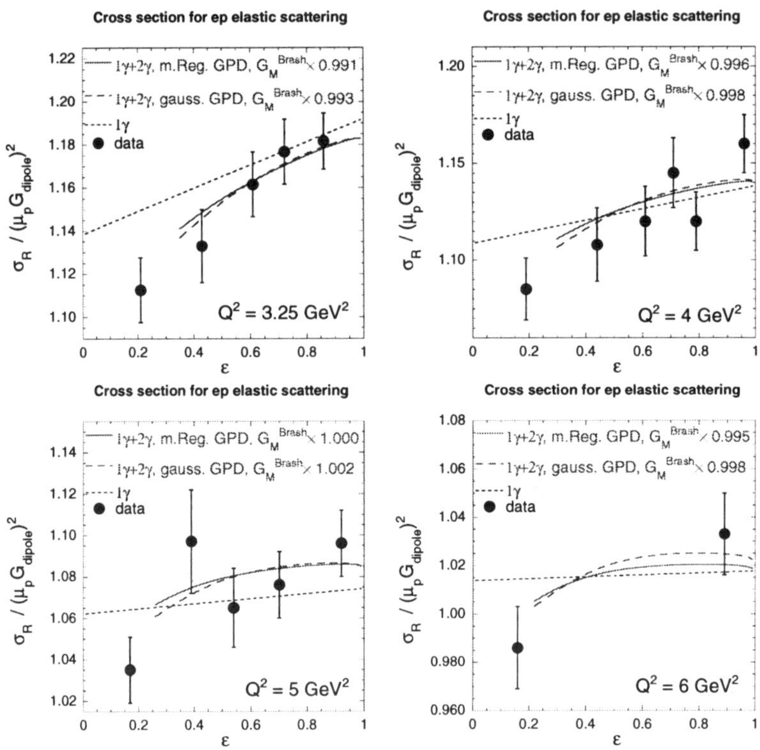

Fig. 1. Rosenbluth plots for elastic ep scattering: R divided by $(\mu_p G_D)^2$, with $G_D = (1 + Q^2/0.71)^{-2}$. Dotted curves: Born approximation. Solid curves: full calculation using the modified Regge GPD. Dashed curves: same as solid curves but using the gaussian GPD. The data are from Ref. [3], and the kinematical range $u > M^2$.

where H^{hard} is evaluated by using $\tilde{f}_1^{hard}$, $\tilde{f}_3$, and $n^\mu = 2(-\eta P^\mu + K^\mu)/\sqrt{M^4 - su}$. H^q, E^q ,$\tilde{H}^q$ are the GPDs for a quark in the nucleon. From Eq. (5), the hard 2γ contribution to $\tilde{F}_3$, $\delta\tilde{G}_M$, and $\delta\tilde{G}_E$ can be evaluated using a GPD parametrization. For the GPD's function input, first we use a gaussian valence model[10] which is unfactorized in x and Q^2 for the GPD's H^q and $\tilde{H}^q$, and adopt a valence parametrization multiplied with $(1-x)^2$ to be consistent with the $x \to 1$ limit for the GPD

E^q. Secondly, we use a different modified Regge model parametrization, which gives a better account of the nucleon form factor constraints.

As for the 1γ exchange form factor input, first we use G_E^p/G_M^p extracted from the polarization transfer experiments[4], for which we assume that the difference between the 1γ and 2γ form factor ratios is small[5]. Then the 1γ form factors G_M^p and G_E^p are obtained from the reduced cross section data, with the slope G_E^p/G_M^p given by polarization transfer method[13]. With the above input, we display the effect of 2γ exchange on the reduced cross section σ_R in Fig.1. The straight lines in fig.1 are the 1γ result, the curved lines are the $1+2\gamma$ results. We see that including the 2γ exchange allows one to reconcile the polarization transfer and Rosenbluth data.

Using the same input, we also find that the corrections to the longitudinal polarization P_l are quite small($< 1\%$), and the corrections to the sideways polarization P_s are smaller than 10 % in the kinematical range $u > M^2$ at large Q^2. Besides P_l and P_l, the single spin asymmetry for a polarized target normal to the scattering plane (A_n) - or the equivalent recoil polarization asymmetry P_n - provides a direct measure of the imaginary part of the 2γ exchange amplitudes. Furthermore, by comparing positron-proton and electron-proton scattering, which have the opposite sign for the two-photon corrections relative to the one-photon terms, one can isolate the two-photon exchange amplitude.

In summary, we have estimated the two-photon exchange contribution to elastic ep scattering at large Q^2 in a partonic calculation, and were able to express this amplitude in terms of the GPDs of the nucleon. We found that the 2γ exchange contribution is able to quantitatively resolve the existing discrepancy between Rosenbluth and polarization transfer experiments. We have also emphasized that there are important experimentally testable consequences of the two-photon amplitude.

References

1. M. E. Christy et al. [E94110 Collaboration], Phys. Rev. C 70, 015206 (2004) .
2. J. Arrington [JLab E01-001 Collaboration], arXiv:nucl-ex/0312017.
3. L. Andivahis et al., Phys. Rev. D 50, 5491 (1994)
4. O. Gayou et al. [Jefferson Lab Hall A Collaboration], Phys. Rev. Lett. 88, 092301 (2002).
5. P. A. M. Guichon and M. Vanderhaeghen, Phys. Rev. Lett. 91, 142303 (2003).
6. M.L. Goldberger, Y. Nambu and R. Oehme, Ann. off Phys. 2, 226 (1957).
7. L. C. Maximon and J. A. Tjon, Phys. Rev. C 62, 054320 (2000).
8. Y. C. Chen, A. Afanasev, S. J. Brodsky, C. E. Carlson and M. Vanderhaeghen, Phys. Rev. Lett. 93, 122301 (2004).
9. A. V. Afanasev, S. J. Brodsky, C. E. Carlson, Y. C. Chen, and M. Vanderhaegen, Phys. Rev. D 72, 013008 (2005) .
10. A.V. Radyushkin, Phys. Rev. D 58, 114008 (1998).
11. M. Diehl, Th. Feldmann, R. Jakob, P. Kroll, Phys. Lett. B 460, 204 (1999); Eur. Phys. J. C 8, 409 (1999).
12. M. Guidal, M. Polyakov, A. Radyushkin, and M. Vanderhaeghen, Phys. Rev. D 72, 054012 (2005).
13. E. J. Brash, A. Kozlov, S. Li and G. M. Huber, Phys. Rev. C 65, 051001 (2002).

Probing the Magnetic Dipole Moment of the $\Delta^+(1232)$ via $\gamma p \to \gamma\pi^0 p$ Reaction

W.T. Chiang and Shin Nan Yang

Department of Physics, National Taiwan University,
Taipei 10617, Taiwan
E-mail: snyang@phys.ntu.edu.tw

M. Vanderhaeghen

Jefferson Laboratory,
Newport News, Virginia 23606, USA
E-mail: marcvdh@jlab.org

D. Drechsel

Institut für Kernphysik, Universität Mainz,
D-55099 Mainz, Germany
E-mail: drechsel@kph.uni-mainz.de

The mangnetic dipole moment μ_{Δ^+} of the $\Delta^+(1232)$ is studied via the radiative pion photoproduction in the Δ-resonance region with a unitary model of the $\gamma p \to \gamma\pi N$, where the πN rescattering is included in an on-shell approximation. The model also satisfies the low-energy theorem exactly. The sensitivity of the $\gamma p \to \gamma\pi N$ at higher final photon energy to μ_{Δ^+} is studied. We find that the final photon asymmetry and a helicity cross section are particularly sensitive to μ_{Δ^+}.

The magnetic dipole moment μ_Δ of the $\Delta(1232)$ is of considerable theoretical interest. However, different theoretical models predict considerable deviation[1] from the SU(6) prediction, $\mu_\Delta = e_\Delta\,\mu_p$, where e_Δ is the Δ electric charge and μ_p the proton magnetic moment. μ_Δ has also been studied by LQCD but still at rather large quark masses[2]. Therefore, it would be helpful to measure μ_Δ through experiment.

$\mu_{\Delta^{++}}$ has been measured by the reaction $\pi^+ p \to \gamma\pi^+ p$ and the PDG quotes the range : $\mu_{\Delta^{++}} = 3.7 - 7.5\,\mu_N$. The large uncertainty is due to large non-resonant contributions to $\pi^+ p \to \gamma\pi^+ p$ because of bremsstrahlung from the π^+ and proton. For the Δ^+, it has been proposed to determine its magnetic moment through measurement of the $\gamma p \to \gamma\pi^0 p$ reaction. Due to the small cross sections for this reaction, a first measurement has only recently been reported by the A2/TAPS collaboration at MAMI[3]. Dedicated experiments are being performed with much higher count rates by using 4π detectors, such as the Crystal Ball detector at MAMI[4]. The analysis of these experiments requires a substantial theoretical effort to minimize the model dependence in the extraction of the μ_{Δ^+} from $\gamma p \to \gamma\pi^0 p$.

A calculation for $\gamma p \to \gamma\pi^0 p$, was recently carried out in Ref. 5. The model starts with an effective Lagrangian which describes $\gamma p \to \pi^0 p$. Then an additional photon is coupled in a gauge invariant way to describe $\gamma p \to \gamma\pi^0 p$. The result is a tree level calculation with part of the final state interaction effects taken into account by the finite width of the Δ. This model was used to analyze the first measurement of the $\gamma p \to \gamma\pi^0 p$ cross sections and a value of $\mu_{\Delta^+} = \left[2.7^{+1.0}_{-1.3}(\text{stat.}) \pm 1.5(\text{syst.}) \pm 3(\text{theor.})\right] \mu_N$ has been extracted in Ref. 3.

Although the tree level model of Ref. 5 gives a qualitatively good description of the data of Ref. 3, a detailed quantitative comparison requires the inclusion of rescattering effects. Such rescattering effects are known to be important in the case of pion photoproduction. Since an accurate theoretical description of the reaction $\gamma p \to \gamma\pi^0 p$ is essential for extracting a reliable value for μ_{Δ^+}, it is imperative to obtain an estimate of the effects of the final-state interaction to the best of our capability. In this talk, we present our recent work[6] in this direction which describes the radiative pion photoproduction by a properly unitarized theory.

We start with a unitary model for pion photoproduction and then extend it to $\gamma p \to \gamma\pi N$ reaction in the $\Delta(1232)$ resonance region, which will then be used as a tool to investigate the size of μ_{Δ^+}.

We employ the dynamical approach to pion photoproduction[7], where the unitarity is built in by explicitly including the final state πN interaction such that the T-matrix is expressed as $t_{\gamma\pi} = v_{\gamma\pi} + v_{\gamma\pi} g_0 t_{\pi N}$, where $v_{\gamma\pi}$ is a transition potential for the reaction $\gamma N \to \pi N$, and $t_{\pi N}$ and g_0 denote the πN scattering matrix and the free propagator, respectively.

In the Δ resonance region, the transition potential $v_{\gamma\pi}$ consists of two terms, $v_{\gamma\pi} = v^B_{\gamma\pi} + v^\Delta_{\gamma\pi}$, where $v^\Delta_{\gamma\pi}$ corresponds to the resonance contribution $\gamma N \to \Delta \to \pi N$ and $v^B_{\gamma\pi}$ describes the background as derived from an effective Lagrangian[8], which includes the Born terms in PV coupling and contribution from $t-$channel (ρ, ω) vector-meson exchange. The resulting T-matrix can be decomposed into two terms[9], $t_{\gamma\pi} = t^B_{\gamma\pi} + t^\Delta_{\gamma\pi}$, where $t^B_{\gamma\pi} = v^B_{\gamma\pi} + v^B_{\gamma\pi}\, g_0\, t_{\pi N}$, and $t^\Delta_{\gamma\pi} = v^\Delta_{\gamma\pi} + v^\Delta_{\gamma\pi}\, g_0\, t_{\pi N}$.

If the on-shell or K-matrix approximation is made, the multipole amplitudes $t^{B,\alpha}_{\gamma\pi}$, for the partial wave α, take the form[7],

$$t^{B,\alpha}_{\gamma\pi} = v^{B,\alpha}_{\gamma\pi} \cos\delta_\alpha\, e^{i\delta_\alpha}\,, \tag{1}$$

where δ_α is the phase shift for πN scattering in the respective partial wave α.

$v^\Delta_{\gamma\pi}$ is obtained with the commonly used interaction Lagrangians[6] for the vertices $\pi N\Delta$ and $\gamma N\Delta$, with Δ treated as a Rarita-Schwinger particle. The $\gamma N\Delta$ vertex contains the magnetic dipole G_M and electric quadrupole G_E couplings. We then approximate the resonance contribution $t^\Delta_{\gamma\pi}$ by

$$t^\Delta_{\gamma\pi} = v^\Delta_{\gamma\pi}(M_\Delta \to M_\Delta - \frac{i}{2}\Gamma_\Delta)e^{i\phi}\,, \tag{2}$$

with the phase ϕ adjusted such that the multipole amplitudes $(M^\Delta_{1+},\ E^\Delta_{1+})$ carry the phase δ_{33}. We further adopt the "complex mass scheme" by substituting $M_\Delta \to$

$M_\Delta - \frac{i}{2}\Gamma_\Delta$ with an energy independent width Γ_Δ, as was suggested by Ref. 10 in order to maintain the gauge invariance of the Δ contribution to the $\gamma N \to \gamma\pi N$ reaction. Since $t_{\gamma\pi}^{B,\alpha}$ of Eq. (1) satisfies the Fermi-Watson theorem separately, the total multipole amplitude will carry the proper πN phase.

The full unitary model described above contains two parameters G_M and G_E. With the choice of $G_M = 3.00$ and $G_E = 0.065$, we obtain very good description of the $\gamma p \to \pi^0 p$ and $\gamma p \to \pi^+ n$ processes through the Δ-resonance region as can be found in Ref. 6.

For $\gamma p \to \gamma\pi N$ reaction, we start from the tree diagrams of the effective Lagrangian for $\gamma p \to \pi N$ as described above and couple a photon to all the charged particles. The resulting diagrams are shown in Fig. 1, where we have also include the diagrams in Fig. 1(f) which arises from the $\pi^0 \to \gamma\gamma$ anomaly as given by the Wess-Zumino-Witten term. The Δ resonance diagrams of Fig. 1(a) can then be sim-

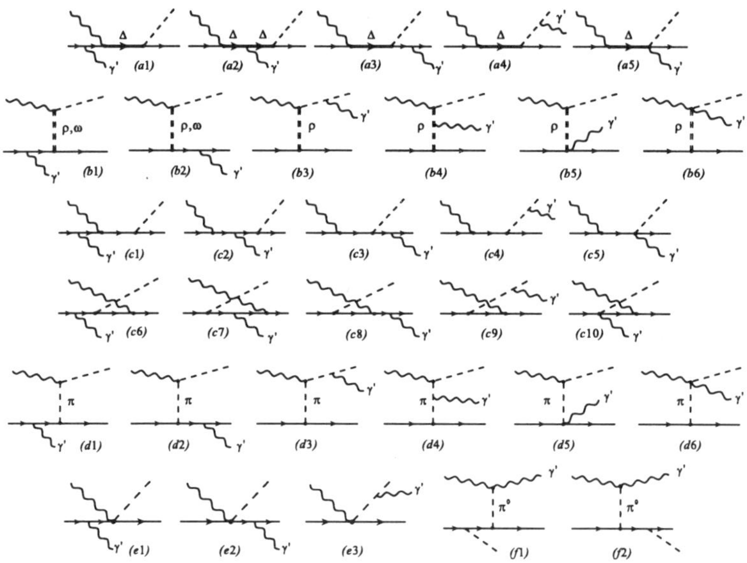

Fig. 1. Tree diagrams considered in the calculation of the $\gamma p \to \gamma\pi N$ reaction in the $\Delta(1232)$ region: Δ resonance (a1-a5), vector-meson exchange (b1-b6), nucleon-pole (c1-c10), pion-pole (d1-d6), Kroll-Rudermann (e1-e3), and anomaly diagrams (f1-f2).

ilarly evaluated as explained earlier except for diagram 1(a2). The latter diagram contains the interaction Lagrangian

$$L_{\gamma\Delta\Delta} = e_\Delta \bar{\psi}_\Delta^{\beta'} \left\{ g_{\beta'\beta} \left(\gamma_\nu A^\nu - \frac{\kappa_\Delta}{2M_\Delta} \sigma_{\nu\lambda} \partial^\lambda A^\nu \right) + \frac{1}{3} \left(\gamma_\beta \gamma_\nu \gamma_{\beta'} - \gamma_\beta g_{\nu\beta'} - \gamma_{\beta'} g_{\nu\beta} \right) A^\nu \right\} \psi_\Delta^\beta, \quad (3)$$

which contains the information on κ_Δ, the anomalous MDM of Δ. Therefore, in comparison with the $\gamma p \to \pi N$ process, the only new parameter entering in the description of the $\gamma p \to \gamma\pi^0 p$ process is κ_{Δ^+}.

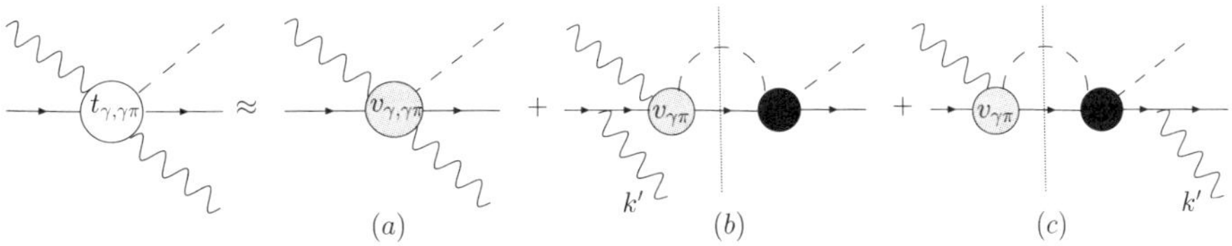

Fig. 2. Model for the T-matrix for the $\gamma p \to \gamma \pi^0 p$ reaction used in this work. The transition potential $v_{\gamma,\gamma\pi}$ (diagram a) corresponds to the diagrams of Fig. 1. The rescattering contributions (diagrams b and c) are evaluated in the soft-photon approximation for the final photon, i.e. $k' \to 0$. The black blob corresponds with the full T-matrix $t_{\pi N}$ for πN scattering. The vertical dotted lines indicate that the πN intermediate state is taken on-shell (K-matrix approximation).

We next turn to the rescattering contribution to the $\gamma p \to \gamma \pi^0 p$ reaction for low energy final photon. We estimate this rescattering in the K-matrix approximation. In the soft-photon limit for the final photon, the T-matrix for the $\gamma p \to \gamma \pi^0 p$ reaction has to be directly proportional to the full T-matrix for $\gamma p \to \pi^0 p$. We construct the full T-matrix for $\gamma p \to \gamma \pi^0 p$ as shown in Fig. 2 by the expression

$$t_{\gamma,\gamma\pi} \approx v_{\gamma,\gamma\pi} + e\,\varepsilon^*_\nu(k',\lambda') \left(\frac{p'^\nu}{p' \cdot k'} - \frac{p^\nu}{p \cdot k'} \right) [t_{\gamma\pi} - v_{\gamma\pi}] \,. \tag{4}$$

The first term in Eq. (4), denoted by the transition potential $v_{\gamma,\gamma\pi}$, is the sum of all tree diagrams shown in Fig. 1 and is gauge invariant by itself. The second term in Eq. (4) is the rescattering contribution. Since we only keep the leading term in the outgoing photon energy for the rescattering term ($k' \to 0$), this amounts to evaluate the $t_{\gamma\pi}$ in the second term of Eq. (4) in soft-photon kinematics. For $t_{\gamma\pi}$, we adopt the unitary model as given by Eqs. (1) and (2). In the soft-photon limit, it is not difficult to verify that the full amplitude for the $\gamma p \to \gamma \pi N$ process as given by Eq. (4) satisfies the low energy theorem, namely,

$$t_{\gamma,\gamma\pi} \stackrel{k'\to 0}{\longrightarrow} e\,\varepsilon^*_\nu(k',\lambda') \left(\frac{p'^\nu}{p' \cdot k'} - \frac{p^\nu}{p \cdot k'} \right) t_{\gamma\pi}. \tag{5}$$

Note that both terms in our model for $t_{\gamma,\gamma\pi}$ in Eq. (4) satisfy gauge invariance with respect to both intial and final photons.

Using the unitary model constructed above in Eq. (4), we investigated several $\gamma p \to \gamma \pi N$ observables. In general, we found good agreement for the existing experimental data of the $\gamma p \to \gamma \pi^0 p$ reaction[6]. We found that there are several polarization observables which are very sensitive to κ_{Δ^+}. In Fig. 3, we show the photon asymmetry for linearly polarized incident photons which changes between 0.35 and 0.15 when κ_{Δ^+} is varied between 0 and 6. The helicity cross sections which can be measured with circularly polarized photon beam and a longitudinally polarized proton target also exhibit strong dependence on κ_{Δ^+}. In Fig. 3, one sees that the helicity cross section $\sigma_{1/2}$ changes by about a factor of 2 when κ_{Δ^+} is varied between 0 and 6.

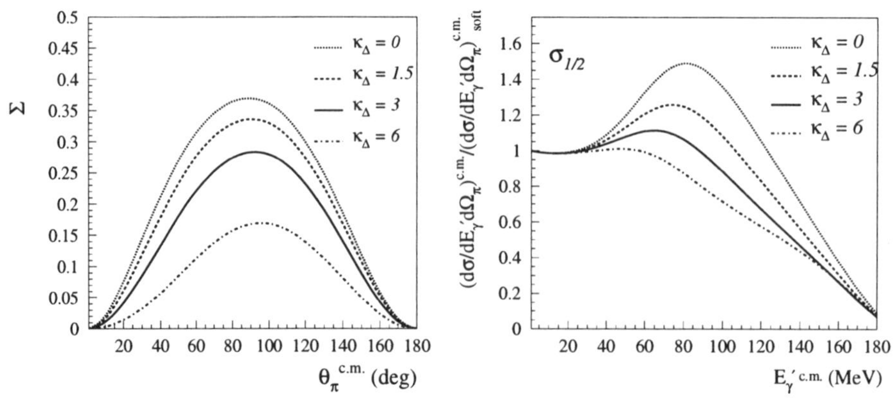

Fig. 3. Left panel : the angular distribution of the emitted pions for the $\gamma p \to \gamma\pi^0 p$ linear photon asymmetry Σ at incident photon lab energy $E_\gamma^{lab} = 400$ MeV and fixed outgoing photon energy $E_\gamma^{\prime c.m.} = 100$ MeV. The sensitivity of the unitary model to different values of κ_{Δ^+} is shown. Right panel : the helicity dependence of the $\gamma p \to \gamma\pi^0 p$ *c.m.* cross section $d\sigma/dE'_\gamma d\Omega_\pi$ for total helicity 1/2, divided by its soft photon value, as function of the outgoing photon energy $E_\gamma^{\prime cm}$, at incident photon *lab* energy $E_\gamma^{lab} = 400$ MeV and pion emission angle $\theta_\pi^{cm} = 90^o$. The curves correspond with the predictions of the unitary model for different values of κ_{Δ^+}.

In summary, we have constructed a unitary model for the reaction $\gamma p \to \gamma\pi^0 p$ which includes the final-state πN resacttering effects in K-matrix approximation. The model is gauge invariant and satisfies the low-energy theorem for the final photon. We find that several polarization observables like photon asymmetry, and the helicity cross sections $\sigma_{1/2}$ and $\sigma_{3/2}$ are strongly dependent on κ_{Δ^+}. The dedicated measurements of the photon asymmetry that are currently underway at MAMI are therefore highly promising for a more quantitative extraction of the Δ^+ MDM. For this purpose the presented model may be compared with a recent chiral effective-field theory calculation [11] of radiative pion photoproduction in the Δ-resonance region, allowing to quantify the model errors in the extraction of the Δ^+ MDM.

References

1. T.M. Aliev, A. Özpineci, and M. Savci, *Nucl. Phys.* **A678**, 443 (2000).
2. D.B. Leinweber, T. Draper, and R.M. Woloshyn, *Phys. Rev.* **D46**, 3067 (1992).
3. M. Kotulla *et al.*, *Phys. Rev. Lett.* **89**, 272001 (2002).
4. R. Beck, B. Nefkens, spokespersons Crystal Ball @ MAMI experiment.
5. D. Drechsel and M. Vanderhaeghen, *Phys. Rev.* **C64**, 065202 (2001).
6. W.T. Chiang *et al.*, *Phys. Rev.* **C71**, 015204 (2005).
7. S.N. Yang, *J. Phys. G* **11**, L205 (1985).
8. M.G. Olsson and E.T. Osypowski, *Phys. Rev.* **D17**, 174, (1978).
9. S.S. Kamalov and S.N. Yang, *Phys. Rev. Lett.* **83**, 4494 (1999).
10. M. El Amiri, G. López Castro, and J. Pestieau, *Nucl. Phys.* **A543**, 673 (1992).
11. V. Pascalutsa and M. Vanderhaeghen, Phys. Rev. Lett. **94**, 102003 (2005).

Baryon-Meson Interactions in Chiral Quark Model*

F. Huang, Z. Y. Zhang and Y. W. Yu

Institute of High Energy Physics, P.O. Box 918-4, Beijing 100049, China

Using the resonating group method (RGM), we dynamically study the baryon-meson interactions in chiral quark model. Some interesting results are obtained: (1) The ΣK state has an attractive interaction, which consequently results in a ΣK quasibound state. When the channel coupling of ΣK and ΛK is considered, a sharp resonance appears between the thresholds of these two channels. (2) The interaction of ΔK state with isospin $I = 1$ is attractive, which can make for a ΔK quasibound state. (3) When the coupling to the ΛK^* channel is considered, the $N\phi$ is found to be a quasibound state in the extended chiral SU(3) quark model with several MeV binding energy. (4) The calculated S-, P-, D-, and F-wave KN phase shifts achieve a considerable improvement in not only the signs but also the magnitudes in comparison with other's previous quark model study.

1. Introduction

Nowadays people still need QCD-inspired models to study the non-perturbative QCD effects in the low-energy region. Among these models, the chiral SU(3) quark model has been quite successful in reproducing the energies of the baryon ground states, the binding energy of deuteron, the NN scattering phase shifts, and the NY (nucleon-hyperon) cross sections. Inspired by these achievements, we try to extend this model to study the baryon-meson systems. In order to study the short-range feature of the quark-quark interaction in the low-energy region, we further extend our chiral SU(3) quark model to include the vector meson exchanges. The OGE that dominantly governs the short-range quark-quark interaction in the original chiral SU(3) quark model is now nearly replaced by the vector-meson exchanges. We use these two models to study the baryon-meson interactions

In this paper, we show the RGM dynamical calculating results of the ΛK, ΣK, ΔK, $N\phi$, and KN states obtained in the chiral SU(3) quark model as well as in the extended chiral SU(3) quark model [1-4].

2. Formulation

The chiral SU(3) quark model and the extended chiral SU(3) quark model have been widely described in the literature [1-4], and we refer the reader to those works for details. Here we just give the salient features of our chiral quark model.

*This work is supported by the National Natural Science Foundation of China No. 10475087.

In the chiral quark model, the total Hamiltonian of baryon-meson systems can be written as

$$H = \sum_{i=1}^{5} T_i - T_G + \sum_{i<j=1}^{4} V_{ij} + \sum_{i=1}^{4} V_{i\bar{5}}, \tag{1}$$

where T_G is the kinetic energy operator for the center-of-mass motion, and V_{ij} and $V_{i\bar{5}}$ represent the quark-quark and quark-antiquark interactions, respectively,

$$V_{ij} = V_{ij}^{OGE} + V_{ij}^{conf} + V_{ij}^{ch}, \tag{2}$$

where V_{ij}^{ch} is the chiral fields induced effective quark-quark potential. In the chiral SU(3) quark model, V_{ij}^{ch} can be written as

$$V_{ij}^{ch} = \sum_{a=0}^{8} V_{\sigma_a}(\boldsymbol{r}_{ij}) + \sum_{a=0}^{8} V_{\pi_a}(\boldsymbol{r}_{ij}), \tag{3}$$

and in the extended chiral SU(3) quark model, V_{ij}^{ch} can be written as

$$V_{ij}^{ch} = \sum_{a=0}^{8} V_{\sigma_a}(\boldsymbol{r}_{ij}) + \sum_{a=0}^{8} V_{\pi_a}(\boldsymbol{r}_{ij}) + \sum_{a=0}^{8} V_{\rho_a}(\boldsymbol{r}_{ij}). \tag{4}$$

$V_{i\bar{5}}$ in Eq. (1) includes two parts: direct interaction and annihilation parts,

$$V_{i\bar{5}} = V_{i\bar{5}}^{dir} + V_{i\bar{5}}^{ann}, \tag{5}$$

with

$$V_{i\bar{5}}^{dir} = V_{i\bar{5}}^{conf} + V_{i\bar{5}}^{OGE} + V_{i\bar{5}}^{ch}. \tag{6}$$

The annihilation interaction $V_{i\bar{5}}^{ann}$ is not included in baryon-meson interactions since they are assumed not to contribute significantly to a molecular state or a scattering process, which is the subject of our study.

Table 1. Model parameters. Λ = 1100 MeV, m_u = 313 MeV, m_s = 470 MeV, g_{ch} = 2.621, b = 0.5 fm for set I and 0.45 fm for sets II and III. The meson masses are taken to be the experimental data except for m_σ.

	m_σ (MeV)	g_{chv}	f_{chv}/g_{chv}	g_u^2	g_s^2	a_{uu}^c (MeV/fm^2)	a_{us}^c (MeV/fm^2)	a_{ss}^c (MeV/fm^2)
I	595	–	–	0.781	0.865	46.1	58.0	98.0
II	535	2.351	0	0.067	0.212	44.1	79.0	162.1
III	547	1.973	2/3	0.143	0.264	36.7	68.6	141.3

The model parameters can be fitted by several conditions, e.g. the energies of baryons, the binding energy of deuteron, the chiral symmetry, and the stability conditions of baryons. We listed them in Table 1, where the first set is for the chiral SU(3) quark model, the second and third sets are for the extended chiral SU(3) quark model. All these three sets of parameters can give a reasonable description of the NN phase shifts [5].

3. Results and Discussion

3.1. ΛK and ΣK States

The nucleon resonance $S_{11}(1535)$ is explained as an excited three quark state in the traditional constitute quark model [6,7]. But on the hadron level, it is explained as a ΛK-ΣK quasibound state [8,9]. A dynamical study on a quark level of the ΛK and ΣK interactions will be useful to get a better understanding of the $S_{11}(1535)$ [1].

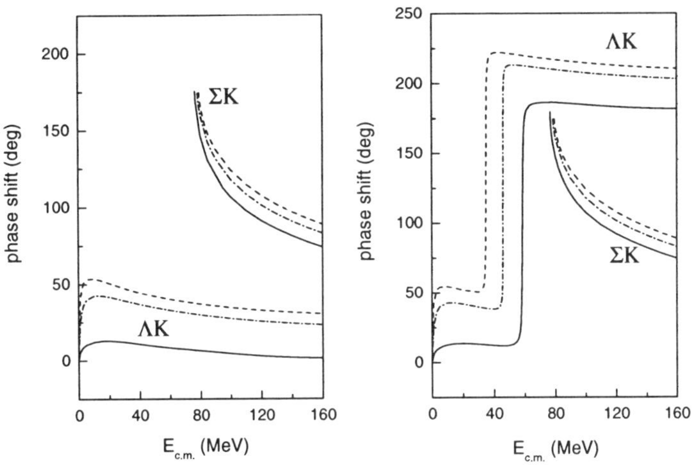

Fig. 1. S-wave ΛK and ΣK phase shifts in one-channel and coupled-channel calculation. The solid curves represent the results obtained in the chiral SU(3) quark model. The dashed and dash-dotted curves show the results from the extended chiral SU(3) quark model by taking f_{chv}/g_{chv} as 0 and 2/3, respectively.

Fig. 1 shows the S-wave ΛK and ΣK phase shifts in one-channel (left) and coupled-channel (right) calculations. The phase shifts show that there is a strong attraction of ΣK, which can make for a ΣK quasibound state, while the interaction of ΛK is comparatively weak. When the channel coupling of ΛK and ΣK is considered, the phase shifts show a sharp resonance between the thresholds of these two channels. The spin-parity is $J^P = 1/2^-$ and width $\Gamma \approx 5$ MeV. The narrow gap of the ΛK and ΣK thresholds, the strong attraction between Σ and K, and the sizeable off-diagonal matrix elements of ΛK and ΣK are responsible for the appearance of this resonance. The final conclusion regarding what is the resonance we obtained and its exact theoretical mass and width will wait for further work where more channels will be considered.

3.2. ΔK State

The ΔK state has first been studied on the hadron level [10]. We perform a RGM dynamical study of the structures of ΔK state within our chiral quark model [2]. Fig.

2 shows the diagonal matrix elements of the Hamiltonian in the generator coordinate method (GCM) calculation, which can describe the ΔK interaction qualitatively.

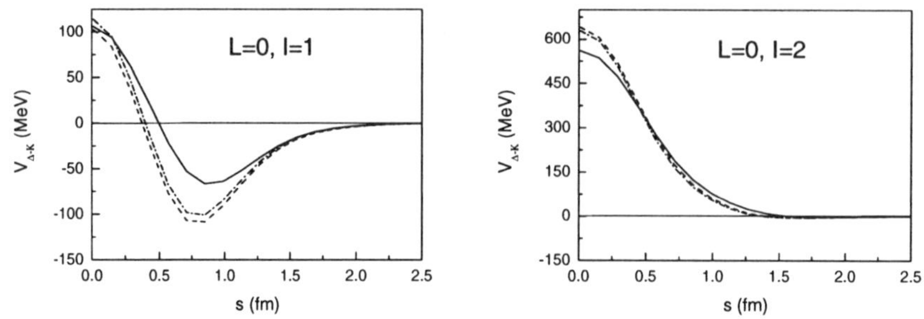

Fig. 2. The GCM matrix elements of the Hamiltonian for ΔK. Same notation as in Fig. 1.

From Fig. 2 one sees that the ΔK interaction is attractive for isospin $I = 1$ channel, while strongly repulsive for $I = 2$ channel. The attraction can result in a ΔK bound state with a binding energy of about 3, 20, and 15 MeV by using three different sets of parameters. To examine if $(\Delta K)_{LSJ=0\frac{3}{2}\frac{3}{2}}$ is possible to be a resonance or a bound state, the channel coupling between $(\Delta K)_{LSJ=0\frac{3}{2}\frac{3}{2}}$ and $(NK^*)_{LSJ=0\frac{3}{2}\frac{3}{2}}$ would be considered in our future work.

3.3. *Nϕ State*

It has been reported that the QCD van der Waals attractive potential is strong enough to bind a ϕ meson onto a nucleon to form a bound state [11]. We try to study the possibility of a $N\phi$ bound state in our chiral quark model. Table 2 shows the calculated binding energy of $N\phi$.

Table 2. The binding energy of $N\phi$.

	One-channel		Coupled-channel	
	$S = 1/2$	$S = 3/2$	$S = 1/2$	$S = 3/2$
I	–	–	–	–
II	1	3	3	9
III	–	–	1	6

From Table 2 one sees that in the one channel study, one can get a bound state only by using the second set of parameters. When the channel coupling to ΛK^* is considered, the $N\phi$ is found to be a bound state in the extended chiral SU(3) quark model with several MeV binding energy. Further the tensor force will be considered in the future work, which would make a bigger binding energy.

For the $N\phi$ system, the two color-singlet clusters have no quark in common. The attractive interaction is dominantly provided by σ exchange. Thus $N\phi$ is an ideal place to test the strength of the coupling of the quark and σ chiral field.

3.4. *KN Scattering*

The KN scattering has aroused particular interest in the past. But most of the works on the quark level cannot give a reasonable description of the KN phase shifts up to $L = 3$. We dynamically study the KN scattering in our chiral quark model [3,2,4]. When m_σ is chosen to be 675 MeV and the mixing of σ_0 and σ_8 is considered, we can get a satisfactory description of the S-, P-, D-, and F-wave KN phase shifts (see Figs. 1-4 in Ref. 4). The results from the chiral SU(3) quark model and the extended chiral SU(3) quark model are quite similar although the short-range interaction mechanisms in these two models are quite different. Compared with the results in other's previous quark model study [12], our theoretical phase shifts achieve correct signs for several partial waves and a considerable improvement in the magnitude for many channels.

4. Summary

We dynamically study the baryon-meson interactions in chiral quark model by using the RGM. Some interesting results are obtained: (1) The ΣK state has an attractive interaction, which consequently results in a ΣK quasibound state. When the channel coupling of ΣK and ΛK is considered, a sharp resonance appears between the thresholds of these two channels. (2) The interaction of ΔK state with $I = 1$ is attractive, which can make for a ΔK quasibound state, while for $I = 2$ channel, the interaction is strongly repulsive. (3) When the coupling to the ΛK^* channel is considered, the $N\phi$ is found to be a quasibound state in the extended chiral SU(3) quark model with several MeV binding energy. (4) The calculated KN phase shifts achieve a considerable improvement in not only the signs but also the magnitudes in comparison with other's previous quark model study.

References

1. F. Huang, D. Zhang, Z.Y. Zhang and Y.W. Yu, *Phys. Rev.* **C71**, 064001 (2005).
2. F. Huang and Z.Y. Zhang, *Phys. Rev.* **C70**, 064004 (2004).
3. F. Huang, Z.Y. Zhang and Y.W. Yu, *Phys. Rev.* **C70**, 044004 (2004).
4. F. Huang and Z.Y. Zhang, *Phys. Rev.* **C72**, 024003 (2005).
5. L.R. Dai, Z.Y. Zhang, Y.W. Yu and P. Wang, *Nucl. Phys.* **A727**, 321 (2003).
6. L.Ya. Glozman and D.O. Riska, *Phys. Rept.* **268**, 263 (1996).
7. N. Isgur and G. Karl, *Phys. Rev.* **D19**, 2653 (1979).
8. N. Kaiser, T. Waas and W. Weise, *Nucl. Phys.* **A612**, 297 (1997).
9. T. Inoue, E. Oset and M.J. Vicente Vacas, *Phys. Rev.* **C65**, 035204 (2002).
10. S. Sarkar, E. Oset and M.J.V. Vacas, *Eur. Phys. J.* **A24**, 287 (2005).
11. H. Gao, T.S.H. Lee and V. Marinov, *Phys. Rev.* **C63**, 022201 (2001).
12. S. Lemaire, J. Labarsouque and B. Silvestre-Brac, *Nucl. Phys.* **A714**, 265 (2003).

The Few-Body Physics of Heavy Quark Systems

Kamal K. Seth

Northwestern University, Evanston, IL 60208, USA.
E-mail: k-seth@northwestern.edu

Heavy–quark systems provide very attractive examples of few–body systems that can be studied with great precision. They can also be understood more directly in terms of the fundamental theory of QCD. Exciting recent experimental developments in the field are reviewed.

1. Introduction

This conference is devoted primarily to the few–body physics of light–quark systems. So, a little introduction about heavy–quark systems is in order. In QCD the only natural scale is $\Lambda(\text{QCD}) \approx 250$ MeV. Quarks are considered light or heavy with respect to Λ. So, u, d, s quarks with masses ≤ 150 MeV are light, and c, b, t quarks are heavy. Light quarks make π, K and other light mesons, and p, n, Δ, N^*, Σ, and other baryons. I will not talk about them. The top quark decays too quickly to form any hadrons. So, heavy quark spectroscopy is the spectroscopy of c and b quarks. I will limit myself to charmonium ($c\bar{c}$) and bottomonium, i.e., when both quark and anti–quark are heavy quarks of the same flavor.

In principle, the mysteries of the strong interaction can be explored in systems of any flavors of quarks, because the QCD interactions are flavour independent. In practice, this is not true for the "effective interaction", primarily because of the differences in the quark masses. The light quarks in hadrons are highly relativistic, $(v/c)^2 \approx E_b/2m_q \geq 0.7$, have a large value of the strong coupling constant of QCD, $\alpha_S \geq 0.6$, and because the masses of the u, d, s quarks are similar, nearly all light–quark structures contain mixtures of all three flavors, leading to a very high density of overlapping states. The result is to require very cumbersome partial wave analysis to identify and classify states.

In contrast, the heavy quarks in $|c\bar{c}>$ charmonium and $|b\bar{b}>$ bottomonium are far less relativistic, with $(v/c)^2 \approx 0.2(c\bar{c})$, $0.1(b\bar{b})$, and have $\alpha_S \approx 0.2 - 0.3$, making perturbative calculations possible. Further the spectra of charmonium and bottomonium are sparse, with well–resolved states of narrow widths, as illustrated in Fig. 1 for the spectrum of charmonium. Clean, precision spectroscopy of charmonium and bottomonium is therefore possible. The only problem which needs to be faced is that $|q\bar{q}>$ formation cross–sections are small, and experimental measurements require high energy, high intensity accelerators, and commensurate detectors.

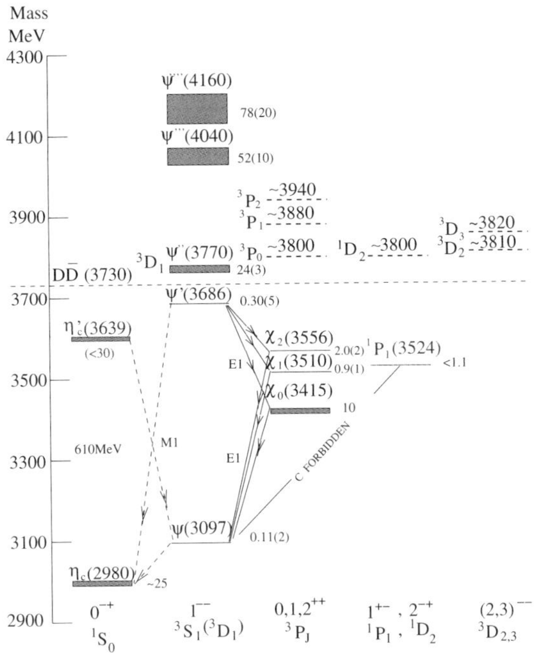

Fig. 1. Spectrum of the states of Charmonium.

2. Charmonium

Let me begin with charmonium (Fig. 1), with discoveries and precision measurements. First a few examples of precision measurements and what they tell us.

We all know that QCD began in earnest in 1974 with the discovery of J/ψ, the spin–triplet $^3S_1(J^{PC} = 1^{--})$ state of charmonium. For ten years the emphasis at SLAC, DESY, and ORSAY was on completing the spectrum of the bound states of the charmonium family and their decays. The measurements often lacked precision. Unparalleled levels of precision have now been reached.

2.1. *Masses and Widths*

Using $p\bar{p}$ annihilation to form charmonium $|c\bar{c}>$ states, Fermilab experiments E760/E835 [1] achieved unprecedented precision in measuring masses and widths of charmonium resonances. Using resonance depolarization techniques in e^+e^- annihilation, Novosibirsk [2] has achieved even greater precision in measuring J/ψ and ψ' ($\psi(2S)$) masses. Examples are:

$$M(J/\psi) = 3096.917 \pm 0.010 \pm 0.007 \text{ MeV, i.e., 3 parts in } 10^7 \text{ (Novosibirsk)}$$

$$\Gamma(\chi_{c1}) = 0.88 \pm 0.05 \text{ MeV (E835)}$$

2.2. *Branching Ratios*

CLEO-c has just started taking data with e^+e^- annihilation at ψ'. Some notable results are the following.

2.2.1. *Hadronic Decays*

At CLEO [3,4], *lepton universality* has been confirmed at a $\pm 1\%$ level in J/ψ decays to e^+e^- and $\mu^+\mu^-$.

$$\mathcal{B}(J/\psi \to e^+e^-)/\mathcal{B}(J/\psi \to \mu^+\mu^-) = 0.997 \pm 0.013.$$

CLEO has also measured

$$\mathcal{B}(\psi' \to J/\psi + \pi^+\pi^-)/\mathcal{B}(\psi' \to J/\psi + \pi^0\pi^0) = 2.030 \pm 0.099,$$

which confirms *isospin conservation* at a $\pm 1.5\%$ level in $c\bar{c}$ systems [4], and

$$\mathcal{B}(\psi' \to J/\psi + \pi^0)/\mathcal{B}(\psi' \to J/\psi + \eta) = 0.040 \pm 0.005,$$

which implies *isospin violation* at a $\pm 0.4\%$ level [4], when due account is taken of the phase space differences.

2.2.2. *Radiative Decays*

CLEO has made precision measurements of branching ratios for radiative transitions from ψ', $\mathcal{B}(\psi' \to \gamma\chi_{c0}, \gamma\chi_{c1}, \gamma\chi_{c2}, \gamma\eta_c)$ [5], and $\mathcal{B}(\chi_{c0}, \chi_{c1}, \chi_{c2} \to \gamma J/\psi)$ [4]. The results are displayed in Fig. 2 and tabulated in Table 1. CLEO results show large differences form the original Crystal Ball (SLAC) results, especially for the χ_{cJ} radiative decays. Notice, for example, that the CLEO result for $\mathcal{B}(\chi_{c2} \to \gamma J/\psi)$ is 60% larger than the Crystal Ball result.

An example of how precision results affect other important measurements is provided by the story of the two–photon decays of charmonium resonances. Two–photon widths of $\chi_{c0}(^3P_0)$ and $\chi_{c2}(^3P_2)$ are pure QED processes akin to the decays of positronium. QCD enters only through radiative corrections and relativistic effects, which can therefore be studied by measuring these decays with precision. Unfortuantely, there has been huge disagreement (factors of ≥ 3) in the results of these decay widths for a long time between measurements of χ_{c2} *formation* by $\gamma\gamma$–fusion (at e^+e^- colliders) and χ_{c2} *decay* into $\gamma\gamma$ (in $p\bar{p}$ annihilation at Fermilab). The latest of these published results are by Belle [6] with $\Gamma_{\gamma\gamma}(\chi_{c2}) = 860 \pm 127$ eV, and by Fermilab [7] with $\Gamma_{\gamma\gamma}(\chi_{c2}) = 270 \pm 59$ eV. Who is right? To answer this question, CLEO [8] has recently made a new measurement of χ_{c2} formation in two–photon fusion. The result is that both measurements are right. What is wrong is the old value of $\mathcal{B}(\chi_{c2} \to \gamma J/\psi)$, shown in Table 1, which was used by both Belle and Fermilab. As shown in Table 2, when the new CLEO result for $\mathcal{B}(\chi_{c2} \to \gamma J/\psi)$ is used, the Belle result for $\Gamma_{\gamma\gamma}(\chi_{c2})$ comes in excellent agreement with the new CELO result (see Fig. 3(left)), and the $p\bar{p}$ result also becomes statistically consistent with the other two.

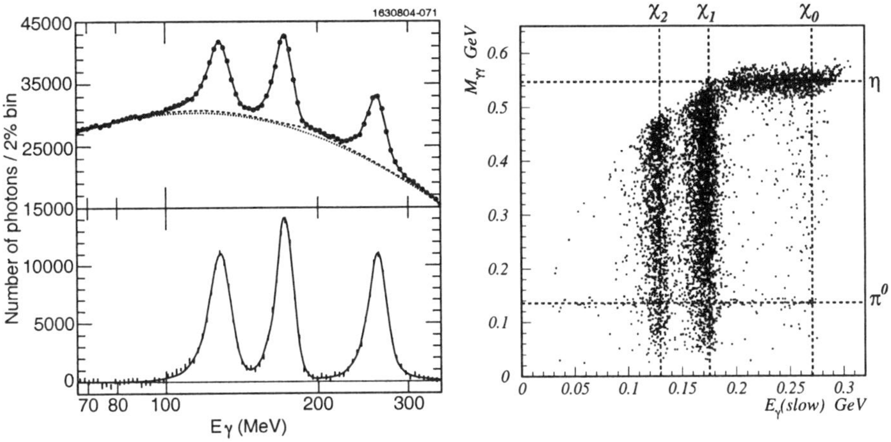

Fig. 2. Radiative transitions of charmonia from CLEO measurements. (Top) $\psi' \to \gamma\chi_{cJ}$ (bottom) $\psi' \to \gamma\chi_{cJ} \to \gamma\gamma J/\psi$.

Table 1. Results of new measurements of radiative decays of $\psi'(^3S_1)$ and $\chi_{cJ}(^3P_J)$ states of charmonium.

	CB(86)%	CLEO(04)% [5,4]
$\mathcal{B}(\psi' \to \gamma\chi_{c0})$	9.9 ± 0.9	9.22 ± 0.47
$\mathcal{B}(\psi' \to \gamma\chi_{c1})$	9.0 ± 0.8	9.07 ± 0.55
$\mathcal{B}(\psi' \to \gamma\chi_{c2})$	8.0 ± 0.8	9.33 ± 0.63
$\mathcal{B}(\psi' \to \gamma\eta)$	0.28 ± 0.06	0.32 ± 0.07
$\mathcal{B}(\chi_{c0} \to \gamma J/\psi)$	0.60 ± 0.18	1.95 ± 0.24
$\mathcal{B}(\chi_{c1} \to \gamma J/\psi)$	28.4 ± 2.1	37.9 ± 2.7
$\mathcal{B}(\chi_{c2} \to \gamma J/\psi)$	12.4 ± 1.5	19.9 ± 1.7

Table 2. The latest experimental results for $\Gamma_{\gamma\gamma}(\chi_{c2})$.

Measurement	$\Gamma_{\gamma\gamma}(\chi_2)$ (eV) (as published)	$\Gamma_{\gamma\gamma}(\chi_2)$ (eV) with new $B(\chi_2 \to \gamma J/\psi)$
CLEO (2005) ($\gamma\gamma \to \chi_2\gamma l^+l^-$)		559 ± 81
Belle (2002) ($\gamma\gamma \to \chi_2\gamma l^+l^-$)	850 ± 127	570 ± 81
E835 (2002) $(p\bar{p} \to \gamma\gamma)/(p\bar{p} \to \gamma J/\psi)$	270 ± 59	384 ± 83

The long-standing controversy is resolved. And now that the controversy is resolved, we are able to obtain a more reliable estimate of the strong coupling constant by comparing the two–photon width of χ_{c2} with its two–gluon width. The CLEO result leads to $\Gamma(\chi_{c2} \to \gamma\gamma)/\Gamma(\chi_{c2} \to gg) = (3.65 \pm 0.04) \times 10^{-4}$, and hence to $\boldsymbol{\alpha_S = 0.29 \pm 0.01}$, using pQCD prediction with first order radiative correction.

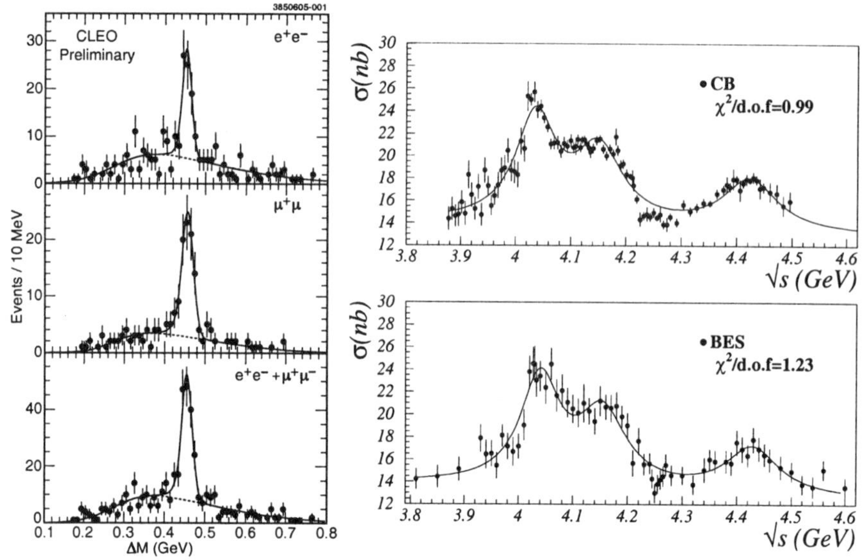

Fig. 3. (left) Mass difference $\Delta M \equiv M(\gamma l^+l^-) - M(l^+l^-)$ for the reaction $\gamma\gamma \to \chi_{c2} \to \gamma J/\psi$, $J/\psi \to l^+l^-$; (right) The data and best fit for the latest R measurements.

2.2.3. *The $\rho - \pi$ Problem, or the 13% Rule*

According to pQCD, since both the leptonic ($\to l^+l^-$) and hadronic ($\to ggg$) decays of J/ψ and ψ' are proportional to their wave functions at the origin,

$$Q \equiv \mathcal{B}(\psi' \to ggg)/\mathcal{B}(J/\psi \to ggg) = \mathcal{B}(\psi' \to e^+e^-)/\mathcal{B}(J/\psi \to e^+e^-) = 0.13 \pm 0.01.$$

This has led to numerous investigations, originally by BES [9], and joined recently by CLEO-c [10], to study the extent to which the ratio behaves as above for individual hadronic decays. The result is that for 2–body decays, Q varies all over the map, with values as small as $Q(\rho\pi) = 0.002 \pm 0.007$, but for multibody decays it appears to range within $Q = 1 - 10$. It appears that it is not reasonable to ask global pQCD predictions to hold for individual decays, which must have their own individual hadronization physics.

2.2.4. *Hadronic Decays Above the $D\overline{D}$ Threshold*

Above $\sqrt{s} = 3.74$ GeV, charmonia can decay into $D\overline{D}$ open–charm pairs, and the vector states acquire large widths. Not much is known above these energies except for the measurements of the quantity R, defined as $R = \sigma(e^+e^- \to \text{hadrons})/\sigma(e^+e^- \to \mu^+\mu^-)$, with $\sigma(e^+e^- \to \mu^+\mu^-) = 4\pi\alpha_{em}^2/3s$. A recent reanalysis [11] of the available data in the region $\sqrt{s} = 3.8 - 4.6$ GeV has shown (see Fig. 3(right)) that the region is dominated by three vector states whose widths are much larger than believed so far. For example, $\Gamma(4426) = 119 \pm 15$ MeV, i.e., three

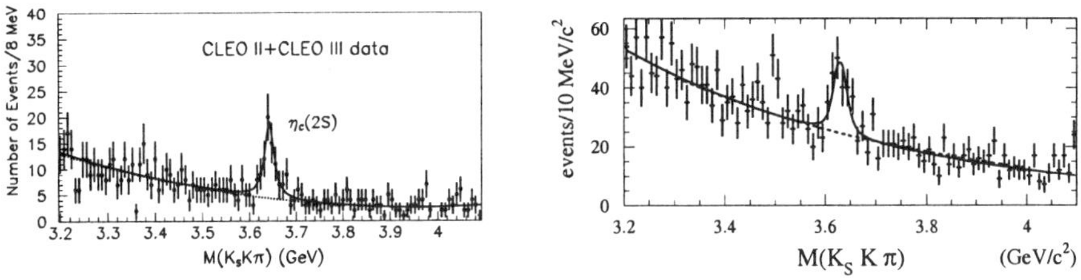

Fig. 4. Observation of η_c' in the reaction $\gamma\gamma \to K_S K\pi$ by CLEO (left) and Babar (right).

times larger than claimed by the 1978 analysis by DASP [12], and adopted by PDG.

2.2.5. *Threshold Resonances*

BES [13] has recently reported the observation of enhancements near the thresholds for J/ψ decays containing $p\bar{p}$, $p\overline{\Lambda}$, $K\overline{\Lambda}$. They have suggested that these enhancements denote sub–threshold or near–threshold resonances in each case. The claim is rather controversial, with other explanations attributing the observations to final state interactions and cusps of various kinds.

2.3. *Discovery of the Long–Lost Singlet States of Charmonium*

As is well known, the central $q\bar{q}$ interaction is well represented by the so–called Cornell potential (or its variants) as the sum of a one–gluon exchange Coulombic potential proportional to $1/r$ and a not–well–understood confinement potential, generally taken as Lorentz scalar and proportional to r. Far less is known about the spin dependence of the $q\bar{q}$ potential which is responsible for the spin–orbit and spin–spin splittings of the $q\bar{q}$ states. The crucial determinant of the spin dependent $q\bar{q}$ interaction is the hyperfine, spin–singlet/spin–triplet, splitting of the states. Unfortunately, despite numerous and valiant efforts, no spin–singlet states have ever been successfully identified in bottomonium, and we have to depend exclusively on charmonium. In charmonium the singlet S–state $\eta_c(1^1S_0)$ and the triplet S–state $J/\psi(1^3S_1)$ have been known for a long time, with the hyperfine splitting, $M(J/\psi) - M(\eta_c) = 117 \pm 1$ MeV. The other two bound singlet states η_c', or $\eta_c(2^1S_0)$, and $h_c(^1P_1)$ have long eluded successful identification. Well, now they have been finally discovered.

2.3.1. *The Radial Excitation of the Charmonium Ground State, η_c', or $\eta_c(2^1S_0)$*

The S-wave radial excitations, ψ' and η_c' sample the confinement potential almost exclusively. As a result, they provide our only insight into how the hyperfine interaction varies in going from $1S$ states (η_c and J/ψ) to $2S$ states (η_c' and ψ'), which

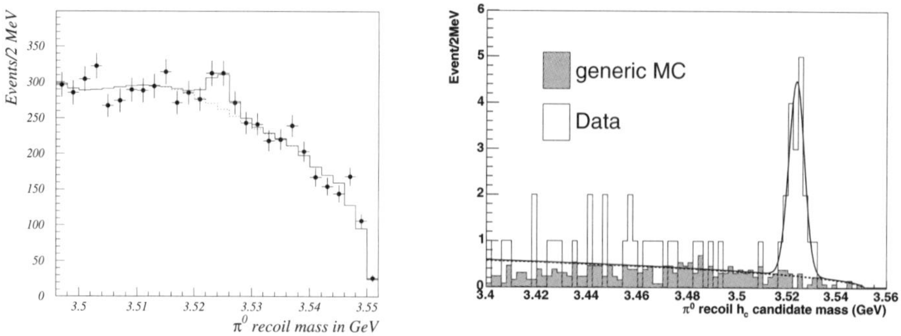

Fig. 5. Observation of $h_c(1^1P_1)$ in (left) inclusive analysis and (right) exclusive analysis at CLEO.

lie in the confinement potential territory.

The mass of ψ' is extremely well known, $M(\psi') = 3686.093 \pm 0.034$ MeV, but $M(\eta_c')$ was unknown. None of the earlier attempts to identify η_c' were successful. The breakthrough came from an unexpected source, Belle [14]. Since then several experiments have confirmed the identification of η_c'. Fig. 4 shows the results of CLEO [15] and BaBar [16] for the reaction $\gamma\gamma \to K_S K \pi$. Using the world average $M(\eta_c') = 3628.3 \pm 2.1$ MeV, we get $\Delta M_{hf}(2S) = 47.8 \pm 2.1$ MeV, which means that $\Delta M(2S)/\Delta M(1S) = 0.41 \pm 0.02$. Most theoretical predictions were that $\Delta M(2S)/\Delta M(1S) \approx 0.65$. So, here we have a result for theory to digest.

2.3.2. *The Singlet P-wave State of Charmonium, $h_c(1^1P_1)$*

The singlet P–wave state $h_c(1^1P_1)$ relates to the question of whether or not there is a long range $\vec{s}_1 \cdot \vec{s}_2$ interaction in the $q\bar{q}$ system. In absence of such an interaction, the lowest order pQCD prediction is that the hyperfine, singlet–triplet splitting is finite only for S–wave states, and is identically zero for all higher L, i.e. $\Delta M_{hf}(L \neq 0) = 0$. In particular, $\Delta M_{hf}(1P) = \langle M(^3P_J)\rangle - M(^1P_1) = 0$. The centroid of the $^3P_J(J = 0, 1, 2)$ states is $\langle M(^3P_J)\rangle = 3525.36 \pm 0.06$ MeV[1]. What is needed is a firm identification of $h_c(1^1P_1)$ and a precise measurement of its mass. CLEO [17] has just reported the unambigous (6σ) observation of h_c in a "tour de force" measurement of the reaction $\psi(2S) \to \pi^0 h_c$, $h_c \to \gamma\eta_c$. Both inclusive and exclusive measurements of $M(h_c)$ in recoils against π^0 have been made. The inclusive and exclusive spectra are shown in Fig. 5. The result is $\Delta M_{hf}(1P) = +1.0 \pm 0.6 \pm 0.4$ MeV.

Two conclusions can be derived from these measurements. The first is that the "naive" pQCD predication $\Delta M_{hf}(1P) = 0$ is not being violated in any substantial manner, as had been feared by many theorists. The second conclusion is that the all–important sign and the small magnitude of $\Delta M_{hf}(1P)$ is not yet pinned down. In the near future, larger ψ' running at CLEO is expected to reduce the errors in the CLEO measurement.

3. The Surprising States

Let me move on to the unexpected states which seem to be popping up all over during the last two years. The veterans among these are X(3872) and the Pentaquark. The newcomers are X(3943), Y(3943), Z(3931), and V(4260). This proliferation is exciting, but also rather baffling. It arises primarily from the fact that huge integrated e^+e^- luminosities ($\geq$ 300 fb^{-1}) are now available at Belle and BaBar, and very weak resonances are showing up. It will be a while before the dust settles down and we really know what is going on.

3.1. *The Sad Story of the Pentaquark*

As is probably well known to most of the audience, when the Pentaquark was born in 2003, it caused great interest. Google tells me that there are 31,300 entries for it by now. There were many reported sightings of pentaquarks of all kinds, and even a greater number of reported failures to find the expected signals. Finally, there is the recent JLab report of the absence of the pentaquark signal in a large statistics repeat of their earlier measurement. This reminds me of a similar history of claimed observations of dozens of dibaryons, which all eventually evaporated [18]. My personal, perhaps biased, conclusion is that the pentaquark is barely alive now.

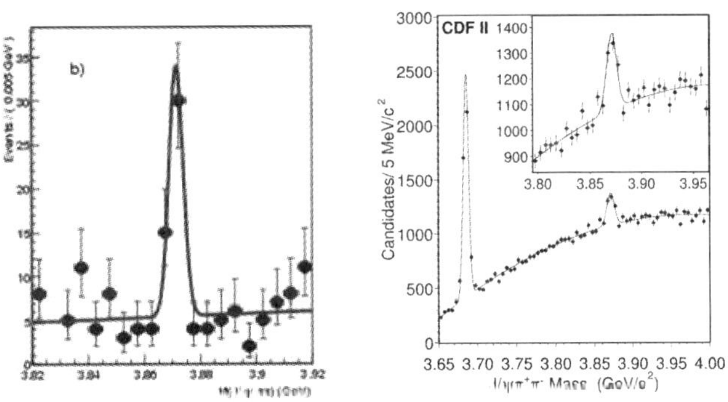

(a) (b)

Fig. 6. Observations of X(3872) by (a) Belle, (b) CDF.

3.2. *The Mystery of X(3872)*

Belle [19] recently announced the discovery of an unexpected state, X(3872). It was quickly confirmed by CDF [20], DØ [21] and BaBar [22] (see Fig. 6). The decay X(3872)$\rightarrow$ $\pi^+\pi^- J/\psi$ is the dominant decay. The average of the masses measured by the four

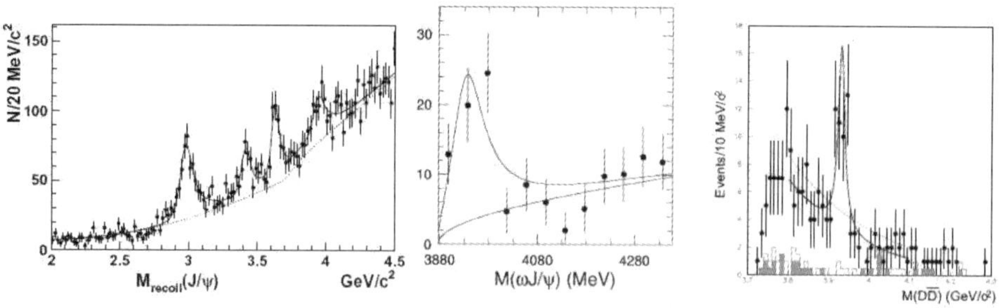

(a) X(3943) (b) Y(3943) (c) Z(3931)

Fig. 7. Observations of the three new states with masses near 3940 MeV by Belle.

Table 3. Summary of properties of three new states observed by Belle.

	M(MeV)	Γ(MeV)	Formed in	Decays in	not in	suggests
X [28]	$3943 \pm 6 \pm 6$	15 ± 10	$e^+e^- \to J/\psi(c\bar{c})$	$D^*\overline{D}$	$D\overline{D}$, $\omega J/\psi$	?
Y [29]	$3943 \pm 11 \pm 13$	87 ± 22	$B \to K(\omega J/\psi)$	$\omega J/\psi$	$D^*\overline{D}$(?)	$c\bar{c}$ hybrid?
Z [30]	$3931 \pm 4 \pm 2$	$20 \pm 8 \pm 3$	$\gamma\gamma$ fusion	$D\overline{D}$		$\chi'_{c2}(2^3P_2)$

experiments is $M(\mathrm{X}) = 3871.5 \pm 0.4$ MeV. Note that $M(D^0)+M(\overline{D^{*0}}) = 3870.3 \pm 2.0$ MeV [6]. The best measurement of the width gives $\Gamma(\mathrm{X}) \leq 2.3$ MeV.

The unique decay, the narrow width, and the closeness of its mass to $M(D^0\overline{D^{*0}})$ have given rise to intense theoretical speculations about the nature of X(3872). Theoretical speculations are that X(3872) is a charmonium state (1^{++}, 2^{--}, 3^{--}), a hybrid (1^{++}), a glueball mixed with vector charmonium (1^{--}), a $D^0\overline{D^{*0}}$ (1^{++}, 0^{-+}) molecule. Perhaps the most provocative of these proposals is the molecular model, because no $|q\bar{q}> \cdot |q\bar{q}>$ molecules have ever been found! To sift through these speculations it is necessary to determine $J^{PC}(\mathrm{X})$. Frantic searches are in progress at Belle and BaBar for decays to establish $J^{PC}(\mathrm{X})$.

CLEO [24], BES [25], and Babar [26] have put the limit, $\mathcal{B}(\mathrm{X} \to \pi^+\pi^- J/\psi)\Gamma(\mathrm{X} \to e^+e^-) < 6$ eV, which might weigh in against X being a vector. However, DØ finds $\mathrm{X} \to \pi^+\pi^- J/\psi$ decays to have all the same characteristics as the vector $\psi(2S)$ decays to $\pi^+\pi^- J/\psi$. Belle [27] has presented arguments against the charge parity of X being negative on the basis of observations of $\mathrm{X} \to \gamma J/\psi$ and $X \to \omega J/\psi$. However the two observations consist of 13.6 ± 4.4 and 12.1 ± 4.1 counts (each in one mass bin), respectively. CDF [20] reports the $\pi^+\pi^-$ mass distribution to be consistent with X being either a $^3S_1$ vector with $C = -1$, or a $C = +1$ object decaying into $\rho J/\psi$. Finally, potential model calculations do not really rule out $^3D_2(2^{--})$ and $^3D_3(3^{--})$ states at 3872 MeV. In view of all this, I personally believe that all options are still open for X(3872).

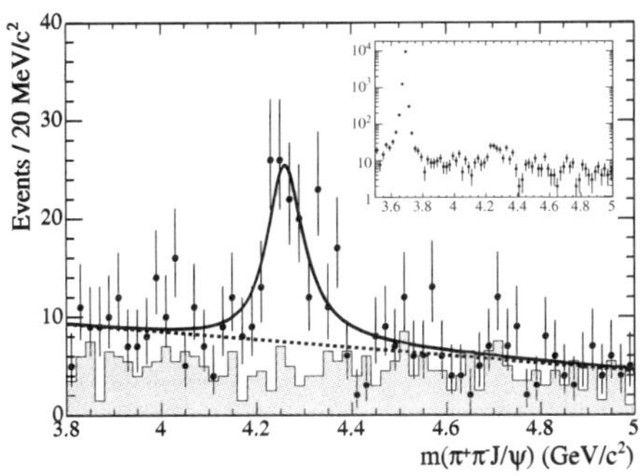

Fig. 8. Observation by Babar of V(4260) enhancement in ISR.

3.3. *The X, Y, Z States at M ~ 3940 MeV, and the V(4260)*

There are three new states reported by Belle with masses which are statistically consistent with being identical [28–30]. The spectra in which they were observed are shown in Fig. 7, and their characteristics are summarized in Table 3. Each is formed in a different reaction and decays dominantly in a different channel. X(3943) is observed in a recoil mass spectrum, in which only $J = 0^{\pm}$, $C = +$ states, $\eta_c(0^{-+})$, $\eta_c'(0^{-+})$, $\chi_{c0}(0^{++})$, are seen. This would suggest $J(\text{X}(3943)) = 0^{\pm}$, $C = +$. The state Z(3931) with width similar to that of X(3943) is produced in $\gamma\gamma$ fusion, which guarentees $C = +$, but is found to have the $D(\bar{D})$ angular distribution characteristic of $J(\text{Z}(3931)) = 2$. The fitted width of Y(3942) is claimed to be ~ 4 times larger than that of X or Z, and its decays are almost opposite to those of X(3943). Despite these differences, it appears to be rather implausible that three distinct states exist within a 10 MeV mass interval. Also, BaBar has yet to weigh in on this story.

BaBar has analyzed ISR events from 211 fb^{-1} of data [26], and reports a broad enhancement in the invariant mass $M(\pi^+\pi^- J/\psi)$ spectrum (Fig. 8). Since production via ISR guarentees a vector, and X, Y, and Z have been overused, I take the liberty of christening this state as V(4260). The parameters of this enhancement are $M(\text{V}) = 4259 \pm 8^{+2}_{-6}$ MeV, $\Gamma(\text{V}) = 88 \pm 23^{+6}_{-4}$ MeV, $N = 125 \pm 23$ events. They suggest that it might be a previously unobserved 1^{--} resonance. This is quite surprising because no vector around this mass is predicted, and the R measurements actually show a dip in this mass region, as illustrated in Fig. 3 (right) [11].

4. Bottomonium

The Bottomonium spectrum is shown in Fig. 9.

The world's largest sample of 21 million $\Upsilon(1S)$, 9 million $\Upsilon(2S)$, and 6 million $\Upsilon(3S)$ comes from CLEO. These data sets have been analyzed to yield interesting new results and improved precision.

- For the first time, a non-$\pi\pi$ hadron transition between bottomonium resonaces

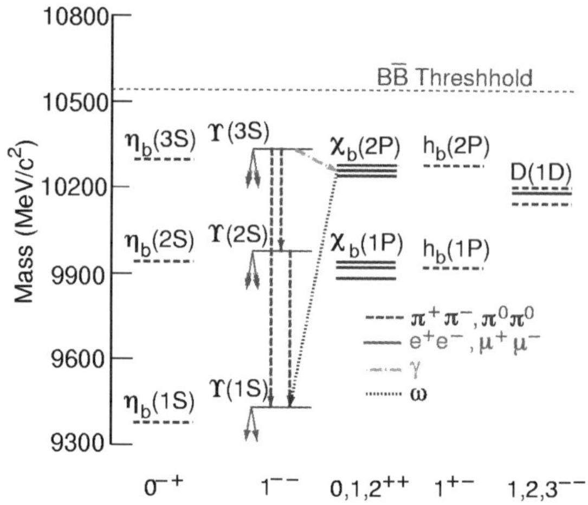

Fig. 9. Spectra of the bound states of Bottomonium.

has been observed [31], with $\mathcal{B}(\chi'_{b1} \to \omega\Upsilon(1S)) = (1.63 \pm 0.38)\%$ and $\mathcal{B}(\chi'_{b2} \to \omega\Upsilon(1S)) = (1.10 \pm 0.34)\%$.

- The 1^3D_2 state of bottomonium has been identified in a 4-step cascade with $M = 10161.1 \pm 0.6 \pm 1.6$ MeV [32].
- $\Upsilon(1S)$ decays to $X + (J/\psi, \psi(2S), \chi_{c1}, \chi_{c2})$ have been measured [33]. The measured branching fraction, $B(\Upsilon(1S) \to J/\psi + X) = (6.4 \pm 0.4 \pm 0.6) \times 10^{-4}$. The measured branching fraction ratios to $B(\Upsilon(1S) \to J/\psi + X)$ for $\Upsilon(1S) \to (\psi(2S), \chi_{c1}, \chi_{c2}) + X$ are $0.41 \pm 0.11 \pm 0.08$, $0.35 \pm 0.08 \pm 0.06$, and $0.52 \pm 0.12 \pm 0.09$, respectively.
- Precision measurements of $\Upsilon(1S, 2S, 3S) \to \mu^+\mu^-$ have been made, with the result that $\mathcal{B}(\Upsilon(2S) \to \mu^+\mu^-)$ and $\mathcal{B}(\Upsilon(3S) \to \mu^+\mu^-)$ are 56% and 32% larger, respectively, than their current PDG values [34].
- The radiative decays of bottomonium 1P and 2P states have been measured with improved precision [35]. While the branching ratios for $\chi_{bJ}(1P)$ states are found to be in good agreement with the current PDG values, those for $\chi_{bJ}(2P)$ are found to be $\sim 30 - 40\%$ larger.

Postscript for bottomonium: No new data taking at the bound bottomonium resonances is expected at any e^+e^- collider, although much remains to be explored. As an example, none of the spin singlet states of bottomonium, not even the ground state $\eta_b(1S)$, have been identified so far.

5. The Timelike Form Factors of Pion, Kaon, and Proton

In recognition of the interests of this conference in systems of light quarks, let me close with report of a unique measurement of form factors.

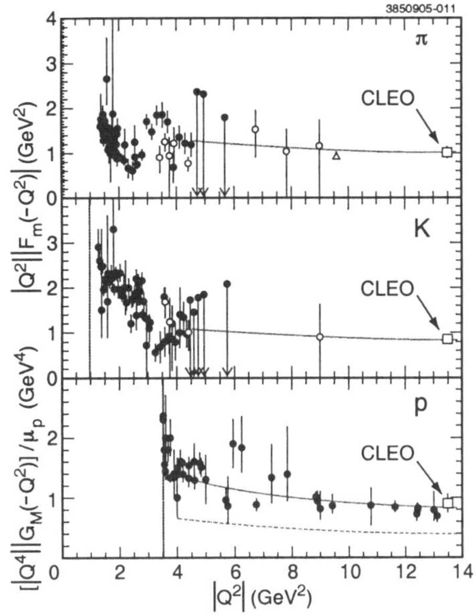

Fig. 10. CLEO results for the timelike form factors at $Q^2 = 13.48$ GeV^2 of the pion (top), kaon (middle), and proton (bottom).

Electromagnetic form factors of hadrons provide deep insight into their quark structure and help define the domain of validity of pQCD. Except for the magnetic form factor of the proton for spacelike momentum transfers, few measurements of spacelike or timelike form factors of any hadrons exist at large enough momentum transfers to shed light on the highly controversial debate on the validity of pQCD at modestly large momentum transfers. Unfortunately, the sparse data which exist for the form factors of pions and kaons (which often could not be separately identified) are essentially limited to $Q^2 \leq 4$ GeV, and have very large (up to 100%) errors. This situation has been remedied recently by a very demanding measurement made at CLEO [36] for $Q^2 = 13.48$ GeV^2. To get an idea of how difficult these measurements are, it is enough to point out that at this momentum, e^+e^- collisions produce about 500 times more muon pairs than pion pairs, and one must distinguish between them. The results of this CLEO measurement with less than $\pm 10\%$ errors are shown in Fig. 10.

References

1. E835 Collaboration, Nucl. Phys. **B 717** (2005) 34.
2. KEDR Collaboration, Phys. Lett. **B 573** (2003) 63.
3. CLEO Collaboration, Phys. Rev. **D 71** (2005) 111103(R).
4. CLEO Collaboration, Phys. Rev. Lett. **94** (2005) 232002.
5. CLEO Collaboration, Phys. Rev. **D 70** (2004) 112002.
6. Belle Collaboration, Phys. Lett. **B 540** (2002) 33.

7. E835 Collaboration, Phys. Rev. **D 62** (2000) 052002.
8. CLEO Collaboration, submitted to Phys. Rev. Lett.
9. BES Collaboration, Phys. Rev. **D 67** (2004) 072004.
10. CLEO Collaboration, Phys. Rev. Lett., **94** (2005) 012005; Phys. Rev. Lett., **95** (2005) 062001.
11. K. K. Seth, Phys. Rev. **D 72** (2005) 017501.
12. DASP Collaboration, Phys. Lett. **B 76** (1978) 361; Z. Physik **C1** (1979) 233.
13. BES Collaboration, Phys. Rev. Lett. **91** (2003) 022001; Phys. Rev. Lett. **93** (2004) 112002.
14. Belle Collaboration, Phys. Rev. Lett. **89** (2002) 102001; hep-ex/0507019.
15. CLEO Collaboration, Phys. Rev. Lett. **92** (2004) 142001.
16. BaBar Collaboration, Phys. Rev. Lett. **92** (2004) 142002; Phys. Rev. **D 72** (2005) 031101.
17. CLEO Collaboration, Phys. Rev. Lett. **95** (2005) 102003.
18. "Dibaryons in Theory and Practice", Kamal K. Seth, Invited Paper International Conference on Medium- and High-Energy Nuclear Physics, Taipei 1988.
19. Belle Collaboration, Phys. Rev. Lett. **91** (2003) 262001.
20. CDF Collaboration, Phys. Rev. Lett. **93** (2004) 072001.
21. DØCollaboration, Phys. Rev. Lett. **93** (2004) 162002.
22. BaBar Collaboration, Phys. Rev. **D 71** (2005) 071103.
23. PDG Collaboration, Phys. Lett. **B 592** (2004) 1.
24. CLEO Collaboration, Phys. Rev. Lett. **94** (2005) 032004.
25. BES Collaboration, Phys. Lett. **B 579** (2004) 74.
26. BaBar Collaboration, Phys.Rev.Lett. 95 (2005) 142001.
27. Belle Collaboration, hep-ex/0505037; hep-ex/0505038.
28. Belle Collaboration, hep-ex/0507019.
29. Belle Collaboration, Phys. Rev. Lett. 94 (2005) 182002.
30. Belle Collaboration, hep-ex/0507033.
31. CLEO Collaboration, Phys. Rev. Lett **92** (2004) 222002.
32. CLEO Collaboration, Phys. Rev. **D 70** (2004) 032001.
33. CLEO Collaboration, Phys. Rev. **D 70** (2004) 072001.
34. CLEO Collaboration, Phys. Rev. Lett **94** (2005) 012001.
35. CLEO Collaboration, Phys. Rev. Lett **94** (2005) 032001.
36. CLEO Collaboration, hep-ex/0510005, submitted to Phys. Rev. Lett.

The Perturbative Chiral Quark Model and Hadron Properties

A. Faessler

Institute of Theoretical Physics
University of Tuebingen
Auf der Morgenstelle 14
72076 Tuebingen
Germany

Quantum Chromo-Dynamics (QCD) is due to the non-linearity in the gluon sector hard to solve exactly. But it has many exact and approximate symmetries which can be built in an effective Lagrangian to investigate low energy properties of hadrons. Chiral perturbation theory (χPT) eliminates gluon and quark degrees of freedom and is based on a chirally symmetric effective Lagrangian on the hadron level. We in Tuebingen [1,2,3] kept the quark degrees of freedom and formulated an effective Lagrangian on the quark level treating hadron properties. This perturbative chiral quark model (PχQM) has only two free parameters. The size of the small relativistic component of the quark wave functions is adjusted to the axial coupling constant $g_A = 1.25$ and a radius parameter R is used to reproduce the correct mean square charge radius of the proton. Here hadron properties are calculated within the $P\chi QM$.

1. Introduction

The Lagrangian of Quantum Chromo-Dynamics (QCD):

$$\mathcal{L}_{QCD} = \sum_{f=1}^{6} \bar{\varphi}_f (i\not{D} - m_f)\varphi_f - \frac{1}{4} F^a_{\mu\nu} F_a^{\mu\nu} \tag{1}$$

is very difficult to solve due to the non-linearities in the gluon sector. But the QCD Lagrangian has several exact and approximate symmetries which can be used to build an effective Lagrangian for low-energy properties.

If some quark masses are zero or very small as for the u, d and s quarks,

$$
\begin{aligned}
\mathcal{L}_{QCD} &= \mathcal{L}_L + \mathcal{L}_R & m_u &\approx 5\,MeV \sim 0 \\
q_L &= \frac{1}{2}\,(1-\gamma^5)\,q & m_d &\approx 9\,MeV \sim 0 \\
q_R &= \frac{1}{2}\,(1+\gamma^5)\,q & m_s &\approx 175\,MeV \sim 0 \\
\bar{\varphi}\varphi &= & \varphi^+\gamma^0 \cdot \varphi &= \bar{\varphi}_R\varphi_L + \bar{\varphi}_L\varphi_R
\end{aligned}
\qquad (2)
$$

chirality for these quarks is a good or almost a good quantum number. Left quarks stay left-handed and right quarks stay right-handed. But a finite mass term breaks this symmetry as indicated in the last line of equation (2). Left and right quarks transform separately.

$$
\begin{aligned}
\varphi_L &\to L_{\varphi_L} = e^{i\vec{\alpha}_L\cdot\vec{\tau}/2}\,\varphi_L \\
\varphi_R &\to R_{\varphi_R} = e^{i\vec{\alpha}_R\cdot\vec{\tau}/2}\,\varphi_R
\end{aligned}
\qquad (3)
$$

The mass term M gets also invariant if we introduce a chiral field U in the following way:

$$
\begin{aligned}
M &= U \begin{pmatrix} m_u & 0 & 0 \\ 0 & m_d & 0 \\ 0 & 0 & m_s \end{pmatrix} = U \cdot \hat{m} \\
U &\to LUR^+
\end{aligned}
\qquad (4)
$$

The chiral field with the required property can be constructed in the non-linear σ model.

$$
\mathrm{SU(2)}: \quad U = e^{i(\vec{\tau}\cdot\vec{\pi})/f_\pi(0)} \qquad (5)
$$

For flavor $SU(3)$, the chiral field $\hat{\pi}$ is given by the expression:

$$
\begin{aligned}
\hat{\pi} &= \sum_{i=1}^{8} \lambda_i \phi_i \\
\lambda_1 &= \begin{pmatrix} 0\,1\,0 \\ 1\,0\,0 \\ 0\,0\,0 \end{pmatrix} \\
\phi_{1/2} &= \frac{\pi^+ \pm \pi^-}{\sqrt{2}} \\
\phi_3 &= \pi^0 \\
\phi_{4/5} &= \frac{K^+ \pm K^-}{\sqrt{2}} \\
\phi_8 &= \eta_8
\end{aligned}
\tag{6}
$$

An expansion of the chiral fields up to second order gives the final Lagrangian.

$$
\begin{aligned}
\mathcal{L} &= \mathcal{L}_0(x) + \mathcal{L}_{int}(x) + 0(\pi^3) \\
\mathcal{L}_0 &= \bar{\varphi}_q(x)\left\{i\partial\!\!\!/ - S(r) - \gamma^0 V(r)\right\}\varphi_q - \frac{1}{2}\vec{\pi}\left\{\Box + m_\pi^2\right\}\vec{\pi}(x) \\
\mathcal{L}_{int} &= -\frac{1}{f_\pi} S(r)\bar{\varphi} i\gamma^5 \left(\vec{\tau}\cdot\vec{\pi}\right)\varphi_q + \frac{1}{2f_\pi^2} S(r)\bar{\varphi}_q \vec{\pi}\cdot\vec{\pi}\varphi_q
\end{aligned}
\tag{7}
$$

The quark fields ϕ_q are quantized to yield particle annihilation and antiparticle creation operators with the corresponding four-component spinors.

This Lagrangian fulfills the Gell-Mann-Oaks-Renner and Gell-Mann-Okubo relations from current algebra.

$$
\begin{aligned}
m_\pi^2 &= 2\hat{m}B \\
m_K^2 &= (\hat{m} + m_s)\,B \\
m_\eta^2 &= \frac{2}{3}(\hat{m} + 2m_s)\,B \\
3m_q^2 + m_\pi^2 &= 4m_K^2
\end{aligned}
\tag{8}
$$

The u and d quark masses are averaged to be $\hat{m} = 7\,MeV$ and the strange quark mass is $m_s = 175\,MeV$.

The quark condensate B is obtained from the Gell-Mann-Oaks-Renner relation (9).

$$\begin{aligned}\hat{m} &= \frac{1}{2}(m_u + m_d) = 7\,MeV \\ m_s/\hat{m} &= 25\,MeV \\ m_s &= 175\,MeV \\ B &= -\frac{\langle 0|\bar{u}u|0\rangle}{f_\pi^2} = \frac{m_\pi^2}{2\hat{m}} = 1.4\,[GeV]\end{aligned} \tag{9}$$

The relativistic quark wave function:

$$u_0(\vec{r}) = N_{exp}\left[-\frac{r^2}{2R^2}\right]\begin{pmatrix} 1 \\ i\varrho\frac{\vec{\sigma}\cdot\vec{r}}{R}\end{pmatrix}\chi_s\chi_f\chi_c \tag{10}$$

can be obtained solving the Dirac equation with a scalar $S(r)$ a vector $V(r)$ confinement harmonic oscillator potential. The quark wave function contains two parameters. ρ gives the size of the small part of the Dirac spinor. It is adjusted to the axial coupling constant $g_A = 1.25$ and one obtains $\rho = (2/13)^{1/2}$. The radius parameter R is adjusted to reproduce the mean square charge radius of the proton.

2. The pion-nucleon σ term

The pion-nucleon σ term can be described on the quark level by the quark expectation value in the nucleon times the average mass of the u and d quarks.

$$\begin{aligned}\sigma^{QCD}{}_{\pi N} &= \hat{m}\,\langle P|\bar{u}u + \bar{d}d|P\rangle \\ &= \hat{m}\,\frac{\partial}{\partial\hat{m}}\,\langle P|\mathcal{H}^{\chi SB}_{QCD} = \hat{m}(\bar{u}u + \bar{d}d)|P\rangle\end{aligned} \tag{11}$$

We have also calculated the scalar formfactor of the nucleon. At small momentum transfer it is dominated by the meson cloud while above $Q^2\,[GeV^2] = 0.22$ the valence quarks get dominating.

3. Electromagnetic properties of baryons

The electromagnetic properties of baryons are calculated [8] according to fig. 4.

The results for the magnetic moments of the protons and the neutrons in units of nuclear magnetons for the electric and magnetic mean square radii of the protons and the neutrons in units of fm^2 are listed in table 2.

4. Strangeness in the nucleon

In the perturbative chiral quark model, strangeness is introduced into nucleons by the diagrams shown in fig. 2.

Table 1. Meson nucleon σ term: the first row gives the pion-nucleon σ term $\sigma_{\pi N}$ decomposed into the contributions from the valence quarks, the pion, the kaon and the η cloud. The total value is 55 MeV. This has to be compared with chiral perturbation theory χPT of Leutwhyler and Gasser of 45 $\pm$ 8 MeV. The probable experimental value lies closer to 60 MeV than to 45 MeV. The quantity Δ_σ gives the difference of the pion-nucleon σ-term at twice the restmass of the pion squared minus the value at zero energy. The third row gives the scalar mean square radius. The main contribution comes from the pion cloud. The table gives further the kaon-nucleon σ-term for isospin $I = 0$ and $I = 1$ and the eta-nucleon σ-term.

	$3q$	π	K	η	Total	χPT
$\sigma_{\pi N}$ $[MeV]$	13	39	2.1	0.1	55	45(8)
Δ_σ $[MeV]$	1	12	.3	.02	14	15(.4)
$\langle r^2\rangle^{scalar}_{\pi N}$ $[fm^2]$	.1	1.4	.04	.002	1.5	1.6
$\sigma^{I=0}_{KN}$ $[MeV]$	85	256	40	4.5	386	395
$\sigma^{I=1}_{KN}$ $[MeV]$	28	0	4.5	0	33	
$\sigma_{\eta N}$ $[MeV]$	4	13	69	9.4	96	

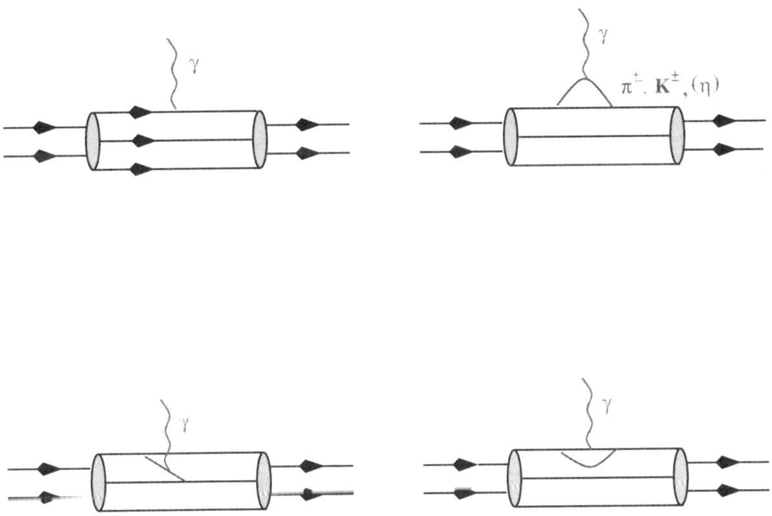

Fig. 1. Interaction of electromagnetic field with baryons. Photon can interact directly with one of the valence quarks or with the meson cloud. The meson can be reabsorbed at the same quark or at a different valence quark. Furthermore the photon can interact with the valence quark while a pseudo-scalar meson is exchanged.

The strange magnetic moment defined by the proton magnetic strange formfactor at zero momentum transfer and the electric and magnetic mean square radii defined by the derivative with respect to the transferred momentum squared at zero with the minus sign and the factor six are given for the perturbative chiral quark

Table 2. Magnetic moments of protons and neutrons in units of nuclear magnetons and the electric E and magnetic M mean square radii of the protons p and neutrons n in units of fm^2.

	$3q$	loops	Total	Exp.
μ_p	1.8	0.80	2.60	2.79
μ_n	-1.2	-0.78	-1.98	-1.91
$\langle r_E^2 \rangle^p$	0.60	0.12	0.72	0.76
$\langle r_E^2 \rangle^n$	0	-.111	-.111	-.116
$\langle r_M^2 \rangle^p$	0.37	0.37	0.74	0.74
$\langle r_M^2 \rangle^n$	0.33	0.61	1.89	1.61

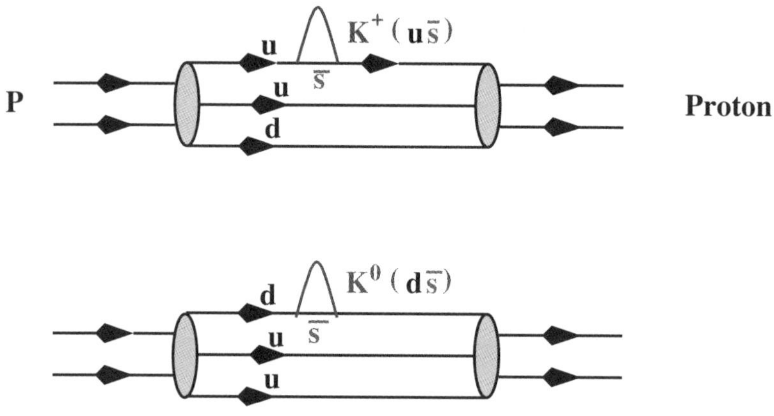

Fig. 2. Diagrams for the strangeness in the proton. Strangeness is admixed by the K-meson cloud.

model and different other approaches in table 3.

Strange formfactors measured by the proton and the deuterium have been performed at BATES (SAMPLE) [11], at JeffersonLab (HAPPEX) [10] and in Mainz (PVA4) [9].

$$G_M^S\,(Q^2 = 0.1\,GeV^2/c^2) = 0.23 \pm 0.76 \tag{12}$$

$$G_E^S\,(0.48\,GeV^2/c^2) + 0.39\,G_M^S = 0.025 \pm 0.034 \tag{13}$$

$$G^S\,(0.23) = G_E^S\,(Q^2 = 0.23\,GeV^2/c^2) + 0.25\,G_M^S \tag{14}$$

$$G^S\,(0.1) = G_E^S\,(Q^2 = 0.108\,GeV^2/c^2) + 0.106\,G_M^S = 0.071 \pm 0.036 \tag{15}$$

Table 3. Strange magnetic moment of the proton in units of nuclear magnetons in the perturbative chiral quark model (PχQM) and in different approaches. Leinweber I and Leinweber II (published after our result) are lattice QCD calculations where the second line is the more reliable one as seen from the error bar. The table contains also lattice QCD results of Dong, chiral perturbation theory (χPT) of Meissner et al., Nambu-Jona-Lasino results of Weigel, chiral quark soliton model results of Goeke and collaborators and results from the constituent quark model by Riska.

Approach	$\mu^s[n.m.]$	$\langle r^2\rangle^s_E[fm^2]$	$\langle r^2\rangle^s_M[fm^2]$
QCD Leinweber I	-0.16 (0.18)		
QCD Leinweber II	-0.051 (0.021)		
QCD Dong	-0.36 (0.20)	-0.16 (0.20)	
χPT Meissner	0.18 (0.34)	0.05 (0.09)	-0.14
NJL Weigel	0.10 (0.15)	-0.15 (0.05)	
χQSM Goeke	0.115	-0.095	0.073
CQM Riska	-0.046		$\sim$ 0.02
PχQM	-0.048 (0.012)	-0.011 (0.003)	0.024 (.003)

Table 4. Strange formfactors in the combinations given in eq. (12) to (15) for the electric and magnetic strange formfactors $G_E^S + \eta\, G_M^S$. The table shows the result of the Tuebingen perturbative chiral quark model $P\chi QM$ and results of chiral perturbation theory (χPT) of Meissner and co-workers, of Goeke and co-workers in the chiral quark soliton model, of Riska and co-workers in the constituent quark model and the experimental data in the last line.

Approach $Q^2[GeV^2/c^2]$	Gs (0.1) SAMP §	Gs (0.48) HAPP §	Gs (0.23) Mainz*
χPT Meissner	0.023 (0.44)	0.023 fit (0.048)	0.007 (0.127)
Skyrme Goeke	0.09	0.087 (0.016)	0.14 (0.03)
Riska	-0.06	-0.08	
$P\chi QM$	-0.04 (0.01)	0.0018 (.0003)	0.00029 (.00005)
EXP	0.23 § (0.76)	.025 § (.034)	*

The measured formfactors for the different momentum transfers Q^2 $[GeV^2/c^2]$ are given in table 4. The theoretical and experimental values are given as linear combinations of the electric and the magnetic strange formfactors of the nucleon as indicated in eq. (12) to (14).

5. Electric and magnetic nucleon polarizabilities

The polarizabilities of the protons and the neutrons are measured and calculated by Compton scattering of photons on nucleons.

6. Conclusion

The starting point of our consideration was the QCD Lagrangian with several exact and approximate symmetries which we build into an effective Lagrangian of the

Table 5. Electric α and magnetic β polarizabilities for the protons $\alpha(pE)$, $\beta(pM)$ and for the neutrons $\alpha(n,E)$, $\beta(n,M)$. The first row lists the experimental value given in the review article by Schumacher in units of $10^{-4}\,fm^3$. The last line gives in the same units the result of the perturbative chiral quark model (PχQM) from Tuebingen while the other results of different chiral perturbation theories (χPT) by Meissner et al., Babusci et al., Hemmert et al. and by Lvov. χPT has several free parameters while the perturbative chiral quark model determines the two free parameters by the charge radius of the protons and the axial coupling constant g_A so that the results are completely parameter free.

	$\alpha(p,E)$	$\beta(p,M)$	$\alpha(n,E)$	$\beta(n,M)$
DATA 10**(−4) fm^3 Schumacher	12.0 (0.6)	1.9 (0.6)	12.5 (1.7)	2.7 (1.8)
χPT Meissner	7.9	-2.3	11.0	-2.0
χPT Babusci	10.5 (2.0)	3.5 (3.6)	13.6 (1.5)	7.8 (3.6)
χPT Hemmert	12.6	1.26	12.6	1.26
χPT Lvov	7.3	-1.8	9.8	-0.9
PχQM Tuebingen	10.9	5.1	10.9	1.15

"Perturbative Chiral Quark Model (PχQM)".

In the PχQM developped in Tuebingen, we eliminate gluons but work on the level of quarks. The Lagrangian of the quarks is invariant concerning the kinetic energy with respect to separate unitary rotations among the left-handed and the right-handed quarks. To guarantee also chiral invariance for scalar terms, one introduces a chiral field U. This field has the property that it transforms on the left by the unitary left-handed rotation and on the right by the unitary right-handed rotation. This guarantees also for scalar terms chiral invariance. Such a chiral field U can be constructed in the non-linear σ model.

One adds in addition a scalar and vector confinement field.

The chiral fields are then expanded up to second order. This yields interaction terms between the quarks and the chiral pseudo-scalar meson fields.

Within this Tuebingen perturbative chiral quark model, the pion-nucleon σ term, electromagnetic properties of the nucleon and the electric and the magnetic spin independent polarizabilities of the protons and the neutrons are studied. Although the model has only two free parameters it yields equally good agreement as the chiral perturbation theories with more quantities to be adjusted.

I would like to thank Thomas Gutsche, Valery Lyubovitskij and Yupeng Yan for their collaboration on the above problems. Further contributions to the above results are due to Dr. Dong and Prof. Shen from the High Energy Institute of the Academia Sinica in Beijing and by the PhD students Kuckei, Cheedket, Pumsa-ard,

Khosonthongkee, Giacosa and Nicmorus.

References

1. V.Lyubovitskij, T.Gutsche and A.Faessler, *Phys. Rev.* **C64**, 065203 (2001)
2. T.Inoue, V.Lyubovitskij, T.Gutsche and A.Faessler, hep-ph/0404051
3. K.Khosongthongkee, V.Lyubovitskij, T.Gutsche, A.Faessler, K.Pumsa-ard, S.Cheedket and Y.Yan, *Phys.* **G30** 793 (2004)
4. S.Cheedket, V.Lyubovitskij, T.Gutsche, A.Faessler, K.Pumsa-ard and Y.Yan, *Eur. Phys. J.* **A20** 317 (2004)
5. S.Weinberg, *Physica* **A96** 327 (1979)
6. T.Inoue, V.Lyubovitskij, T.Gutsche and A.Faessler, *Phys. Rev.* **C69** 035207 (2004)
7. J.Gasser and H.Leutwhyler, *Phys. Rept.* **87** 779 (1982)
8. T. Gutsche, V.Lyubovitskij, A. Faessler et al., *Phys. Rev.* **C68** 015205 (2003) AND *Phys. Rev.* **C69** 035207 (2004)
9. K.Pumsa-ard, V.Lyubovitskij, T.Gutsche, A.Faessler and S.Cheedket, *Phys. Rev.* **C68** 015205 (2003)
10. S.Kox, these Proceedings
11. E.J.Beise et al., *Prog. Part. Nucl. Phys.* **54** 289 (2005)
12. F.E.Maas et al., *Phys. Rev. Lett.* **94** 152001 (2005)
13. M. Schumacher, *Prog. Part. Nucl. Phys.*, to be published
14. Y.Dong, A.Faessler, T.Gutsche, J.Kuckei, V.Lyubovitskij, K.Pumsa-ard and P.Shen, hep-ph/0507277 to be published

Multi-Quark Study*

Fan Wang

Center for Theoretical Physics, Nanjing University, Nanjing 210093, P.R. China

Jialun Ping

Physics Department, Nanjing Normal University, Nanjing, 210097 P.R. China

The Quark Delocalization, Color Screening Model has successfully been applied to 6-quark systems to describe baryon-baryon potentials, phase shifts and deuteron properties. It is applied here to $q^3-q\bar{q}$, $qq-qq-\bar{s}$ (Jaffe-Wilczek), $qq-qq\bar{s}$ (Karliner-Lipkin) and tetrahedral structures proposed for the reported pentaquark state, Θ^+. In this model, the negative parity state is always found at a lower mass than the positive parity one. It is possible to obtain the mass of pentaquark close to the observation for the Jaffe-Wilczek and tetrahedral structure with the correction for the physical masses of kaon. The results depend largely on how to model the interaction among the quarks with different color structures.

1. Introduction

The success of quark models: Isgur-Karl's gluon exchange model, Glozman-Riska's Goldstone boson exchange model, Bonn instanton exchange model etc. on hadron spectroscopy and baryon-baryon interaction (where two color singlet clusters are good approximation) tell us that the constituent quark might be a good effective degree of freedom for low energy QCD. The wavefunctions and hamiltonian used are good approximations for baryon (q^3) and meson ($q\bar{q}$). However no any prediction about multi-quark state based on the above models has been verified so far. It suggests that the hamiltonian and/or wavefunctions may not be a good approximation for multi-quark system, due to the abundance of color structure of multi-quark system.

The "discovery" of pentaquark [1] really provide a good chance to test our understanding of the low energy behavior of QCD. So far no model with the constraint that given the hadron spectroscopy and baryon-baryon interaction can obtain the mass of pentaquark Θ^+ consistent with the observation [2]. If the pentaquark is confirmed, then every model should be modified accordingly. Even the pentaquark disappears, the model still has to answer the question, why is there no multi-quark system in nature except the deuteron, which is a rather weak bound state of two nucleon than six-quark system?

*This work is supported by National Science Foundation of China

Comparing to the unique color structure of q^3 and $q\bar{q}$ system, the multi-quark system has more color structures. Whether the two-body interactions, which employed in the constituent quark models and worked well, can be used here is an open question. Generally the multi-quark calculation is an multi-body interaction and multi-channel coupling one. It would be quite involved. To find a way to make the calculation trackable is an important object of constituent quark model. Quark delocalization, color screening model was developed which aims at multi-quark system. First, the quark-delocalization (similar to the percolation of electrons in atoms) was introduced to take into account the contribution of orbital excitation. Secondly, different parametrization of the confinement interaction is assumed for the quark pairs in different states. The parametrization is trying to account for the various color structure for the multi-quark system. The main advantage of QDCSM is that it allows the multi-quark system to choose its most favorable configuration (by variation the energy of the system to delocalization parameter, which have the meaning that the self-consistent of quark distribution and gluon distribution is taken into account) through its own dynamics. This model reproduces the existing baryon-baryon (bound state deuteron and NN, $N\lambda$, $N\Sigma$ scattering) interaction data well [3,4]. It is very interesting to apply QDCSM to study the pentaquark system. Some generalizations are needed here: the quark can delocalize among clusters and the color confinement between quarks in different clusters is screened. The cluster may be colorless or colorful.

2. QDCSM and Pentaquark

The detail of QDCSM can be found in ref.[3]. The parameters are determined by baryon spectroscopy and deuteron properties. The calculation method is also given in ref.[4]. In the following we only give a our results for different configurations.

2.1. q^3–$q\bar{q}$ *configuration*

The channels included were KN, K^*N, K_8N$_8$, K_8^*N$_8$ and K_8^*N$_8^*$, where the "8" indicates the quarks were coupled to a color octet instead of a singlet in that cluster ("hidden color" combinations).

Under the adiabatic approximation, the effective potential between a meson containing a strange antiquark and a nucleon is calculated. The effective potential at separation S_0 is defined as

$$V_{eff}(\vec{S_0}) = E(S_0) - E(\infty). \tag{1}$$

If there is a resonance, the mass can be obtained [3]:

$$M_{q^5} = M_{q^3} + M_{q\bar{q}} + V_{eff} + \frac{3\hbar}{4\mu R_0^2}. \tag{2}$$

The lowest energy state here is KN state. The effective potential of isospin 1 KN state is fully repulsive in the naive quark model and QDCSM. For $I = 0$ S-wave KN

state, although the naive quark model gives repulsive potential, there is a rather strong attraction ~ -70 MeV in QDCSM. This result contradicts to the former phase-shift analysis of KN scattering [5]. However it is compatible with the latest analysis of experimental data [6]. By taking into account of the double scattering of K^+ and the neutron Fermi momentum in the deuteron, which were ignored in the ref.[5], the new analysis showed that the phase-shifts are positive around $P_{lab} = 0.6$ GeV/c. We know the phase-shifts for $I = 0$ are obtained from K^+p and K^+n scattering, whereas the phase-shifts of K^+n are derived from K^+d scattering. To get reliable K^+n scattering amplitude from K^+d scattering data is very hard because it is a small component in comparison to the big K^+p amplitude.

We find a state energy of 1615 MeV, which includes a zero point energy (~ 250 MeV) of the fluctuations in separation between the two clusters. The delocalization is complete at the minimum energy and the quark wavefunctions are symmetric between the two orbitals. However it is about 100 MeV above the reported mass of the Θ^+.

In the P-wave (for the strange antiquark), there is again negligible attraction for $\mu = 0$, and the effective potential develops a $\sim$ 180 MeV minimum for for $\mu = 1$ fm^{-2}. However, this occurs at a hadron separation of about 0.4 fm and this much smaller distance effectively prevents delocalization from producing significant reduction of the energy of the state. The mass of this positive parity state is $\sim$ 1820 MeV, and so is too high to match with experiment even taking a threshold correction into account.

Similar results are found with the strange quark mass adjusted in the meson to give the correct mass for the kaon.

2.2. *qq–qq-$\bar{s}$ (Jaffe-Wilczek) configuration*

In this configuration, the five quarks are separated into three clusters and form an isosceles triangle with the two strongly correlated pairs of u, d sitting at the bottom corners with separation S and the $\bar{s}$ at top with the height T. The two diquarks can each be in several spatial-color-isospin-spin states, which we label by integers, $i = 1 - 4$:

$$1 \equiv [2][2]10; \quad 2 \equiv [2][2]01; \quad 3 \equiv [2][11]11; \quad 4 \equiv [2][11]00;$$

consistent with Pauli statistics. There are then seven four-quark combinations also shown in Table 1, where $i = 1, 2, 3, 4$ and the bar indicate symmetrization and the tilde represents antisymmetrization.

Table 1. Quantum number correlations for diquark pairs.

k	1	2	3	4	5	6	7
(ij)S_4	$(\overline{13})1$	$(\overline{24})1$	$(33)1$	$(\widetilde{13})1$	$(\widetilde{24})1$	$(33)0$	$(44)0$

It is true that the diquark with the symmetry $i = 4$ has the lowest energy, this makes Jaffe-Wilczek's channel ($k = 7$) is the ground state of the positive parity channels, 1780 MeV with $S = 1.4$ fm and $T = 0.1$ fm (a linear pattern), the diquark wavefunctions do not delocalize into each other's orbitals at all, but diquark and $\bar{s}$-quark do develop a relative amplitude of ~ 0.4 intruding into each other's orbitals. The channel coupling can introduce more attraction, 1710 MeV. However, the P-wave orbital between the two diquarks and the color-magnetic interaction between quarks in different diquarks make the total energy of the 5-quark system higher than that of the channel $k = 2$, where S-wave orbital is permitted. In QDCSM, the energy of the lowest channel (The parity is negative) is 1650 MeV with $S = 0.7$ fm and $T = 0.6$ fm (a triangular pattern), the delocalizations are similar to the positive parity states, but the $\bar{s}$-quark delocalizes strongly (~ 0.99) into the diquark orbitals. And the channel coupling almost has no effect on it. The reason for this can be seen from the orbital symmetry of the four nonstrange quarks: for negative parity, the lowest channel has $[\nu_4] = [4]$ and other channels have $[\nu_4] = [22]$, whose energy is much higher than the channel with $[\nu_4] = [4]$, whereas the four channels with positive parity all have $[\nu_4] = [31]$, where the channel coupling lower the energy considerably. So it appears that it is possible to obtain a negative parity pentaquark with triangular pattern in QDCSM, although the mass is a little high, but with the correction for the discrepancy with the physical threshold, this closely approximates the reported mass of the Θ^+.

The states (both positive and negative parity) are again too high in mass to match experiment when color screening is turned off ($\mu = 0$, conventional quark model). Delocalization and channel coupling have small effects ($\sim 5 - 10$ MeV reduction in mass). Only color screening introduce much attraction. The behavior is different from the dibaryon case, where only the combining of the two effects introduces substantial attraction in certain channels. Clearly the difference is due to the color structure of the clusters used. In dibaryon case, two colorless clusters are used whereas the colorful clusters are used for pentaquark.

2.3. *qq–$qq\bar{s}$ (Karliner-Lipkin) configuration*

In the Karliner-Lipkin configuration, We again found the positive parity states to be significantly higher in mass, at 1770 MeV even with channel coupling, which lower the energy about 100 MeV, and at cluster separation of order 1 fm. The clusters in the negative parity state found an energy minimum at smaller separation, $\sim 0.5 - 0.6$ fm. Here channel coupling made a marked difference in the delocalization parameters, which were small without coupling, but order one for both the diquark and triquark clusters with coupling. However, the effect on the final mass of the state was negligible, with a value of ~ 1690 MeV. This seems somewhat high, even after correcting for the model threshold being 120 MeV above the physical KN threshold. The mass can reach 1586 MeV with the strange quark mass adjusted in the meson to give the correct mass of the kaon.

2.4. *Tetrahedral configuration*

In this configuration, Each light quark was assigned an orbital centered on one of the corners of a regular tetrahedron and the strange antiquark orbital was centered at the geometrical center of the tetrahedron. In this case, the positive parity state is higher in mass, at 1740 MeV, similar to other configurations. For the negative parity state, we found, with channel coupling of all color configurations, a state with mass $\sim$ 1630 MeV. The minimum energy configuration occurred for a tetrahedron side length of 0.8 fm and with u and d delocalization amplitude parameters of $\sim$ 0.1 (as always, relative to unity in the initial orbital). The strange antiquark was slightly more delocalized, with $\epsilon \sim 0.2$ at the minimum configuration. Again, if one allow for a $\sim$ 120 MeV correction between the model and physical KN thresholds, the mass of the Θ^+ is accounted for in this molecule-like configuration also. The calculation with lower strange quark mass, which give the correct KN threshold, really obtain 1530 MeV for negative parity state. It has the further advantage of being very different in structure from a meson-baryon configuration, which suggests the decay width from this structure will be suppressed relative to normal hadronic expectations.

3. Summary

The constituent quark model, which is successfully applied to the $q\bar{q}$ and qqq system, cannot obtain the mass of pentaquark Θ^+, even with the various refinements. However all of these refinements emphasized the residual interactions between quarks. The QDCSM, which aims at the multi-quark system, can obtain the mass of pentaquark Θ^+ with Jaffe-Wilczek and tetrahedral structure with the correction for physical threshold. The molecular structure, the tetrahedron for pentaquark, might be the lowest configuration. How to model the interaction among quarks which form different color and/or geometry structure is crucial to obtain the pentaquark.

References

1. Nakano T et al 2003 *Phys. Rev. Lett.* **91** 012002;
   Kabana S 2005 *J. Phys.* **G31** S1155
   Hicks K 2005 *Prog. Part. Nucl. Phys.* (hep-ex/0504027)
   and references therein
2. Zhu S L 2004 hep-ph/0410002
   Maltman K 2004 hep-ph/0408144
   Goeke K *et al.* 2005 *Prog. Part. Nucl. Phys.* **55** 350 and references therein.
3. Wang F, Wu G H and Teng L J 1992 *Phys. Rev. Lett.* **69** 2901
   Ping J L, Wang F and Goldman T 2001 *Nucl. Phys* A **688** 871
   Ping J L, Wang F and Goldman T 2002 *Phys. Rev.* C **65** 044003
4. Wu G H, Teng L J, Wang F and Goldman T. 1996 *Phys. Rev.* C **53** 1161
   Wu G H, Ping J L, Teng L J, Wang F and Goldman T 2000 *Nucl. Phys* A **673** 279
   Wang F, Ping J L, Teng L J and Goldman T 1995 *Phys. Rev.* C **51** 3411
5. Arndt R A, Strakovsky I I and Workman R L 2003 *Phys. Rev.* C **68** 042201
6. Gibbs W R 2004 *Phys. Rev.* C **70** 045208

EXOTIC HADRONS AND ATOMS

1^{-+} Exotic on the Lattice with FLIC Fermions

J. N. Hedditch*, B. G. Lasscock, D. B. Leinweber, A. G. Williams
Special Research Centre for the Subatomic Structure of Matter
and
Department of Physics, University of Adelaide, 5005, Australia
E-mail: jhedditc@physics.adelaide.edu.au

W. Kamleh
School of Mathematics, Trinity College
Dublin 2, Ireland

J. M. Zanotti
John von Neumann-Institut für Computing NIC,
Deutches Elektronen-Synchrotron DESY,
D-15738 Zeuthen, Germany

Using hybrid interpolating fields, we investigate the mass of the 1^{-+} exotic meson. Access to light quark masses approaching 25 MeV is enabled through use of the Fat-Link Irrelevant Clover (FLIC) fermion action, and large ($20^3 \times 40$) lattices. Our results yield a 1^{-+} mass consistent with that of the $\pi_1(1600)$ candidate.

1. Introduction

The exotic mesons comprise a rare vehicle for the elucidation of the relatively unexplored role of gluons in QCD. The Particle Data Group[1] reports two candidates for the 1^{-+} exotic, the $\pi_1(1400)$ at 1.376(17)GeV, and the $\pi_1(1600)$ at 1.596^{+25}_{-14} GeV.

Michael[2] provides a good summary of lattice results in this field up to 2003, concluding that the light-quark exotic is predicted by lattice studies to have a mass of 1.9(2) GeV.

2. Lattice Simulations

We use local interpolating fields, coupling colour-octet quark bilinears to chromo-electric and chromo-magnetic fields. In this work we focus on two exotic interpolating fields, $\chi_2 = i\epsilon_{jkl}\bar{q}^a\gamma_k B_l^{ab} q^b$, and $\chi_3 = i\epsilon_{jkl}\bar{q}^a\gamma_4\gamma_k B_l^{ab}$, both of which couple large-large and small-small spinor components and provide the strongest signal for the 1^{-+} state.

*Talk given at APFB05

Table 1. 1^{-+} Exotic Meson mass (GeV) obtained from interpolating fields χ_2 and χ_3. Associated χ^2/dof are indicated for each correlation function.

	χ_2		χ_3	
m_π^2	m	χ^2/dof	m	χ^2/dof
0.693(3)	2.15(12)	0.69	2.20(15)	0.45
0.595(4)	2.11(12)	0.77	2.18(16)	0.46
0.488(3)	2.07(12)	0.85	2.15(17)	0.41
0.381(3)	2.01(12)	0.91	2.14(19)	0.29
0.284(3)	1.97(13)	0.78	2.27(29)	0.0001
0.215(3)	1.92(14)	0.78	2.25(31)	0.02
0.145(3)	1.85(17)	0.57	2.26(37)	0.02
0.102(4)	1.80(23)	0.13	2.46(58)	0.03

To permit the use of an $\mathcal{O}(a^4)$ improved field strength tensor[3], from which we construct our hybrid operators, we employ a variant of APE smearing[4], whereby the smeared links do not involve averages which include links in the temporal direction. In this way we preserve the notion of a Euclidean 'time' and avoid overlap of the creation and annihilation operators. In this study, the smearing fraction $\alpha = 0.7$ (keeping 0.3 of the original link) and the process of smearing and $SU(3)$ link projection is iterated four times[5].

Propagators are generated using the fat-link irrelevant clover (FLIC) fermion action[6,7] where the irrelevant Wilson and clover terms of the fermion action are constructed using APE-Smeared links[4], while the relevant operators use the untouched (thin) gauge links. This improves the chiral properties of the FLIC action and mitigates the problem of exceptional configurations encountered with clover actions[8]. The effect of renormalization on the action improvement terms[9] is also reduced, giving a new form of nonperturbative $\mathcal{O}(a)$ improvement[6,9] where near-continuum results are obtained at finite lattice spacing.

We use quenched-QCD gauge fields Collaboration with the $\mathcal{O}(a^2)$ mean-field improved Lüscher-Weisz plaquette plus rectangle gauge action[10] using the plaquette measure for the mean link. The configurations are generated using the Cabibbo-Marinari pseudo-heat-bath algorithm[11] using a parallel algorithm with appropriate link partitioning[12], with some further modifications to improve ergodicity[13,14].

These results are obtained from $20^3 \times 40$ lattices at $\beta = 4.53$, which provides a lattice spacing of $a = 0.128(2)$ fm set by the Sommer parameter $r_0 = 0.49$ fm. Eight quark masses are considered in the calculations. The analysis is based on a sample of 345 configurations, with the error analysis performed via third-order single-elimination jackknife, where the χ^2 per degree of freedom (χ^2/dof) is obtained via covariance matrix fits.

3. Results

Table 1 summarizes our results for the mass of the 1^{-+} meson, with the squared pion-mass provided as a measure of the input quark mass. The agreement between the interpolators is significant, as we expect them to posess considerably different

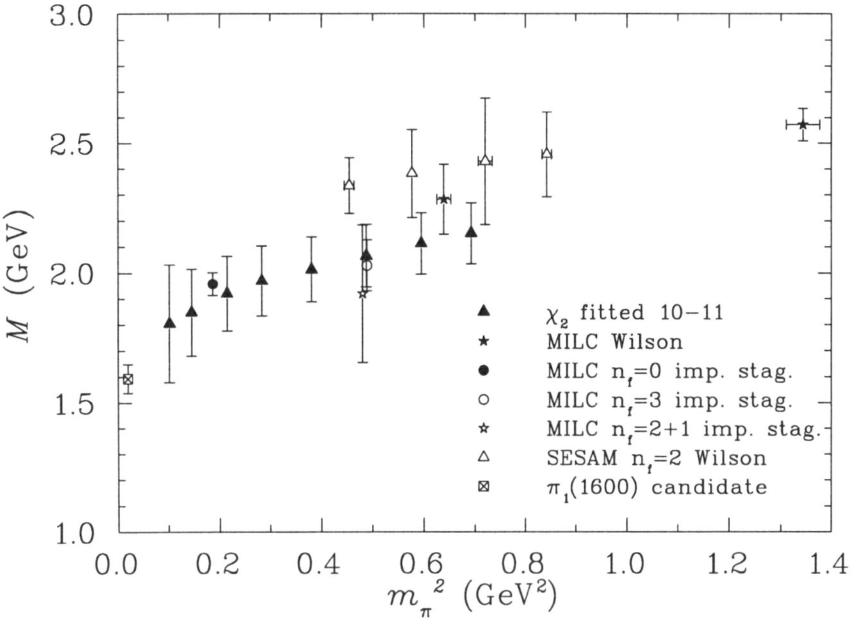

Fig. 1. A survey of results in this field. Open and closed symbols denote dynamical and quenched simulations respectively. Solid triangles are the results from this study. The leftmost point is the $\pi_1(1600)$ candidate..

excited-state contributions, based on experience with pseudoscalar interpolators[15].

Figure 1 shows this work in the context of previous lattice studies. MILC results[16] ($Q^4, 1^{-+} \to 1^{-+}$ results from $t = 3$ to $t = 11$) and SESAM results[17] are included. The illustration shows the novel features of our calculation, including new evidence of curvature close to the chiral limit.

We perform a linear fit to the 1^{-+} mass using the four lightest quark masses and a quadratic form to all 8 masses. A third-order single-elimination jackknife error analysis yields masses of 1.74(24) and 1.74(25) GeV for the linear and quadratic fits, respectively. A discussion of chiral nonanalytic curvature in the extrapolation is presented elsewhere[18,19].

4. Conclusion

We have found a compelling signal for the $J^{PC} = 1^{-+}$ exotic meson at very light quark masses, from which we can extrapolate a physical mass of 1.74(24) GeV. Thus for the first time in lattice studies, we find a 1^{-+} mass in agreement with the $\pi_1(1600)$ candidate, but excluding the $\pi_1(1400)$.

Acknowledgments

We thank Doug Toussaint for sharing his collection of results for the 1^{-+} mass. We acknowledge supercomputer support from the Australian Partnership for Advanced

Computing (APAC), and the South Australian Partnership for Advanced Computing (SAPAC). This work was supported by the Australian Research Council.

References

1. S. Eidelman *et al.* [Particle Data Group], Phys. Lett. B **592**, 1 (2004).
2. C. Michael, [arXiv:hep-ph/0308293].
3. S. O. Bilson-Thompson, D. B. Leinweber and A. G. Williams, Annals Phys. **304**, 1 (2003) [arXiv:hep-lat/0203008].
4. M. Falcioni, M. L. Paciello, G. Parisi and B. Taglienti, Nucl. Phys. B **251**, 624 (1985); M. Albanese *et al.* [APE Collaboration], Phys. Lett. B **192**, 163 (1987).
5. F. D. R. Bonnet, P. Fitzhenry, D. B. Leinweber, M. R. Stanford and A. G. Williams, Phys. Rev. D **62**, 094509 (2000) [arXiv:hep-lat/0001018].
6. J. M. Zanotti, B. Lasscock, D. B. Leinweber and A. G. Williams, Phys. Rev. D **71**, 034510 (2005) [arXiv:hep-lat/0405015].
7. J. M. Zanotti *et al.* [CSSM Lattice Collaboration], Phys. Rev. D **65**, 074507 (2002) [arXiv:hep-lat/0110216].
8. S. Boinepalli, W. Kamleh, D. B. Leinweber, A. G. Williams and J. M. Zanotti, Phys. Lett. B **616**, 196 (2005) [arXiv:hep-lat/0405026].
9. D. B. Leinweber, *et al.* Eur. Phys. J. A **18**, 247 (2003) [arXiv:nucl-th/0211014]
10. M. Luscher and P. Weisz, Commun. Math. Phys. **97**, 59 (1985) [Erratum-ibid. **98**, 433 (1985)].
11. N. Cabibbo and E. Marinari, Phys. Lett. B **119**, 387 (1982).
12. F. D. Bonnet, D. B. Leinweber and A. G. Williams, J. Comput. Phys. **170**, 1 (2001) [arXiv:hep-lat/0001017].
13. F. D. R. Bonnet, D. B. Leinweber, A. G. Williams and J. M. Zanotti, Phys. Rev. D **65**, 114510 (2002) [arXiv:hep-lat/0106023].
14. D. B. Leinweber, A. G. Williams, J. B. Zhang and F. X. Lee, Phys. Lett. B **585**, 187 (2004) [arXiv:hep-lat/0312035].
15. A. Holl, A. Krassnigg, P. Maris, C. D. Roberts and S. V. Wright, [arXiv:nucl-th/0503043].
16. C. W. Bernard *et al.* [MILC Collaboration], Phys. Rev. D **56**, 7039 (1997) [arXiv:hep-lat/9707008].
17. P. Lacock and K. Schilling [TXL collaboration], Nucl. Phys. Proc. Suppl. **73**, 261 (1999) [arXiv:hep-lat/9809022].
18. J. N. Hedditch, W. Kamleh, B. G. Lasscock, D. B. Leinweber, A. G. Williams and J. M. Zanotti, arXiv:hep-lat/0509106
19. A. W. Thomas and A. P. Szczepaniak, Phys. Lett. B **526**, 72 (2002) [arXiv:hep-ph/0106080].

Accurate Evaluation of $\overline{p}D$ Atoms*

Y. Yan, C. Kobdaj and P. Suebka

School of Physics, Suranaree University of Technology,
111 University Avenue, Nakhon Ratchasima 30000, Thailand

The $\overline{p}D$ atomic state problem is studied in the latest version of the Paris $\overline{N}N$ potential. The small energy shifts and decay widths resulting from the short range strong interaction of the $\overline{p}D$ atomic states are accurately derived by solving the dynamical equations of the system in a suitable numerical approach based on Sturmian functions, which is able to account for both the strong short range nuclear potential (for both the local and nonlocal) and the long range Coulomb force and to provide directly the wave function of $\overline{p}D$ atoms with complex eigenvalues $E = E_R - i\frac{\Gamma}{2}$.

1. Introduction

The second simplest antiprotonic atom is the antiprotonic deuteron atom $\overline{p}D$, consisting of an antiproton and a deuteron bound mainly by the Coulomb interaction but distorted by the short range strong interaction. The study of the $\overline{p}D$ atom is much later and less successful than for other exotic atoms like the protonium and pionium. Experiments were carried out at LEAR just in very recent years to study the properties of the $\overline{p}D$ atom[1,2]. There have been some theoretical works[3–5] studying the $\overline{p}D$ atomic states, where over simplified $\overline{p}D$ interactions were employed. The predictions of all those works are not in good agreement with the experimental data.

In the theoretical sector, one needs to overcome at least two longstanding difficulties in the study of the $\overline{p}D$ atom. First, the interaction between the antiproton and the deuteron core should be derived from realistic $\overline{N}N$ interactions, for example, the Paris $\overline{N}N$ potentials[6–8]. This has never been done. Even if a reliable $\overline{p}D$ interaction is in hands, the accurate evaluation of the energy shifts and decay widths (stemming for the strong $\overline{p}D$ interactions) and especially of the nuclear force distorted wave function of the atom is still a challenge. It should be pointed out that the methods employed in the works[3–5] are not accurate enough for evaluating the wave functions of the $\overline{p}D$ atoms.

In the present work we study the $\overline{p}D$ atom problem employing a properly adapted numerical method based on Sturmian functions[9]. The method accounts for both the strong *short* range nuclear potential (local and non-local) and the *long*

*This work is supported in part by the Suranaree University of Technology grant SUT 1-105-48-36-12 range Coulomb force and provides directly the wave function of the $\overline{p}D$ system with complex eigenvalues $E = E_R - i\frac{\Gamma}{2}$. The protonium and pionium problems have been successfully investigated[10,11] in the numerical approach. The numerical method is much more powerful, accurate and much easier to use than all other methods applied to the exotic atom problem in history. The $\overline{p}D$ interaction in the work is derived from a realistic $\overline{N}N$ potentioal, the latest Paris $\overline{N}N$ potential (Paris04)[8] which is state-dependent. The work is organized as follows. The $\overline{p}D$ interaction is derived in Sec. II from Paris04. In Sec. III we solve the $\overline{p}D$ dynamical equation to obtain the energy shifts and decay widths of the atom.

2. $\overline{p}D$ Interactions in Terms of $\overline{N}N$ Interactions

We start from the Schrödinger equation of the antiproton-deuteron system in coordinate space

$$\left[\frac{P_\rho^2}{2M_\rho} + \frac{P_\lambda^2}{2M_\lambda} + V_{12}(\vec{\rho}) + V_{13}(\vec{\lambda} + \vec{\rho}/2) + V_{23}(\vec{\lambda} - \vec{\rho}/2)\right] \Psi(\vec{\lambda}, \vec{\rho}) = E\Psi(\vec{\lambda}, \vec{\rho}) \quad (1)$$

where $\vec{\lambda}$ and $\vec{\rho}$ are the Jacobi coordinates of the system, defined as

$$\vec{\lambda} = \vec{r}_3 - \frac{\vec{r}_1 + \vec{r}_2}{2}, \quad \vec{\rho} = \vec{r}_2 - \vec{r}_1, \quad (2)$$

and $M_\lambda = M/2$ and $M_\rho = 2M/3$ are the reduced masses of the system. Here we have assigned, for simplicity, the proton and neutron the same mass M. One may expect that the nuclear distortion of the deuteron core would be negligible and that the wave function of the $\overline{p}D$ system would be rewritten as $\Psi(\vec{\lambda}, \vec{\rho}) = \psi(\vec{\lambda})\psi_D(\vec{\rho})$, where $\psi_D(\vec{\rho})$ is the wave function of deuteron core. Without nuclear distortion of the deuteron core, we obtain

$$\left[\frac{P_\lambda^2}{2M_\lambda} + V(\vec{\lambda}, \vec{\rho})\right] \Psi(\vec{\lambda}, \vec{\rho}) = E_{\overline{p}D}\Psi(\vec{\lambda}, \vec{\rho}) \quad (3)$$

with

$$V(\vec{\lambda}, \vec{\rho}) = V_s(\vec{\lambda}, \vec{\rho}) + V_c(\vec{\lambda}, \vec{\rho}) \quad (4)$$

$$V_s(\vec{\lambda}, \vec{\rho}) = \frac{1}{4}[V^0(\vec{r}_3 - \vec{r}_1) + V^0(\vec{r}_3 - \vec{r}_2)] + \frac{3}{4}[V^1(\vec{r}_3 - \vec{r}_1) + V^1(\vec{r}_3 - \vec{r}_2)] \quad (5)$$

$$V_c(\vec{\lambda}, \vec{\rho}) = \frac{1}{2}[V_c(\vec{r}_3 - \vec{r}_1) + V_c(\vec{r}_3 - \vec{r}_2)] \quad (6)$$

where V_s and V_c stand respectively the nuclear interaction and Coulomb interaction of the $\overline{p}D$ system. V^0 and V^1 are the isospin 0 and 1 antinucleon-nucleon nuclear interactions, respectively. Note that the coordinate $\vec{\rho}$ is no longer dynamical in Eq. (3).

In this work we investigate the $\overline{p}D$ atoms in the latest version of the Paris nucleon-antinucleon potential where the nucleon-antinucleon interactions are interpreted in the basis of the angular momenta and isospin of the nucleon-antinucleon states. It is necessary and also convenient to express the interactions V in eq. (3)

in the basis of the $\overline{p}D$ $|JMLS\rangle$ states which can be projected onto the basis of the angular momenta of the nucleon-antinucleon states. Detailed calculations lead to the general expression for the antiproton-deuteron interactions in Eq. (3) in the $\overline{p}D$ $|JMLS\rangle$ states

$$\langle K|\,V(\vec{\lambda},\vec{\rho})\,|K'\rangle = \sum_{k,k'}\sum_{\alpha\alpha'} D(K;k;\alpha)W(\lambda,\rho;k,k';\alpha,\alpha')D(K',k';\alpha') \tag{7}$$

where $|K\rangle \equiv |JMLS\rangle$ and $|K'\rangle \equiv |JML'S\rangle$ stand for the antiproton-deuteron states, and $|k\rangle \equiv |jmls\rangle$ and $|k'\rangle \equiv |jml's\rangle$ for the antinucleon-nucleon ones. $W(\lambda,\rho;k,k';\alpha,\alpha')$ in Eq. (7) is the only part depending on the radial variables λ and ρ, taking the form, for example, for the term $V(\vec{r}_3-\vec{r}_1)$

$$W(\lambda,\rho;k,k';\alpha,\alpha') = \left(\frac{\rho}{2\lambda}\right)^{\alpha+\alpha'}\frac{1}{2}\int_{-1}^{1} dx\,\frac{V(k,k';r_{13})}{X_{13}^{\alpha+\alpha'}}P_\alpha(x)P_{\alpha'}(x) \tag{8}$$

where $P_k(x)$ are Legendre polynomials, $r_{13}=\sqrt{\lambda^2+\rho^2/4+\rho\lambda x}$, $X_{13}=r_{13}/\lambda$, and

$$V(k,k';r_{ij}) \equiv \langle k\,|V(\vec{r}_{ij})|\,k'\rangle \tag{9}$$

are simply the interaction of nucleon-antinucleon systems in the $|k\rangle \equiv |jmls\rangle$ basis. $D(K;k;\alpha)$ in Eq. (7) reflect the angular momentum couplings of the antiproton-deuteron and antinucleon-nucleon states.

In the approximation that the deuteron core is in the S-state, we have for the lowest $\overline{p}D$ states $^2S_{1/2}$ and $^4S_{3/2}$

$$\begin{aligned}\langle\,^2S_{1/2}|\,V(\vec{\lambda},\vec{\rho})\,|\,^2S_{1/2}\rangle &= \frac{3}{4}\,W(^1S_0)+\frac{1}{4}\,W(^3S_1)\\ \langle\,^4S_{3/2}|\,V(\vec{\lambda},\vec{\rho})\,|\,^4S_{3/2}\rangle &= W(^3S_1)\end{aligned} \tag{10}$$

with

$$\begin{aligned}W(\,^1S_0) &= \frac{1}{2}\int_{-1}^{1} dx\,V(\,^1S_0;r_{13})\\ W(\,^3S_1) &- \frac{1}{2}\int_{-1}^{1} dx\,V(\,^3S_1;r_{13})\end{aligned} \tag{11}$$

where $^1S_0$ and $^3S_1$ are antinucleon-nucleon states, and hence $V(^1S_0;r_{13})$ and $V(^3S_1;r_{13})$ are corresponding antinucleon-nucleon interactions. For the higher $\overline{p}D$ states like $^2P_{1/2}$, $^2P_{3/2}$, $^4P_{1/2}$, $^4P_{3/2}$, and $^4P_{5/2}$, the results are not so simple since terms with non-zero α and α' will have contributions.

3. Results and Discussions

It is not a simple problem to accurately evaluate the energy shifts and decay widths, especially wave functions of exotic atoms like protonium, pionium and antiproton deuteron atoms, which are mainly bound by the Coulomb interaction, but also effected by the short range strong interaction. In this work we study the $\overline{p}D$ atoms in the Sturmian function approach employed in our previous works[10,11]. Employed

Table 1. The energy shifts and widths of the $1s\ \bar{p}D$ atomic states.

	Paris04		Experimental Data	
	ΔE(eV)	Γ(eV)	ΔE(eV)	Γ(eV)
$^2S_{1/2}$	1337.88	2606.87	–	–
$^4SD_{3/2}$	1546.00	990.16	–	–
$^4S_{3/2}$	1102.72	3476.82	–	–
$\overline{^2S_{1/2},\, ^4SD_{3/2}}$	1181.11	1529.06	1050 ± 250	1100 ± 750

for the $\overline{N}N$ interaction is the latest version of the Paris $\overline{N}N$ potential (Paris04)[8], which is believed the most realistic. In this preliminary study we are restricted to the lowest order approximation that the deuteron core is in the S-state and its size is negligible. The study which properly considers the wave function of the deuteron core and the deuteron D-wave contribution is under way. Listed in Table 1 are the energy shifts and widths of the $\bar{p}D$ $1s$ atomic states in the lowest order approximation. It is found that the theoretical results for both the energy shift and decay width averaged over the $^2S_{1/2}$ and $^4SD_{3/2}$ are in good agreement with the experimental data[1]. The considerable difference between the $^4SD_{3/2}$ and $^4S_{3/2}$ decay widths indicates that the tensor force, which leads to the S- and D-wave coupling, is very important.

The study of the $2p$ $\bar{p}D$ atomic states, however, reveals that the size of the deuteron core must be properly considered for the P-wave $\bar{p}D$ atoms. In the approximation of the size of the deuteron core being zero, both the theoretical energy shifts and decay widths (not shown in the work) of the $2p$ $\bar{p}D$ atomic states are not consistent with the experimental data[2].

References

1. M. Augsburger, et al., Phys. Lett. B461 (1999) 417.
2. D. Gotta, et al., Nucl. Phys. A660 (1999) 283.
3. S. Wycech, A.M. Green, and J.A. Niskanen, Phys. Lett. B152, 308 (1985).
4. G.P. Latta, and P.C. Tandy, Phys. Rev. C42 (1990) R1207.
5.] G.Q. Liu, J.-M Richard, and S. Wycech, Phys. Lett. B2602 (1991) 15.
6. M. Pignone, M. Lacombe, B. Loiseau, and R. Vinh Mau, Phys. Rev. C **50**, 2710 (1994).
7. B. El-Bennich, M. Lacombe, B. Loiseau, and R. Vinh Mau, Phys. Rev. C **59**, 2313 (1998).
8. M. Lacombe, B. Loiseau, R. Vinh Mau, and S. Wycech, "The Paris $N\overline{N}$ potential constrained by recent $\bar{n}p$ total cross section and antiprotonic atom data", in preparation.
9. M. Rotenberg, Adv. At. Mol. Phys. **6**, 233 (1970).
10. Y. Yan, R. Tegen, T. Gutsche and A. Faessler, Phys. Rev. C56 (1997) 1596.
11. P. Suebka, and Y. Yan, Phys. Rev. C70, 034006 (2004).

Cascade Calculation of Kaonic Nitrogen Atoms Involving the Electron Refilling Process

T. Koike

Advanced Meson Science Laboratory, RIKEN
(The Institute of Physical and Chemical Research)
E-mail: tkoike@riken.jp

The cascade calculation of kaonic nitrogen atoms is performed and the electron population during the cascade is investigated in order to estimate the amount of the electron screening effect on the kaonic x-ray energy.

1. Introduction

The Particle Data Group assigned 493.677 ± 0.013 MeV to the charged kaon mass as a world average[1]. However, there exists a serious disagreement between the most two recent mass measurements using kaonic atom x-rays, in which the deduced kaon masses differ 60 keV although their uncertainties are about 10 keV. Therefore, the kaon mass within the precision of 10 keV order is still an open problem. In order to settle this discrepancy, new precise charged kaon mass measurement using kaonic nitrogen atom x-rays in a gaseous target is planned at the DAΦNE[2,3].

For the kaon mass determination, the kaonic x-ray energy must be calculated with sufficient accuracy. Among the various corrections on the x-ray energy, the electron screening effect is difficult to estimate correctly, because it needs the knowledge of the electron population at the moment of x-ray emission, which depends on the balance between Auger electron emission and electron refilling during the atomic cascade process. The correct evaluation of the electron screening effect becomes crucial issue in order to reach the desired accuracy.

Our purpose is to develop the cascade code for kaonic nitrogen atoms which enables us to determine the electron fraction at each kaon atomic level during the cascade, preparing for the forthcoming kaon mass measurement at DAΦNE, Some results of our cascade calculation are published in Ref.[2] together with the experimental results of the kaonic nitrogen atom x-ray yields. In this paper, we report our new results using the improved cascade model. The present model is different from the previous one in the following points; (1) the electron 2s- and 2p-orbit are distinguished each other in the L-shell, and (2) the KLL-Auger transition (Auger process between electrons, not electron and kaon) is explicitly included as the electron rearrangement process.

2. Cascade Model Involving Electron Refilling Process

The atomic cascade starts from kaon principal quantum number $n_{\text{init.}} \sim \sqrt{M^*/m_e} \sim 30$. The deexcitation of kaon takes place in competition mainly between the radiative transitions and the internal Auger process between kaon and electron. After emitting the electron by Auger process, the hole in the electron orbit is refilled by either of the radiative transition, KLL-Auger process and the electron pick-up from outside, depending on the electron configurations. The cascade process ends by the absorption into the nuclei by the strong interaction between kaon and nucleus, which mainly occurs at 3d-state in the case of kaonic nitrogen. In our cascade model, we include the following processes; as for the kaon sector,

- $(n, \ell) \to (n', \ell \pm 1)$ radiative transitions,
- $(n, \ell) \to (n', \ell \pm 1)$ Auger transitions for ejecting 1s-, 2s- and 2p-electron, respectively,
- nuclear absorption by the strong interaction, calculated by using the phenomenological optical potential[4],
- weak decay of kaon,

and for electron sector,

- 2p→1s radiative transition,
- KLL-Auger transition between two 2s-electrons, one 2s- and one 2p- electron, and two 2p-electrons, respectively.
- electron pick-up from outside to L-shell.

The rate equations describing these processes are constructed and numerically solved. Since the formation process involving the break-up of nitrogen *molecules* into atoms is not well-understood yet, several free-parameters are introduced for both of the kaon and electron initial distributions and adjusted to reproduce the experimental x-ray yields observed at DAΦNE[2]. As a results of χ^2-fit to the experimental data, the initial kaon angular momentum distribution becomes just statistical one, $P(l) \propto (2l+1)$ at $n = 30$, and the initial electron population prefers that two or three electrons in L-shell are lost at the start of the cascade. For more detail of our cascade model is found in ref.[5].

It should be stated that there remains the ambiguity in the evaluation of electron pick-up rate from outside, because the velocity distribution of kaonic nitrogen atoms during the cascade is unknown. Here, the electron pick-up rate is estimated by assuming the thermalized velocity, although there is no evidence of the complete thermalization during the cascade. As long as our concern is restricted to the gaseous target at the density $\rho = 3.4\rho_{\text{NTP}}$ (DAΦNE experimental condition), this uncertainty does not affect final results so much.

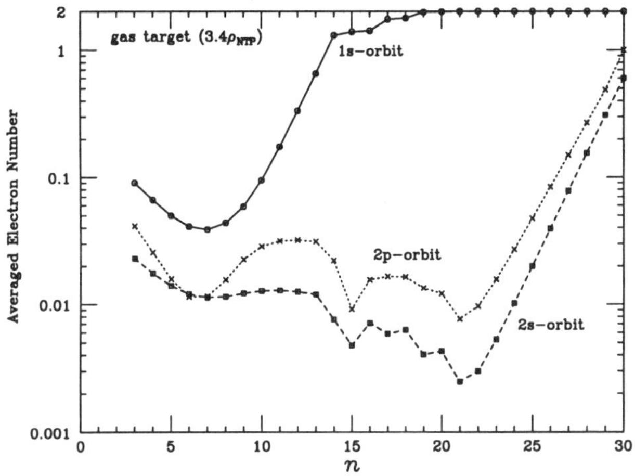

Fig. 1. Average electron number for 1s-, 2s- and 2p-orbit for kaonic nitrogen atom in gaseous target (density $\rho = 3.4\rho_{\rm NTP}$) as a function of kaon principal quantum number n.

3. Results and Discussion

Fig.1 shows the averaged electron number for each electron orbit during the cascade in the gaseous target at a density $\rho = 3.4\rho_{\rm NTP}$, as a function of kaon principal quantum number. At $n \geq 20$, the K-shell Auger transition is energetically forbidden. When the K-shell Auger process is allowed, the K-shell population rapidly falls off by successive electron emission before the electron refilling is completed. It is found that the 1s-electron population becomes minimum around the x-ray observed region, and the averaged 1s-electron number at $n = 6$ is estimated to be 0.04. This is consistent with our previous results[2]. In addition, the present cascade model allows us to examine more details of the electron configurations. Fig.2 separately shows the probabilities that two, one and no electron remain in the 1s-orbit. At $n = 6$, one electron remains in the 1s-orbit with the probability of 4% and the probability that two electrons remain is almost zero. This result may be helpful for the future kaon mass measurement using kaonic nitrogen atom x-rays at DAΦNE.

In order to confirm the results more quantitatively, the velocity distribution of kaonic nitrogen atoms during the cascade remains as a matter to be discussed further. It is also related to the acceleration by "Coulomb explosion" at the formation stage, whose evidence is reported for the pionic nitrogen atoms[6]. In the case of dense target such as liquid or solid, the velocity distribution affects the x-ray yields through the electron pick-up rate. Fig.3 shows the calculated the x-ray yields as a function of the target density. The slope of the decrese of the x-ray yields at higher density depends on the electron pick-up rate. Therefore, the density dependence of the x-ray yields has the information on the velocity distribution and it is important subject to be investigated for the better understanding of the cascade process, apart from the mass measurement.

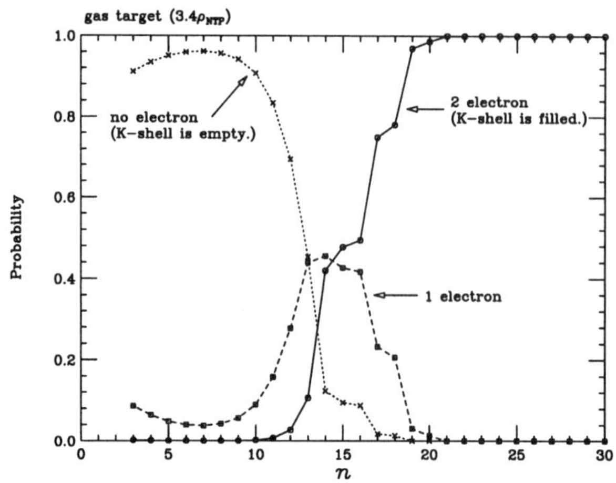

Fig. 2. Probabilities that two, one and no electrons remain in K-shell during the cascade kaonic nitrogen atom in gaseous target (density $\rho = 3.4\rho_{\text{NTP}}$) as a function of kaon principal quantum number n.

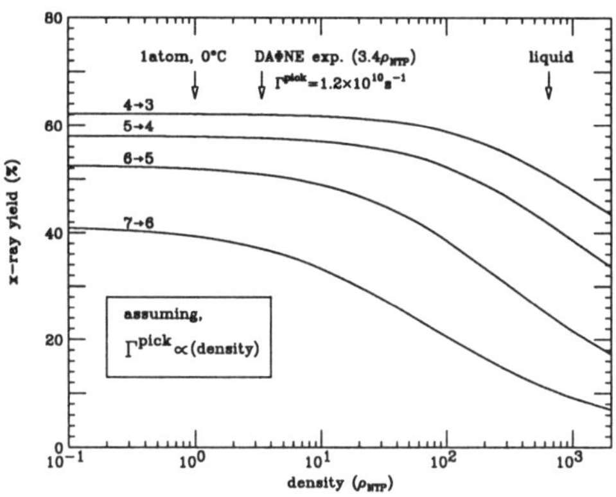

Fig. 3. The calculated x-ray yields of kaonic nitrogen atoms as a function of the target density, assuming that the electron pick-up rate is $1.2\times10^{10}\text{s}^{-1}$ at a density $\rho = 3.4\rho_{\text{NTP}}$.

Acknowledgments

The author is supported by the Special Postdoctoral Researchers Program from RIKEN.

References

1. S. Eidelman *et al.* (Particle Data Group), *Phys. Lett.* **B592**, 1 (2004).
2. T. Ishiwatari *et al.*, *Phys. Lett.* **B593**, 48 (2004).
3. G. Beer *et al.*, *Phys. Lett.* **B535**, 52 (2002).
4. C.J. Batty, E. Friedman and A. Gal, *Phys. Rep.* **287**, 385 (1997).
5. T. Koike, *Genshikaku Kenkyu (Nuclear Studies)* Vol.49 No.6, 159 (2005).
6. T. Siems *et al.*, *Phys. Lett.* **84**, 4573 (2000).

B Decays into $\pi\pi K$ and $K\bar{K}K$: Long Distance an Final-State Effects

B. Loiseau

Laboratoire de Physique Nucléaire et de Hautes Énergies, Groupe Théorie,*
Univ. P. & M. Curie, 4 Pl. Jussieu, F-75252 Paris, France
E-mail: loiseau@lpnhep.in2p3.fr

A. Furman, R. Kamiński, L. Leśniak

Department of Theoretical Physics,
The Henryk Niewodniczański Institute of Nuclear Physics,
Polish Academy of Sciences, 31-342 Kraków, Poland

The interplay of strong and weak decay amplitudes for $B \to \pi\pi K$ and $B \to K\bar{K}K$, with the $\pi\pi$ and $K\bar{K}$ pairs interacting in isospin-0 S-wave, is analyzed for $\pi\pi$ effective mass from threshold to 1.2 GeV. To improve agreement with experiment of a factorization approach with some QCD corrections, addition of long-distance contributions, called charming penguins is necessary.

1. Introduction

Hadronic two- and three-body B-meson weak decays are a rich source of information to study CP violation within the Standard Model (and beyond). On Dalitz plots, many resonances are distinguishable but with entangled interference patterns. Moreover the understanding of the final state interactions is weighty to attain accurate values of the Cabibbo-Kobayashi-Maskawa (CKM) matrix elements. We report here on our recent study[1] (FKLL) of $B \to \pi\pi K$ and $B \to K\bar{K}K$ decays We consider the $\pi\pi$ or $K\bar{K}$ pairs, denoted by $(\pi\pi)_S$ or $(K\bar{K})_S$, to interact in isospin-0 S-wave in the $m_{\pi\pi}$ effective mass range from threshold to 1.2 GeV. We extend to the $f_0(980)$ case the approach of Gardner and Meißner[2] (GM) who have examined the effect of the $f_0(600)$ (or σ) on the $B^0 \to \pi^+\pi^-\pi^0$ decays. The decay amplitudes are calculated within the QCD factorization approach. Due to cancellations between penguin amplitudes, the $B \to f_0(980)K$ branching fractions are too small compared to data. We add long-distance contributions, called charming penguins. Expressions for the B decay amplitudes and description of the final state interactions, are given in section II. Comparison of our results to the data and some conclusions are presented in section III.

*Unité de Recherche des Universités Paris 6 et Paris 7, associée au CNRS

2. B Decay Amplitudes and Final State Interactions

From the effective weak Hamiltonian, H_{eff}, and the operator product expansion, the B decay amplitude into mesons M_1 and M_2 is[3]

$$\langle M_1 M_2 | H_{eff} | B \rangle = \frac{G_F}{\sqrt{2}} \sum_{q=u,c} V_q^{CKM} \left[\sum_{k=1}^{10} C_k(\mu) \langle M_1 M_2 | O_k^q(\mu) | B \rangle \right] . \quad (1)$$

In (1) $G_F = 1.66 \times 10^{-5}$ GeV^{-2} is the Fermi constant, V_q^{CKM} represents the CKM matrix-element factors and μ is a renormalization scale. The perturbative Wilson coefficients $C_k(\mu)$ characterize the short-distance physics. The non-perturbative hadronic matrix elements $\langle M_1 M_2 | O_k^q(\mu) | B \rangle$ describe the long-distance physics. The local operators $O_k^q(\mu)$ *"effectively"* govern the decay. The main task of the theory is to compute these hadronic matrix elements in a reliable way. Beneke, Buchalla, Neubert and Sachrajda[4] have developed the following QCD factorization formula,

$$\langle M_1 M_2 | O_k(\mu) | B \rangle \sim \langle M_1 | J_1 | 0 \rangle \langle M_2 | J_2 | B \rangle \left[1 + \sum_n r_n \alpha_s^n + O(\Lambda_{QCD}/m_b) \right] . \quad (2)$$

In (2) J_1 and J_2 are bilinear quark currents, $\langle M_1 | J_1 | 0 \rangle$ is the meson M_1 decay constant and $\langle M_2 | J_2 | B \rangle$ represents the transition form factor of the meson B into the meson M_2. The radiative corrections r_n have been calculated to order $n = 1$, α_s being the strong coupling constant. The term $O(\Lambda_{QCD}/m_b)$ in Eq. (2) represents power corrections and if set to zero, together with r_n, one recovers the naive factorization formula as applied by Bauer, Stech and Wirbel[5].

The possible quark line diagrams for the $B^- \to (\pi\pi)_S K^-$ and $B^- \to (K\bar{K})_S K^-$ are shown in Fig. 1 of FKLL[1]. For the B^0 decays there is no tree diagram (a), only penguin diagrams similar to (b) or (c) diagrams exist. Within the QCD factorization framework, just reminded above, the $B^- \to (\pi^+\pi^-)_S K^-$ decay amplitude is

$$\langle (\pi^+\pi^-)_S K^- | H_{eff} | B^- \rangle = \frac{G_F}{\sqrt{2}} \sqrt{\frac{2}{3}} \Big\{ \chi \left[P(m_{\pi\pi})(U_a + U_b) + C(m_{\pi\pi}) \right] \Gamma_1^{n*}(m_{\pi\pi}) + \\ + \left[Q(m_{\pi\pi}) U_c + \chi C(m_k) \right] \Gamma_1^{s*}(m_{\pi\pi}) \Big\} . \quad (3)$$

The constant χ, to be fitted to reproduce the averaged branching fraction $\mathcal{B}[B^\pm \to f_0(980) K^\pm]$, is estimated to 30 GeV^{-1} from the $f_0(980)$ decay properties[1]. The non-strange and strange pion scalar form factors $\Gamma_1^n(m_{\pi\pi})$ and $\Gamma_1^s(m_{\pi\pi})$ describe the final-state interactions. In (3) $P(m_{\pi\pi})$ corresponds to the product of the kaon decay constant f_K times the transition form factor of B into $(\pi\pi)_S$, $F_0^{B\to(\pi\pi)_S}(M_K^2)$,

$$P(m_{\pi\pi}) = f_K (M_B^2 - m_{\pi\pi}^2) F_0^{B\to(\pi\pi)_S}(M_K^2). \quad (4)$$

The function $Q(m_{\pi\pi})$ is proportional to the product of the $(\pi\pi)_S$ "decay constant" times the transition form factor of B into K, $F_0^{B\to K}(m_{\pi\pi}^2)$, and reads[1,2],

$$Q(m_{\pi\pi}) = \frac{2\sqrt{2} B_0}{m_b - m_s} \left(M_B^2 - M_K^2 \right) F_0^{B\to K}\left(m_{\pi\pi}^2 \right) , \quad (5)$$

with $B_0 = -\langle 0|q\bar{q}|0\rangle/f_\pi^2 = m_\pi^2/(m_u+m_d)$.

The functions U_a, U_b and U_c, given in terms of the coefficients[4,6–8] a_i of the decay operators, are calculated from the Wilson coefficients. They correspond to the contributions of Figs. 1(a), 1(b) and 1(c) of FKLL[1], respectively. One has $U_a = V_{ub}V_{us}^* a_1$, $U_b = V_{ub}V_{us}^*(a_4^u - a_6^u r) + V_{cb}V_{cs}^*(a_4^c - a_6^c r) = V_{ub}V_{us}^*[a_4^u - a_4^c + (a_6^c - a_6^u)r] + V_{tb}V_{ts}^*(a_6^c r - a_4^c)$ and $U_c' = -V_{ub}V_{us}^* a_6^u - V_{cb}V_{cs}^* a_6^c = V_{ub}V_{us}^*(a_6^c - a_6^u) + V_{tb}V_{ts}^* a_6^c$. We have used the unitarity of the CKM matrix elements, $V_{ub}V_{us}^* + V_{cb}V_{cs}^* + V_{tb}V_{ts}^* = 0$. In the reduction of the four-quark operators, split into product of two-matrix elements (Eq. (2)), a Fierz transformation is needed for the penguin diagrams in order to match the flavor quantum numbers of the quark currents to those of the hadrons[6]. The chiral factor r is equal to $2M_K^2/[(m_b+m_u)(m_s+m_u)]$. Here and in Eq. (5), quark masses appear as one makes use of the Dirac equation. The a_1, $a_{4,6}^{u,c}$ coefficients at next-to-leading order in α_s are given in Eq. (35) of Beneke and Neubert[7] (BN). In the numerical calculations we take the values of table III of de Groot, Cottingham and Whittingham[8] (GCW) at the scale $\mu = 2.1$ GeV. We do not include the small contributions from hard spectator interactions[7], annihilation and electroweak diagrams.

To improve their fit to charmless hadronic two-body B decays with pseudoscalar-pseudoscalar and pseudoscalar-vector decays, GCW[8], following Ciuchini et al.[9], introduce the charming penguin term $C(m)$, the parametrization of which is reminded in Eq. (6) of FKLL[1]. Expressions for the other $B \to (\pi\pi)_S K$ and for the $B \to (K\bar{K})_S K$ amplitudes are given in FKLL[1]. If one neglects the penguin contraction terms (see Eq. (35), (39) and (41) of BN[7]) then $a_{4,6}^u = a_{4,6}^c = a_{4,6}$ and our amplitude (3) is similar to that of the $B^- \to \sigma\pi^-$ decay[2] (see their Eq. (25)). Note the near cancellation[1] in the $b \to s$ transition of the two penguin contributions as $r \sim 1$ and $a_4^c \sim a_6^c$.

The $\Gamma_1^{n,s}(m)$ obey unitary $\pi\pi$ and $K\bar{K}$ coupled channel equations[1]. We use the solution A of the two-body scattering matrix of Kamiński, Leśniak and Loiseau[10]. Below the $K\bar{K}$ threshold $\Gamma_1^{n,s^*}(m) = R_1^{n,s}(m)\cos\delta_{\pi\pi}(m)e^{i\delta_{\pi\pi}(m)}$. $|\Gamma_1^{n,s}|$ is maximum for $\delta_{\pi\pi} \simeq 180°$, which corresponds to the $f_0(980)$ production. The $R_1^{n,s}(m)$, depicting the formation of mesons prior to rescattering, have been obtained by Meißner and Oller[11].

3. Results and Conclusions

Our results with the charming-penguin $C(m)$ terms of GCW[8], model I, and of Ciuchini, Franco, Martinelli, Masiero, Pierini and Silvestrini[12] (CFMMPS), model II, are compared to the data[13–16] in Table 1. Values of all parameters used are given in FKLL[1]. The effective $m_{\pi\pi}$ mass distribution for $B^\pm \to \pi^+\pi^-K^\pm$ of our model II is compared with Belle[13] in Fig. 1. The sharp maximum near 1 GeV is from the $f_0(980)$. Near 0.5 GeV the broad bump is related to the $f_0(600)$. Events near 0.8 GeV are due to the $B^\pm \to \rho^0 K^\pm$ decay not considered in our model. Similar agreement with $m_{\pi\pi}$ effective mass distributions of BaBar and Belle for charged

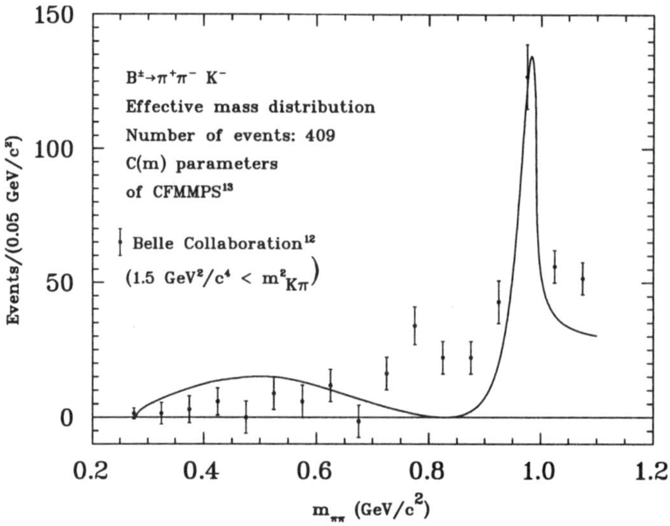

Fig. 1. Comparison of our model II with Belle data[13]

Table 1. Average branching fractions $\mathcal{B}$ in units of 10^{-6}, direct asymmetries $\mathcal{A}_{CP}$, $\mathcal{A}$ and $\mathcal{S}$ parameters of our model compared to data. Model errors come from $C(m)$ parameters fitted by GCW[8], model I, and CFMMPS[12], model II. For a qualitative comparison we give in parenthesis the experimental $B \to K\bar{K}K$ decay observables obtained by considering the full range of the $K\bar{K}$ invariant mass rather than the upper limit[1] of 1.1 GeV. For data[14,16] we only quote statistical errors.

B decay mode		Average HFAG's values[15]	Model I $\chi = 33.5$ GeV^{-1}	Model II $\chi = 23.5$ GeV^{-1}
$B^{\pm} \to f_0(980)K^{\pm}$,	$\mathcal{B}$	$8.49^{+1.35}_{-1.26}$	8.49 (fit)	8.46 (fit)
$f_0 \to \pi^+\pi^-$	$\mathcal{A}_{CP}$	-0.02 ± 0.07	-0.52 ± 0.12	0.20 ± 0.20
$B^0 \to f_0(980)K^0$,	$\mathcal{B}$	6.0 ± 1.6	5.9 ± 1.6	5.8 ± 2.8
$f_0 \to \pi^+\pi^-$	$\mathcal{A}$	-0.14 ± 0.22	0.01 ± 0.10	0.0004 ± 0.0010
	$\mathcal{S}$	-0.39 ± 0.26	-0.63 ± 0.09	-0.77 ± 0.0004
$B^{\pm} \to (K^+K^-)_S K^{\pm}$	$\mathcal{B}$	< 2.9[13]	1.8 ± 0.4	1.7 ± 0.7
	$\mathcal{A}_{CP}$	no data	-0.44 ± 0.12	0.29 ± 0.21
$B^0 \to (K^+K^-)_S K^0_S$	$\mathcal{B}$	no data	1.1 ± 0.3	1.2 ± 0.5
	$\mathcal{A}$	(-0.09 ± 0.10)	0.01 ± 0.10	0.001 ± 0.001
	$\mathcal{S}$	(-0.55 ± 0.22)[14] (-0.74 ± 0.27)[16]	-0.64 ± 0.09	-0.77 ± 0.0006
$B^0 \to (K^0_S K^0_S)_S K^0_S$	$\mathcal{B}$	no data	1.1 ± 0.3	1.2 ± 0.5
	$\mathcal{A}$	(0.41 ± 0.21)	0.01 ± 0.10	0.001 ± 0.001
	$\mathcal{S}$	(-0.26 ± 0.34)	-0.64 ± 0.09	-0.77 ± 0.0006

and neutral B decays can be seen in the figures shown in FKLL[1].

To conclude, our effective mass distributions and branching fractions compare well with data. The long distance contribution $C(m)$ is important. If $C(m) = 0$,

$\mathcal{B}(B^0 \to \pi^+\pi^- K_S^0)$ is too small by a factor ~ 18 and $\mathcal{B}(B^\pm \to \pi^+\pi^- K^\pm)$ by a factor ~ 4. If charming penguins are included the agreement with data is good with a χ compatible with the $f_0(980)$ properties. The recent determination[15] of $\mathcal{A}_{CP} = -0.02 \pm 0.07$ for $B^\pm \to f_0(980)K^\pm$, $f_0 \to \pi^+\pi^-$, should help to determine better the charming-penguin $C(m)$ parameters. In progress we are adding the P-wave contribution which will allow us to describe the ρ-meson region.

Acknowledgments

We acknowledge good advices from B. El-Bennich. This work benefits from IN2P3-Polish Laboratories Convention (project N^0 99-97).

References

1. A. Furman, R. Kamiński, L. Leśniak and B. Loiseau, Phys. Lett. **B622**, 207 (2005), *Long-distance effects and final state interactions in $B \to \pi\pi$ and $B \to K\bar{K}K$ decays.*
2. S. Gardner and U.-G. Meißner, Phys. Rev. **D65**, 094004 (2002), *Rescattering and chiral dynamics in $B \to \rho\pi$ decay.*
3. *The BABAR Physics Book: Physics at an Asymmetric B Factory* by BABAR collaboration, P. F. Harrison and H. R. Quinn, Editors, report SLAC-R-504, Oct. 1998 (see in particular chapters 2 and 10).
4. M. Beneke, G. Buchalla, M. Neubert and C. T. Sachrajda, Phys. Rev. Lett. **83**, 1914 (1999), *QCD factorization for $B \to \pi\pi$ decays: strong phases and CP violation in the heavy quark limit;*
5. M. Bauer, B. Stech and M. Wirbel, Z. Phys. **C34**, 103 (1987), *Exclusive non-leptonic decays of $D-$, D_s- and $B-$ mesons.*
6. A. Ali, G. Kramer and C.-D. Lü, Phys. Rev. **D58**, 094009 (1998), *Experimental tests of factorization in charmless non leptonic two-body B decays.*
7. M. Beneke, M. Neubert, Nucl. Phys. **B675**, 333 (2003), *QCD factorization for $B \to PP$ and $B \to PV$ decays.*
8. N. de Groot, W. N. Cottingham and I. B. Whittingham, Phys. Rev. **D68**, 113005 (2003), *Factorization fits and the unitarity triangle in charmless two-body B decays.*
9. M. Ciuchini, E. Franco, G. Martinelli and L. Silvestrini, Nucl. Phys. **B501**, 271 (1997), *Charming penguins in B decays.*
10. R. Kamiński, L. Leśniak and B. Loiseau, Phys. Lett. **B413**, 130 (1997), *Three channel model of meson-meson scattering and scalar meson spectroscopy.*
11. U.-G. Meißner and J . A. Oller, Nucl. Phys. **A679**, 671 (2001), *$J/\psi \to \phi\pi\pi(K\bar{K})$ decays, chiral dynamics and OZI violation.*
12. M. Ciuchini, E. Franco, G. Martinelli, A. Masiero, M. Pierini and L. Silvestrini, talk at Rencontre de Moriond, March 2004, hep-ph/0407073, *Two-Body noleptonic B decays in the Standard Model and beyond.*
13. A. Garmash *et al.*, Belle Collaboration, Phys. Rev. **D71**, 092003 (2005), *Dalitz analysis of the three-body charmless decays $B^+ \to K^+\pi^+\pi^-$ and $B^+ \to K^+K^+K^-$.*
14. B. Aubert *et al.*, BaBar Collaboration, Phys. Rev. **D71**, 091102(R) (2005), *Measurement of CP asymmetries in $B^0 \to \phi K^0$ and $B^0 \to K^+K^-K_S^0$ decays.*
15. J. Alexander *et al.*, The Heavy Flavor Averaging Group (HFAG), http://www.slac.stanford.edu/xorg/hfag/index.html.
16. K.-F. Chen *et al.*, Belle Collaboration, Phys. Rev. **D72**, 012004 (2005), *Time-dependent CP-violating asymmetries in $b \to s\bar{q}q$ transitions.*

The Search for the Pentaquark

Wolfgang Eyrich

Physical Institute, University of Erlangen-Nuremberg,
Erwin-Rommel-Str. 1,
91058 Erlangen, Germany
E-mail: eyrich@physik.uni-erlangen.de

In the last two years, starting with the LEPS collaboration[1] several experiments reported evidence for a manifestly exotic narrow state with a mass of about 1530 MeV/c^2. The state was found to decay into K^0p and K^+n. This object with strangeness $S = +1$ was named Θ^+ and identified with the lightest exotic antidecuplet baryon predicted in the soliton model[2]. Many experiments have scanned their data for a pentaquark signal with varying results. Some searches resulted in evidence for the Θ^+ while others fail to produce any narrow structure in the region of interest. Currently, a number of high statistics experiments are being evaluated with the goal to confirm or refute the existence of the Θ^+. In this contribution the experimental status and further prospects will be discussed.

1. Introduction

As presently understood, QCD does not forbid the existence of states other than quark-antiquark and three-quark systems as long as they form colour singlets. In fact already in the early phase of QCD motivated models systems consisting of more than three quarks, in particular five quark systems have been discussed[3]. Pentaquark states consist of four quarks and one antiquark. Among these the so called exotic pentaquarks having an antiquark with different flavour than their quarks (e.g. $uudd\bar{s}$) are of most interest. Experimental searches starting in the sixties reported some evidence on such states which, however, were not confirmed. In more recent theoretical publications the possible existence of such pentaquark states has been worked out based on specific assumptions and production scenarios, also including heavy quarks (e.g.[4,5] and others). One of the most cited publications[2] is based on the soliton model assuming an antidecuplet as third rotational excitation in a three flavour system. The corners of this antidecuplet are occupied by exotic pentaquark states with the lightest state having a mass of $\approx 1530\,MeV/c^2$, strangeness +1, spin $\frac{1}{2}$ and isospin 0. This state, originally known as the Z^+, has more recently been renamed Θ^+. In this model the mass of the Θ^+ is fixed by the N^* resonance at $1710\,MeV/c^2$, which is assumed to be a member of the antidecuplet. The most striking property of the Θ^+ resonance is the predicted narrow width of $\Gamma < 15\,MeV/c^2$. With the predicted quark content of $uudd\bar{s}$ this pentaquark resonance is expected to decay into the channels K^+n and K^0p.

2. Experimental Evidence on Θ^+

The first report on the discovery of a narrow resonance in the expected mass region came from the LEPS collaboration at SPring8[1] where in the γK^- mass spectrum of the reaction $\gamma n \rightarrow K^+K^-n$ on ^{12}C a narrow resonance was observed at $1.54 \pm 0.01\, GeV/c^2$ with a significance of 4.6 σ and an upper limit for the width of $\Gamma = 25\, MeV/c^2$. This was confirmed by the DIANA collaboration at ITEP[6] which observed a narrow resonance with a mass of $1539 \pm 2\, MeV/c^2$ and a width of $\sigma = 3\, MeV/c^2$ in the K^0p invariant mass spectrum in reanalysed data from the reaction $K^+Xe \rightarrow K^0pXe'$ with a quoted evidence of 4.4 σ. In the meantime several other experiments have presented successful observations in the mass region between 1526 and 1555 MeV/c^2. The CLAS collaboration[7] reported on a narrow peak in the K^+n system produced in the reaction $\gamma d \rightarrow pnK^+K^-$ at a mass of $1542 \pm 2\, MeV/c^2$ and a width of $\Gamma < 21\, MeV/c^2$ and a narrow peak around $1555\, MeV/c^2$ from the reaction $\gamma p \rightarrow nK^+K^-\pi^+$ [8]. A further γ-induced measurement of the reaction $\gamma p \rightarrow n\bar{K}^0K^+$ has been reanalysed by the SAPHIR collaboration[9]. The observed mass of $1540 \pm 4 \pm 2\, MeV/c^2$ and the width of $\Gamma < 25\, MeV/c^2$ is in agreement with the recent experiments. Moreover, evidence comes from neutrino-scattering on nuclei investigating the pK_s^0 system at ITEP[10], where a peak was found at 1533 MeV/c^2. Positive results were also reported from the HERMES collaboration[11], where a narrow baryon state was found at 1528 MeV/c^2 in quasi-real photo production on a deuterium target in the decay channel pK_s^0, and from the SVD collaboration[12], where in the pA interaction a peak shows up at 1526 MeV/c^2 in the $pK_s^0 \rightarrow p\pi^+\pi^-$ system. The ZEUS collaboration[13] used e^+p collisions at $\sqrt{s} \approx 300$ GeV and observed a peak in the pK_s^0 invariant mass spectrum at 1522 MeV. Moreover, they observed for the first time evidence also for the $\bar{\Theta}$ resonance. Evidence was also found by the COSY-TOF[14] experiment. Here a peak at 1530 MeV was observed in the pK_s^0 invariant mass spectrum using the reaction $pp \rightarrow \Sigma^+K_s^0p$ very close to threshold.

Within the errors practically all experiments measure a Θ^+ width consistent with the experimental resolution. From the non-observation of the Θ^+ in previous experiments and the analysis of KN- scattering data a width in the order of 1 MeV or less is estimated[15]. A high precision measurement of the reaction $p(K^+, \pi^+)$ which is underway e.g. at KEK[16] should provide strongly improved information on the width of the Θ^+.

There is an ongoing discussion whether the observed spreading and the difference between the mean values of the reported Θ^+ masses for the two decay channels is acceptable considering the individual errors. However this picture was changed significantly by the preliminary results of the new experiments discussed in the next chapter. A detailed review of the experimental situation of pentaquarks is given e.g. by K. Hicks[17].

3. Searching for the Θ^+ - Second Round

It is obvious that higher statistics is necessary in order to confirm or refute the pentaquark observations and to get a better understanding of the production mechanism. In 2004 a second round of experiments started in this direction. The preliminary output reported up to now is, however still confusing. Already at the pentaquark04 conference the LEPS collaboration[18] reported evidence in the reaction $\gamma d \rightarrow K^- pX$. After correction of Fermi motion they observed a peak in the γK^- mass spectrum around 1530 MeV/c^2. And very recently they reported evidence[19] from the same reaction with the special condition $M(K^-p) = \Lambda^*(1520)$. Again a peak was found at a mass around 1530 MeV/c^2 with a significance of about 5 standard deviations. On the other hand, very recently the CLAS collaboration reported negative results[20,21] from high statistics experiments in the photo production measurement of the two reaction channels $\gamma p \rightarrow K_s K^+ n$ and $\gamma d \rightarrow K^- p K^+ n$. Whereas the first negative result is in conflict with the published positive result of SAPHIR the second result has to be confronted with the positive evidence in the CLAS data of the first measurement. It was shown that a corrected background reduces the significance of the published signal to a value of about three standard deviations.
Very recently SVD-2 reported new results[22] of an improved analysis of pA-interactions. In the $K^0 p$ spectrum again a signal was found around 1525 MeV/c^2. Compared to their published data they strongly improved statistics and now claim evidence of 8.0 standard deviations which is the highest value of all reported evidences. There are ongoing analyses of other experiments (e.g. COSY-TOF) which measured very recently to improve the precision of their data.

4. Search at COSY-TOF

The COSY-TOF experiment is discussed a little more in detail here, since the author of this article is member of the COSY-TOF collaboration. Using the wide angle TOF detector at the COSY storage ring the hadronic reaction $pp \rightarrow \Sigma^+ K^0 p$ was measured in two runs (in 2000 and 2002) exclusively at a beam momentum of 2.95 GeV/c. As shown in Fig. 1 a narrow peak at $1530 \pm 5\, MeV/c^2$ was observed in the invariant mass spectrum of the $K^0 p$ subsystem with a significance of 4-6 standard deviations, depending on background assumptions. The data cover the full phase space. This enabled COSY-TOF as the first experiment to search for the Θ^+ in the corresponding Dalitz plot.
End of 2004 a beam time was performed with the goal to get more precise information on the existence or non-existence of the exotic Θ^+ state. Again, the reaction $pp \rightarrow \Sigma^+ K^0 p$ was studied. To separate the region of interest around 1.53 GeV more from the kinematical limit of the $K^0 p$ mass spectrum a slightly higher beam momentum of $p_{beam} = 3.05$ GeV/c was chosen. To improve the reconstruction efficiency the experiment was upgrade by a new fiber hodoscope of the inner detector. From MonteCarlo simulations an increase of the reconstruction efficiency of more than 50% is expected for the reaction channel of interest. The expected overall gain

for the number of events in the reaction channel of interest $pp \to \Sigma^+ K^0 p$ is at least a factor of five compared to the published data. This can be estimated from the reaction channel $pp \to K^+ \Lambda p$ which has the same trigger condition and a similar event pattern with a delayed decay and which was measured simultaneously. This channel has a higher reconstruction efficiency in our detector. The much larger event sample of the new measurement will allow to investigate in detail not only the invariant $K^0 p$ spectrum but also the Dalitz plot. To control all steps of the analysis chain in an optimal way and to minimize systematic errors, independent analyses at several institutes are performed. These analyses are based on a common calibration database but use different codes which are partly emphasising different aspects of the detector. Moreover improved MonteCarlo simulations will be used, including background modelling.

5. Other Pentaquark States

Apart from the positive Θ^+ results evidence is also reported for other pentaquark states. Preliminary evidence for a Θ^{++} partner of the Θ^+ is reported by the STAR collaboration[23], where a peak is seen in the pK^+ and $\bar{p}K^-$ invariant mass at 1.53 GeV/c^2. The NA49 experiment[24] has observed in pp reactions the pentaquark candidates Ξ^{--}, Ξ^0 and the corresponding antiparticles at around 1862 MeV/c^2. Assuming this result will be confirmed a consequence would be that the spacing in the antidecuplet has to be reduced significantly and accordingly the $N^*(1710)$ could not be a member of this scheme. Preliminary evidence for a narrow resonance at 1670 MeV which would roughly fit into the modified spacing is reported by the GRAAL experiment[25]. Moreover, evidence for a charmed pentaquark at a mass of about 3.10 GeV/c^2 is reported by the H1 experiment[26]. However, all these results suffer from the fact that each of them is seen in only one experiment and other experiments claiming to have similar significance do not confirm these evidences.

6. Summary and Outlook

The first round of observations of pentaquark candidates looked on the one hand very promising, but on the other hand they all suffer from the low statistical significance. Moreover, there is a list of experiments which did not find any narrow structure in the region of interest. This holds for the Θ^+, and even more for the other pentaquark candidates which have the additional problem that they have been observed in only one experiment each. Very recently a series of high statistics experiments and refined analyses has been started to clarify the situation. The actual result is still confusing. Whereas two CLAS measurements show negative results giving strong arguments against the existence of the Θ^+ there is improved evidence from LEPS and SVD-2. The analysis of other experiments as, e.g., COSY-TOF, ZEUS and CLAS is still not completed. It is obvious that this puzzle has to be resolved in a common experimental and theoretical effort as soon as possible.

References

1. LEPS Collaboration, T. Nakano et al. Phys. Rev. Lett. **91** (2003) 012002
2. D. Diakonov, V. Petrov and M. V. Polyakov, Z. Phys. A **359** (1997) 305
3. R.L. Jaffe, Proc. Topical Conference on Baryon Resonances, Oxford, July 1976, SLAC-PUB-1774
4. C. Gignoux, B. Silvestre-Brac, and J.M. Richard, Phys. Lett. B **193** (1987) 323
5. H. J. Lipkin, Phys. Lett. B **195** (1987) 484
6. DIANA Collaboration,V. V. Barmin et al., hep-ex/0304040 V.V. Barmin et al. Phys. Atom. Nucl. **66** (2003) 1715; Yad. Fiz. **66** (2003) 1763
7. CLAS Collaboration, S. Stepanyan et al., hep-ex/0307018 S. Stepanyan et al., Phys. Rev. Lett. **91** (2003) 252001
8. CLAS Collaboration, V. Kubarovsky et al., Phys. Rev. Lett. **92** (2004) 032001
9. SAPHIR Collaboration, J. Barth et al., Phys. Lett. B **572** (2003) 127
10. A.E. Asratyan, A.G. Dolgolenko and M.A. Kubantsev, hep-ex/0309042
11. HERMES Collaboration, A. Airapetian et al., hep-ex/0312044
12. SVD Collaboration, A. Aleev et al., hep-ex/0401024
13. ZEUS Collaboration, S. Chekanov et al., Phys. Lett. B **591** (2004) 7
14. COSY-TOF Collaboration, M. Abdel-Bary et al., Phys. Lett. B **595** (2004) 127
15. R. A. Arndt et al., Phys. Rev. C **68** (2003) 042201
16. K. Imai et al., experiment E559 at KEK
17. K. H. Hicks, Prog. Part. Nucl. Phys. **55** (2005) 647
18. T. Nakano et al.; Contr. Pentaquark04, SPring8, Japan, 2004
19. T. Nakano et al.; Contr. QCD2005, Beijing, China, 2005
20. R. de Vita, Contr. APS, Tampa, USA, 2005
21. V. Burkert, Contr. LP2005, Uppsala, Sweden, 2005
22. SVD-2 Collaboration, A. Aleev et al., hep-ex/0509033
23. H. Huang, Contr. QCD2005, Beijing, China, 2005
24. NA49 Collaboration, C. Alt et al., Phys. Rev. Lett. **92** (2004) 042003
25. V. Kouznetsov et al., Contr. workshop Pentaquarks, Trento (2004)
26. H1 Collaboration, A. Aktas et al., Phys. Lett. B **588** (2004) 17

From Spectroscopy of Mesonic Atoms to a Search for Deeply Bound Kaonic States

M. Iwasaki

DRI, RIKEN, Wako-shi, Saitama, 351-0198, Japan

Recently we reported the formation of a so-called strange tribaryon $S^0(3115)$ in the proton spectrum from the negative kaon reaction at rest in a helium target. The experiment was triggered by a recent theoretical prediction of a deeply bound kaonic state, as well as a series of exotic atom experimental results. It is still not clear that the strange tribaryon is due to the formation of a deeply bound kaonic state. In the present paper, we overview the present situation of the experimental studies.

1. Introduction

Our picture of the strong interaction in the low energy region is somewhat strange in many ways. Finite sized particles, baryons $\langle qqq \rangle$ and mesons $\langle q\bar{q} \rangle$, consist of quarks and exchange gluons. The quarks in these hadrons cannot be separated because of the color-confinement mechanism. On the other hand, nuclei are composed of baryons (nucleons) glued together by the exchange of virtual mesons. In this system, baryons stay in nuclei as if they keep their identity (mass, spin, *etc.*), and conserve their own volume in nuclei ($radii \propto A^{1/3}$). At the high density, a totally new phase, the "quark gluon plasma (QGP)" is expected to arise in heavy nuclear collisions at RHIC energies, and an extensive study is in progress. Is there any other way to detect *quark degree-of-freedom* in the nucleus?

2. Mesonic Atoms

Meson properties in the nuclear medium are one of the most interesting subjects. The typical experimental approach to study the *meson-nucleon* interaction at the lowest (threshold) energy is to observe the energy shift and width of the low-lying states of mesonic atoms. The conventional method to form a mesonic atom is to stop mesons in a target and feed them into an atomic orbit by the Auger process. However, it was believed for long time that the meson would be instantly absorbed and annihilated if it came close to the nucleus. Actually, this method does not work for the study of low-lying atomic orbits in a heavy nucleus, because the mesons react with nuclei much before mesons reach these states by the cascade process.

2.1. *Pionic atoms*

For pions, the situation has been changed by a novel technique established in a GSI experiment. In the experiment [1], the formation of a pionic-atom ground state in heavy nuclei has been clearly seen for the first time, as shown in Fig. 1. The

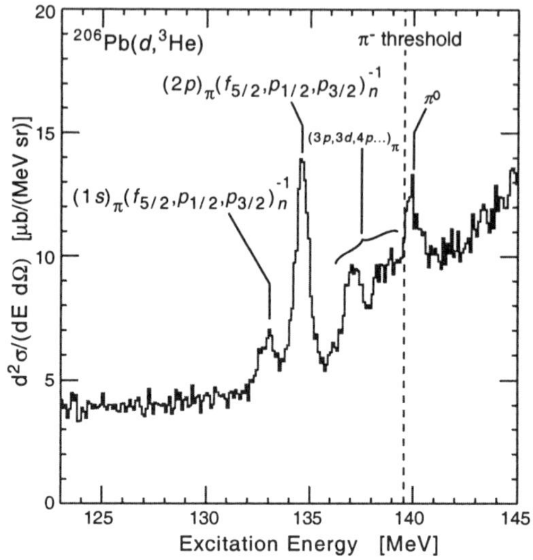

Fig. 1. Deeply bound pionic atom signal. The yield from the ^{206}Pb(d,^{3}He) reaction is plotted as a function of the excitation energy. Atomic orbits for the pion are labeled together with possible excited neutron-hole states in the nucleus. A peak marked with π^0 is for the energy calibration using the reaction $p(d,^3\text{He})\pi^0$.

spectrum is obtained by the (d,^{3}He) reaction on a ^{206}Pb target in which a neutron is picked up as a proton, leaving behind a negative pion. The key to forming such a deeply bound pionic atom is the pion production at the surface of the nucleus by a nucleon picking-up. Because ^{3}He is heavier than d by $\sim$ 1 GeV/c^2, a recoilless condition can be realized for produced particles with mass below that (*e.g.* pion).

The radius of the pionic 1s orbit is comparable with that of the nucleus. The calculation by Hirenzaki *et al.* [2] shows that the pion wave function is pushed away from the nucleus because of the s-wave repulsion in this system, but it has finite overlap at the surface of the nucleus. A further experimental study of deeply bound pionic states in Sn isotopes was performed to deduce pion properties in the nuclear medium [3].

2.2. *Kaonic atoms*

Conventional techniques is the only way to form kaonic atoms because of the strangeness. Thus the 1s level is observed only in hydrogen. The first reliable data reported in the KEK experiment [4] (shown in Fig. 2).

Kaonic atom experiments are more difficult because of the much shorter lifetime of the kaon than the pion, and the huge x-ray background due to kaon absorption reactions in the target nuclei. Also, because the kaonic hydrogen atom is neutral, the Stark effect is very strong, resulting in a very low x-ray yield. This requires the use of a lower-density gaseous hydrogen target even for the high momentum kaons at KEK.

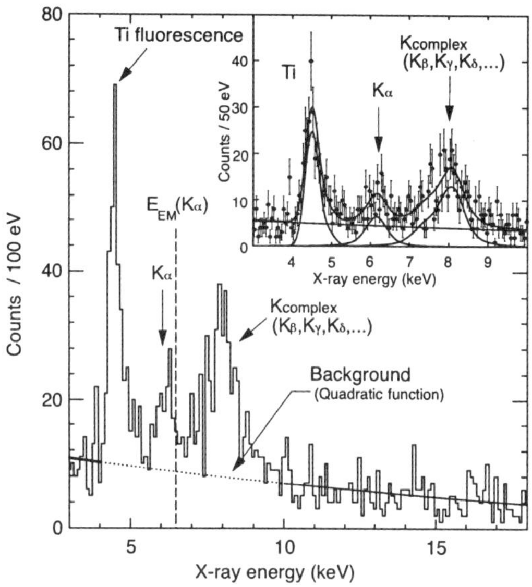

Fig. 2. Kaonic hydrogen x-ray spectrum. Kaonic hydrogen x rays together with Ti electronic K_α lines for energy calibration are seen. To reduce the background originating from the π^0 (or γ) originated x-ray background, and to select a fiducial cut on kaons stopping in the gaseous hydrogen target, a two-charged-pion signal from the K^-p reaction is required.

A recent experiment at INFN utilized the low momentum kaons from ϕ decay and thin layered CCD x-ray detector [5]. Both reported a repulsive energy shift of the 1s level. This repulsive shift is due to the presence of the lower-lying $\Lambda(1405)$ resonance, which pushes the kaonic hydrogen 1s level upward in energy (repulsive shift does not mean repulsive strong-interaction between $\overline{K}N$).

3. Search for Kaonic Nuclei

There are two possible ways to to interpret the present experimental data on kaonic hydrogen; i) conventional optical-model calculations give a consistent result with the experimental data (excluding kaonic helium x-ray data) using an attractive but strongly absorptive interaction, or ii) a coupled channel calculation by Akaishi and Yamazaki [6], in which the $\Lambda(1405)$ is treated as a bound state between the K^- and the proton, suggests that the $\overline{K}N$ interaction is extremely attractive and less absorptive, thus stable K-mesonic state formation is allowed in light nuclei.

In the latter framework, a kaon can survive longer than the typical time scale of the strong interaction, without loosing its identity as a meson in the nucleus. It is predicted that the negative kaon forms a meta-stable state in ^{3}He with total isospin $T = 0$ at a binding energy as large as ~100 MeV with an absorption width of ~20MeV. A high density baryon system is formed due to the strong attraction of the kaon.

The discovery of such a system will open a new research area to study hyper-dense matter in objects such as "compact stars", with densities greater than neutron stars, in the laboratory framework. In such a system, one may study the precursor effect toward the totally new $\langle\bar{q}q\rangle$-condensation phase, color super-conductivity, which may realize at the high density and low temperature limit. Because the constituent-quark or hadron mass is expected to be a function of $\langle\bar{q}q\rangle$-condensation strength, one may obtain a hint as to how hadronic mass is realized after the big bang.

Very recently, we performed an experimental search for the predicted deeply-bound kaonic state by kaon absorption at rest in a liquid helium target. In the experiment, we observed nucleon emission from the reaction, and discovered distinct mono-energetic proton formation [7] as shown in Fig. 3. This can be produced by

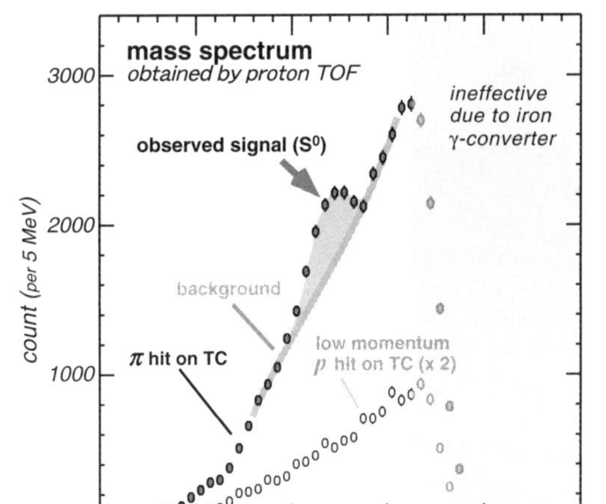

Fig. 3. Missing-mass spectrum of the semi-inclusive proton events. In the experiment, we required a charged particle to perform reaction-vertex analysis, in addition to nucleon production. Closed and open circles are for rough PID for the charged particle.

the two-body reaction of a proton together with the unknown object, S^0, whose mass is about 3115 MeV/c^2, namely,

$$(K^- \, ^4\text{He})_{\text{atomic}} \rightarrow S^0(3115) + p. \quad (1)$$

The observed system should have baryon number $B = 3$, charge $Z = 0$, isospin $T = 1$ and strangeness $S = -1$. The discovery of peak structure in the proton spectrum was quite astonishing to us, because we were searching for mono-energetic neutron formation with $T = 0$. It is pointed out that the observed peak may correspond to

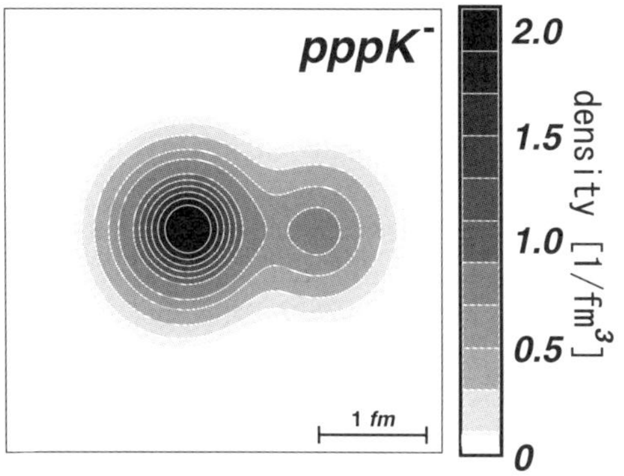

Fig. 4. Central nucleon density plot of $ppp K^-$ system, which could be a isospin partner of the observed state, calculated by antisymmetric molecular dynamics (AMD) method.

the isospin partner of the $pppK^-$ calculated by Doté et al. [8], whose isospin $T = 1$ and spin $J = 3/2$.

The nature of these peaks should be examined in the forthcoming experiments. There are two experimental programs scheduled in 2005 at KEK before the PS shutdown. One is for more detailed study of the tribaryon $S^0(3115)$ by inclusive proton data using a proton tracking device with better energy resolution. The other is to measure the kaonic helium x-ray transition from $3d$ to $2p$ to determine the $\overline{K}N$ interaction in more detail.

Acknowledgement

The author is grateful to all the collaborators of all the experiments mentioned in the paper.

References

1. H. Geissel et al., *Phys. Rev. Lett.* **88**, 122301 (2002).
2. S. Hirenzaki and H. Toki, *Phys. Rev.* **C55**, 215 (1997).
3. K. Suzuki et al., *Phys. Rev. Lett.* **92**, 072302 (1995).
4. M. Iwasaki et al., *Phys. Rev. Lett.* **78**, 3067 (1997).
5. G. Beer et al., *Phys. Rev. Lett.* **94**, 212302 (2005).
6. Y. Akaishi and T. Yamazaki, *Phys. Rev.* **C65**, 044005 (2002).
7. T. Suzuki et al., *Phys. Lett.* **B597**, 263 (2004).
8. A. Doté, H. Horiuchi, Y. Akaishi and T. Yamazaki, *Phys. Rev.* **C70**, 044313 (2004).

Search for the Θ^+ in Photoproduction on the Deuteron*

K. H. Hicks† for the CLAS Collaboration

Department of Physics and Astronomy
Athens, OH 45701, USA
E-mail: hicks@ohio.edu

A high-statistics experiment on a deuterium target was performed using a real photon beam with energies up to 3.6 GeV at the CLAS detector of Jefferson Lab. The reaction reported here is for $\gamma d \to pK^-K^+n$ where the neutron was identified using the missing mass technique. No statistically significant narrow peak in the mass region from 1.5-1.6 GeV was found. An upper limit on the elementary process $\gamma n \to K^-\Theta^+$ was estimated to be about 4-5 nb, using a model-dependent correction for rescattering determined from $\Lambda(1520)$ production. Other reactions with less model-dependence are being pursued.

1. Introduction and Results

The search for pentaquarks, made from four quarks and one antiquark, has captured the interest of the nuclear-particle physics community since the announcement of a possible experimental signal by the LEPS Collaboration[1]. Since then, there have been many results published, some positive and some null, and the reader is referred to a recent review[2] for more details.

Here, we focus on the reaction $\gamma d \to pK^-K^+n$ which was previously published[3] but with low statistics. The present results are for a high-statistics experiment, known as "g10", carried out using the same detector, the CEBAF Large Acceptance Spectrometer (CLAS) at Jefferson Lab. The experimental setup is the same as Ref. 3, known as "g2a", except for two items: (1) the beam energy was increased, allowing photons from 0.9-3.6 GeV; (2) the target was moved upstream by 25 cm to increase the acceptance for negative particles. The data analysis and event selection cuts used in the present analysis are the same as Ref. 3, and the photon energy range has been restricted (by software) to match as closely as possible the conditions of Ref. 3. In this sense, the analysis is a "blind" analysis, so that no bias was introduced in the high-statistics result.

In the g10 experiment, the data was taken at two magnetic field settings of the CLAS torus coils. At both field settings, a luminousity of about 25 pb^{-1} was collected, which is nearly 10 times the luminousity of the previous g2a data[3]. Only the high-field data will be presented, which matches the conditions of the g2a data,

*The g10 experiment at Jefferson Lab: Hicks and Stepanyan, co-spokesmen.
†This work is supported in part by the National Science Foundation

although the results from the low-field setting are found to be similar. After restricting the photon energy range to be the same as g2a, the g10 experiment had 5.9 times the luminousity of the g2a experiment. A comparison of the two experiments is shown in Fig. 1, where the missing mass of the pK^- system (equal to the mass of the nK^+ system, from possible Θ^+ decay) is plotted. The vertical scale shows the number of counts in the published g2a data[3] and the g10 data have been scaled by the luminosity, shown by the solid histogram.

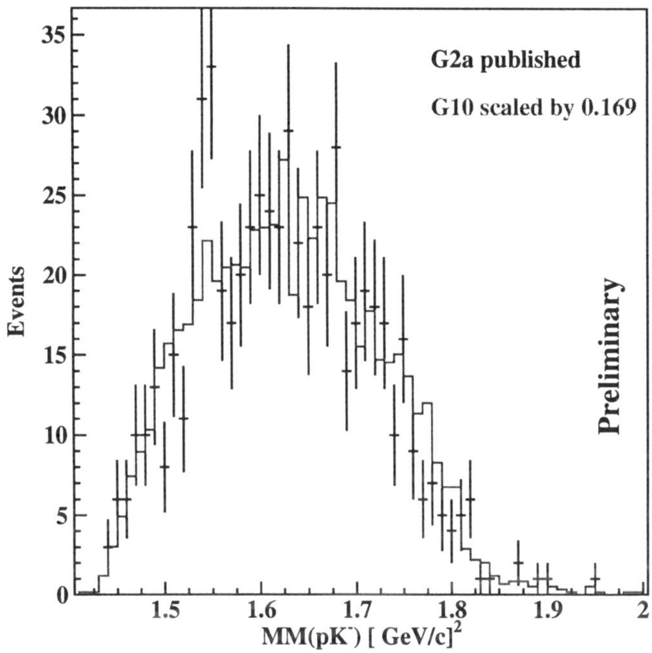

Fig. 1. Missing mass of the pK^- system for the reaction $\gamma d \rightarrow pK^- K^+ n$ measured at CLAS. The points with error bars is from Ref. [3], and the solid histogram shows the present (high-statistics) results, scaled down by the factor shown.

Clearly, the peak seen in the g2a data is not reproduced by the g10 data. Using the g10 data as a guide to the background shape, the probability of a fluctuation of the amount seen at 1.54 GeV in the g2a data is found to be about 3-σ (three standard deviations). The claim of a 5-σ statistical significance in Ref. 3 was due to a lower estimate of the background. We note that the g2a data fluctuate downward from the g10 shape on either side of the 1.54 GeV "peak". In hindsight, we see that the evidence for the Θ^+ claimed in Ref. 3 is due to a combination of an underestimate of the background shape and a statistical fluctuation in the region of 1.54 GeV.

These results show the importance of high statistics, along with a "blind" analysis procedure where the event selection criteria are determined before the experiment is done.

It is now a straight-forward procedure to fit the g10 mass spectra with an overall background shape (using a third order polynomial). Using a fixed background and fitting a gaussian (with a 6 MeV width, equal to the CLAS resolution) across the mass spectrum, an upper limit on the number of counts in the mass region of 1.54 GeV is found. Using the luminosity, along with and the gaussian fit results and a detector acceptance from Monte Carlo, an upper limit on the *measured* reaction on deuterium has been calculated. We assume a uniform angular distribution for Θ^+ production, even though the CLAS detector does not measure particles at forward angles (the angle is momentum-dependent but roughly 15°-20° lab for K^- and roughly 8°-10° lab for the K^+). This may not be a valid assumption if the Θ^+ is produced primarily at forward angles, as suggested by the LEPS data[1].

An upper limit on the cross section for the elementary reaction $\gamma n \rightarrow K^-\Theta^+$ is desired. Of course, the reaction we measured was on deuterium, not a free neutron. In order to convert from the measured result to the elementary reaction, a theoretical model must be used. The model is complicated by the fact that the proton is detected in CLAS, which requires it to have a momentum of $>$ 350 MeV to exit the liquid deuterium target. Ideally, the proton in the deuterium target would be a spectator to the elementary reaction, having nearly zero momentum (smeared by Fermi momentum). In the g10 experiment, the proton must gain momentum by final state (rescattering) reactions. In order to estimate the rescattering correction, we look to the mirror reaction $\gamma p \rightarrow K^+\Lambda(1520)$. In this mirror reaction, the neutron would be a spectator, and its momentum is found in g10 by the missing momentum. By cutting on the neutron momentum above 350 MeV, the rescattering probability of the mirror reaction is found to be about 0.10 ± 0.01. Assuming a similar correction for rescattering in Θ^+ production on deuterium, the cross section for the elementary process is estimated at 4-5 nb.

The model dependence in the above cross section estimate is undesirable but unavoidable. One could imagine other ways to do the rescattering correction, such as using the tail of the Fermi momentum above 350 MeV for the proton in deuterium. In this case, the upper limit is increased by a factor of 5, to 20-25 nb. Both estimates, one using the $\Lambda(1520)$ model and the other using the Fermi tail, are shown as a function of the nK^+ mass in Fig. 2. Other models might suggest a bigger rescattering probability, thus reducing the upper limit. Clearly, a measurement without a rescattering correction would be better. For example, the reaction $\gamma d \rightarrow K^-\Theta^+ p$ where the proton is not detected, and the decay $\Theta^+ \rightarrow K^0 p$ is measured, has less model dependence to deduce the elementary reaction cross section. Further analysis of the g10 data is in progress and more results are expected soon.

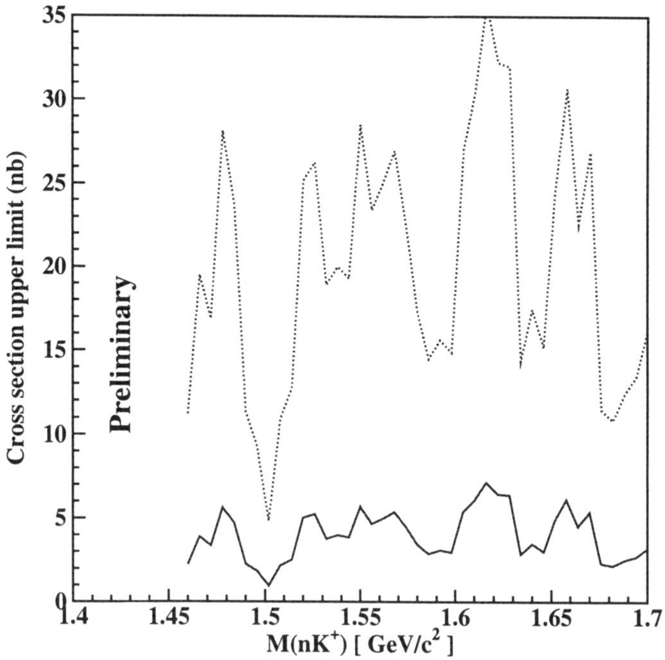

Fig. 2. Preliminary upper limit for the elementary reaction $\gamma n \rightarrow K^- \Theta^+$ using the $\Lambda(1520)$ for the rescattering correction (lower line) and the Fermi momentum tail for the correction (dotted upper line). All curves come from fitting a fixed-width gaussian on top of a fixed polynomial background, using the same mass spectrum from the g10 experiment.

2. Summary

The exclusive reaction $\gamma d \rightarrow pK^- K^+ n$ was measured at CLAS with high statistics. No evidence for a narrow peak in the nK^+ mass spectrum was observed, contrary to earlier low-statistics results[3]. A *model dependent* upper limit for the cross section in the elementary reaction $\gamma n \rightarrow K^- \Theta^+$ was estimated to be about 4-5 nb, using the $\Lambda(1520)$ as a model for the rescattering.

References

1. T. Nakano *et al.*, (LEPS), Phys. Rev. Lett. **91**:012002 (2003); hep-ex/0301020.
2. K.H. Hicks, Prog. Part. Nucl. Phys. **55** (2005) 647; hep-ex/0504027.
3. S. Stepanyan *et al.*, (CLAS), Phys. Rev. Lett. **91**:25001 (2003); hep-ex/0307018.

EFFECTIVE FIELD THEORY IN FEW-BODY PHYSICS

Bose-Einstein Condensation in Neutron Stars

Debades Bandyopadhyay
Saha Institute of Nuclear Physics
Kolkata-700064
India.

We have constructed equations of state for neutron star matter involving Bose-Einstein condensates of K^- and $\bar{K}^0$ mesons within relativistic mean field models. Here baryon-baryon and antikaon-baryon interactions are mediated by exchange of mesons. In this calculation, we consider both K^- as well as $\bar{K}^0$ condensation. Later those equations of state are exploited to investigate the mass-radius (M-R) relationship of neutron stars. We also discuss a new family of superdense stars containing antikaon condensates beyond the neutron star branch.

1. Introduction

There is a growing interplay between the physics of dense matter formed in laboratories and the physics of dense matter in compact stars. The study of dense matter formed in relativistic heavy ion collisions is enriching our knowledge about medium modification of hadron properties, strange matter including hyperons etc. On the other hand, the matter density at the core of compact stars could exceed by a few times normal nuclear matter density.Several novel phases with large strangeness fraction such as, hyperon formation and Bose-Einstein condensation may appear there [1].

Here we are interested to understand (anti)kaon properties in dense matter formed in relativistic heavy ion collisions as well as in strangeness condensation in a neutron star core. As (anti)kaons are produced at an early stage of relativistic heavy ion collisions, the in-medium properties of (anti)kaons might be studied through the measurements of collective flow and particle spectra in those reactions. The FOPI collaboration at SIS(GSI) observed antiflow for kaons in reaction Ni+Ni at $E_{beam} = 1.93A$ GeV [2]. Also, the directed flow of K_s^0 mesons has been observed in Au+Au collisions at 6A GeV [3]. The KaoS collaboration measured the azimuthal angular distributions of kaons for Au+Au reaction at 1A GeV [4] and found a large K^- production cross section in Ni+Ni collisions at (0.8-1.8)A GeV [5]. All these experimental results suggest that the in-medium kaon-nucleon interaction is repulsive whereas it is attractive for antikaon-nucleon [6,7]. Also, informations about antikaon-nucleon interaction in medium may be obtained by studying the K^--atomic data. The analysis of K^--atomic data using a phenomenological density dependent potential showed that the real part of antikaon optical potential could be as large as -180 ± 20 MeV [8]. Recently, Suzuki et al. [9] discovered tightly bound strange tribaryon $S^0(3115)$. It has been argued that this discovery of the 'strange nugget' might provide strong support for strangeness condensation in neutron star interior [10].

It was first demonstrated by Kaplan and Nelson within a chiral $SU(3)_L \times SU(3)_R$ model that K^- meson may undergo Bose-Einstein condensation in dense matter formed in heavy ion collisions [11] because of the strongly attractive K^--baryon interaction in dense matter. Later, K^- condensation in the core of neutron stars was studied by other groups using chiral models [12,13]. In this article, we investigate $\bar{K}$ (K^- and $\bar{K}^0$) meson condensation in neutron star matter in meson exchange models.

2. Equation of State

In this work, we investigate condensation of both K^- and $\bar{K}^0$ mesons in compact stars. Here K^- condensation is treated as first order phase transition whereas $\bar{K}^0$ condensation is considered as a second order phase transition. For first order K^- condensation, we have hadronic phase, antikaon condensed phase and a mixed of those two phases. In the hadronic phase, the constituents of matter are all the species of the baryon octet, electrons and muons whereas the antikaon condensed phase is composed of baryons of the octet embedded in the condensate, electrons and muons. We adopt relativistic field theoretical models to describe hadronic and antikaon condensed phase. Here baryon-baryon [14–17] and baryon-(anti)kaon [16–18] interactions are mediated by the exchange of scalar and vector mesons. The model is also extended to include hyperon-hyperon interaction through two additional strangeness meson $f_0(975)$ and $\phi(1020)$. Both hadronic phase and antikaon condensed phase are to satisfy beta-equilibrium and charge neutrality conditions. The mixed phase of antikaon condensed matter and hadronic matter is governed by Gibbs phase rules and global conservation laws [19]. The energy density and pressure in hadronic and antikaon condensed phases are given as in Ref. [17].

In neutron star interior, strangeness changing processes such as, $N \rightleftharpoons N + \bar{K}$ and $e^- \rightleftharpoons K^- + \nu_e$ may occur. Here $N \equiv (n,p)$ and $\bar{K} \equiv (K^-, \bar{K}^0)$ denote the isospin doublets for nucleons and antikaons, respectively. The requirement of chemical equilibrium yields

$$\mu_n - \mu_p = \mu_{K^-} = \mu_e \ , \tag{1}$$

$$\mu_{\bar{K}^0} = 0 \ , \tag{2}$$

where μ_{K^-} and $\mu_{\bar{K}^0}$ are respectively the chemical potentials of K^- and $\bar{K}^0$. The above conditions determine the onsets of antikaon condensations.

3. Results and Discussion

In this calculation, we adopt GM1 parameter set [20] where nucleon-meson coupling constants are determined from the nuclear matter saturation properties. The vector meson coupling constants for (anti)kaons and hyperons are determined from the quark model [14]. The scalar meson coupling constants for hyperons and antikaons are obtained from the potential depths of hyperons and antikaons in normal nuclear matter [14,17]. The phenomenological fit to the K^- atomic data yielded the real part of antikaon potential as $U_{\bar{K}} = -180 \pm 20$ MeV [8]. On the other hand, the recent microscopic calculations predict a shallow attractive potential at the saturation density [21]. We perform this calculation with an antikaon optical potential of -160 MeV at normal nuclear matter density ($n_0 = 0.153 fm^{-3}$). The coupling constants for strange mesons with (anti)kaons are given as $g_{f_0 K} = 2.65$ and $\sqrt{2} g_{\phi K} = 6.04$ [14]. The coupling constants for scalar strange meson-hyperons are calculated by fitting them to the potential depth for a hyperon in hyperon matter at the saturation density [14,17].

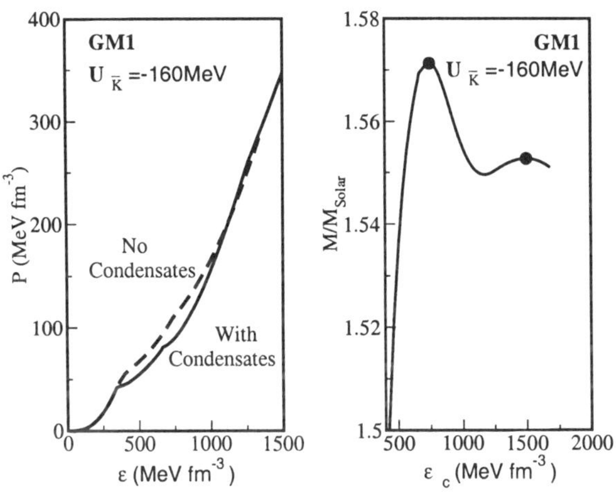

Fig. 1. The equation of state with and without antikaon condensates are shown in the left panel and the compact star mass sequences are plotted in the right panel.

We investigate the role of K^- and $\bar{K}^0$ condensation on the composition of β-equilibrated and charge neutral matter. In the pure hadronic phase, abundances of nucleons, electrons and muons increase with density. Here, charge neutrality is maintained among protons, electrons and muons. With the onset of K^- condensation, the mixed phase begins at $2.23n_0$. We find that Λ hyperon is the first strange baryon to appear in the mixed phase at $2.51n_0$. As soon as K^- condensate is formed, it rapidly grows with density and replaces electrons and muons. In the lowest energy state, K^- mesons are energetically more favourable to maintain charge neutrality than any other negatively charged particles. Consequently, the proton density becomes equal to the density of K^- condensate. Once the mixed phase is over, $\bar{K}^0$ condensation occurs at $4.06n_0$. With the appearance of $\bar{K}^0$ condensate, neutron and proton abundances become equal. The density of $\bar{K}^0$ condensate increases with baryon density uninterruptedly and even becomes larger than the density of K^- condensate. As soon as negatively charged hyperons - Ξ^- and Σ^- appear at higher densities, the density of K^- condensate is observed to fall drastically. This is quite expected because it is energetically favourable for particles carrying conserved baryon numbers to achieve charge neutrality in the system.

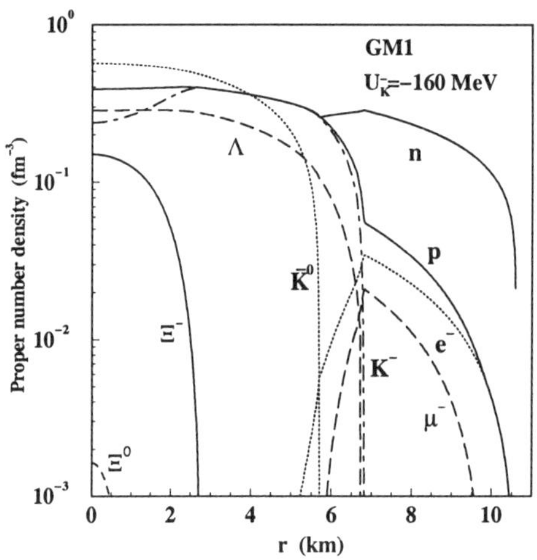

Fig. 2. The composition of a superdense star in the third family branch is plotted with Schwarzschild radius.

We study the high density behaviour of the equation of state (EoS) with and without antikaon condensates which are exhibited in the left panel of Figure 1. The curve indicating the overall EoS with hyperons and antikaon condensates (solid line) is softer compared with the EoS with hyperons and no condensate (dashed line). The kinks on the lower curve mark the beginning and end of the mixed phase. These kinks may lead to discontinuity in the velocity of sound. Here, we calculate the structure of compact stars calculated using Tolman-Oppenheimer-Volkoff equations [1] and the EoS with and without $\bar{K}$ condensates. The compact star mass sequences are shown with central energy density in the right panel of Figure 1. For the EoS with $\bar{K}$ condensates, it is found that after the positive slope neutron star branch, there is an unstable region followed by another positive slope compact star branch called the third family [22]. From the study of fundamental mode of radial vibration, we find that the third family branch is a stable one. However, there is no third family solution for the EoS without $\bar{K}$ condensate (not shown in the figure). The maximum masses of neutron star and third family branch are $1.571M_{\odot}$ and $1.553M_{\odot}$ corresponding to radii 12.8 km and 10.7 km. The compact star in the third family branch has a smaller radius than its counterpart in the neutron star branch. It was demonstrated that "non-identical" stars of same mass, but different radii and compositions, could exit because of partial overlapping mass regions of the neutron star branch and the third family branch. These pairs are known as "neutron star twins" [23]. The third family of compact stars were also investigated by other authors [17,23–27].

The distribution of particles in a compact star in the third family branch is shown as a function of Schwarzschild radius in Figure 2. The superdense star has a mass of $1.5526M_{\odot}$, radius of 10.7 km and central density $7.93n_0$. The mixed phase exists in the region between 5.74 km and 6.84 km. In the inner core of the star, the matter is dominated by $\bar{K}^0$ condensate and nucleons. The density of K^- condensate decreases with the appearance of Ξ^- hyperon.

4. Summary and Conclusion

We focus on the role of antikaon condensates on the composition, equation of state and structure of compact stars. We obtain a new class of superdense stars beyond the neutron star branch. Future experiments on K^--nuclei at J-PARC might provide important informations about the connection between 'strange nuggets' in laboratories and strangeness condensation in neutron stars.

References

1. N. K. Glendenning, Compact stars, (Springer, New York, 1997).
2. J.R. Ritman et al., Z. Phys. A352 (1995) 355; D. Best et al., Nucl. Phys. A625 (1997) 307; Y. Shin et al., Phys. Rev. Lett. 81 (1998) 1576.
3. P. Chung et al., nucl-ex/0112002.
4. Y. Shin et al., Phys. Rev. Lett. 81 (1998) 1576.
5. R. Barth et al., Phys. Rev. Lett. 78 (1997) 4007.
6. G.Q. Li, C.-H. Lee and G.E. Brown, *Phys. Rev. Lett.* **79**, 5214 (1997).
7. S. Pal, C.M. Ko, Z. Lin and B. Zhang, *Phys. Rev.* **C62**, 061903 (R) (2000).
8. E. Friedman, A. Gal, J Mareš and A Cieplý *Phys. Rev.*, **C60** 024314 (1999).
9. T. Suzuki et al., *Phys. Lett.* **B597**, 263 (2004).
10. G.E. Brown, C-H. Lee, H-J Park, M. Rho, *nucl-th* **0504029**
11. D.B. Kaplan and A.E. Nelson, *Phys. Lett.* **B175**, 57 (1986); A.E. Nelson and D.B. Kaplan, *Phys. Lett.* **B192**, 193 (1987).

12. M. Prakash, I. Bombaci, M. Prakash, Paul J. Ellis, J. M. Lattimer and R. Knorren, *Phys. Rep.* **280**, 1 (1997).
13. M. Hanauske, D. Zschiesche, S. Pal, S. Schramm, H. Stöcker and W. Greiner, *Astrophys. J.* **537**, 958 (2000).
14. J. Schaffner and I.N. Mishustin, *Phys. Rev.* **C53**, 1416 (1996).
15. S. Pal, D. Bandyopadhyay and W. Greiner, *Nucl. Phys.* **A674**, 553 (2000).
16. S. Banik and D. Bandyopadhyay, *Phys. Rev.* **C63**, 035802 (2001).
17. S. Banik and D. Bandyopadhyay, *Phys. Rev.* **C64**, 055805 (2001); S. Banik and D. Bandyopadhyay, *J. Phys.* **G28**, 1949 (2002).
18. N.K. Glendenning and J. Schaffner-Bielich, *Phys. Rev. Lett.* **81**, 4564 (1998); N.K. Glendenning and J. Schaffner-Bielich, *Phys. Rev.* **C60**, 025803 (1999).
19. N.K. Glendenning, *Phys. Rev.* **D46**, 1274 (1992).
20. N.K. Glendenning and S.A. Moszkowski, *Phys. Rev. Lett.* **67**, 2414 (1991).
21. J. Schaffner-Bielich, V. Koch and M. Effenberger, *Nucl. Phys.* **A669**, 153 (2000).
22. U.H. Gerlach, *Phys. Rev.* **172**, 1325 (1968).
23. N. K. Glendenning and C. Kettner, *Astron. Astrophys.* **353**, L9 (2000).
24. K. Schertler, C. Greiner, J. Schaffner-Bielich and M.H. Thoma, *Nucl. Phys.* **A677**, 463 (2000).
25. J. Schaffner-Bielich, M. Hanauske, H. Stöcker and W. Greiner, *Phys. Rev. Lett.* **89**, 171101 (2002).
26. E.S. Fraga, R.D. Pisarski and J. Schaffner-Bielich, *Phys. Rev.* **D63**, 121702 (2001).
27. S. Banik and D. Bandyopadhyay, *Phys. Rev.* **D67**, 123003 (2003).

Elastic Properties of Argon under High Pressure

S. Gupta, D. Gupta, P. Gupta and S.C. Goyal
Department of Physics, Agra College
Agra-282002.INDIA
E-mail:suresh_c_goyal@rediffmail.com

Elastic properties and their pressure dependence of Ar are calculated using the many body potential. The second order elastic constants, Cauchy's deviation and Zener anisotropy ratio are investigated as a function of pressure up to 75 GPa. The variation of pressure derivative of bulk and shear moduli with pressure are also computed for the first time. The computed results are fairly in agreement with the experimental results.

1. Introduction

Rare gas solids are an important class of materials[1,2] to provide an ideal system allowing fruitful comparisons between the experiments and the theoretical calculation. A considerable amount of efforts has been expended in attempts to determine the interatomic potentials for rare gas solids, including possible effects of many body forces[3–6]. The elastic properties of dense Ar are of fundamental interest as the hydrostatic pressure medium for high-pressure research in a diamond anvil cell (DAC) and as a component in rocks and the atmosphere of planetary bodies..At ambient pressure, Ar liquefies at 87.3 K and solidifies in the fcc phase at 83.8 K. At 300K, the liquid Ar crystallizes into the same phase at about P=1.3 GPa[7,8]. With increasing pressure, a single crystal Ar grown in the DAC always recrystallizes to a small group of single crystals at a pressure of about 4 GPa[7–9]. This crystalline behavior of solid Ar needs careful analysis above P=4 GPa in order to obtain the reliable elastic properties, because the acoustic velocities are sensitive to the crystal direction[8]. Gewurtz and Stoicheff[9] measured the Brillouin scattering of the crystalline Ar at 82.3K and 1 atm. and determined three adiabatic elastic constants. Grimsditch et al.[9] determined the pressure dependence of three elastic constants up to 33 GPa at 300 K by Brillouin spectroscopy... It was the first time that such a complete data set exists for any material at ultrahigh pressure and it has provided an ideal case for investigating the validity of the pair potential concept, which is usually employed in calculations. The elastic constants through the Cauchy relation indicate that a complete description of the elastic constants is not feasible using a pair potential model. The fact that noncentral forces must be incorporated to produce a complete description is also confirmed by the calculation, which attempts to describe dense Ar. There is no point in trying to refine an effective two-body potential. The first attempts to incorporate many body effect into the binding energy calculations of rare gas solids came from Axilrod and Teller[10] who approximated the three body non-additivity by the first term in the multipole expansion of the third order dispersion energy, called the ATM or triple dipole (ddd) term. Recently many authors[11,14,15] have successfully explained the second order elastic constants and other related properties of Ar at different pressure using ab initio method. These studies are limited to explain only the harmonic properties of Ar at different pressure. However they have not explained the variation of dK/dP, dCS/dP and dC_{44}/dP with pressure. In the present study we are trying to explain both harmonic and anharmonic elastic properties of Ar at different pressure. The various workers[16,17] in explaining the different properties of rare gas solids have used Lundqvist potential[18] extensively. Therefore, we have used this three-body potential [18] to derive the expression for the second-order elastic constants and the first-order pressure derivative of bulk and shear moduli of rare gas solids, including the effect of pressure.

2. Theory

The following relations for the potential energy for the rare gas solids[16] is considered.

$$\phi_{(r)} = \frac{A'\lambda[\lambda + 12f(r)]}{r^6} + Nw(r) + \frac{9}{8}K_B\theta_D \tag{1}$$

where the terms and symbols are defined in[6]. The following expression for the pressure (P) is derived using the equation of state[19] and the above potential (11) and is given below:

$$P = \frac{2}{a}[0.4516\frac{\alpha'}{a^8} - B + \frac{9}{32}\frac{K_B\theta_D\gamma}{a^2}] \tag{2}$$

The equations for SOE constants C_{11}, C_{12} and C_{44} at different pressure P can be derived by using the potential energy of rare gas solids[16] in the homogeneous deformation theory[20]. Following the procedure[6], one may derive the expressions for the first order pressure derivatives $dK/dP, dCs/dP, dC_{44}/dP$ and of bulk and shear module.

3. Result and Discussion

The calculated values of the parameters are used to compute the three elastic constants i.e. C_{11}, C_{12} and C_{44} of Ar up to 75 GPa pressure. The values of these three SOE constants (C_{ij}) are increasing as the pressure increases. Shows the pressure dependence of the elastic stiffness coefficient C_{11}, C_{12} and C_{44}. The elastic stiffness coefficients increase linearly with increasing pressure and variationis is in agreement with the results obtain by first principles method by Litaka & Ebisuzaki[11] and with

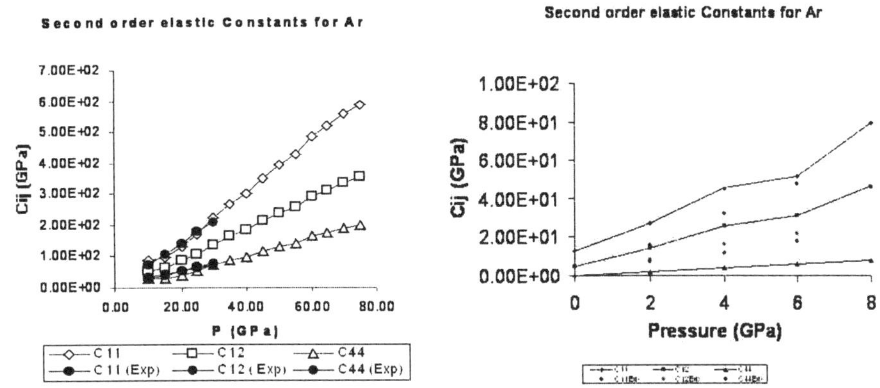

Fig. 1. The graph between Cij and Pressure

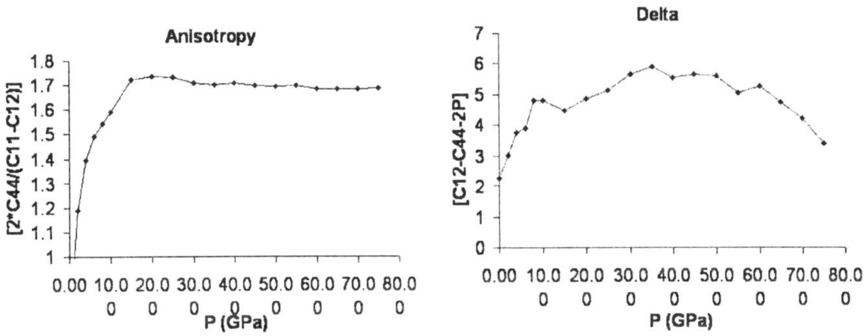

Fig. 2. The graph between anisotropy (A) and Pressure. & The graph between Cauchy deviation (δ) and Pressure.

the experimental values by Shimizu et al.[4]& Grimsditch et al.[9]. In the lower range of pressure the variation of elastic constants is not clearly linear but changing in irregular manner up to 5Gpa[9]. The calculated values also are indicating that the velocity of longitudinal wave changes more than the velocity of transverse wave as the pressure increases.

It is interesting to note from fig. (3) of Litaka & Ebisuzaki[11] that their experimental value of anisotropy[4–9] and the theoretical value[11] of anisotropy are different from each other. However, the present calculated values of anisotropy are in agreement with the experimental values of Grimsditch et al.[9]. AZ can characterize the elastic anisotropy of a cubic crystal. Several interesting effects can be noted by comparison of the ratio of two independent shear moduli C_{44} and $(C_{11} - C_{12})$ along the [110] and [100] directions in the [100] plane. For isotropic elasticity the ratio of two shear moduli (Zener ratio) i.e. $AZ = 2C_{44}/(C_{11} - C_{12})$ is equal to 1. It is encouraging to note from fig. (2) that the value of elastic anisotropy (AZ) varies in irregular way with increase in pressure. Such trend seems to be characteristic of the rare gas solids. The deviation from Cauchy relation with pressure is computed and is show in fig (3). It is to be noted that the values of the Cauchy deviation is changing with pressure. The fig. (2) Shows the values of σ is increasing with pressure. It is quite apparent from the experimental values of Grimsdtich et al.[9] that the values of σ is increasing in irregular way up to 5GPa and remains constants in the higher limit of pressure. This may be due to the reason that the deformation of the atomic orbital is increasing with increase of pressure. We have also computed the values of dK/dP, dCs/dP and dC_{44}/dP of Ar at different pressure with the help of equations (19-21). It is interesting to note from fig's.(3-5) that the variation in the values of dK/dP, dCs/dP and dC_{44}/dP with increasing pressure (P) is almost the same in nature. The values of these constants sharply fall in the lower limit of pressure up to 30 GPa and then become nearly constants. Such variations clearly suggest that the values the atomic orbital is increasing with increase of pressure. . This is consistent with the variation of the elastic constants C_{ij} as they are showing the same linear variation see fig.(1) with the increase of pressure (P). Moreover, the value of dK/dP at $P = 0$ (fig. (3)) is equal to 6.5, which is very close with the value taken in the study of equation of state28

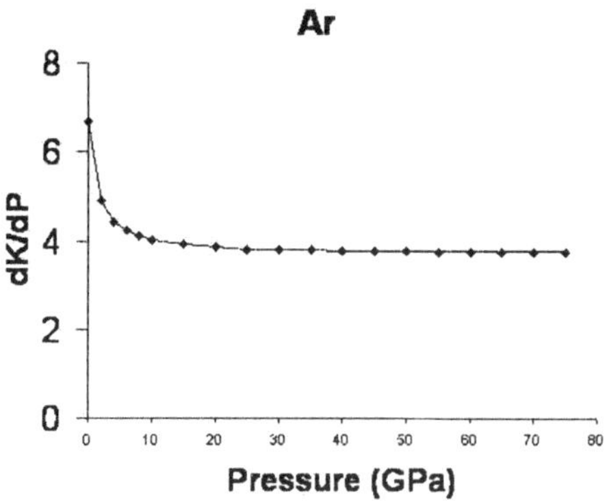

Fig. 3. The graph between pressure derivative

The computed values of C_{ij} at different pressure may also be useful in order to calculate the longitudinal and shear wave velocity, which are most important sources of information to know about the composition of earth's interior. Thus these curves fig's.(3-5) will be helpful in analyzing the experimental values.

Acknowledgment: The Authors are grateful to Dr. S.M. Sharma, Head, SRS division, BARC Mumbai for useful discussions and are also thankful to University Grant Commission, Delhi (India) for providing the financial grant through Major Research Project Scheme. One of the author Seema Gupta is grateful to DST (Delhi) for providing the financial assistance through Research Project.

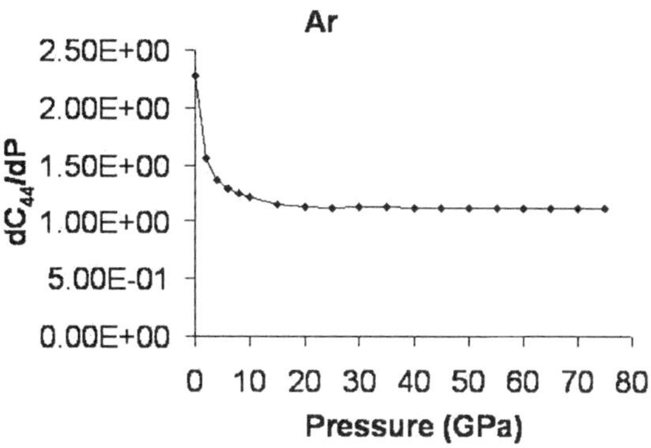

Fig. 4. The graph between shear moduli (dC_{44}/dP) and Pressure.

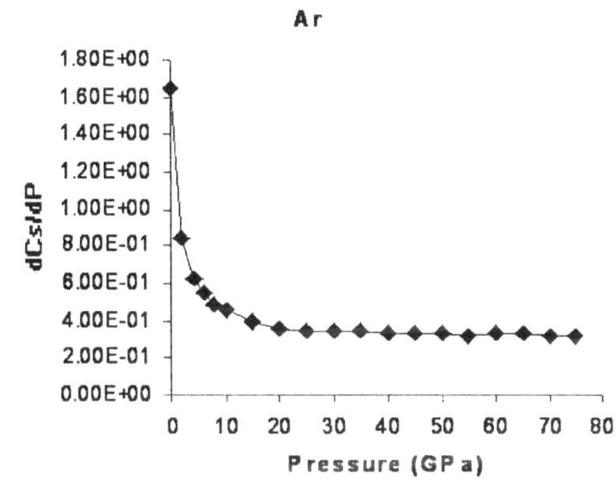

Fig. 5. The graph between shear moduli (dCs/dP) and Pressure.of bulk moduli and Pressure.

References

1. W. Reisdorf et al., *Phys. Lett.* **B595**, 118 (2004). R.J. Bell and I. J. Zucker " Rre gas solids" Ed. M.L. Klein and J.A. Venables, (Academic, New York) 1 (1976) 2.
2. M.I. Eremets et al. Phys. Rev. Lett. 85 (2000) 2797.
3. K. Rosciszewski et al. Phys. Rev. B 62 (2000) 5482.
4. H. Shimizu, H. Tashiro, T. Kume and S. Sasaki, Phys. Rev. Lett. 86 (2001) 4568.
5. T.Tsuchiya and K. Kawamura, J. Chem. Phys. 117 (2002) 5859.
6. S.Gupta and S.C.Goyal, Physica B 352 (2004) 24.
7. L.W. Finger et al., Appl. Phys. Lett. 39 (1981) 892.
8. F. Datchi, P. Loubeyre and R.LeToullec, Phys. Rev.B 61 (2000) 6535.
9. M. Grimsditch, P.Loubeyre and A. Polian, Phys. Rev. B. 33 (1986) 7192.
10. H. Shimizu et al., Phys. Rev. B. 53 (1996) 6107.
11. S. Gewurtz and B.P. Stoicheff, Phys. Rev. B. 10 (1974) 3487.

12. B.M. Axilrod and E. Teller, J. Chem. Phys. 11(1943) 299.
13. T. Iitaka et al., Phys. Rev. B, 65 (2001) 012103.
14. T. Tsuchiya and K. Kawamnura, J. Chem. Phys. 117 (2002) 5859.
15. J.S. Tse, Sol. Stat. Commun. 122 (2002) 557.
16. S. Lahari and M.P. Verma, Phys. Stat. Solidi (b) 98 (1980) 789.
17. R.K. Singh and D.K. Neb, Phys. Stat. Sol. (b) 112 (1982) 735.
18. S.O. Lundqvist, Ask. Fys. 8 (1956) 263.
19. M. Born and K. Huang, Dynamical Theory of Crystal Lattices (Oxford University, Oxford, 1985) 129.
20. M.P. Tosi, "Low Temperature Solid State Physics", Clarendon Press Oxford, 16 (1964) 1.

Electroweak Processes of the Deuteron in Effective Field Theory*

Shung-ichi Ando
Theory Group, TRIUMF, 4004 Wesbrook Mall
Vancouver, B.C. V6T 2A3, Canada
E-mail: sando@triumf.ca

We review our recent calculations of electroweak processes involving the deuteron, based on pionless effective field theory with dibaryon fields. These calculations are concerned with neutron-neutron fusion and $np \to d\gamma$ at BBN energies.

1. Introduction

The study of electroweak processes plays an important role in few-body physics. Effective field theory (EFT) provides a systematic way of calculating the transition amplitudes for those processes. It can also establish, through the symmetry of QCD, useful relations between the amplitudes for weak- and strong-interaction processes. Some of the important processes, *e.g.*, neutron β-decay[1] and the electroweak processes involving the deuteron, have been studied in the framework of EFT[2]. In this talk, we review two recent studies on neutron-neutron fusion[3] and $np \to d\gamma$ for big-bang nucleosynthesis (BBN)[4]; these studies employ pionless EFT with dibaryon fields (dEFT)[5].[a] As regards nn-fusion, we pay particular attention to the consequences of uncertainties in the existing experimental data on the *neutron-neutron* scattering length and effective range. As for the $np \to d\gamma$ cross section at BBN energies, a Markov Chain Monte Carlo (MCMC) is adapted to analyze the relevant experimental data and determine the low energy constants (LECs) in dEFT.

2. Neutron-Neutron Fusion, $nn \to de^-\bar{\nu}_e$

Ultra-high-intensity neutron-beam facilities are currently under construction at, e.g., the Oak Ridge National Laboratory and J-PARC and are expected to bring great progress in high-precession experiments concerning the fundamental properties of the neutron. Besides these experiments that focus on the properties of a single neutron, one might consider processes that involve the interaction of two free neutrons, which allow the model-independent determination of the neutron-

*This work is supported by Natural Science and Engineering Research Council of Canada.
[a]We refer to it as "dibaryon EFT" (dEFT) in this talk.

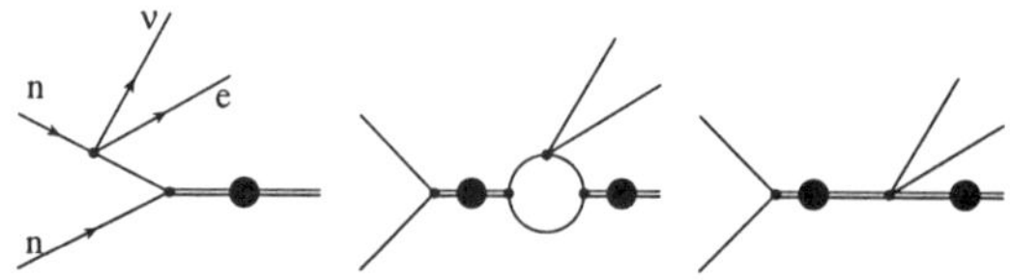

Fig. 1. Diagrams for the nn fusion process up to NLO in dEFT.

neutron scattering length and effective range, a_0^{nn} and r_0^{nn}. In this talk, we first consider the nn-fusion process for neutrons of very low energies such as the ultra-cold neutrons and thermal neutrons. It is worth noting that, for very low energy neutrons, the maximum energy E_e^{max} of the outgoing electrons from nn-fusion is $E_e^{max} \simeq B + \delta_N \simeq 3.52$ MeV, where B is the deuteron binding energy and $\delta_N = m_n - m_p$. The value of E_e^{max} is significantly larger than the maximum energy of electrons from neutron β-decay, $E_{e,\beta\text{-}decay}^{max} \simeq \delta_N \simeq 1.29$ MeV, and thus the nn-fusion electrons with energies larger than δ_N are in principle distinguishable from the main background electrons of neutron β-decay.

Diagrams for the nn-fusion process up to next-to leading order (NLO) are shown in Fig. 1, from which the cross section is calculated[3]. We also include the Fermi function and α-order radiative corrections pertaining to the one-body interaction[1] to ensure accuracy better than 1 % in the cross section. The two low-energy constants (LECs), e_V^R and l_{1A}, appear in our calculation. Using the formula for neutron β-decay[1] and the recent values of G_F, V_{ud}, g_A, and the neutron lifetime τ in the literatures, we deduce $\frac{\alpha}{2\pi} e_V^R = (2.01 \pm 0.40) \times 10^{-2}$. The LEC, l_{1A}, which also contributes to other processes, *e.g.*, *pp*-fusion and ν-d reactions, can in principle be fixed from the tritium β-decay data. However, there has been no attempt to include the weak current into the three-nucleon system in dEFT. So we make use of the result from the pionful EFT[6], and obtain $l_{1A} = -0.33 \pm 0.03$. Hence the uncertainties due to the errors in these LECs and higher order terms should be less than 1%. The prime uncertainty in the cross section comes from a_0^{nn} and r_0^{nn}, [7]

$$a_0^{nn} = -18.5 \pm 0.4[\text{fm}]\,, \quad r_0^{nn} = 2.80 \pm 0.11[\text{fm}]\,. \tag{1}$$

We are now in a position to carry out numerical calculations of the electron spectrum and the cross section. Since the nn-fusion cross section obeys the $1/v$ law, where v is the relative velocity between the two neutrons, we may concentrate on a particular value of the incident neutron energy. We consider here a head-on collision of two ultra-cold neutrons (UCN) ($v_{UCN} \simeq 5$m/sec), and thus $v = 2v_{UCN} \sim 10$m/sec. In Fig. 2, we plot the calculated electron spectrum, $d\sigma/dE_e$, as a function of E_e. As mentioned, the electrons with $E_e > \delta_N = 1.29$ MeV are in principle distinguishable from the electrons coming out of neutron β-decay. The total cross section σ is calculated to be

$$\sigma = (38.6 \pm 1.5) \times 10^{-40}[\text{cm}^2]\,. \tag{2}$$

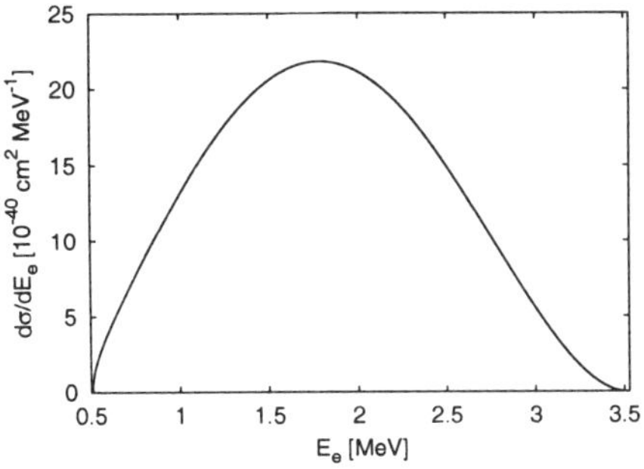

Fig. 2. Spectrum of the electrons from neutron-neutron fusion, $nn \to de\nu$.

We find that the significant uncertainty (~4%) in the cross section comes solely from the current experimental errors of a_0^{nn} and r_0^{nn}. Since the cross section obtained here is very small, the experimental observation of this reaction does not seem to belong to the near future.

3. $np \to d\gamma$ at the BBN Energies

Primordial nucleosynthesis processes take place between 1 and 10^2 seconds after the big bang at temperatures ranging from $T \simeq 1$ MeV to 70 keV. Predictions of primordial light element abundances, D, ^{3}He, ^{4}He, and ^{7}Li, and the comparison of them with observations are a crucial test of the standard big bang cosmology. The uncertainties in these predictions are dominated by the nuclear physics input for the reaction cross sections. Reaction databases are continuously updated[8], with more attention now paid to the error budget. The cross section of the $np \to d\gamma$ process at the BBN energies has been thoroughly studied by using pionless EFT up to N^{3}LO by Chen and Savage[9], and up to N^{4}LO by Rupak[10]. In this part of talk, we present an estimation of the cross section employing a new method, *i.e.*, a combination of dEFT up to NLO and an MCMC analysis with the aid of the relevant experimental data. We find that this method leads to a result comparable with that obtained by Rupak, and we discuss that the estimated $np \to d\gamma$ cross section at the BBN energies is reliable to within 1%.

Diagrams for the $np \to d\gamma$ process up to NLO in dEFT are shown in Fig. 3. From these diagrams we calculate the amplitudes for the $S(^1S_0$ and $^3S_1)$- and P-waves of

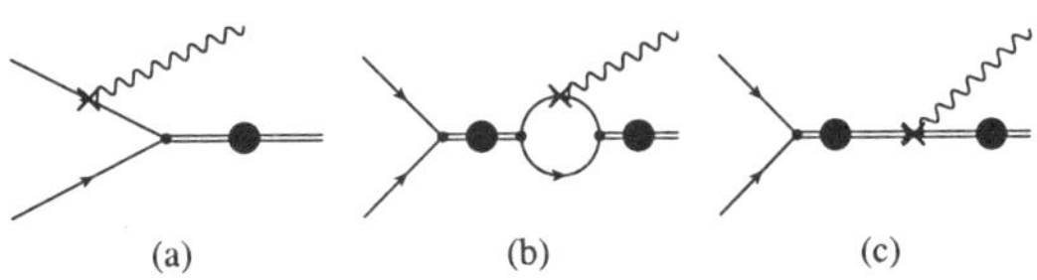

Fig. 3. Diagrams for the $np \to d\gamma$ process up to NLO in dEFT.

Table 1. Values of parameters

	MCMC	Prev. Method
a_0	-23.7426 ± 0.0081	-23.749 ± 0.008
r_0	2.783 ± 0.043	2.81 ± 0.05
ρ_d	1.7460 ± 0.0072	1.760 ± 0.005
l_1	0.798 ± 0.029	0.782 ± 0.022

the initial two-nucleon. We note that since the $^3S_1$ amplitude is highly suppressed due to the orthogonality of the scattering and bound $^3S_1$ states, we neglect it in our calculations. Using these amplitudes, we can easily calculate the cross section for $np \to d\gamma$.

Five parameters, a_0, r_0, γ, ρ_d, and l_1, appear in the amplitudes. We determine the values of the four parameters, a_0, and r_0, ρ_d, and l_1, by the MCMC analysis of the relevant low energy experimental data; the total cross section of the np scattering at the energies $\leq$5 MeV (2124 data) from the NN-OnLine web page, the $np \to d\gamma$ cross section from Suzuki *et al.*[11] and Nagai *et al.*[12] including two thermal capture data[13], the $d\gamma \to np$ cross section from Hara *et al.*[14] and Moreh *et al.*[15], and the photon analyzing power from Tornow *et al.*[16] and Schreiber *et al.*[17]. Meanwhile, we constrain γ from the accurate value of B. In Table 1 we give our estimates of the parameters obtained from the present MCMC analysis along with the values obtained in the previous method ("Prev. Method").[b] We find small differences ($\leq$2%) between the values of the parameters for the two cases; we will come back to this later.

In Table 2 the theoretical estimates of the $np \to d\gamma$ cross section at BBN energies are given as a function of the initial two-nucleon energy E in the center of mass (CM) frame. The column labeled "dEFT(MCMC)" gives our preliminary results for the mean values and standard deviations obtained in MCMC. Table 2 also shows the results of four other methods: "dEFT(Prev. Meth.)" based on the parameter set "Prev. Method" in Table 1, pionless EFT up to N^{4}LO by Rupak, a high-precision potential model calculation including the meson-exchange current by Nakamura, and an R-matrix analysis by Hale. Good agreement is found among the different approaches except that the results of "dEFT(Prev. Meth.)" at E = 0.5 and 1 MeV and those of Hale exhibit some deviations, which are ~0.6% in the former and go up to 4.5 % in the latter. The ~0.6% difference at E = 0.5 and 1 MeV between "dEFT(MCMC)" and "dEFT(Prev. Meth.)" is significant compared to the small ~0.3% statistical errors obtained here. This difference can be accounted for by higher order terms that are not included in the amplitudes of dEFT up to NLO. By including the higher order terms associated with the P-wave scattering volumes[9], we can reproduce the "dEFT(MCMC)" results at E = 0.5 and 1 MeV in "dEFT(Prev. Meth.)". This implies that the values fitted by MCMC mimic the

[b]The values of the effective ranges, a_0, r_0, and ρ_d, are taken from Ref.[18], and the value of l_1 is obtained from the averaged value of the two thermal capture rates[13].

Table 2. Theoretical estimates of the $np \to d\gamma$ cross sections at the BBN energies. E is the initial two-nucleon energy in CM frame. See the text for more details.

E(MeV)	dEFT(MCMC)	dEFT(Prev. Meth.)	Rupak	Nakamura	Hale
1.265×10^{-8}	333.8(4)	333.7(15)	334.2(0)	335.0	332.6(7)
5×10^{-4}	1.667(2)	1.666(8)	1.668(0)	1.674	1.661(7)
1×10^{-3}	1.171(1)	1.171(5)	1.172(0)	1.176	1.167(2)
5×10^{-3}	0.4979(6)	0.4976(21)	0.4982(0)	0.4999	0.4953(11)
1×10^{-2}	0.3321(4)	0.3319(14)	0.3324(0)	0.3335	0.3298(9)
5×10^{-2}	0.1079(1)	0.1079(4)	0.1081(0)	0.1084	0.1052(9)
0.100	0.06341(7)	0.0634(2)	0.06352(5)	0.06366	0.0605(10)
0.500	0.03413(8)	0.0343(1)	0.0341(2)	0.03416	0.0338(8)
1.00	0.03502(10)	0.0352(2)	0.0349(3)	0.03495	0.0365(8)

roles of the higher order terms. Since our results "dEFT(MCMC)" agree quite well with those of Rupak and Nakamura's calculations, and since in the N^4LO pionless EFT calculation by Rupak, various corrections due to the higher order terms have been studied, we infer that the estimated $np \to d\gamma$ cross section at the BBN energies should be reliable within 1% accuracy. A dEFT calculation provides a systematic perturbation scheme and a simple model-independent expression for the amplitudes in terms of a finite number of LECs. As demonstrated above, the combination of a dEFT calculation and an MCMC analysis of available experimental data would be a useful method to deduce reliable cross sections for other few-body processes.

The author would like to thank K. Kubodera, R. H. Cyburt, S. W. Hong, and C. H. Hyun for collaboration.

References

1. S. Ando *et al.*, Phys. Lett. B 595 (2004) 250.
2. S. R. Beane *et al.*, nucl-th/0008064; P. F. Bedaque and U. van Kolck, Ann. Rev. Nucl. Part. Sci. 52 (2002) 339; K. Kubodera and T.-S. Park, Ann. Rev. Nucl. Part. Sci. 54 (2004) 19; D. R. Phillips, J. Phys. G 31 (2005) S1263.
3. S. Ando and K. Kubodera, nucl-th/0507048.
4. S. Ando, R. H. Cyburt, S. W. Hong, and C. H. Hyun, in preparation.
5. See, e.g., S. R. Beane and M. J. Savage, Nucl. Phys. A 694 (2001) 209; S. Ando and C. H. Hyun, Phys. Rev. C 72 (2005) 014008.
6. T.-S. Park *et al.*, Phys. Rev. C 67 (2003) 055206.
7. See, e.g., G. F. de Téramond and B. Gabioud, Phys. Rev. C 36 (1987) 691.
8. See, e.g., R. H. Cyburt, Phys. Rev. D 70 (2004) 023505.
9. J.-W. Chen and M. J. Savage, Phys. Rev. C 60 (1999) 065205.
10. G. Rupak, Nucl. Phys. A 678 (2000) 405.
11. T. S. Suzuki *et al.*, Astrophys. J. Lett. 439 (1995) L59.
12. Y. Nagai *et al.*, Phys. Rev. C 56 (1997) 3173.
13. A. E. Cox *et al.*, Nucl. Phys. 74 (1965) 497; D. Cokinos and E. Melkonian, Phys. Rev. C 15 (1977) 1636.
14. K. Y. Hara *et al.*, Phys. Rev. D 69 (2003) 072001.
15. R. Moreh, T. J. Kennett and W. V. Pretwich, Phys. Rev. C 39 (1989) 1247.
16. W. Tornow *et al.*, Phys. Lett. B 574 (2003) 8.
17. E. C. Schreiber *et al.*, Phys. Rev. C 61 (2000) 061604(R).
18. L. Koester and W. Nistler, Z. Phys. A 272 (1975) 189.

Nuclear Forces and Chiral Symmetry

R. Higa

Thomas Jefferson National Accelerator Facility,
1200 Jefferson Avenue, Newport News, VA 23606, USA
E-mail: higa@jlab.org

M. R. Robilotta

Instituto de Física, Universidade de São Paulo,
C.P. 66318, 05315-970, São Paulo, SP, Brazil
E-mail: robilotta@if.usp.br

C. A. da Rocha

Núcleo de Pesquisa em Bioengenharia, Universiade São Judas Tadeu,
Rua Taquari, 546, 03166-000, São Paulo, SP, Brazil
E-mail: carlos.rocha@usjt.br

We review the main achievements of the research programme for the study of nuclear forces in the framework of chiral symmetry and discuss some problems which are still open.

The research programme for the study of nuclear forces, based on the idea that long and medium range interactions are dominated by one and two-pion exchanges, was formulated more that fifty years ago in a seminal paper by Taketani, Nakamura and Sasaki[1]. Nevertheless, only in the last fifteen years it has achieved its full strength, due to the consistent use of chiral symmetry. This led to a considerable improvement in our knowledge of the basic mechanisms underlying nuclear interactions. For a comprehensive discussion of the subject, the reader is directed to the recent review produced by Epelbaum[2]. Here we describe the main topics in which progress has been made towards clarifying dynamics and outline some problems which still remain open, in a perspective biased by the work done by our group[3].

The works by Weinberg[4] in the early nineties motivated the systematic use of chiral symmetry in the study of nuclear forces. The rationale for this approach is the fact that nuclear interactions are dominated by low-energy processes involving the quarks u and d, which have small masses. This allows one to work with a two-flavor version of QCD and to treat these masses as a perturbation in a chiral symmetric massless lagrangian. The procedure for the systematic inclusion of the effects associated with the quark masses is known as chiral perturbation theory (ChPT). In order to be able to perform chiral expansions, one uses a typical scale q, set by either pion four-momenta or nucleon three-momenta, such that $q < 1$ GeV.

Chiral symmetry is especially suited for dealing with multipion processes. Hence, in the case of the one-pion exchange potential ($OPEP$), it becomes relevant only when form factors are taken into account. On the other hand, it is essential to the accurate description of the two-pion exchange potential ($TPEP$), which is closely related to the πN scattering amplitude. In the sequence, we concentrate on this component of the force.

The leading contribution to the $TPEP$ is $O(q^2)$ and, at present, there are two independent expansions of the potential up to $O(q^4)$ in the literature, based on either heavy baryon[5] or covariant[3] ChPT. These results allow one to put the problem in perspective and note that the following aspects of the problem have been tamed:

• Quite generally, asymptotic (large r) expressions for the potential have the status of theorems and are written as sums of chiral layers, with little model dependence.

• The minimal realization of the symmetry is implemented by just pions and nucleons, but realistic potentials require other degrees of freedom, either hidden within the-low energy constants (LECs) of effective lagrangians or represented as explicit deltas.

• The dynamical content of the $TPEP$ is associated with three families of diagrams, shown in the figure below. Diagrams of family I correspond to the minimal realization of chiral symmetry and involve only the πN coupling constants g_A and f_π. Family II describes effects associated with pion-pion correlations, whereas the interactions in family III depend on the LECs, represented by black dots.

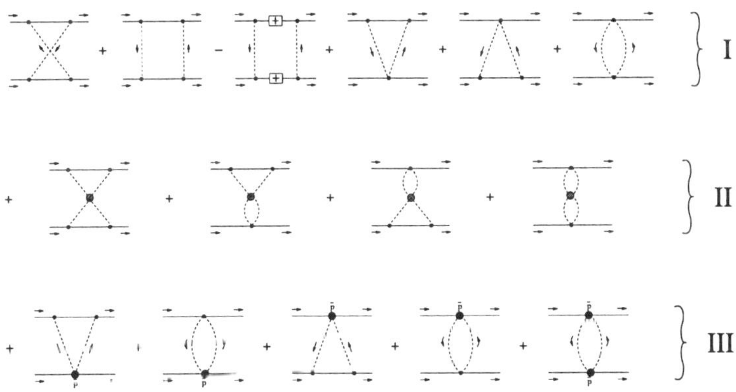

• The relation of these diagrams with πN scattering is well understood and may be used to fix the (LECs) in family III. When this procedure is adopted, the $TPEP$ does not contain any free parameters and becomes fully determined.

• There is no room for scalar mesons, such as the σ, in the long and medium range parts of the $TPEP$.

• Relativistic effects are visible in the final form of the *TPEP* and arise from the proper covariant treatment of loop integrals. Therefore they are present even when the external nucleon momenta are small.

• Due to the treatment of loop integrals, heavy baryon and relativistic derivations of the potential do not coincide. This problem is conceptually important, since it is related with the form of the asymptotic chiral theorems. From the point of view of internal theoretical consistency, the covariant procedure is favored. On the other hand, heavy baryon calculations have the advantage of producing analytical results. As far as numerical applications are concerned, the differences between both approaches are small.

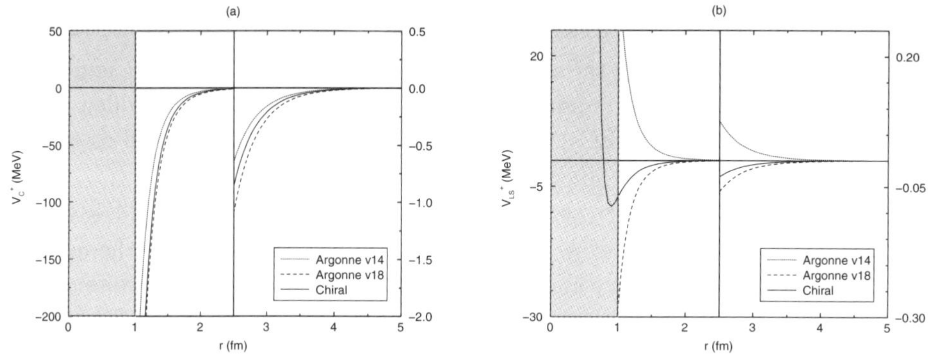

• The chiral picture is well supported by partial wave analyses[6]. In the figure above we compare the chiral TPEP V_C^+ and V_{LS}^+ components with results from the Argonne group[7].

• The dominant features of the isospin independent component of the central potential are directly related with the QCD vacuum through the nucleon scalar form factor[8].

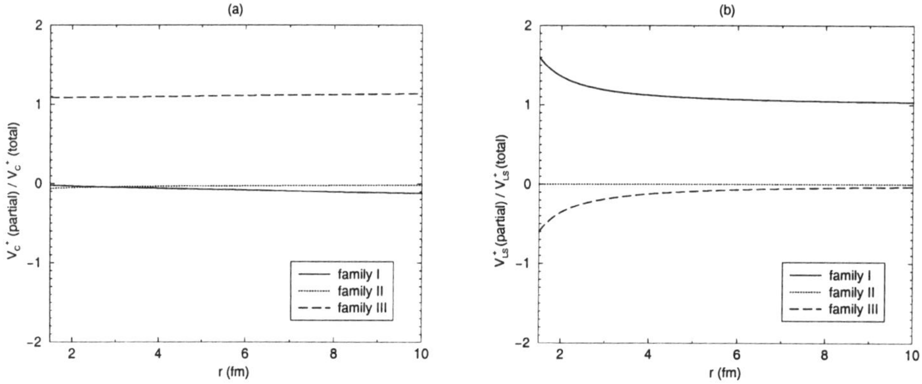

• As far as dynamics is concerned, the various channels of the potential are clearly dominated by isolated contributions arising from either family I or III and intermediate $\pi\pi$ scattering processes in family II are almost completely irrelevant. This is in sharp contrast with older models for the $TPEP$, which were not based on chiral symmetry. As typical instances, in the preceding figure we show the ratios of the individual contributions from families I, II and III by their sum, in the case of the components V_C^+ and V_{LS}^+.

• The relative importances of $O(q^2)$, $O(q^3)$ and $O(q^4)$ terms in all the components of the potential has been assessed. At distances of physical interest, they are consistent with converging series, with the exception of the isospin independent central potential. In the figure that follows we display the relative contribution of each chiral order to the $TPEP$ for V_C^+ and V_{LS}^+. The black dots in the curves correspond to the points where the ratio is 0.5.

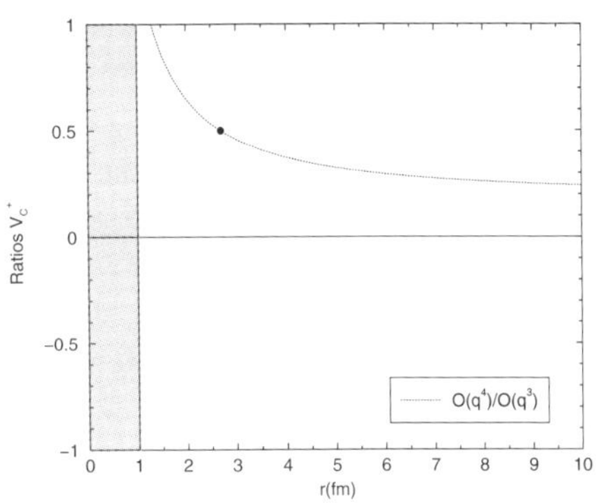

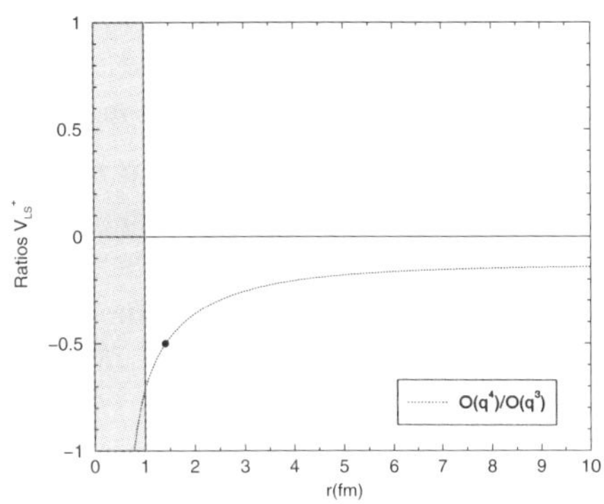

The clear picture of the $TPEP$ dynamics promoted by chiral symmetry allows one to identify some problems that remain open and deserve being tackled in order to put the potential in a yet firmer basis.

• The construction of the potential involves loop integrals, which must be regularized. At present, the best regularization procedure for chiral symmetry in the baryon sector is the infrared scheme[9], which gives rise to power counting. Even if indications are that the influence of the regularization scheme is restricted to distances smaller than 1 fm, the $TPEP$ problem remains in the want of a full calculation based on the infrared method.

• A rather puzzling aspect of the chiral $TPEP$ is that its leading terms are formally predicted to be $O(q^2)$, whereas the all important central isospin independent component V_C^+ begins at $O(q^3)$. This may be associated with the poor convergence of the chiral series for this term, as shown in the preceding figure. The numerical reasons for its odd behavior can be traced back to the large size of the LECs that are used in the first two diagrams of family III. These LECs, in turn, are dynamically generated by processes involving delta intermediate states. Therefore the explicit inclusion of delta degrees of freedom in a covariant calculation could prove useful in shedding light into this problem.

• The relation of the potential with data is very important. At present, one has good indications that the chiral $TPEP$ is able to reproduce well empirical phase shifts. However, a problem that occurs in this kind of testing is that the theoretical potential can only be directly used in the study of peripheral waves, which are small and carry large uncertainties. In order to study a larger set of waves and energies, theoretical expressions have to be corrected at short distances by means of cutoffs or form factors, which also influence numerical results. As this problem cannot be avoided, a proper assessment of the merits of the $TPEP$ could be obtained by mapping in detail the influence of cutoffs and form factors over numerical results.

• A related problem concerns the determination of the numerical values of the LECs present in the effective lagrangians. Many of the LECs relevant to the NN interaction also contribute to elastic πN scattering and it would be useful to know whether values extracted from these two processes are compatible. In doing this comparison, it is important to bear in mind that the numerical values for the LECs depend on the chiral order of the expansion one is working with.

• In the long run, it would be interesting to consider the extension of the chiral picture to potentials used in many-body calculations which, for technical reasons, tend to be more schematic. Usually, they rely heavily on scalar-isoscalar interactions inspired in the linear σ model. However, the chiral $TPEP$ does not support this assumption, especially as far as the $O(q^3)$ nature of the central potential is concerned.

References

1. M. Taketani, S. Nakamura and M. Sasaki, Progr. Theor. Phys. **VI**, 581 (1951).
2. E. Epelbaum, peprint nucl-th/0509032.
3. R. Higa, M.R. Robilotta and C. A. da Rocha, Phys. Rev. C **69**, 034009 (2004); R. Higa and M.R. Robilotta, Phys. Rev. C **68**, 024004 (2003); J-L. Ballot, M. R. Robilotta and C. A. da Rocha, Phys. Rev. C **57**, 1574 (1998); M. R. Robilotta an C. A. da Rocha, Nucl. Phys. A **615**, 391 (1997); M. R. Robilotta, Nucl. Phys. A **595**, 171 (1995); C. A. da Rocha and M. R. Robilotta, Phys. Rev. C **49**, 1818 (1994).
4. S. Weinberg, Phys. Lett. B **251**, 288 (1990); Nucl. Phys. B **363**, 3 (1991).
5. N. Kaiser, R. Brockman and W. Weise, Nucl. Phys. A **625**, 758 (1997); N. Kaiser, Phys. Rev. C**64**, 057001 (2001); N. Kaiser, Phys. Rev. C**65**, 017001 (2001); E. Epelbaum, W. Glöckle and Ulf-G. Meissner, Nucl. Phys. A **637**, 107 (1998); Nucl. Phys. A **671**, 295 (2000); D.R. Entem and R. Machleidt, Phys. REv. C **66**, 014002 (2002).
6. http://nn-online.org; http://gwdac.phys.gwu.edu.
7. R. B. Wiringa, R. A. Smith and T. L. Ainsworth, Phys. Rev. C **29**, 1207 (1984); R. B. Wiringa, V. G. J. Stocks, and R. Schiavilla, Phys. Rev. C **51**, 38 (1995).
8. M. R. Robilotta, Phys. Rev. C **63**, 044004 (2001).
9. P. J. Ellis and H-B. Tang, Phys. Rev. C **57**, 3356 (1998); K. Torikoshi and P. Ellis, Phys. Rev. C **67**, 015208 (2003); T. Becher and H. Leutwyler, Eur. Phys. J. C **9**, 643 (1999); J. High Energy Phys. **106**, 17 (2001).

ELECTROMAGNETIC AND WEAK PROCESSES IN FEW-BODY SYSTEMS

A Practical Method to Solve the Cut-Off Coulomb Problem in the Lippmann-Schwinger RGM Formalism*

Y. Fujiwara

Department of Physics, Kyoto University, Kyoto 606-8502, Japan
E-mail: fujiwara@ruby.scphys.kyoto-u.ac.jp

A practical method to solve the cut-off Coulomb problem in the Lippmann-Schwinger Resonating-Group Method (RGM) formalism is given. The RGM equation is solved in the momentum representation, using the cut-off Coulomb potential at the constituent level. The obtained phase shifts are transformed to the local phase shifts at some distance $R_{\rm in}$, by solving the asymptotic waves from far away distance. The nuclear phase shifts are then obtained through the matching condition at $R_{\rm in}$, which is a standard procedure proposed by Vincent and Phatak. Some examples are shown for the $\alpha\alpha$ RGM and the scattering problems by the quark-model baryon baryon interactions.

1. Introduction

In the momentum representation, the incorporation of the long-range Coulomb force always poses problems. In particular, the three-body scattering problems with the Coulomb force included are still under intensive investigations. Here we consider a much simpler problem, in which the Coulomb force is included in the two-cluster Resonating-Group Method (RGM) solved in the Lippmann-Schwinger (LS) formalism in the momentum representation. In this particular case, the longest range direct potential consists of a nuclear direct potential and the long-range Coulomb potential in the error function form, if simple harmonic-oscillator shell-model cluster wave functions are employed. We introduce a sharp cut-off radius R_c for the Coulomb force acting between constituent particles. We can solve the LS equations and obtain the T-matrix in the standard procedure. A problem is how to extract the correct nuclear phase shifts from this T-matrix or the asymptotic phase shifts. Here we propose a simple method, taking examples of the $\alpha\alpha$ RGM and scattering problems of the quark-model baryon baryon interactions.

2. The Basic Idea

The essential part of this treatment is based on the method by Vincent and Phatak [1] for the sharp cut-off Coulomb problem with asymptotically local potentials. They

*This work is supported by the Grant-in-Aid for Scientific Research (C) from the Japan Society for the Promotion of Science (JSPS), No. 15540270.

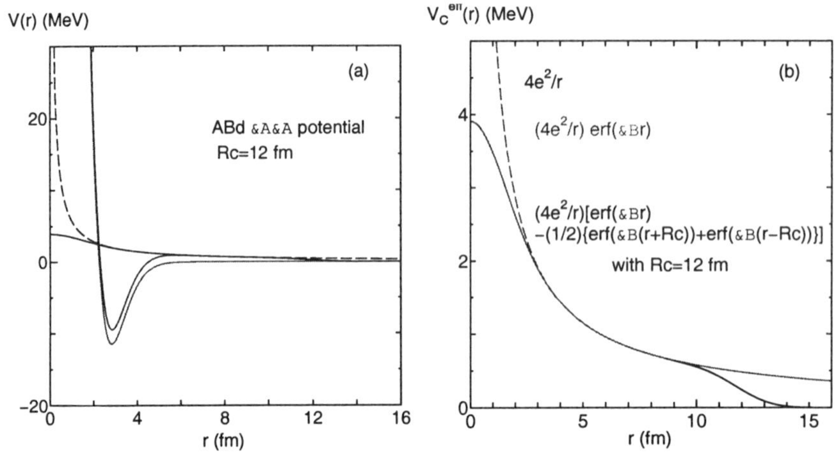

Fig. 1. (a): S-wave Ali-Bodmer potential ABd for the $\alpha\alpha$ system. The dashed curve is the simple Coulomb potential $V_C(r) = 4e^2/r$. The cut-off radius R_c of the cut-off Coulomb force at the nucleon level is assumed to be $R_c = 12$ fm. (b): The enlarged profiles of (a) for the various Coulomb potentials. The solid curve is the screening Coulomb direct potential Eq. (3) with the error function form.

solved the $^{12}C + \pi^{\pm}$ scattering in the momentum representation, by introducing

$$V_C^{R_c}(r) = \frac{Z_1 Z_2 e^2}{r}\theta(R_c - r) , \tag{1}$$

where R_c should be taken beyond the nuclear interaction range. The nuclear phase shift is obtained by smoothly connecting the asymptotic waves with and without the Coulomb effect at the radius R_c.

$$\tan\,\delta_\ell^N = -\frac{[F_\ell, u_\ell]_{R_c} + \tan\,\delta_\ell^{Rc}\,[F_\ell, v_\ell]_{R_c}}{[G_\ell, u_\ell]_{R_c} + \tan\,\delta_\ell^{R_c}\,[G_\ell, v_\ell]_{R_C}} \; . \tag{2}$$

Here $\delta_\ell^{R_c}$ is the phase shift obtained from the LS equation, and F_ℓ, G_ℓ and u_ℓ, v_ℓ are the Coulomb and Riccati Bessel, Neumann functions, respectively. Also, $[F_\ell, u_\ell]_{R_c}$ implies the Wronskian at R_c. It should be noted that $\delta_\ell^{R_c}$ approaches to $\sigma_\ell - \eta \log 2kR_c + \delta_\ell^N$, when R_c becomes large.[2] The connection condition Eq. (2) is exact for a finite R_c, which is an advantage of this method.

In the $\alpha\alpha$ RGM, the direct potential has no longer sharp cut-off, but is given by

$$V_C^{R_c}(r) = \frac{4e^2}{r}\left\{\mathrm{erf}\,(\beta r) - \frac{1}{2}\left[\mathrm{erf}\,(\beta(r + R_c) + \mathrm{erf}\,(\beta(r - R_c)\right]\right\} \; . \tag{3}$$

The phenomenological Ali-Bodmer potential for the two α system uses this folding potential, and gives a good testing ground for our Coulomb method. Figure 1 shows the potential profile of the type ABd. Since the Coulomb potential tail remains beyond the cut-off radius R_c, we cannot assume the phase shift $\delta_\ell^{R_c}$ in Eq. (2) as it is. Instead, we have to introduce the local phase shift at $R_{\rm in}$, which is a little smaller than R_c, based on the Calogero's variational phase approach.[3]

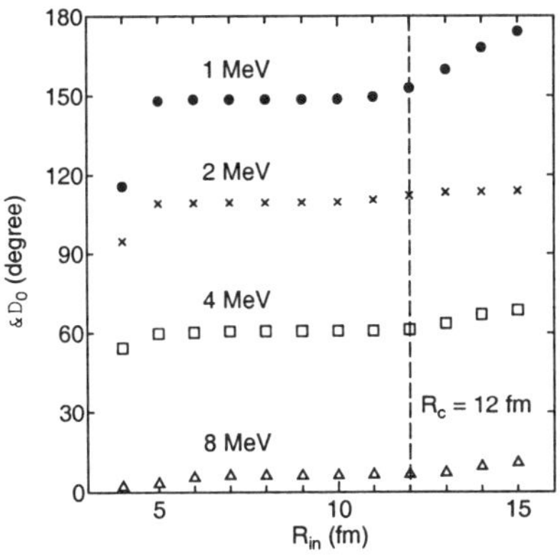

Fig. 2. $R_{\rm in}$ dependence of the S-wave phase shifts at $E_{\rm c.m.}=1$, 2, 4 and 8 MeV, predicted by $\alpha\alpha$ RGM. The Volkov No. 2 force is used. The other parameters are $R_{\rm in}-R_c-R_{\rm out}=R_{\rm in}-12-16$ fm. The flat part is a good candidate for the correct value.

We can also equivalently modify the Riccati Bessel and Neumann wave functions, $u_\ell(kr)$ and $v_\ell(kr)$ in Eq. (2) with the solutions of the simple potential problem by Eq. (3) in the interval $[R_{\rm in},\infty)$, by just keeping $\delta_\ell^{R_c}$ as it is and connecting it at $R_{\rm in}$. We therefore need three ranges, $R_{\rm in}<R_c<R_{\rm out}$, which are characterized as follows:

$R_{\rm out}$: the smallest radius beyond which $u_\ell(kr)$ and $v_\ell(kr)$ become the solution of Eq. (3),

R_c : the smallest cut-off Coulomb radius in which the coupling between the error function Coulomb and the full nuclear potential is sufficiently taken into account,

$R_{\rm in}$: the smallest radius beyond which $F_\ell(kr,\eta)$ and $G_\ell(kr,\eta)$ become the solution of Eq. (3).

The necessary buffer radius around R_c depends on the range parameter β in Eq. (3), which is related to the size of clusters. In the $\alpha\alpha$ case, we needed at least 2 fm. In the following calculations, we choose $R_{\rm in}-R_c-R_{\rm out}=8-12-16$ fm.

3. Application to $\alpha\alpha$ RGM

Accuracy of the present method is examined in three steps. We first neglect the nuclear potential and solve a simple potential problem of Eq. (3). Since the pure Coulomb repulsion is reduced, the nuclear phase shift is not zero but weakly attractive. It monotonically increases up to about 11° at $E_{\rm c.m.}=15$ MeV for the S-state, 0.5° for D-state etc. The obtained phase shifts are compared with the direct solutions by the Runge-Kutta-Gill (RKG) method, and the difference was found to be less than 0.001°. The second step is to solve the phase shifts of the Ali-Bodmer potential (Abd) with Eq. (3) both in the LS formalism and the standard RKG method. Here we again obtained good agreement less than about 0.02°. The agreement deteriorates at the resonance region and reaches 0.1°. However, these differences mainly originate from the inaccuracy of solving the LS equation due to the numerical angular momentum projection for the oscillating plane wave matrix elements of Eq. (3):

$$\langle e^{i\boldsymbol{q}_f\cdot\boldsymbol{r}}|V_C^{R_c}(r)|e^{i\boldsymbol{q}_i\cdot\boldsymbol{r}}\rangle=4e^2 2\pi {R_c}^2\left(\frac{\sin\frac{kR_c}{2}}{\frac{kR_c}{2}}\right)^2 e^{-\frac{1}{4}\left(\frac{k}{\beta}\right)^2}\,, \tag{4}$$

with $k=|\boldsymbol{q}_f-\boldsymbol{q}_i|$. This was confirmed by directly solving $\delta_\ell^{R_c}$ in the RKG method. After these preparations, we have solved $\alpha\alpha$ RGM by using the Volkov No. 2 and

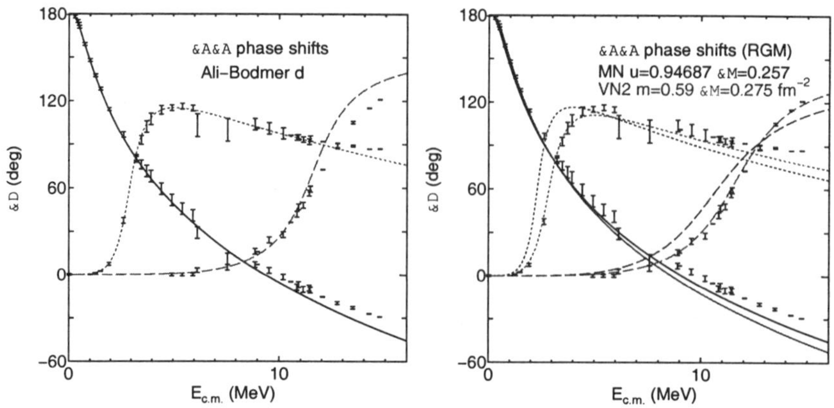

Fig. 3. (a): S-, D- and G-wave $\alpha\alpha$ phase shifts predicted by the Ali-Bodmer potential ABd. (b): The same as (a), but for $\alpha\alpha$ RGM with Volkov No. 2 (VN2) and Minnesota 3-range (MN) potentials, the latter result is in better agreement with experiment.

3-range Minnesota forces. We find almost no dependence on R_c and R_{out}, as long as $R_c \geq 12$ fm. However, some dependence is found on R_{in}. Figure 2 shows the S-wave phase shift values at several energies as a function of R_{in}, where $R_c = 12$ fm and $R_{\text{out}} = 16$ fm are fixed. They slowly increase in the range of $R_{\text{in}} = 6$ - 10 fm, but the increment is less than 0.1°. We therefore expect the accuracy of the present calculation is less than 0.05°, by taking the middle point of this range, e.g., $R_{\text{in}} = 8$ fm. In Fig. 3, we show the $\alpha\alpha$ nuclear phase shifts for S-, D- and G-waves, obtained by the Ali Bodmer potential ABd and by the RGM calculations using the Volkov No. 2 and the 3-range Minnesota forces.

4. Application to the Quark-Model Baryon Baryon Interactions

We have applied the present method to the recent quark-model baryon baryon interactions [4,5] in the particle basis. The so-called pion-Coulomb corrections are incorporated. The Coulomb force is introduced at the quark level and the Coulomb exchange kernel is exactly evaluated. In the simplest pp scattering, the nuclear and Coulomb interference part of the differential cross sections around $\theta_{\text{c.m.}} = 10°$ - $20°$ is nicely reproduced over the wide energy range $E_{\text{cm}} = 50$ - 400 MeV. In the strangeness $S = -1$ sector, accurate determination of the nuclear phase shifts is very important for the evaluation of the low-energy parameters, including the inelastic capture ratio at rest for $\Sigma^- p$ scattering. We find that this ratio has the almost equal values whatsoever the Coulomb force is included or not. However, this conclusion is obtained after the divergent factors by the Coulomb attraction are correctly treated. Another example is the $\Xi^- p$ and $\Xi^0 p$ total cross sections shown in Fig. 4. These are to be compared with the previous result, Fig. 9 of Ref. 6, calculated in the isospin basis without the Coulomb force. Quite naturally, the $\Xi^0 p$ total cross sections are almost unchanged, but the $\Xi^- p$ cross sections have appreciable change from the

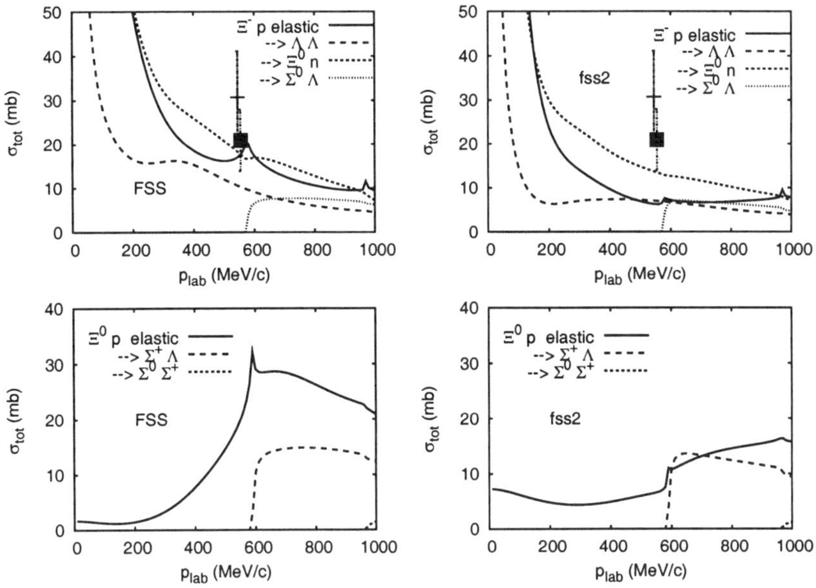

Fig. 4. ΞN total cross sections when the Coulomb force is included in the particle basis.

previous result over the whole energy region depicted.

5. Summary

We have proposed a simple and accurate method to solve the Lippmann-Schwinger RGM (resonating-group method) equations, with the cut-off Coulomb force included at the constituent level. The direct potential becomes a screened Coulomb force in the error function form. Although the cut-off is no longer sharp, the standard procedure by Vincent and Phatak [1] can still be used by solving the asymptotic waves down to $R_{\text{in}} \ll R_c$, where the full Coulomb force is acting. We applied this method to the $\alpha\alpha$ RGM and the quark-model baryon baryon interactions, and found that accuracy with less than 0.1° is maintained unless the energy is extremely too low.

References

1. C. M. Vincent and S. C. Phatak, *Phys. Rev.* **C10**, 391 (1974).
2. J. R. Taylor, *Nuovo Cinmento* **23B**, 318 (1975).
3. F. Calogero, *The Variable Phase Approach to Potential Scattering*, (New York, N.Y., 1967)
4. Y. Fujiwara, T. Fujita, M. Kohno, C. Nakamoto and Y. Suzuki, *Phys. Rev. C* **65**, 014002 (2002).
5. Y. Fujiwara, M. Kohno, C. Nakamoto and Y. Suzuki, *Phys. Rev. C* **64**, 054001 (2001).
6. Y. Fujiwara, C. Nakamoto, Y. Suzuki, M. Kohno and K. Miyagawa, *Prog. Theor. Phys. Suppl.* No. 156, 17 (2004).

Deuteron Spin Observables from Electron Scattering at Intermediate Energies

R. Alarcon*

THE BLAST COLLABORATION

Department of Physics and Astronomy
Arizona State University
Tempe, AZ 85287, USA
E-mail: ralarcon@asu.edu

Double-polarization asymmetries have been measured for the $(\vec{e}, e'p)$ electrodisintegration from vector polarized deuterium over a range of four-momentum squared, Q^2 from 0.1 to 0.7 $(GeV/c)^2$, and a missing momentum range from 0 to 475 (MeV/c). Simultaneously, single polarization asymmetries have been measured from tensor polarized both in the elastic channel as well as in the $(e, e'p)$ reaction channel. These sets of polarization observables have been obtained with great statistical precision and low systematical errors. They are compared with theoretical calculations and provide a new way to test the ground state wave function and the electromagnetic currents that connect to the continuum np system.

1. Introduction

The deuteron is an ideal system for testing nuclear theory. Due to its simple structure reliable calculations can be performed in both nonrelativistic[1,2] and relativistic frameworks.[3,4] Such calculations use phenomenological nucleon-nucleon (NN) potentials, and the resulting ground-state wave function is dominated by the S-state, especially at low missing momentum p_M. Because of the tensor part of the NN interaction a D-state component is generated. The models predict that the S- and D-state components strongly depend on p_M and are sensitive to the repulsive core of the NN interaction at short distances.[2] Traditionally, the spin structure of the deuteron has been studied through measurements of the tensor analyzing power T_{20} in elastic electron-deuteron scattering,[5] and more recently by electrodisintegration studies in the region of quasielastic scattering.[6] The cross section for the $^2\vec{H}(\vec{e}, e'p)n$ reaction, in which longitudinally polarized electrons are scattered from a polarized deuterium target, can be written as[7]

$$\sigma = \sigma_0\{1 + \sqrt{\frac{3}{2}}P_z A_d^V + \sqrt{\frac{1}{2}}P_{zz}A_d^T + h(A_e + \sqrt{\frac{3}{2}}P_z A_{ed}^V + \sqrt{\frac{1}{2}}P_{zz}A_{ed}^T)\} \quad (1)$$

*Work supported by grant PHY-0354878 of the US National Science Foundation.

where σ_0 represents the unpolarized cross section, h the polarization of the electrons, and $P_z(P_{zz})$ the vector (tensor) polarization of the target. The beam analyzing power is denoted by A_e, with $A_d^{V/T}$ and $A_{ed}^{V/T}$ the vector and tensor analyzing powers and spin-correlation parameters, respectively.

The Bates Large Acceptance Spectrometer Toroid (BLAST) is a detector designed to measure spin observables from few-body nuclei. With BLAST the A_{ed}^V double-polarization asymmetries have been measured for the $(\vec{e}, e'p)$ electrodisintegration from vector polarized deuterium over a range of four-momentum squared, Q^2 from 0.1 to 0.7 $(\text{GeV/c})^2$, and p_M values up to 475 (MeV/c). Simultaneously, single polarization asymmetries A_d^T have been measured from tensor polarized both in the elastic channel as well as in the $(e, e'p)$ reaction channel.

2. The BLAST Experiment

The MIT-Bates linear accelerator consists of a 500 MeV linac with recirculator to produce electrons with energies up to 1 GeV and polarizations of $\sim$ 70%. The BLAST experiment is situated on the South Hall Storage Ring. Injected currents of $\sim$ 200 mA with lifetimes of $\sim$ 25 minutes are typical. A Siberian snake maintains the longitudinal polarization of the stored electron beam and a Compton polarimeter is used to monitor the stored beam polarization.

The BLAST detector is based on an eight sector, toroidal magnet with a maximum field of $\sim$ 3800 G. The detectors in the left and right, horizontal sectors nominally subtend 20°–80° in polar angle and ±15° azimuthally. The detector is roughly symmetric with each sector containing: wire chambers for charged particle tracking, Čerenkov detectors for electron identification, and time of flight scintillators to measure the relative timing of scattered particles. The neutron detectors are somewhat asymmetric with the right sector having a greater thickness.

An atomic beam source is used to produce highly polarized targets of pure hydrogen or deuterium. Combinations of magnets and RF transition units populate and transport the desired spin states into a 15 mm diameter, 60 cm long, thin aluminium cylinder open at either end through which the electron beam passes. For deuterium, a target thickness of 6×10^{13} atoms/cm^2 with polarizations of 72% vector or 68% tensor are typical. A magnetic holding field of $\sim$ 500 G defines the target spin angle. This is typically horizontal, at 32° into the left sector so an electron scattering to the left (right) corresponds to a momentum transfer roughly perpendicular (parallel) to the spin angle.

The data acquisition uses the CODA and trigger supervisor systems from TJLAB. This allows multiple triggers to be defined so data can be accumulated for elastic, quasi-elastic, inclusive, and production reactions at the same time. Data are collected reversing the beam helicity each fill and randomly cycling through the target spin states every five minutes.

3. Results

Fig. 1 shows preliminary results for the T_{20} and T_{21} single polarization asymmetries obtained from elastic *ed* scattering as a function of the momentum transfer Q. The two lowest points have been normalized to the latest parameterization of the deuteron form factors. Also shown is the world data on these observables as well as a number of different theoretical calculations. Unfolding of the deuteron from factors to include the new BLAST data is in progress.

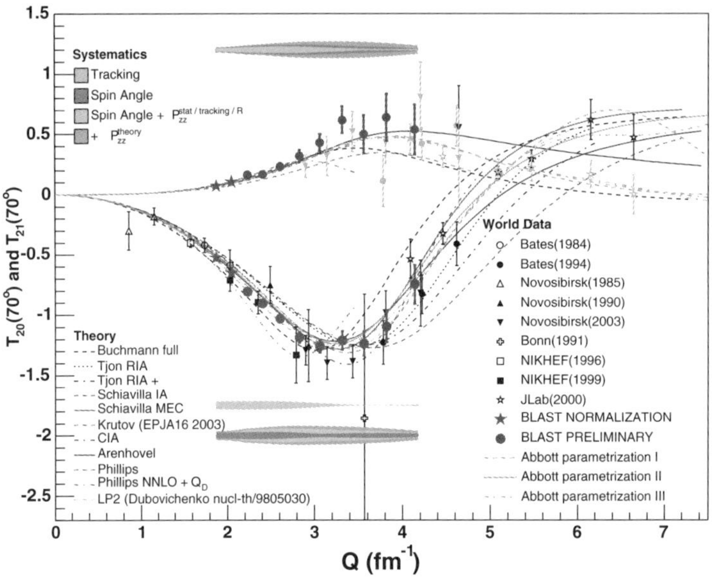

Fig. 1. Preliminary results for the T_{20} and T_{21} single polarization asymmetries obtained from elastic *ed* scattering as a function of the momentum transfer Q

By studying quasi-elastic *ep* scattering from deuterium the D-state contribution and hence tensor force can be studied. In the Born approximation the tensor asymmetry for quasi-elastic scattering should be zero in the absence of D-state contributions. As seen in Fig. 2 for both perpendicular and parallel kinematics a significant asymmetry is seen as p_M increases indicating the onset of D-state contributions. Similar effects are seen in the vector asymmetries (see Fig. 3). At $p_M \leq$ 100 MeV/c, the theoretical results[1] for A_{ed}^V neither depend on the choice of the NN potentials nor on the models for the reaction mechanism. This shows that in this specific kinematic region the deuteron can be used as an effective neutron target. Thus, these data were normalized to the calculations and yielded an accurate determination of hP_z for our measurement of the charge form factor of the neutron (see M. Kohl, these proceedings). For increasing p_M, both the data and predictions for the asymmetry reverse sign. This reflects an increasing contribution from the D-state component in the ground-state wave function of the deuteron.

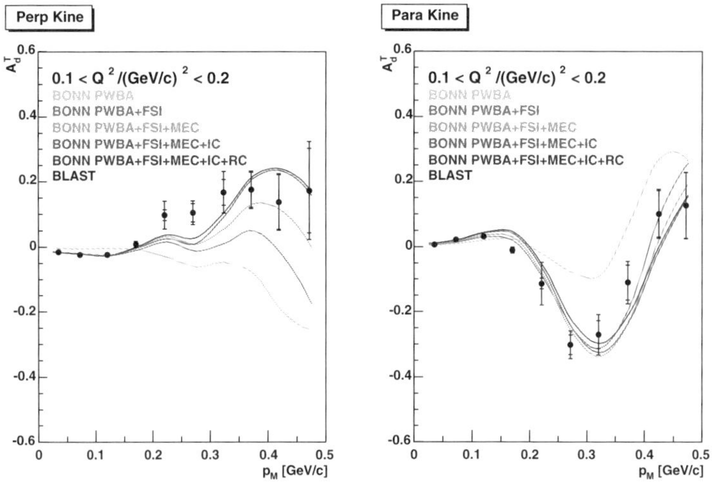

Fig. 2. Tensor single spin asymmetry A_d^T as a function of p_M for the lowest Q^2 bin.

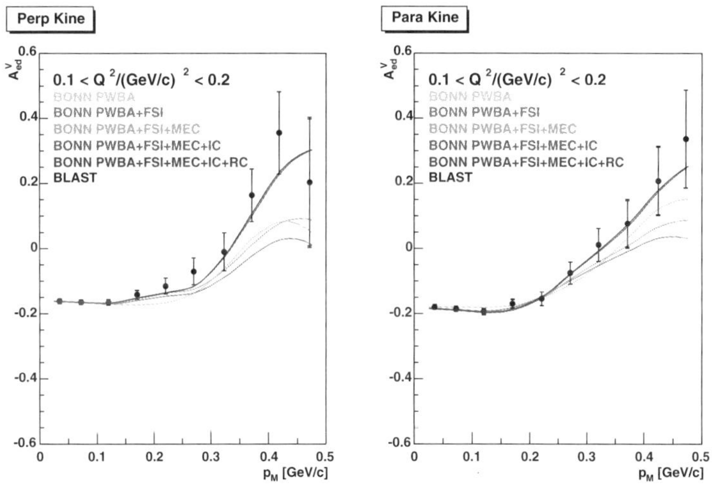

Fig. 3. Vector double spin asymmetry A_{ed}^V as a function of p_M for the lowest Q^2 bin.

References

1. F. Ritz, H. Göller, Th. Wilbois and H. Arenhövel, *Phys. Rev.* **C55**, 2214 (1997).
2. J.L. Forest *et al.*, *Phys. Rev.* **C54**, 646 (1996).
3. E. Hummel and J.A. Tjon, *Phys. Rev.* **C49**, 21 (1994).
4. J.W. Van Orden, N. Devine, and F. Gross, *Phys. Rev. Lett.* **75**, 4369 (1995).
5. D. Abbott *et al.*, *Phys. Rev. Lett.* **84**, 5053 (2000) and references therein.
6. I. Passchier *et al.*, *Phys. Rev. Lett.* **88**, 102302 (2002).
7. H. Arenhövel, W. Leidemann, and E.L. Tomusiak, *Phys. Rev.* **C46**, 455 (1992).

Isobar Model for Photoproduction of $K^+\Sigma^0$ and $K^0\Sigma^+$ on the Proton

T. Mart

Departemen Fisika, FMIPA,
Universitas Indonesia, Depok 16424, Indonesia

Kaon photoproductions on the proton $\gamma p \to K^+\Sigma^0$ and $\gamma p \to K^0\Sigma^+$ have been simultaneously analyzed by using isobar models and new SAPHIR data. The result shows that isobar models such as KAON MAID require more resonances in order to explain the data.

1. Introduction

It has been widely known that pion nucleon scattering provides an excellent tool for investigation of nucleon resonances. Indeed, this process has become a longstanding source of information on nucleon resonances properties, such as their masses, decay widths, electromagnetic and hadronic coupling constants, as well as their spins. However, the exclusive use of pion nucleon scattering would bias the information on the existence of certain resonances. Modern quark model studies predict a much richer resonance spectrum [1,2] than has been observed in $\pi N \to \pi N$ scattering experiments. Where have all these resonances gone? The answer also comes from these quark models. They have suggested that those "missing" resonances may couple strongly to other channels, such as the $K\Lambda$ and $K\Sigma$ channels [3] or other final states involving vector mesons. Since performing kaon-nucleon or hyperon-nucleon scattering experiments is a daunting task, kaon photoproduction on the nucleon appears as the best solution.

The new generation of electron and photon facilities have paved the way to kaon photoproduction experiments of unprecedented accuracies. Such experiments have been performed at ELSA in Bonn, JLab in Newport News, SPring8 in Osaka, and ESRF in Grenoble. In this paper we will only focus on the newly published data by SAPHIR collaboration of ELSA [4,12] for the $\gamma p \to K^+\Sigma^0$ and $\gamma p \to K^0\Sigma^+$ channels. These data could be very interesting, since they indicate that isobar models (e.g., KAON MAID [7]) require more resonances in order to reproduce them.

2. Isobar Model

We start with the KAON MAID model [7]. The model consists of the standard Born terms, along with the K^* and K_1 exchanges as the background part, while the resonance part includes the isospin 1/2 nucleon resonances $S_{11}(1650)$, $P_{11}(1710)$,

Table 1. Masses and widths of resonances from the three models.

Resonance	Mass or Width	Original value [6]	Model 1	Model 2
$S_{11}(1650)$	M	1650	2167	1795
	Γ	150	186	158
	M	–	–	2112
	Γ	–	–	400
$P_{11}(1710)$	M	1710	1690	1680
	Γ	100	100	100
$P_{13}(1720)$	M	1720	2133	2141
	Γ	150	256	279
$S_{31}(1900)$	M	1900	1920	1900
	Γ	200	355	329
$P_{31}(1910)$	M	1910	1936	1800
	Γ	250	399	400
χ^2/N		4.14	2.44	1.76

and $P_{13}(1720)$, as well as the isospin 3/2 deltas $S_{31}(1900)$ and $P_{31}(1910)$. A brief discussion of the model can be found in Refs. [8,9].

3. Result and Discussion

As shown by the two panels of Fig. 1 KAON MAID obviously cannot reproduce the new data, especially in the $K^0\Sigma^+$ channel. This shortcoming is understandable, since KAON MAID was fitted to previous SAPHIR data. From this figure it is also clear that the new data do not exhibit a certain resonance structure and, therefore, an arbitrarily inclusion of new resonances in the model is not advocated. However, a close inspection to the $K^+\Sigma^0$ total cross section reveals that a significant discrepancy between the prediction of KAON MAID and new data exists at $W \simeq 2.1$ GeV. This discrepancy originates from the previous data, where no indication of resonances required by the model at this energy as well as due to the large error-bars.

To investigate this phenomenon we refit the original KAON MAID coupling constants by including only new data in our database. The obtained χ^2 per degrees of freedom is shown in the third column of Table 1, which is clearly far from satisfactory. In the second step, we allow the masses and widths of nucleon resonances to be determined by the fit. As shown by the fourth column of Table 1, the result is quite interesting. Besides significantly reducing χ^2, the fit shifts both S_{11} and P_{13} masses to higher values (2167 MeV and 2133 MeV, respectively), while other resonances seem to be more stable. From the experience in the multipoles study of kaon photoproduction [13], we suspect that such behavior could be an indication for the existence of similar resonances, but with relatively different masses. Therefore, in the next step we put two S_{11} resonances and leave their masses and widths to be determined by the fit. As shown by the fifth column of the same Table, we obtain two S_{11}s with masses 1795 MeV and 2112 MeV, which is seemingly to support the finding of Ref. [13]. The result clearly indicates that more resonances are required to

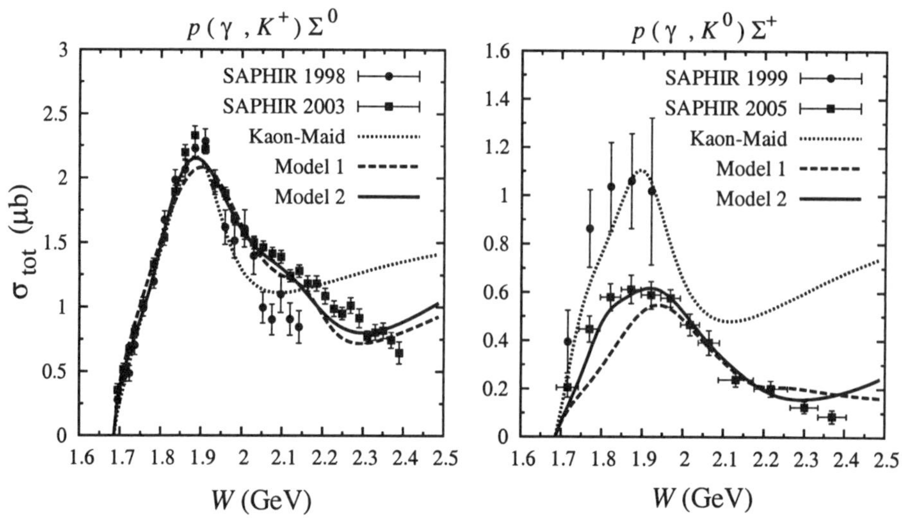

Fig. 1. Total cross section for $\gamma p \to K^+\Sigma^0$ and the $\gamma p \to K^0\Sigma^+$ channels. Experimental data are taken from Refs. [4,10–12].

explain kaon photoproduction process, a point which should be addressed in future studies.

In the $K^+\Sigma^0$ channel the difference between the last two models does not clearly show up, but this is not the case of the $K^0\Sigma^+$ channel. As shown by Fig. 1, it is obvious that Model 1 cannot reproduce the $K^0\Sigma^+$ total cross section data at energies below 1900 MeV due to the lack of resonances with $M \approx 1800$ MeV. In Model 2 the fitted mass of the first S_{11} and that of the P_{31} are in this region. We also note that in the case of KAON-MAID the divergent behavior of the total cross section at high energies is attributed to the large value of the hadronic form factor cut-off (0.82 GeV). At this region a slight increment in total cross section is also observed in Model 1, but not in Model 2.

Figure 2 shows differential cross sections obtained from all models compared with experimental $K^+\Sigma^0$ data. In this case, similar to the case of the total cross section, only KAON-MAID substantially deviates from experimental data.

Figure 3 compares the differential cross sections for the $K^0\Sigma^+$ channel. It is evident from this figure that KAON-MAID is unable to reproduce the shape as well as the magnitude of differential cross sections. In contrast to this, both Model 1 and Model 2 can fairly describe these new data up to some structures shown, e.g., at low energies and backward angles. The main difference from these models can be seen at forward directions, where Model 1 tends to produce more forward peaking cross sections at high energies.

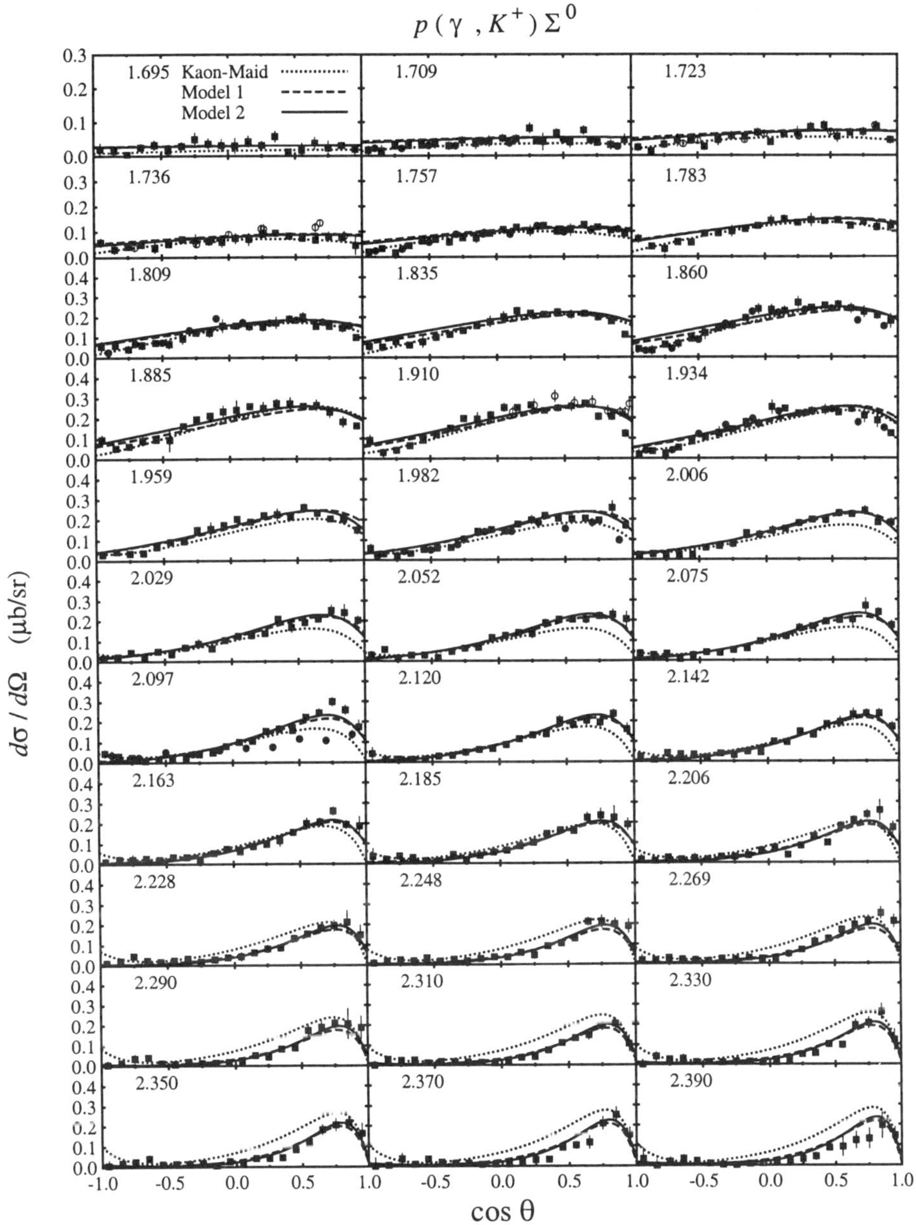

Fig. 2. Differential cross section for the $\gamma p \rightarrow K^+\Sigma^0$ channel. Experimental data are taken from Ref. [12]. Some older measurements are also shown for comparison.

4. Conclusion

We have shown that new SAPHIR data provide a stringent constrain to isobar models such as KAON MAID. New resonances are required by the models in order

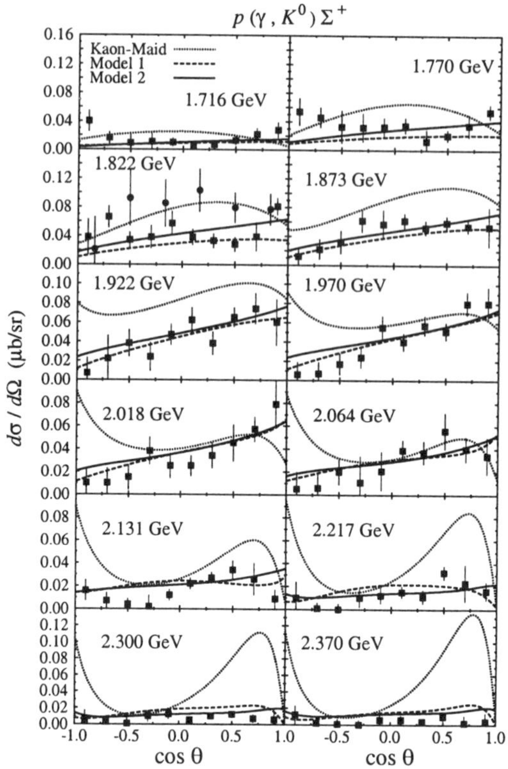

Fig. 3. Differential cross section for the $\gamma p \to K^0\Sigma^+$ channel. Experimental data are taken from Ref. [4] (solid squares) and Ref. [10] (solid circles).

to explain the data. Whether or not they are "missing resonances" should be checked in the future by the coupled channels studies.

References

1. N. Isgur and G. Karl, Phys. Lett. B **72**, 109 (1977); Phys. Rev. D **23**, 817 (1981); R. Koniuk and N. Isgur, Phys. Rev. D **21**, 1868 (1980).
2. S. Capstick and W. Roberts, Phys. Rev. D **49**, 4570 (1994).
3. S. Capstick and W. Roberts, Phys. Rev. D **58**, 074011 (1998).
4. R. Lawall *et al.*, Eur. Phys. J. A **24**, 275 (2005).
5. H. Haberzettl, C. Bennhold, T. Mart and T. Feuster, Phys. Rev. C **58**, 40 (1998).
6. Review of Particle Physics (S. Eidelman *et al.*), Phys. Lett. B **592**, 1 (2004).
7. KAON-MAID: http://www.kph.uni-mainz.de/MAID/kaon/kaonmaid.html.
8. T. Mart and C. Bennhold, Phys. Rev. C **61**, 012201 (2000).
9. T. Mart, Phys. Rev. C **62**, 038201 (2000).
10. S. Goers *et al.*, Phys. Lett. B **464**, 331 (1999).
11. M. Q. Tran *et al.*, Phys. Lett. B **445**, 20 (1998).
12. K.-H. Glander *et al.*, Eur. Phys. J. A **19**, 251 (2004).
13. T. Mart, A. Sulaksono and C. Bennhold, arXiv:nucl-th/0411035.

New Few-Body Model for α-Decay and Cluster Radioactivity*

Zhongzhou Ren and Chang Xu

Department of Physics, Nanjing University,
Hankou Road, Nanjing 210093, China
E-mail: zren@nju.edu.cn

A new cluster model (Density-dependent cluster model (DDCM)) has been proposed for α-decay and cluster radioactivity. In DDCM, the nuclear potential between α particle (or cluster) and daughter nucleus is obtained from a double-folding procedure where the M3Y effective interaction is folded with the density distributions of α particle (or cluster) and daughter nucleus. The present cluster model is well-grounded in physics because the M3Y interaction is derived from the G-matrix element of the Reid potential. The theoretical half-lives from DDCM are in good agreement with available experimental data for both α-decay and cluster radioactivity.

1. Introduction

In recent years, the study of radioactive decay has received particular attention in nuclear physics[1,2]. Alpha-decay and cluster radioactivity are two important nuclear decay modes for unstable nuclei. It is well known that both α-decay and cluster radioactivity share the same quantum tunnelling mechanism: the process in which a radioactive nuclide spontaneously emits a ^{4}He particle or a cluster of neutrons and protons. Study on nuclear decays can provide detailed information for nuclear structure properties[3,4]. The development of new microscopic models for nuclear decays is very useful for future experiments.

Recently, we have proposed a new cluster model (DDCM) for both α-decay and cluster radioactivity[5–7], in which the nuclear potential between α-particle and daughter nucleus is obtained from the double-folding integral of the renormalized M3Y nucleon-nucleon interaction with the matter density distributions of α-particle and daughter nucleus[8,9], and the Coulomb potential is obtained from the double-folding integral of the proton-proton Coulomb interaction with the charge density distributions of α-particle and daughter nucleus. The nuclear and Coulomb potentials are obtained in a natural way where important physics related to α-decay and cluster radioactivity is included. We perform systematic calculations on half-lives of α-decay and cluster radioactivity in the framework of DDCM. The numerical

*This work is supported by the National Natural Science Foundation of China (Grant No. 10125521), by the 973 National Major State Basic Research and Development of China (Grant No. G2000077400).

results obtained by DDCM are analyzed and discussed.

2. Density-Dependent Cluster Model (DDCM)

In DDCM, the ground state of parent nucleus is assumed to be an α-particle (or cluster) orbiting the daughter nucleus[4–7]. The α- (or cluster-) core potential is the sum of the nuclear potential, the Coulomb potential and the centrifugal potential.

$$V(R) = V_N(R) + V_C(R) + \frac{\hbar^2}{2\mu}\frac{(L+\frac{1}{2})^2}{R^2} \tag{1}$$

where $V_N(R)$ and $V_C(R)$ are the double-folding nuclear and Coulomb potentials, respectively. The coordinates system used in the double-folding potential is plotted in Fig.1.

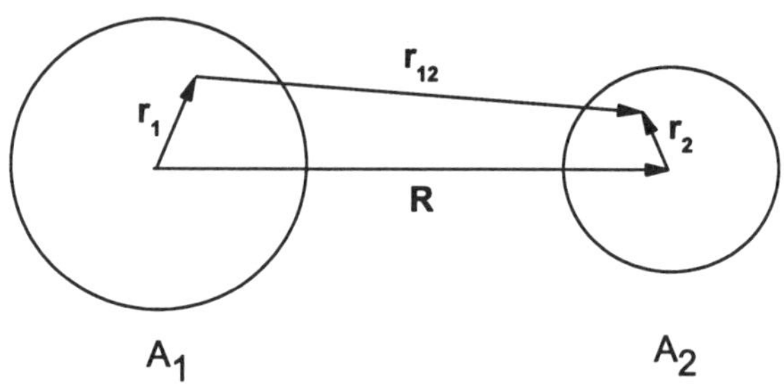

Fig. 1. Coordinates system used in the double-folding potential.

The renormalized factor λ in $V_N(R)$ is determined separately for each decay by applying the Bohr-Sommerfeld quantization condition.

$$V_N(R) = \lambda \int d\mathbf{r}_1\, d\mathbf{r}_2 \rho_1(\mathbf{r}_1)\rho_2(\mathbf{r}_2)\, g(E, |\mathbf{s}|) \tag{2}$$

The M3Y nucleon-nucleon interaction is given by two direct terms with different ranges, and by an exchange term with a delta interaction[8,9].

$$g(E, |\mathbf{s}|) = 7999\frac{\exp(-4s)}{4s} - 2134\frac{\exp(-2.5s)}{2.5s} + J_{00}\delta(\mathbf{s}), \tag{3}$$

$$J_{00} = -276(1 - 0.005\, E_\alpha/A_\alpha) \tag{4}$$

The Coulomb potential is obtained from the double-folding integral of the proton-proton Coulomb interaction with the charge density distributions of α particle and daughter nucleus.

$$V_C(R) = \int d\mathbf{r}_1\, d\mathbf{r}_2 \frac{e^2}{|\mathbf{R} + \mathbf{r}_2 - \mathbf{r}_1|}\rho_1'(\mathbf{r}_1)\rho_2'(\mathbf{r}_2) \tag{5}$$

In quasiclassical approximation, the α-decay width Γ is given by[4-7]

$$\Gamma = P_\alpha \frac{\hbar^2}{4\mu} \frac{exp[-2\int_{R_2}^{R_3} dR\, K(R)]}{\int_{R_1}^{R_2} dR \frac{1}{K(R)} \cos^2(\int_{R_1}^{R} dR' K(R') - \frac{\pi}{4})}. \tag{6}$$

The α-decay half-life is then related to the width by[4-7]

$$T_{1/2} = \hbar \ln 2/\Gamma. \tag{7}$$

3. Numerical Results and Discussions

In DDCM, the M3Y interaction is derived from the G matrix of the Reid potential and the parameterized form of the M3Y potential by Satchler and Love is used for calculations[8,9]. It is known that the M3Y interaction represents an average of the nucleon-nucleon interaction for densities from zero to normal nuclear matter (ρ_0). The behavior of the nucleon-nucleon interaction at the $(1/3)\rho_0$ is correctly included in the M3Y interaction. The nucleon-nucleon correlation is also taken into account by the G-matrix of the Reid potential. Therefore important physics related to the α decay and cluster radioactivity is included in DDCM. Detailed framework of DDCM is plotted in Fig.2.

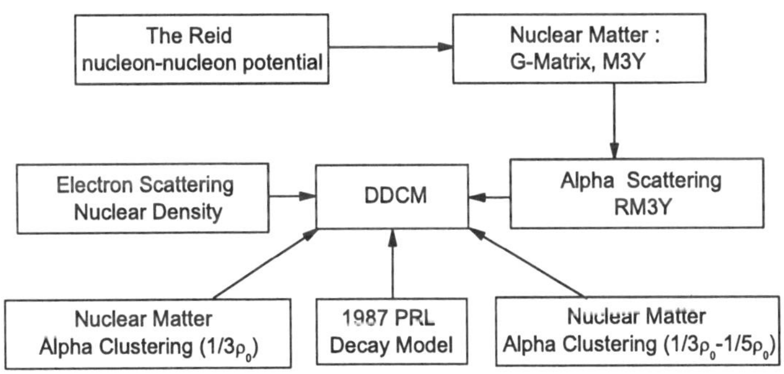

Fig. 2. The framework of the density-dependent cluster model.

Through a systematic calculation on the half-lives of α emitters, we have found that DDCM reproduces the experimental α-decay half-lives to within a factor of 3 or better in the whole mass region[6,7]. For the superheavy nuclei, the agreement between model and data is also satisfactory. In Fig.3, we compare the theoretical α-decay half-lives of DDCM with available experimental data for the superheavy nuclei with $Z \geq 106$. We define the ratio of experimental half-life and theoretical one as a hindrance factor HF= T_α(expt.)/T_α(calc.). It is seen from Fig.3 that theoretical half-lives are very close to experimental ones. For even-even superheavy nuclei the ratio between experimental half-lives and theoretical ones is approximately within a factor of 3. For odd-A nuclei and odd-odd nuclei the agreement of experimental half-lives and theoretical ones are generally good. For nucleus $^{287}114$ there is a large deviation. It is unclear whether it means a new structure change around this nucleus (N=173) or whether the assumption of the favored transition is suitable.

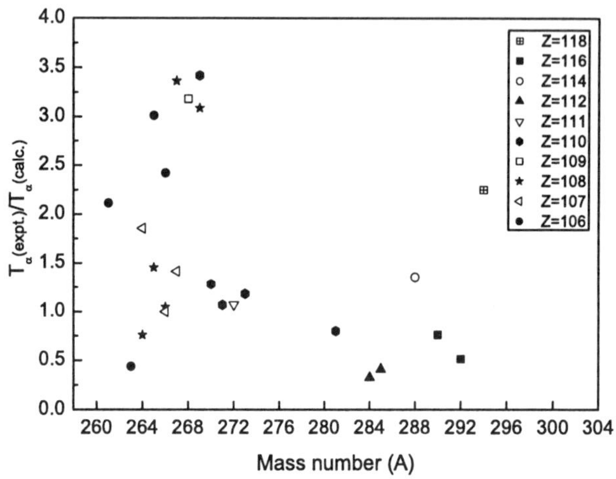

Fig. 3. The variation of the hindrance factor (HF) of α-decay with nucleon for superheavy nuclei with Z=106-116 where the double-folding Coulomb potential is taken into account.

In table 1, we give the theoretical α-decay half-lives for the newly discovered Z=115 and Z=113 α-decay chains. The experimental α-decay energies and half-lives are taken from Ref.[10] and Ref.[11]. It is seen from table 1 that the experimental half-lives are generally well reproduced by DDCM. The deviation between model and data is slightly large for $^{279}111$, $^{274}111$, and ^{270}Mt. Overall, the agreement between experiment and model is satisfactory.

Now we discuss the calculations of half-lives of complex cluster radioactivity. In 1984, Rose and Jones first observed a new kind of radioactivity: ^{14}C from ^{223}Ra[12]. Later, other kinds of heavier cluster radioactivity (such as ^{20}O, ^{24}Ne, ^{28}Mg, ^{34}Si) were also observed in experiments. Although the data of cluster radioactivity from ^{14}C to ^{34}Si have been accumulated in past years, there is few systematic analysis on these data. It is very useful for experimentalists to analyze the data of cluster radioactivity. We systematically investigate these data in the framework of DDCM[5]. In many cases, the experimental data agree with the theoretical results of DDCM within a factor of 1-3[5]. For one or two nuclei the agreement between model and data is within a factor of 4-5[5]. This successful description shows that DDCM can give reliable predictions for future experiments by combing it with any reliable structure

Table 1. Experimental and theoretical half-lives for newly discovered α-decay chains of Z=115 and Z=113.

Nuclei	Q_α(MeV)	$T_{1/2}$(Expt.)	$T_{1/2}$(Theo.)
$^{288}115$	10.657	87^{+105}_{-30}ms	160ms
$^{287}115$	10.789	32^{+155}_{-14}ms	44ms
$^{284}113$	10.191	$0.48^{+0.58}_{-0.17}$s	0.63s
$^{283}113$	10.313	100^{+490}_{-45}ms	177ms
$^{278}113$	11.728	344μ	125μ
$^{280}111$	9.938	$3.6^{+4.3}_{-1.3}$s	0.7s
$^{279}111$	10.568	170^{+810}_{-80}ms	8.6ms
$^{274}111$	11.197	9.26ms	0.51ms
^{276}Mt	9.899	$0.72^{+0.87}_{-0.25}$s	0.19s
^{275}Mt	10.528	$9.7^{+46}_{-4.4}$ms	2.5ms
^{270}Mt	10.076	7.16ms	74.9ms
^{272}Bh	9.199	$9.8^{+11.7}_{-3.5}$s	4.2s
^{266}Bh	9.125	2.47s	8.56s

model or nuclear mass model.

4. Summary

To conclude, we have systematically investigate the half-lives of both α-decay and cluster radioactivity in the framework of density-dependent cluster model. The cluster model generally reproduces the experimental half-lives to within a factor of 3 and will have a good predicting ability for the decay energies and half-lives of unknown mass range.

References

1. S. Hofmann and G. Münzenberg, *Rev. Mod. Phys.* **72**, 733 (2000).
2. Yu. Ts. Oganessian *et al.*, *Nature* **400**, 242 (1999).
3. S. A. Gurvitz and G. Kalbermann, *Phys. Rev. Lett.* **59**, 262 (1987).
4. B. Buck *et al.*, *Atomic Data and Nuclear Data Table* **54**, 53 (1993).
5. Zhongzhou Ren, Chang Xu and Zaijun Wang, *Phys. Rev.* **C70**, 034304 (2004).
6. Chang Xu and Zhongzhou Ren, *Nucl. Phys.* **A753**, 174 (2005).
7. Chang Xu and Zhongzhou Ren, *Nucl. Phys.* **A760**, 303 (2005).
8. G. F. Bertsch, J. Borysowicz, H. Mcmanus and W. G. Love, *Nucl. Phys.* **A 284**, 399 (1977).
9. G. R. Satchler and W. G. Love, *Phys. Reps.* **55**, 183 (1979).
10. Yu. Ts. Oganessian *et al.*, *Phys. Rev.* **69**, 021601(R) (2004).
11. K Morita *et al.*, *Journal of the Physical Society of Japan* **73**, 2593 (2004).
12. H. J. Rose and G. A. Jones, *Nature*, **307**, 245 (1984).

Nucleon Strange Form Factors from Parity Violation Experiments at JLab

Serge Kox
LPSC Grenoble, IN2P3/CNRS-UJF,
38026 Grenoble Cedex, France
E-mail: kox@in2p3.fr

In this contribution we present a new set of measured parity-violating asymmetries in elastic electron-proton scattering over the range of momentum transfers $0.12 \leq Q^2 \leq 1.0$ GeV^2. These asymmetries, arising from interference of the electromagnetic and neutral weak interactions, are sensitive to strange quark contributions to the currents of the nucleon. The measurements were made at JLab in two experiments HAPPEX and G^0. The results indicate non-zero strange quark contributions, and provide new information beyond that obtained previously.

1. Introduction

At short distance scales, bound systems of quarks have relatively simple properties and QCD is successfully described by perturbation theory. On size scales similar to that of the bound state itself, ~ 1 fm, the QCD coupling constant becomes large and the effects of the color fields cannot anymore be calculated accurately, even in lattice QCD. In addition to valence quarks, e.g., *uud* for the proton, there is a sea of gluons and quark-antiquark ($q\bar{q}$) pairs that plays an important role at these distance scales. The strange quarks are exclusively part of the sea, because there are no strange valence quarks in the nucleon, but are light enough to be frequently produced and thus contribute to the nucleon properties. There have been already signals about this strange quarks contribution to nucleon properties (mass, spin ...) but those are strongly model dependent.

The data presented in this contribution are related to a different sector of the nucleon matrix, i.e. the one related to the vector currents carried by the quarks. Parity-violation (PV) experiments in electron-nucleon scattering have become over the years a reliable and powerful technique to extract information on the nucleon internal structure[1] allowing the measurement and the separation of the neutral weak form factors G_E^Z, G_M^Z and the effective axial current of the nucleon. Combined with the known electromagnetic form factors, these new information permit in particular to isolate the contribution of the s quarks to the charge and magnetization densities of the nucleon. Two experimental programmes (G0[2] and HAPPEX[3]) are underway at the Jefferson Laboratory using polarized electron beams provided by the CEBAF accelerator. Both have reached important milestones presented here.

2. The Physics Case

The separation of the strange quark contributions to nucleon currents in the context of the neutral weak interaction was developed by Kaplan and Manohar[4]. As the coupling of both photons and Z bosons to point-like quarks is well defined, it is possible, to separate the contributions of the various flavors by comparing the corresponding currents[5]. Assuming that the proton and neutron are related by a simple exchange of u and d quarks (and the corresponding anti-quarks), and neglecting the contribution of the heaviest quarks (c, b, t) in the sea, it is possible to express the strange quark contribution to the electromagnetic structure of the nucleon via the relation :

$$G^s_{E,M} = (1 - 4sin^2\theta_W)G^{\gamma,p}_{E,M} - G^{\gamma,n}_{E,M} - G^{Z,p}_{E,M} \tag{1}$$

The ordinary electromagnetic form factors ($G^{n,p}_{E,M}$) are measured in elastic electron scattering are measured elsewhere. Then the isolation of $G^s_{E,M}$ requires only the measurements of the weak interaction form factors of the proton $G^{Z,p}_{E,M}$ which is the topic of this contribution.

3. The Experiments at the Jefferson Laboratory

Fig. 1. G0 experimental set-up, forward angle configuration installed in Hall C.

In order to isolate the small contribution of the neutral weak interaction in the elastic electron-proton scattering, the parity-violating asymmetry is measured for longitudinally polarized (R and L) electrons[5]

$$A = \frac{d\sigma_R - d\sigma_L}{d\sigma_R + d\sigma_L} = -\frac{G_F Q^2}{4\sqrt{2}\pi\alpha} \frac{\varepsilon G^\gamma_E G^Z_E + \tau G^\gamma_M G^Z_M - (1 - 4\sin^2\theta_W)\varepsilon' G^\gamma_M G^e_A}{\mathcal{D}} \tag{2}$$

where

$$\tau = \frac{Q^2}{4M_p^2}, \ \varepsilon = \left(1 + 2(1+\tau)\tan^2\frac{\theta}{2}\right)^{-1},$$

$$\mathcal{D} = \varepsilon(G^\gamma_E)^2 + \tau(G^\gamma_M)^2, \text{ and } \varepsilon' = \sqrt{\tau(1+\tau)(1-\varepsilon^2)},$$

Q^2 is the squared four-momentum transfer ($Q^2 > 0$), G_F and α the usual weak and electromagnetic couplings, M_p the proton mass and θ the laboratory electron scattering angle. The three new form factors in this asymmetry, G_E^Z, G_M^Z and G_A^e can be separated by measuring elastic scattering from the proton at forward and backward angles, and quasi-elastic scattering from the deuteron at backward angles[5]. To reach that goal, measurements of tiny PV asymmetries (10^{-5} to 10^{-6}) have to be performed with an overall relative uncertainty smaller than 10 %.

Both G0 [2] and HAPPEX experiments[3] are performed at the Jefferson Laboratory. G0 is implanted in hall C and a very specific set-up has been developed by the collaboration (see figure 1), whereas the HAPPEX collaboration makes use of the standard HRS spectrometers of Hall A. Both experiments are using the highly polarized (75-85 %) electron beams of CEBAF with large intensities 40-100 μA. Beam polarization was measured at the 1-2 % precision level by Moeller or Compton polarimeters. For G0 experiment, the electron beam has a very specific 32 ns pulse timing (rather than the standard 2 ns used for HAPPEX). This allows for Time of Flight measurement and background rejection in off-line analysis. Beam energy were 3.03 GeV for G0 and 3.3 GeV for HAPPEX. An important feature for these experiments is that the beam properties should remain unchanged for the two helicity states. Helicity-correlated current and position changes were then corrected with active feedback to levels of a fraction of part-per-million (ppm) and a few nm, respectively. Corrections to the measured asymmetry were applied via linear regression for residual helicity-correlated beam current, position, angle and energy variations. The polarized electrons scattered from extended high power cryogenic

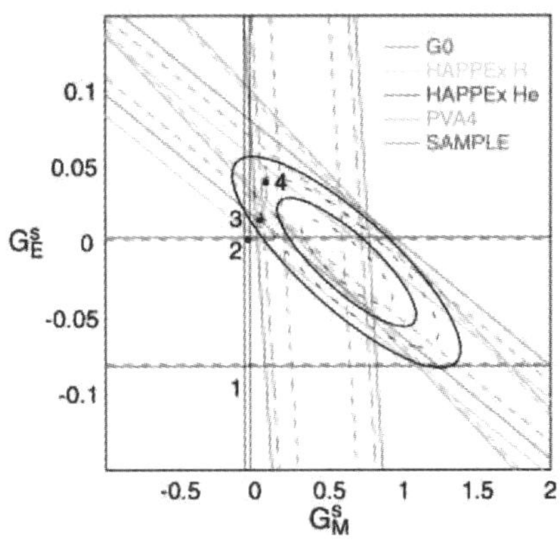

Fig. 2. Combinations of world data at Q^2 of 0.1 $(GeV/c)^2$.

targets (liquid hydrogen or deuterium, gaseous He). The recoil product of the elastic scattering (proton for G0 forward angle and electrons for HAPPEX) are selected with magnetic spectrometer and detected with dedicated systems. By detecting protons, the G^0 forward angle experiment is able to perform the measurement over the wide range of momentum transfer $0.12 \leq Q^2 \leq 1.0$ GeV^2 with a single beam energy.

4. The Results

The figure 2 shows the combination of the world data at Q^2=0.1 $(\mathrm{GeV/c})^2$ from SAMPLE[6], PVA4[8], Happex I[3] and II[9] and finally G^0. The agreement between data sets is good, the overlap being given by the various ellipses. For comparison, predictions of some recent theoretical models[11,12] are also displayed. From the figures a positive and rather large value of G_M^s seems favored by the data in contradiction with the theory. Such a conclusion will be tested wit improved statistics of the Happex experiment which running is underway. The figure 3 shows the specific results of the G^0 forward angle experiment which are shown as a function of momentum transfer. From the difference between the experimental asymmetry and the "no-vector-strange" asymmetry A_{NVS}, the quantity:

$$G_E^s + \eta G_M^s = \frac{4\sqrt{2}\pi\alpha}{G_F Q^2} \frac{\mathcal{D}}{\varepsilon G_E^\gamma} \left(A_{phys} - A_{NVS}\right), \tag{3}$$

(where $\eta\left(Q^2\right) = \tau G_M^p / \varepsilon G_E^p)$ is determined. The value of A_{NVS} is calculated using the electromagnetic form factors of Kelly [13]. Also shown in the figure 3 is

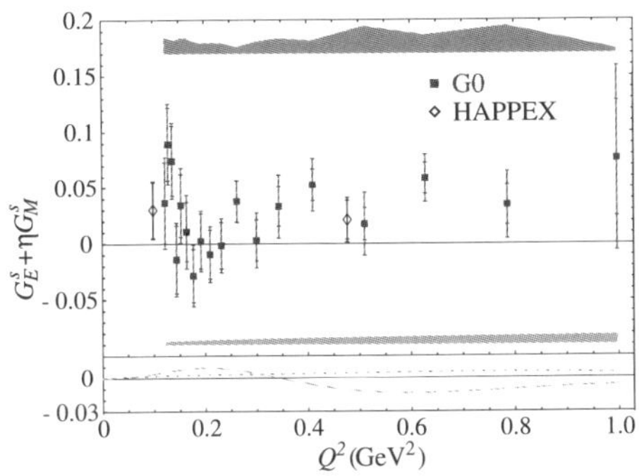

Fig. 3. The combination $G_E^s + \eta G_M^s$ for the present measurement. See text for explanation of errors and curves.

the excellent agreement with the HAPPEX measurements[3,9] made at nearly the same kinematical condition than G0. The error bars include the statistical uncertainty (inner) and statistical plus point to point systematic uncertainties added in quadrature (outer). The error bands represent, for the G0 experiment [7], the global systematic uncertainties: from the measurement (upper) and from the uncertainties in the quantities entering A_{NVS} (lower). The sensitivity of the result to electromagnetic form factors is shown separately in the lower panel for the alternative form factor parameterizations(FW[14] (dashed), Arrington[15]). Note that $G_E^s+\eta G_M^s = 0$ at $Q^2 = 0$ and that $\eta \sim 0.94Q^2$ for our kinematics. To characterize these results with a single number, the hypothesis $G_E^s + \eta G_M^s = 0$ was first tested and the conclusion was that the non-strange hypothesis is disfavored with 89% confidence. Secondly, and even more important is the intriguing Q^2 dependence of the data. The initial rise from zero to about 0.05 is consistent with the $G_M^s(Q^2 = 0.1\ \mathrm{GeV}^2) \sim 0.5$ found from the SAMPLE[6], PVA4 [10] and HAPPEX [9] measurements. Because η increases linearly throughout, the apparent decline of the data in the intermediate region up to $Q^2 \sim 0.3$ GeV^2 suggests that G_E^s may be *negative* in this range. There is also some support for this conclusion from the combination of G0 and PVA4[8] results at $Q^2 = 0.23$ GeV^2. There is then a significant trend, consistent with HAPPEX[3], to positive values of $G_E^s + \eta G_M^s$ at higher Q^2. Experiments planned for Jefferson Lab, including G0 measurements at backward angles, and MAMI (Mainz) should provide precise separations of G_E^s and G_M^s over a range of Q^2 to address this situation.

5. Acknowledgments

I gratefully thank organizer of the conference for their invitation and hospitality. I also thank the G^0 and HAPPEX collaborations to have provided me their results.

References

1. Proceedings of the PAVI04 Conference, Published in Eur. Phys. J. A., **24** (2005)
2. P. Roos *et al.* (Collaboration G0) Eur. Phys. J. **A24**, s2 59(2005)
3. K. A. Aniol *et al.* (collaboration HAPPEX), Phys. Rev. **C69**, 065501 (2004).
4. D. Kaplan and A. V. Manohar, Nucl. Phys. **B310**, 527 (1988).
5. D.H. Beck and R. D. McKeown, Ann. Rev. Nucl. Part. Sci. **51**, 189 (2001); K. Kumar and P. A. Souder, Prog. Part. Nucl. Phys. **45**, S333 (2000).
6. D. T. Spayde *et al.* (Collaboration SAMPLE), Phys. Lett. **B583**, 79 (2004).
7. D. Armstrong *et al.* (collaboration G0), Phys. Rev. Lett. **95**, 092001 (2005).
8. F. E. Maas *et al.* (Collaboration PVA4), Phys. Rev. Lett. **93**, 022002 (2004).
9. K. A. Aniol *et al.* (Collaboration HAPPEX-II), nucl-ex/0506011 and nucl-ex/0506010.
10. F. E. Maas *et al.* (Collaboration PVA4), Phys. Rev. Lett. **94**, 152001 (2005).
11. For a recent review, see D.H. Beck and B.R. Holstein, Int. J. Mod. Phys. **E10**,1 (2001).
12. D. B. Leinweber *et al.*, Phys. Rev. Lett. **94**, 212001 (2005).
13. J. J. Kelly, Phys. Rev. **C70**, 068202 (2004).
14. J. Friedrich and T. Walcher, Eur. Phys. J. **A 17**, 607 (2003).
15. J. Arrington, Phys. Rev. **C69**, 022201 (2004).

Parity Violation in pp Scattering and Vector-Meson Weak-Coupling Constants

C. H. Hyun

Institute of Basic Science, Sungkyunkwan University, Suwon 440-746, Korea
E-mail: hch@meson.skku.ac.kr

C.-P. Liu

KVI, Zernikelaan 25, Groningen 9747 AA, The Netherlands
E-mail: C.P.Liu@KVI.nl

B. Desplanques

Laboratoire de Physique Subatomique et de Cosmologie,
(UMR CNRS/IN2P3-UJF-INPG), F-38026, Grenoble Cedex, France
E-mail: desplanq@lpsc.in2p3.fr

We calculate the parity-nonconserving longitudinal asymmetry in the elastic $\vec{p}p$ scattering at the energies where experimental data are available. In addition to the standard one-meson exchange weak potential, the variation of the strong-coupling constants and the non-standard effects such as form factors and 2π-exchange description of the ρ-exchange potential are taken into account. With the extra effects, we investigate the compatibility of the experimental data and the presently-known range of the vector-meson weak-coupling constants.

1. Introduction

The measurements of the parity-nonconserving (PNC) asymmetry in $\vec{p}p$ scattering at both low energies (13.6 and 45 MeV) and at high energy (221 MeV) have been expected to determine the ρNN and ωNN coupling constants of a one-meson exchange model of the PNC NN force. The analysis has been recently performed by Carlson *et al.*[1], and they obtain a positive ωNN PNC coupling. Though their results do not disagree with the largest range provided by Desplanques, Donoghue and Holstein (DDH)[2], updated calculations of the weak-coupling constants do not give much room for positive values of the ωNN weak coupling[3,4].

With the DDH "best-guess" values of PNC meson-nucleon couplings, the contribution of PNC effects in pp scattering is dominated by the ρ-meson exchange. When the low-energy data are well reproduced, it gives a result suppressed roughly by a factor of 2 at high energy. In this work, we concentrate on the ρ-meson exchange with various effects such as strong-coupling constants of the ρ and ω mesons, form factor in the strong or weak vertices[5,6] and the two-pion resonance nature of the ρ

meson[7]. We investigate the role of these effects and whether they can give results that satisfy all the low and high-energy data simultaneously with a negative ωNN weak-coupling constant. We illustrate the most important features of the work only, and the details will be given elsewhere[8].

2. Extra Effects

Coupling constants

The concerned PNC potential reads

$$V_\rho(\boldsymbol{r}) = -\frac{g_{\rho NN}}{m_N}\left[h^0_\rho\,\boldsymbol{\tau}_1\cdot\boldsymbol{\tau}_2 + \frac{1}{2}h^1_\rho(\tau^z_1+\tau^z_2) + \frac{h^2_\rho}{2\sqrt{6}}\,(3\tau^z_1\tau^z_2 - \boldsymbol{\tau}_1\cdot\boldsymbol{\tau}_2)\right]\times$$
$$\Big((\boldsymbol{\sigma}_1-\boldsymbol{\sigma}_2)\cdot\{\boldsymbol{p},\,f_{\rho+}(r)\} - (\boldsymbol{\sigma}_1\times\boldsymbol{\sigma}_2)\cdot\hat{r}f_{\rho-}(r)\Big). \tag{1}$$

In the one-meson-exchange description, $f_{\rho+}$ is the usual Yukawa function, $\mathrm{e}^{-m_\rho r}/4\pi r$, and $f_{\rho-}(r)$ is its derivative with respect to r (with the factor $(1+\kappa_V)$). The relevant PNC coupling constants are $h^{pp}_\rho \equiv h^0_\rho + h^1_\rho + h^2_\rho/\sqrt{6}$ and $h^{pp}_\omega \equiv h^0_\omega + h^1_\omega$ and their "best-guess" values are -15.5 and -3.04, respectively. We consider the variations on the strong-coupling constants $g_{\rho NN}$ and κ_V as well as $g_{\omega NN}$ and κ_S. Our choices of the strong-coupling constants are summarized in Tab. 1.

Table 1. Sets of the strong-coupling constants. The cutoffs Λ_ρ and Λ_ω are in units of GeV.

	$g_{\rho NN}$	$g_{\omega NN}$	κ_V	κ_S	Λ_ρ	Λ_ω
S1	2.79	8.37	3.70	−0.12	-	-
S2	2.79	8.37	6.10	0	-	-
S3	2.79	8.37	3.70	−0.12	1.31	1.50
Cal	3.25	15.58	6.10	0	1.31	1.50

Form factor at the strong vertex

In Tab. 1, the set S3 introduces the cutoffs in the strong meson-nucleon vertices in the PNC potential. With a monopole-type form factor, the normal Yukawa function is modified as

$$\frac{\mathrm{e}^{-mr}}{r} \to \frac{\mathrm{e}^{-mr}}{r} - \frac{\mathrm{e}^{-\Lambda r}}{r}\left[1+\frac{1}{2}\Lambda r\left(1-\frac{m^2}{\Lambda^2}\right)\right]. \tag{2}$$

Contributions with 2π and N^* intermediate states

In order to account for the two-pion resonance nature of the ρ-meson, we rely on the work presented in Ref. 7, based on dispersion relations. In this formalism, only stable particles are involved and the ρ-meson appears indirectly in the transition amplitude $N\bar{N} \to \pi\pi$ through its propagator. To satisfy unitarity, the width of the ρ-meson has to be accounted for. A background contribution involving the exchange of the nucleon and the Δ or N^* resonances in the t-channel includes the three lowest-lying resonances, $\Delta(1232)$, $N(1440)$ and $N(1520)$[7].

Table 2. Sensitivity of the PNC asymmetry, $A_L(\times 10^7)$, to different choices of strong-coupling constants or to monopole form factors, and comparison with experiment.

Strong	S1	S2	S3	Cal	Exp.
13.6	−0.96	−1.33	−0.66	−1.13	-0.95 ± 0.15
45	−1.73	−2.39	−1.16	−2.00	-1.50 ± 0.23
221	0.43	0.75	0.25	0.52	0.84 ± 0.29

Correction at the weak vertex

As a result of a specific dynamics, the weak meson-nucleon interaction may acquire a momentum dependence that cannot be reduced to a monopole form factor. This could affect in particular the isoscalar ρNN coupling[6]. In momentum space, the corresponding NN interaction here represented by the meson propagator, $(\boldsymbol{q}^2 + m_\rho^2)^{-1}$, in absence of form factor, can be approximately parametrized as

$$\tilde{f}_\rho(\boldsymbol{q}) = \left(1 - 2\frac{\boldsymbol{q}^2}{\boldsymbol{q}^2 + \Lambda'^2}\right)\frac{1}{\boldsymbol{q}^2 + m_\rho^2}. \tag{3}$$

The Fourier transformation to configuration space gives

$$\tilde{f}_{\rho+}(r) = \frac{\Lambda'^2 + m_\rho^2}{\Lambda'^2 - m_\rho^2} f_\rho(r) - \frac{2\Lambda'^2}{\Lambda'^2 - m_\rho^2} f_{\Lambda'}(r). \tag{4}$$

3. Results and Discussions

Coupling constants and form factor at the strong vertex

In Tab. 2, we show our results with various strong-coupling constants and form factors. DDH "best-guess" values are employed for the weak couplings. The resulting asymmetry is sensitive to the strong couplings as well as the form factor, but results do not fall within experimental error bars simultaneously.

Corrections with 2π and N^*

The effect of the $2\pi + N^*$ contribution is investigated with the strong parameter sets S1 and S2, and DDH "best-guess" values for the weak-coupling constants. The results are summarized in Tab. 3. In the column for $2\pi + N^*$, the numbers in the parentheses represent the ratios $(2\pi + N^*)/(\text{bare}\,\rho)$. The $2\pi + N^*$ contribution gives a relatively larger enhancement at 13.6 MeV than at the remaining two energies, but as a whole, the ratios are similar. Since the magnitudes of the asymmetry at 13.6 and 45 MeV are larger than that at 221 MeV, similar ratios give more increase of the asymmetry at low energies. As a result, $2\pi + N^*$ with S1 gives asymmetries out of the error bars at all the three energies. For S2, the $2\pi + N^*$ contribution worsens the situation at low energies, while keeping it at 221 MeV.

Correction at the weak vertex

The results with the form factor given by the chiral-soliton model at the isoscalar weak ρNN vertex are given in Tab. 4. The set S1 is used for the strong parameters and DDH "best-guess" values for the weak couplings. The asymmetry is also sensitive to the values of the cutoff Λ', but in this case again, the correction at the PNC

Table 3. Sensitivity of the PNC asymmetry, $A_L\,(\times 10^7)$, to the effect of the finite ρ-width correction of the weak potential.

	S1		S2		
	bare ρ	$2\pi + N^*$	bare ρ	$2\pi + N^*$	Exp.
13.6	−0.96	−1.22 (1.27)	−1.33	−1.70 (1.28)	-0.95 ± 0.15
45	−1.73	−2.14 (1.24)	−2.39	−2.97 (1.24)	-1.50 ± 0.23
221	0.43	0.53 (1.23)	0.75	0.93 (1.24)	0.84 ± 0.29

Table 4. Sensitivity of the PNC asymmetry, $A_L\,(\times 10^7)$, to the effect of a specific correction of the isoscalar PNC ρNN-vertex.

Λ' (GeV)	bare ρ	3	1.31	0.771
13.6	−0.96	−1.04	−1.33	−1.69
45	−1.73	−1.88	−2.38	−2.92
221	0.43	0.47	0.61	0.67

vertex does not change the trends we have been observing in other cases.

4. Conclusion

We calculated the PNC asymmetry in $\vec{p}p$ scattering at the energies 13.6, 45 and 221 MeV. We investigated the role of the effects such as different strong-coupling constants, cutoffs in the regularization of the potential, long–range contributions to the ρ-exchange PNC potential and PNC form factors of the isoscalar ρNN vertex. The effects we considered in this work are not helpful to solving the problem raised in the introduction. With this observation, we'd like to suggest the following issues: The first one is that the value of the ωNN coupling, its sign in particular, is correct. This implies that present estimates are missing important contributions. The second issue is the existence of large corrections to the PNC single-meson exchange potential. The last issue concerns the experiment, especially at the highest energy of 221 MeV. Whatever the issue, they are quite interesting problems to be studied in the future.

References

1. J. Carlson, R. Schiavilla, V. R. Brown and B. F. Gibson, Phys. Rev. C **65**, 035502 (2002).
2. B. Desplanques, J. F. Donoghue and B. R. Holstein, Ann. Phys. (N.Y.)
3. B. Desplanques, Nucl. Phys. **A335**, 147 (1980).
4. G. B. Feldman, G. A. Crawford, J. Dubach and B. R. Holstein, Phys. Rev. C **43**, 863 (1991).
5. B. Desplanques, Phys. Rept. **297**, 1 (1998).
6. N. Kaiser and U. G. Meissner, Nucl. Phys. **A510**, 759 (1990).
7. M. Chemtob and B. Desplanques, Nucl. Phys. **B78**, 139 (1974).
8. C.-P. Liu, C. H. Hyun and B. Desplanques, in preparation.

The Electromagnetic Potential of the Neutron and its Applications

M. Nowakowski and N. G. Kelkar

Departamento de Fisica,
Universidad de los Andes,
Bogota, Colombia
E-mail: mnowakos@uniandes.edu.co, nkelkar@uniandes.edu.co

T. Mart

Departemen Fisika,
Universitas Indonesia,
Depok 16424, Indonesia
E-mail: tmart@fisika.ui.ac.id

The charge distribution inside the neutron leads to a non-zero electromagnetic form-factor F_1^n. As a consequence the Fourier transform of the non-relativistic amplitude for $e^- n \to e^- n$ yields a spin independent electromagnetic potential of the interaction of the neutron with the electron. The latter might be replaced by any charged particle with charge Ze. We use the neutron potential to calculate electromagnetic corrections to the the binding energies of the deuteron and one-neutron halo nuclei. These corrections turn out to be surprisingly large in the case of halo nuclei.

1. Newtonian Potentials from Quantum Field Theory

It is not often that in a single step, one can connect Newtonian Mechanics with the latest state of the art of a physical framework in which nowadays we formulate the fundamental laws of nature, i.e., the relativistic Quantum Field Theory (QFT)[1]. The above mentioned single step consists simply in taking the Fourier transform of the non-relativistic elastic amplitude $M_{\rm NR}$ over the momentum transfer to obtain the Newtonian potential $V(r)$[2]. In the non-relativistic limit, the momentum transfer $t = q^2$ can be approximated as $q^2 \simeq -\mathbf{Q}^2$. If the amplitude depends on the magnitude of $\mathbf{Q}$, i.e. on Q only, the potential can be calculated as[3]

$$V(r) = \frac{1}{2\pi^2}\frac{1}{r}\int_0^\infty dQ Q M_{\rm NR}(Q) \sin Qr\,. \tag{1}$$

Standard examples are the massless particle exchange with $1/Q^2$ propagator giving a $1/r$ potential, or a massive intermediate particle exchange resulting into the Yukawa potential e^{-mr}/r. Since QFT offers many possibilities to construct an elastic amplitude, one expects that 'new' Newtonian forces can be now derived from equation (1). Indeed, this is possible. One can not only get relativistic corrections to the Coulomb potential (contained in the so-called Breit equation), but also higher order loop corrections to the electromagnetic force[2]. A double particle exchange like two photon exchange or the more exotic Feinberg-Sucher force based on the exchange of two neutrinos[4] lead to new long-range forces[5]. Quantum Field Theory offers also the possibility of temperature (T) dependent forces in the framework of finite temperature formalism where the propagators become T dependent. An application of the temperature dependent potentials would be to consider particles in a neutrino thermal heat bath (e.g. cosmic relic neutrinos) which now modifies the Feinberg-Sucher force[6]. In this paper we will consider another application of the formalism to calculate $V(r)$. We will compute the electromagnetic potential of nucleons taking into account the fact that the nucleons are extended objects (finite size corrections). The elastic amplitude we have in mind is one-photon exchange transition matrix element of $AN \to AN$ with N being the proton or neutron and A a particle with a charge Ze. The nucleon-photon-nucleon vertex will contain all electromagnetic form-factors necessary to include the finite size corrections whereas the other vertex is supposed to describe the interaction of the photon with a point-like particle with charge Ze. In the case of the neutron the assumption of A being point-like is not really necessary since the first order non-zero result of a potential is only possible if we include the form-factors for the neutron (in other words the structure effects of A would be a next order effect). As long as we restrict ourselves to a spin independent potential and neglect possible substructure of A the only effect of the $\gamma - A - \gamma$ vertex will be the factor Ze entering the potential. The relevant $\gamma - N - \gamma$ vertex leading to spin independent $V^N(r)$ is[7]

$$F_1^N(q^2)\gamma^\mu \tag{2}$$

where $F_1^p(0) = e$ and $F_1^n(0) = 0$. The Q dependence of $M_{\rm NR}$ is now simply $F_1^N(Q^2)/Q^2$.

2. Electromagnetic Form-Factors and $V^n(r)$

To compute $V^N(r)$ we need a reliable parameterization of the electromagnetic form-factors. Recently, new measurements (using polarization observables)[8] of the form-factors seemingly revealed a discrepancy with older measurements. However, it became apparent that the inclusion of two photon corrections could reconcile the two types of experiments to determine the form-factors[9]. We will therefore make use of the older existing parameterizations. These are mostly given in terms of Sachs form-factors G_E^N and G_M^N. For our purposes here it suffices to quote the expression of F_1^N in terms of the Sachs form-factors given as

$$F_1^N(t) = \frac{4M^2 G_E^N(t) - tG_M^N(t)}{4M^2 - t} \tag{3}$$

where $t = q^2$ is the momentum transfer and M the nucleon's mass. Consistent with data within error bars of 20% is the dipole parameterization for the neutron[10]

$$G_M^n(t) = \mu_n \left(1 - \frac{t}{m^2}\right)^{-2}, \quad G_E^n(t) \simeq 0 \tag{4}$$

with $\mu_n = -1.91$ and $m^2 = 0.71\text{GeV}^2$. Together with the formalism outlined in section 1 this parameterization leads to the following electromagnetic potential of the neutron

$$V^n(r) = \frac{Ze}{4\pi}\frac{|\mu_n|}{[\kappa^2-1]^2}\left[\frac{1}{r}\left(e^{-\kappa mr} - e^{-mr}\right) + \frac{m}{2}(\kappa^2-1)e^{-mr}\right] \tag{5}$$

where we defined $\kappa = 2M/m$. Note that according to the l'Hospital rule the potential has a finite value at $r = 0$.

3. Applications

The recently revived interest in the nucleon's electromagnetic form-factors prompted us to look for effects of the form-factors besides the elastic cross section for $eN \to eN$. With the potential in (5) it is now possible to calculate electromagnetic corrections to the binding energies (i.e. $\Delta E =< \Psi_0|V^n|\Psi_0 >$) of bound states consisting of the neutron n and a charged particle A. Two such systems can be examined in detail. The first one is the deuteron ($A = p$) for which we use a parameterized form of the Paris potential[11] unperturbed wave function, $u(r) = \sum_{j=1}^{13} C_j e^{-m_j r}/r$, for the dominant s-wave. The result as compared to the deuteron's binding energy[12] is

$$E_b^{\text{deut.}} = 2224.6\,\text{keV}, \quad \Delta E_{\text{em}}^{deut.} = 10.4\,\text{keV} \tag{6}$$

Of course, the binding energy for the deuteron is a precisely known number and it is this fact which makes the above electromagnetic correction relevant. The other systems which can be studied are one-neutron halo nuclei in the neutron-core model. For ^{11}Be the core is ^{10}Be and plays here the role of A. The wave function for ^{11}Be is taken from a coupled channel calculation which uses a deformed Woods-Saxon potential to take the ^{10}Be core excitation into account[13]. Our numerical values show now a large electromagnetic correction of the order of 10% to the ^{11}Be binding energy[14], namely

$$E_b^{^{11}\text{Be}} = 505\,\text{keV}, \quad \Delta E_{\text{em}}^{^{11}\text{Be}} = 58.2\,\text{keV} \tag{7}$$

Needless to say, that such sizable electromagnetic corrections to strong interaction physics are unusual. This underlines the importance of knowing precisely the electromagnetic form-factors of the neutron. We expect that with different parameterizations of F_1^n the results in equations (6) and (7) will change giving us the possibility to bring each of the parameterization (or even better, measurement) into direct contact with a physical observable. We do not expect, however, that the order of magnitude of these electromagnetic corrections will change with the different parameterizations. For instance, using one of the fits of G_E^n in Ref. 10 (this is the fit in Ref. 10 constrained by thermal neutron data), we get, $\Delta E_{\text{em}}^{^{11}\text{Be}} = 20\,\text{keV}$. For the other one-neutron halo, ^{19}C, we would suspect that the corrections are even larger.

Acknowledgments

We thank F. Nunes for providing us with the ^{11}Be wave function.

References

1. M. Nowakowski, *Long range forces from quantum field theory at zero and finite temperature*, hep-ph/0009157 (2000).
2. V. B. Berestetskii, E. M. Lifshitz and L. P. Pitaevskii, *Quantum Electrodynamics*, Landau-Lifshitz Course on Theoretical Physics, Vol.4, 2nd edition, Oxford: Butterworth-Heinemann
3. F. Ferrer and M. Nowakowski, *Phys. Rev.* **D59**, 075009 (1999).
4. G. Feinberg and J. Sucher, *Phys. Rev.* **160**, 1638 (1968).
5. E. G. Adelberger et al., *Phys. Rev.* **D68**, 062002 (2003).
6. F. Ferrer, J. A. Grifols and M. Nowakowski, *Phys. Rev.* **D61**, 057304 (2000); *ibid, Phys. Lett.* **B446**, 111 (1999).
7. For a general form-factor decomposition and discussion see M. Nowakowski, E. A. Paschos and J. M. Rodriquez, *Eur. J. Phys.* **26**, 545 (2005).
8. M. K. Jones at al., *Phys. Rev. Lett.* **84**, 1398 (2000); O. Gayou et al., *Phys. Rev. Lett.* **88**, 092301 (2002); J. Arrington, *Phys. Rev* **C69**, 022201 (2004).
9. Y. C. Chen, *Phys. Rev. Lett.* **93**, 122301 (2004); P. Cuichon at al., *Phys. Rev. Lett.* **91**, 142304 (2003).
10. P. E. Bosted, *Phys. Rev.* **C51**, 409 (1995)
11. M. Lacombe et al., *Phys. Lett.* **B101**, 139 (1981).
12. G. L. Greene *et al.*, *Phys. Rev. Lett.* **56**, 819 (1986); C. Van Der Leun and C. Alderliesten, *Nucl. Phys.* **A380**, 261 (1982).
13. F. M. Nunes, I. J. Thompson and R. C. Johnson, *Nucl. Phys.* **A596**, 171 (1996); F. M. Nunes, J. A. Christley, I. J. Thompson, R. C. Johnson and V. D. Efros, *Nucl. Phys.* **A609**, 43 (1996).
14. A. H. Wapstra and G. Audi, *Nucl. Phys.* **A432**, 55 (1985).

The Charge Form Factor of the Neutron at Low Q^2 *

M. Kohl, R. Milner, and V. Ziskin

MIT-Bates Linear Accelerator Center and Laboratory for Nuclear Science
Massachusetts Institute of Technology, Cambridge, MA 02139, USA
E-mail: kohlm@mit.edu

R. Alarcon and E. Geis

Arizona State University, Tempe, AZ 85287, USA

THE BLAST COLLABORATION

Measurement of the charge form factor of the neutron G_E^n presents a sensitive test of nucleon models and QCD-inspired theories. In particular, the pion cloud is expected to play a dominant role in the low-momentum transfer region of G_E^n. At the MIT-Bates Linear Accelerator Center, G_E^n has been measured by means of quasielastic scattering of polarized electrons from vector-polarized deuterium, $^2\vec{H}(\vec{e},e'n)$. The experiment used the longitudinally polarized stored electron beam of the MIT-Bates South Hall Ring along with an isotopically pure, highly vector-polarized internal atomic deuterium target provided by an atomic beam source. The measurements have been carried out with the symmetric Bates Large Acceptance Spectrometer Toroid (BLAST) with enhanced neutron detection capability. From the beam-target double polarization asymmetry A_{ed}^V with the target spin oriented perpendicular to the momentum transfer the form factor G_E^n is extracted over a range of four-momentum transfer Q^2 between 0.12 and 0.70 $(\text{GeV/c})^2$ with minimized model dependencies.

1. Introduction

The electromagnetic form factors of the nucleon provide basic information on nucleon structure. At low momentum transfer, the pion cloud of the nucleon is expected to play a significant role in the quantitative description of the form factors, in particular for the electric form factor of the neutron G_E^n in the absence of a net charge. Thus, the low-Q^2 region of G_E^n is an ideal testing ground for QCD- and pion-cloud inspired and other effective nucleon models. In the nonrelativistic framework, the Fourier transform of G_E^n can be interpreted as the charge distribution of the neutron in the Breit frame. The spatial distribution indicates that the neutron consists of a positively-charged core surrounded by a negatively charged cloud.

Among the four non-strange nucleon electromagnetic form factors, the electric form factor of the neutron G_E^n is experimentally the least known one with uncertainties of typically 15-20%. Significant improvement of the experimental uncertainty

*This work is supported by DOE under Cooperative Agreement DE-FC02-94ER40818

is highly desirable and is setting strong constraints for nucleon models. A precise knowledge of G_E^n at low Q^2 is also essential to reduce the systematic errors of parity violation experiments.

In the past, experimental access to G_E^n was hampered by the absence of free neutron targets, and the extraction of G_E^n from elastic electron-deuteron scattering appeared to be largely model-dependent. This difficulty has been overcome in recent years with the advent of polarized beams and targets which minimize both the systematic errors and the model dependency. This work reports about a new measurement of G_E^n over a range of four-momentum transfer Q^2 between 0.12 and 0.70 $(\mathrm{GeV/c})^2$ with the BLAST experiment at the MIT-Bates Linear Accelerator Center. The technique makes use of quasielastic scattering of polarized electrons from vector-polarized deuterium in the $^2\vec{\mathrm{H}}(\vec{\mathrm{e}},\mathrm{e'n})$ reaction. The asymmetry is sensitive to the electric to magnetic form factor ratio of the neutron G_E^n/G_M^n in a kinematics where the target spin is oriented perpendicular to the momentum transfer vector $\vec{q}$. In Plane Wave Impulse Approximation (PWIA) the beam-vector asymmetry is given by

$$\frac{1}{hP_z}A_{ed}^V = \frac{a\,{G_M^n}^2\cos\theta^* + b\,G_E^n G_M^n\,\sin\theta^*\cos\phi^*}{c\,{G_E^n}^2 + {G_M^n}^2} \approx a\,\cos\theta^* + b\,\frac{G_E^n}{G_M^n}\,\sin\theta^*\cos\phi^*, \quad (1)$$

where θ^* and ϕ^* describe the target spin orientation with respect to the momentum transfer direction and a, b, c are kinematical factors. The term hP_z denotes the product of beam and target polarization.

2. The BLAST Experiment

The BLAST experiment has been designed to measure spin-dependent electron scattering at intermediate energies from polarized targets in the elastic, quasielastic and resonance region. Based on the internal target technique BLAST optimizes the use of a longitudinally polarized electron beam stored in the South Hall Ring of the MIT-Bates Linear Accelerator Center, in combination with an isotopically pure, highly-polarized internal target for both hydrogen or deuterium. In case of deuterium the target was both vector and tensor polarized. The polarized target is provided by an atomic beam source (ABS). The ejected gas molecules are first dissociated into atoms before they pass sections of sextupole magnets and RF transition units to populate the desired single spin states through Stern-Gerlach beam splitting and induced transitions between hyperfine states (see left hand side of Fig. 1). This selection process is highly efficient and thus provides nuclear polarizations of more than 70%. The spin state selection was altered every five minutes in a random sequence to minimize systematics. The polarized atoms are injected into a 60 cm long cylindrical target cell with open ends through which the stored electron beam passes. As there are no target windows the experiment is very clean with negligibly small background rates of only a few percent in the prominent channels.

The direction of the target spin can be freely chosen within the horizontal plane using magnetic holding fields. During BLAST data taking, the spin direction pointed

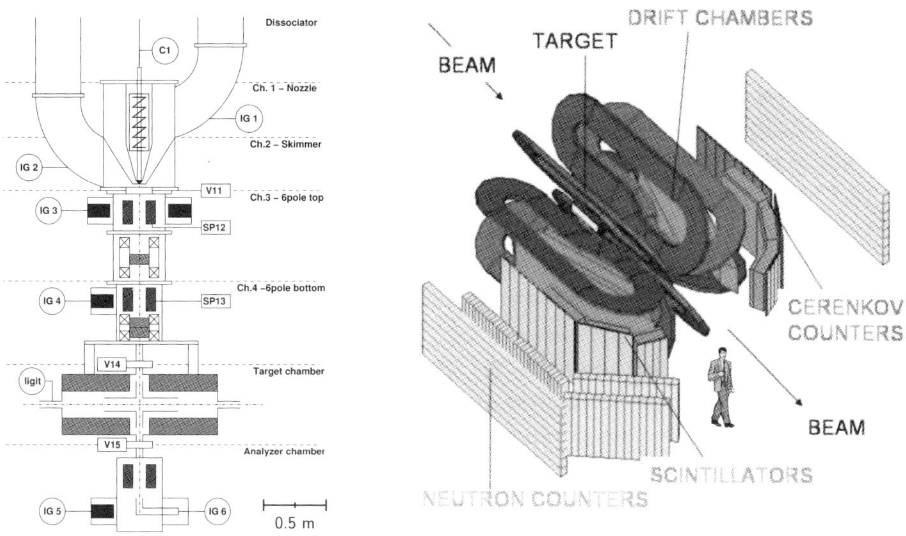

Fig. 1. Schematics of the Atomic Beam Source (left) and the BLAST detector (right).

at 32° and 47° to the left side of the beam axis in the 2004 and 2005 runs, respectively. At Bates beam currents of up to 225 mA were stored in the ring at 65% polarization and beam lifetimes of 20-30 minutes. The electron beam energy was 850 MeV throughout the BLAST program. The relatively thin target in combination with the high beam intensity yields a luminosity of about $5\times10^{31}/(\mathrm{cm}^2\,\mathrm{s})$ at an average current of 175 mA. The Bates storage ring contains a Compton polarimeter to monitor the longitudinal beam polarization in real time and without affecting the beam. The electron spin precession is compensated with a spin rotator (Siberian snake) in the ring section opposite of BLAST. The helicity of the beam was flipped once before every ring fill.

The BLAST detector is schematically shown in the right hand side of Figure 1. It was built as a toroidal spectrometer consisting of eight normal conducting copper coils producing a maximum field of 3800 G. The two in-plane sectors opposing each other are symmetrically equipped with drift chambers for the reconstruction of charged tracks, aerogel-Čerenkov detectors for e/π discrimination and 1" thick plastic scintillators for timing, triggering and particle identification. The angular acceptance covers scattering angles between 20° and 80° as well as ±15° out of plane. The symmetric detector core is surrounded by thick large-area walls of plastic scintillators for the detection of neutrons using the time-of-flight method. The thin scintillators in combination with the volumimous wire chambers in front of the neutron detectors were used as a highly efficient veto for charged tracks, making the selection of (e,e'n) events extremely clean. The setup allows to simultaneously measure the inclusive and exclusive channels (e,e'), (e,e'p), (e,e'n), (e,e'd) elastic or quasielastic, respectively, as well as (e,e'π) in the excitation region of the Δ

resonance. By measuring many reaction channels at the same time over a broad range of momentum transfer, the systematic errors are minimal.

The neutron detectors are enhanced in the right sector with $\approx$ 30% neutron detection efficiency ($\approx$ 10% in the left sector). The reason for this is because of the choice of the target spin orientation. For electrons scattered into the left sector of BLAST, the momentum transfer vector is approximately perpendicular to the target spin direction where the sensitivity to the neutron electric to magnetic form factor ratio is maximal, as can be seen from Eq. (1). In the opposite case with electrons scattered into the right sector, the spin angle is approximately parallel to $\vec{q}$ which serves as a calibration process from which the product of beam and target polarization hP_z can be extracted. A more precise value of the polarization product is however obtained from evaluating the beam-vector asymmetry A^V_{ed} of the $^2\vec{\mathrm{H}}(\vec{\mathrm{e}},\mathrm{e'p})$ reaction at low Q^2 and low missing momentum (see contribution by R. Alarcon in these proceedings). For the 2004 run of BLAST, a value of $hP_z = 0.558 \pm 0.009(stat.) \pm 0.006(sys.)$ has been determined corresponding to a deuteron vector polarization of 86%.

3. Preliminary Results

From the measured (e,e'n) yields in each beam-target spin state combination normalized to the collected deadtime-corrected beam charge the experimental double spin asymmetry A^V_{ed} is formed. For five bins in Q^2, the experimental asymmetry as a function of missing momentum is compared with the full BLAST Montecarlo result based on deuteron electrodisintegration cross section calculations by H. Arenhövel[1] with consistent inclusion of reaction mechanism und deuteron structure effects. In the quasielastic limit of the $^2\vec{\mathrm{H}}(\vec{\mathrm{e}},\mathrm{e'n})$ reaction, the deviation from PWIA is dominated by final-state interaction which increases towards lower Q^2 but is reliably calculable in the whole Q^2 range covered by the experiment. Relativistic effects are accounted for as first-order corrections which are very small in the low-Q^2 region. The electric form factor of the neutron is varied as an input parameter to the Montecarlo simulation and its measured value is extracted by a χ^2 minimization for each Q^2 bin.

Figure 2 (l.h.s.) shows the preliminary result for G^n_E from the 2004 run of BLAST along with the world data from polarization experiments[2]. Also shown is the parameterization by Galster *et al.*[3] (G) $G^n_E = 1.91\tau/(1+5.6\tau)G_{Dipole}$ where $\tau = Q^2/(4M_n^2)$. The excess of the data over the Galster curve at high and at low Q^2 is better accounted for by the more recent parameterization by Friedrich and Walcher[2,4] (FW), who describe all four nucleon form factors as sums of a smooth and a bump part, where the latter is attributed to the role of the pion cloud around the nucleon. The new preliminary BLAST data is quite consistent with both the bulk of existing data as well as with the parameterizations shown in Fig. 2, of which the FW parameterization appears slightly favored. Note that the BLAST data is preliminary and based on only about half of the statistics that were acquired in the

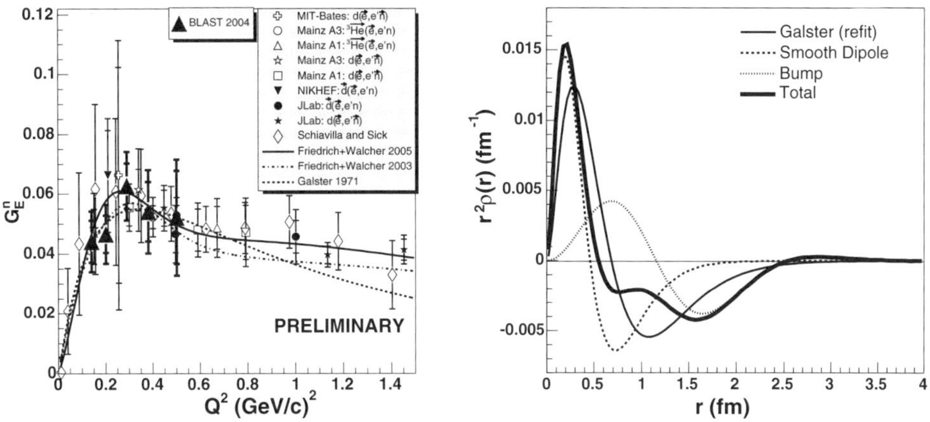

Fig. 2. L.h.s: Electric form factor of the neutron from polarization experiments[2] along with preliminary results from BLAST. The curves are the original parameterization by Galster[3] *et al.* and the recent parameterizations by Friedrich and Walcher[2,4]. R.h.s.: Fourier transforms of G_E^n of the refitted parameterizations after including preliminary data from BLAST. The excess at low Q^2 accounted for by the new parameterization corresponds to a spatial distribution that extends to larger radii than the one obtained by the Galster parameterization.

total run in 2004 and 2005.

The form factor as given in momentum space can be Fourier-transformed to obtain spatial distributions. Although recently the interpretation of the nucleon form factors in r-space has been under debate, in the Breit frame the result can be interpreted as the distribution of the charge density of the neutron, which is particularly valid at low Q^2 where the Breit and the neutron rest frames are very close. Fitting the G and FW parameterizations to the world data from polarization measurements[2] including the BLAST results, yields the spatial distributions shown in the right hand side of Fig. 2. The bump in the form factor distribution at low Q^2 causes the negative part of the charge distribution to extend out to larger radii than suggested by the previous Galster parameterization.

The preliminary results shown here comprise parts of a PhD thesis[5] based on the BLAST data taken in 2004. Analysis of the full 2004-2005 dataset is in progress[6].

References

1. H. Arenhövel, W. Leidemann, and E.L. Tomusiak, *Eur. Phys. J.* **A23**, 147 (2005).
2. D.I. Glazier *et al.*, *Eur. Phys. J.* **A24**, 101 (2005) and references therein.
3. S. Galster *et al.*, *Nucl. Phys.* **B32**, 221 (1971).
4. J. Friedrich and Th. Walcher, *Eur. Phys. J.* **A17**, 607 (2003).
5. V. Ziskin, PhD thesis, Massachusetts Institute of Technology (2005).
6. E. Geis, PhD thesis, Arizona State University, in preparation.

Photo- and Electro-Productions of the Nucleon Resonances in the Point Form Relativistic Quantum Mechanics*

Y. B. Dong

Institute of High Energy Physics, The Chinese Academy of Sciences, Beijing 100049, P. R. China

E-mail: dongyb@mail.ihep.ac.cn

The point form relativistic quantum mechanics is employed to calculate the photo- and electro-productions of the nucleon resonances. Both the transverse and longitudinal transition amplitudes are computed based on the constituent quark model with the relativistic framework. The discrepancies between the results of the relativistic approach and of the non-relativistic framework are shown.

1. Introduction

Dirac [1] first proposed the three equivalent forms of relativistic dynamics. They are instant form, light-front form, and point form. In the instant form, the interactions are involved both in P_0 (the time component of four-momentum) and in Lorentz boost operators: J_{01}, J_{02}, and J_{03}. Therefore, the main difficulty is that the manifest Lorentz covariance is lost. In the point form relativistic quantum mechanics, however, the four momenta $P_\mu(\mu = 0, 1, 2, 3)$ are interaction-dependent. They are the Hamiltonians of the system. Other dynamical operators, like the angular momentum and Lorentz boost operators, are interaction free. Thus, the advantage of the point form is that all the Lorentz transformations remain purely kinematic and the theory is manifestly Lorentz covariant. It is well-known that the instant and light-front forms became more popular than the point form in the past several decades. In fact, the point form has been detailed discussed by Keister and Polyzou [2] and recently been carefully and systematically studied by Klink [3]. It was employed in the calculations of the nucleon form factors [4], nucleon resonance strong decays [10] and some other aspects of hadron physics [5-6], like the form factors of π, ρ and K mesons. The results in the literature show the importance of the relativistic description of the system, particularly, when the momentum transfer Q^2 is at a moderate region of $\sim 1GeV^2$. The strong decay widths predicted by this framework are remarkably different from the non-relativistic constituent quark model calculation. So far, how well the point form relativistic quantum mechanics in understanding the hadron properties is still under investigation [7]. Here, we'll present our calculation

*This work is supported by the National Science Foundations of China

of the invariant electromagnetic transverse and longitudinal transition amplitudes of the low-lying nucleon resonances in the point form relativistic quantum mechanics. Moreover, we will compare the results of the non-relativistic framework to the values of the point form.

2. Calculations in the Point Form

Since in the point form the Lorentz transformations remain purely kinematic, the so-called velocity state is usually introduced as follows:

$$\begin{aligned} \mid v; \vec{k}_1, \vec{k}_2, \vec{k}_3; \mu_1, \mu_2, \mu_3 > &= U_{B(v)} \mid k_1, k_2, k_3; \mu_1, \mu_2, \mu_3 > \\ &= \prod_{i=1}^{3} D^{1/2}_{\sigma_i \mu_i}[R_W(k_i, B(v))] \mid p_1, p_2, p_3; \sigma_1, \sigma_2, \sigma_3 >, \end{aligned} \tag{1}$$

where k_i, $i = 1-3$ are the quark momenta in the center-of-mass frame ($\sum_i \vec{k}_i = 0$). $B(v)$ is a Lorentz boost with four-velocity v. In Eq. (1) $p_i = B(v)k_i$, and $U_{B(v)}$ is a unitary representation of $B(v)$. $D^{1/2}(R_W)$ is the spin-1/2 representation matrix of the Wigner rotation $R_W(k_i, B(v)) = B^{-1}(B(v)k_i)B(v)B(k_i)$. It has been proved that all the Wigner rotations of a canonical boost of a velocity state are the same. Therefore, the spins can thus be coupled together to a total spin state as in the non-relativistic framework as well as in the center-of-mass frame. This is the practical advantage of using the velocity state in the point form relativistic mechanics.

To calculate the transition amplitudes of a nucleon resonance, we simply employ the point form spectator approximation in the electromagnetic interaction. The conserved electromagnetic current operator J_μ contains both the one-body current j^1_μ and the dynamically determined current j^{DD}_μ. j^1_μ in the point form spectator approximation has the usual form of a point-like Dirac particle

$$< p'_i, \lambda'_i \mid j^1_\mu \mid p_i, \lambda_i >= e_i \bar{u}(p'_i, \lambda_i) \gamma_\mu u(p_i, \lambda_i), \tag{2}$$

where $u(p_i, \lambda_i)$ is the Dirac spinor with momentum p_i and spin λ_i for the i-th struck quark. The matrix element of the dynamically determined current is [8]

$$< p', \lambda'_i \mid j^{DD}_\mu \mid p, \lambda >= -\frac{q_\mu + q^\perp_\mu}{q^2} q^\nu < p, \lambda'_i \mid j^1_\nu \mid p, \lambda >, \tag{3}$$

where, q_μ is the four-momentum of the incoming photon and $q^\perp_\mu$ is a four-vector perpendicular to q and is determined by the requirement that there should be no pole at $q^2 = 0$. It is clear that the dynamically determined current does not affect the transverse current. Moreover, the gauge invariant constraint condition $q^\mu J_\mu = 0$ is satisfied.

The predictions of the fully relativistic point form framework are reference frame independent. The amplitude is

$$
\begin{aligned}
A_{\mu',\mu} &= \xi < f,\mu' \mid H^T_{em} \mid i,\mu >= -\sqrt{\frac{2\pi\alpha_E}{\omega}}\xi < f,\mu' \mid \vec{J}\cdot\vec{\epsilon} \mid i,\mu > \\
&= 3\xi \int \frac{d^3p_1}{(2\pi)^3}\frac{d^3p_2}{(2\pi)^3}\frac{d^3p_3}{(2\pi)^3}\frac{d^3p'_1}{(2\pi)^3}\frac{d^3p'_2}{(2\pi)^3}\frac{d^3p'_3}{(2\pi)^3} \times (2\pi)^6\delta^3(\vec{k}_1+\vec{k}_2+\vec{k}_3)\delta^3(\vec{k}'_1+\vec{k}'_2+\vec{k}'_3) \\
&\times \psi^*_{J',\mu'}(\vec{p}\,'_\rho,\vec{p}\,'_\lambda;\mu'_1,\mu'_2,\mu'_3)\psi_{1/2,\mu}(\vec{p}_\rho,\vec{p}_\lambda;\mu_1,\mu_2,\mu_3) \\
&\times D^{*1/2}_{\lambda'_3\mu'_3}[R_W(k'_3,B(v_{out}))] < p'_3,\lambda'_3 \mid -\sqrt{\frac{2\pi\alpha_E}{\omega}}\vec{J}_+ \mid p_3,\lambda_3 > D^{1/2}_{\lambda_3\mu_3}[R_W(k_3,B(v_{in}))] \\
&\times D^{1/2}_{\mu'_1\mu_1}[R_W(k_1,B^{-1}(v_{out})B(v_{in}))]D^{1/2}_{\mu'_2\mu_2}[R_W(k_2,B^{-1}(v_{out})B(v_{in}))] \\
&\times \delta^3(k'_1-B^{-1}(v_{out})B(v_{in})k_1)\delta^3(k'_2-B^{-1}(v_{out})B(v_{in})k_2), \qquad (4)
\end{aligned}
$$

where $\alpha_E = \frac{1}{137}$ and $J_+ = -\frac{1}{\sqrt{2}}(J_x + iJ_y)$. In Eq. (4) $p'_i = B(v_{out})k'_i$, and $p_i = B(v_{in})k_i$. μ and μ' are the projections of the angular momenta of the initial nucleon (1/2) and its final resonance (J'), respectively. The two wave functions are the intrinsic wave functions of the initial nucleon and final resonance with the momentum conjugates p_ρ and p_λ of intrinsic Jacobi coordinates. ξ is the sign of the pionic decay of the resonance. Due to the symmetry of the wave functions it is sufficient to consider only one case where quark 3 is struck by the incoming photon, while the other two are spectators, and to multiply the results by a factor of 3. The conventional transition amplitude $A_{1/2}$ or $A_{3/2}$ is determined from the matrix element of Eq. (4) by setting the quantum numbers of the initial nucleon (μ) and of the final resonance (μ') to be -1/2 and 1/2, or 1/2 and 3/2, respectively.

In a similar way, one can calculate the longitudinal electromagnetic transition amplitude based on the relativistic point form. The amplitude $S_{1/2}$ is defined in terms of the matrix element of the longitudinal electromagnetic interaction H^L_{em}

$$
\begin{aligned}
S_{1/2} &= \xi < f,\mu' \mid H^L_{em} \mid i,\mu >= \xi < f,\mu' \mid \sqrt{\frac{2\pi\alpha_E}{\omega}}\epsilon^L_\mu \cdot J^\mu \mid i,\mu > \\
&= 3\xi \int \frac{d^3p_1}{(2\pi)^3}\frac{d^3p_2}{(2\pi)^3}\frac{d^3p_3}{(2\pi)^3}\frac{d^3p'_1}{(2\pi)^3}\frac{d^3p'_2}{(2\pi)^3}\frac{d^3p'_3}{(2\pi)^3} \times (2\pi)^6\delta^3(\vec{k}_1+\vec{k}_2+\vec{k}_3)\delta^3(\vec{k}'_1+\vec{k}'_2+\vec{k}'_3) \\
&\times \psi^*_{J',1/2}(\vec{p}\,'_\rho,\vec{p}\,'_\lambda;\mu'_1,\mu'_2,\mu'_3)\psi_{1/2,1/2}(\vec{p}_\rho,\vec{p}_\lambda;\mu_1,\mu_2,\mu_3) \\
&\times D^{*1/2}_{\lambda'_3\mu'_3}[R_W(k'_3,B(v_{out}))] < p'_3,\lambda'_3 \mid \sqrt{\frac{2\pi\alpha_E}{\omega}}\frac{\sqrt{Q^2}}{q}J_0 \mid p_3,\lambda_3 > D^{1/2}_{\lambda_3\mu_3}[R_W(k_3,B(v_{in}))) \\
&\times D^{1/2}_{\mu'_1\mu_1}[R_W(k_1,B^{-1}(v_{out})B(v_{in}))]D^{1/2}_{\mu'_2\mu_2}[R_W(k_2,B^{-1}(v_{out})B(v_{in}))] \\
&\times \delta^3(k'_1-B^{-1}(v_{out})B(v_{in})k_1)\delta^3(k'_2-B^{-1}(v_{out})B(v_{in})k_2). \qquad (5)
\end{aligned}
$$

This definition of the longitudinal transition amplitude is consistent with that of the transverse transition amplitude in Eq. (4). Usually, the polarization vector of the longitudinal polarized photon is selected to be $\epsilon^L_\mu = (\frac{q}{\sqrt{Q^2}}, 0, 0, \frac{\omega}{\sqrt{Q^2}})$, so that the constraint condition $q^\mu\epsilon^L_\mu = 0$ is satisfied. The gauge invariant condition $q^\mu J_\mu = 0$ gives that $< f \mid H^L_{em} \mid i >= \frac{\sqrt{Q^2}}{q} < f \mid J_0 \mid i >$. Clearly, the matrix element of the longitudinal electromagnetic interaction in Eq. (5) is proportional to $\sqrt{Q^2}$, and it vanishes in the real photon limit $Q^2 = 0$. The interaction H^L_{em} is a Lorentz scalar. Thus, the results of the longitudinal transition amplitude in the fully relativistic point form are frame-independent. It should be mentioned that the longitudinal transition amplitude defined in the conventional nonrelativistic constituent quark model is $S^{non}_{1/2} = \xi < f, \mu' \mid \sqrt{\frac{2\pi\alpha_E}{\omega}} J_0 \mid i, \mu >$. It is not a Lorentz invariant amplitude. The two definitions of the longitudinal transition amplitude are different by a factor of $\frac{\sqrt{Q^2}}{q}$. In addition, the results of Eq. (5) in the non-relativistic framework are still reference frame-dependent due to the lack of the Lorentz covariance.

In our numerical calculations, the conventional harmonic oscillator wave functions are employed for the nucleon and its resonances. The two parameters are selected to be $0.16 GeV^2$ for harmonic oscillator constant, and $m_q = M_N/3$ for quark mass. Here, we do not consider any configuration mixing effect from hyperfine interactions for simplicity. In Figs. 1 and 2, our results (solid line) for the transverse transition amplitudes $A_{1/2}$ and $A_{3/2}$ of the $\Delta(1232)$ resonance are plotted. The results of the non-relativistic constituent quark model (dotted-dashed line) with the relativistic corrections are also displayed in the Breit reference frame for comparison. Moreover, in Figs. 3 and 4, the results for the transverse and longitudinal amplitudes of the resonance $S_{11}(1535)$ are shown.

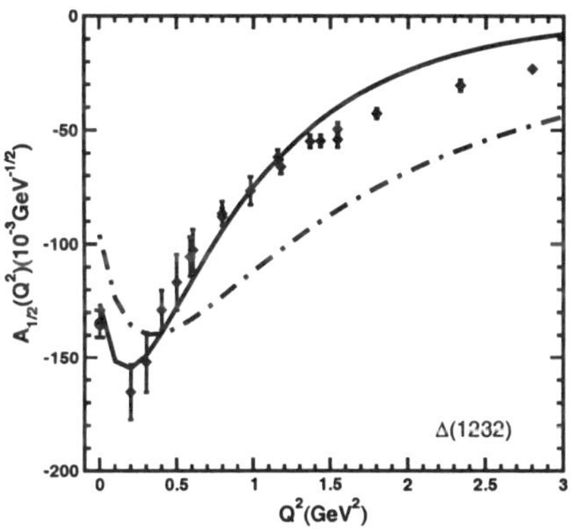

Fig. 1. $A_{1/2}(Q^2)$ of $\Delta(1232)$.

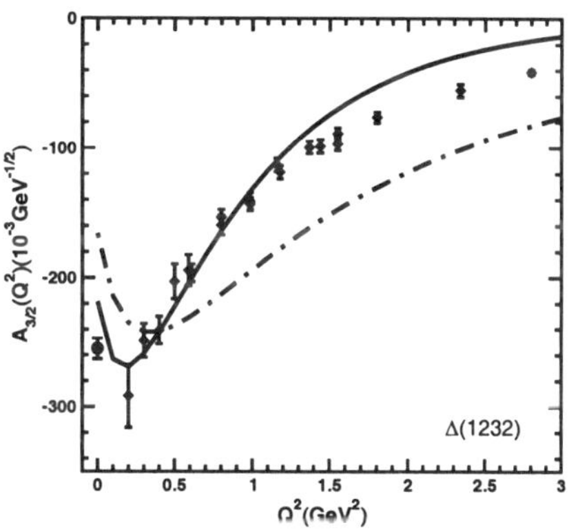

Fig. 2. $A_{3/2}(Q^2)$ of $\Delta(1232)$.

We may conclude that the point form relativistic quantum mechanics can provide a convenient way to study hadron properties, particularly the transitions to resonances. Since the relativistic boost is correctly considered in this framework, the results for the Q^2-dependences of the transition amplitudes are expected to be reasonable. This conclusion can be clearly seen from the point form predictions of the transverse transition amplitudes of the $\Delta(1232)$ resonance as shown in Figs. 1

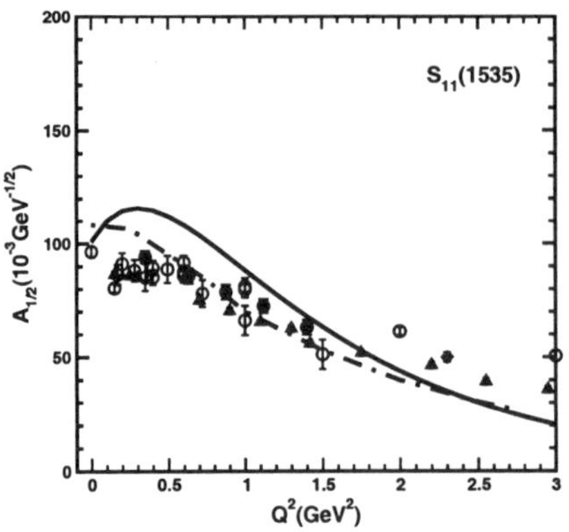

Fig. 3. $S_{1/2}(Q^2)$ of $S_{11}(1535)$.

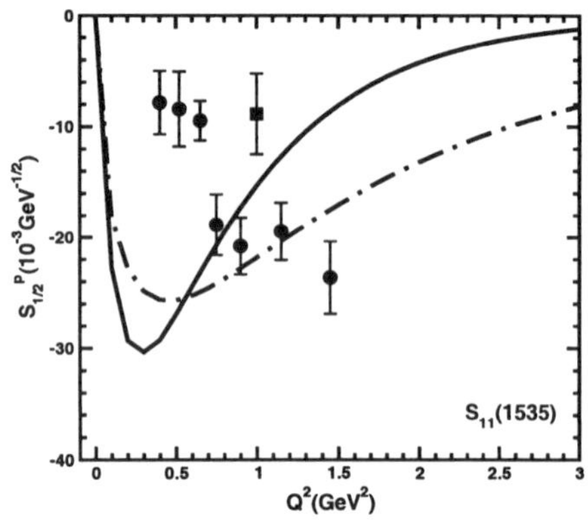

Fig. 4. $S_{1/2}(Q^2)$ of $S_{11}(1535)$

and 2 and of the $S_{11}(1535)$ in Fig. 3. Our work shows that the reasonable results for the transition amplitudes of $A_{1/2,3/2}(Q^2)$ of the $\Delta(1232)$ resonance, of $A_{1/2}$ of $S_{11}(1535)$ and $D_{13}(1520)$ resonances can be achieved. Moreover, there are sizable discrepancies between the point form results and the non-relativistic constituent quark model predictions with the relativistic corrections. Since the harmonic oscillator wave functions are employed in the two frameworks simultaneously, the discrepancies clearly indicate that the relativistic effect needs to be correctly taken into account.

References

1. P. A. M. Dirac, Reviews of Modern Physics, **21**, 392 (1949).
2. B. D. Keister and W. N. Polyzou, Advanced Nuclear physics, Edited by J. W. Negele and E. W. Vogt (Pleneum, New York, 1991), Vol **21**, P. 225.
3. W. H. Klink, Phys. Rev. C**58**, 3587 (1998); *ibid*, 3605 (1998); *ibid*, 3617 (1998).
4. L. Ya. Glozman, M. Radici, R. F. Wagenbrunn, S. Boffi, W. Klink and W. Plessas, Phys. Lett. B**516**, 183 (2001); F. Wagenbrunn, S. Boffi, W. Klink, W. Plessas and M. Radici, Phys. Lett. B**511**, 33 (2001).
5. T. Melde, R. F. Wagenbrunn and W. Plessas, Few Body Syst. Suppl. **14**, 37 (2003); T. Melde, W. Plessas and R. F. Wagenbrunn, hep-ph/0505198.
6. T. W. Allen and W. H. Klink, Phys. Rev. C**58**, 3670 (1998); T. W. Allen, W. H. Klink and W. N. Polyzou, Phys. Rev. C**63**, 034002 (2001).
7. B. Desplanques, L. Theussl, and S. Noguera, Phys. Rev. C**65**, 038202 (2002), A. Amghar, B. Desplanques, and L. Theussl, Phys. Lett. B**574**, 201 (2003); Jun He, B. Julia-Diaz, and Y. B. Dong, Phys. Lett. B**602**, 212 (2004).

Energy-Dependent Phase-Shift Analyses for Elastic *pp* Scattering in COSY-Data Region

J. Nagata

Faculty of Informatics, Hiroshima Kokusai Gakuin University,
6-20-1 Nakano, Aki-ku, 739-0321 Hiroshima, Japan
E-mail: j.nagata@hkg.ac.jp

H. Yoshino

Faculty of Health Sciences, Hiroshima International University,
555-36 Kurose Gakuendai, 724-0695 Higashi-Hiroshima, Japan
E-mail: yoshino@hs.hirokoku-u.ac.jp

M. Matsuda

Kurashiki Sakuyo University,
3515 Tamashima Nagao, 710-0292 Kurashiki, Japan
E-mail: masa@hiroshima-u.ac.jp

We have carried out the energy-dependent phase-shift analyses for elastic *pp* scattering including new COSY-data. The parametrizations of phase shifts and reflection parameters are given by polynomial expansions. The preliminary results are discussed.

Phase-shift analyses (PSA) of experimental data on various spin observables have been playing an important role for the study of few-body systems in hadron physics. Especially in nucleon-nucleon system, they have a very long history since the beginning of nuclear scattering experiments. The theoretical models to describe the few-body interactions are usually constructed so as to reproduce the partial wave amplitudes determined by PSA.

Broard and narrow structures have been found in Nucleon-Nucleon system in the intermediate energy region, and possible resonance states have been argued. Based on the results of PSA, spin angular momentum states were determined. Hyperon production processes, $pp \to KYN$[1] etc, are also interested in studying the Hyperon-Nucleon interacion, where the scattering amplitudes of *pp* system are used as initial channel.

Precise data on A_y for elastic *pp* scattering at $T_L = 0.5 - 2$ GeV have been measured at KEK[2]. Saclay group have measured a rich amount of spin observables on *pp* and *np* scatteirng at $T_L = 0.3 - 2.7$ GeV[3–5]. JINR group have also measured a forward observable $\Delta\sigma_L$ for *np* scattering at $T_L = 1.2 - 3.6$ GeV[6]. All of these data have been used for our single-energy PSA so far.

In last decade, COSY experiments have provided precise data on $d\sigma/d\Omega$, A_N and A_{ij} for elastic pp scattering at the intermediate energies[7–9]. These data have been analysed by several groups[10–13]. Recently, the results of COSY experiments have been updated[14–16]. These data were used for searching the narrow structures related to resonace states in nucleon-nucleon system. Moreover, it is very important to determine the scattering amplitude in several GeV region for revealing the transition from the hadron to the quark-gluon phase.

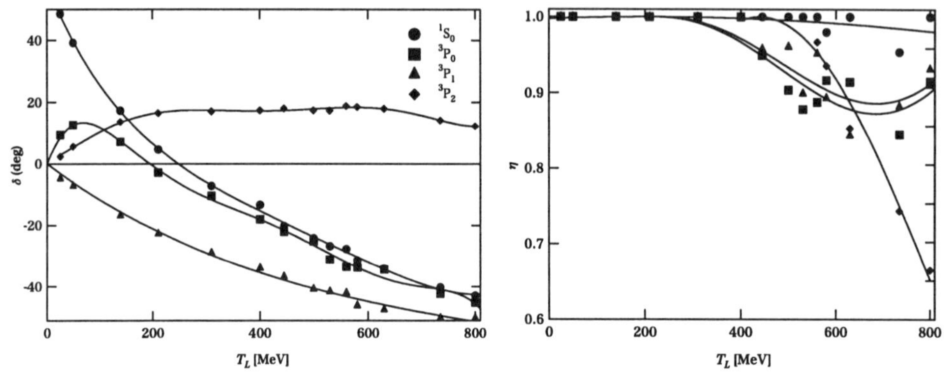

Fig. 1. Solutions of single and energy-dependent PSA. Solutions of ED-PSA are preliminary ones.

In contrast to our previous PSA, we have carried out the energy dependent PSA using polynomial expansions for phase shifts and reflection parameters. The S-matrix for uncoupled waves with the orbital and the total angular momentum ℓ and J are given by,

$$S_{\ell,J} = \eta_{\ell,J} \exp\left(2i\delta_{\ell,J}\right) \tag{1}$$

where $\delta_{\ell,J}$ and $\eta_{\ell,J}$ are the phase shifts and the reflection parameters, For coupled waves,

$$S_J = \begin{bmatrix} \sqrt{1-|\rho_J|^2}\eta_- \exp(2i\delta_-) & i\rho_J\sqrt{\rho_-\rho_+}\exp\{i(\delta_-+\delta_+)\} \\ i\rho_J\sqrt{\rho_-\rho_+}\exp\{i(\delta_-+\delta_+)\} & \sqrt{1-|\rho_J|^2}\eta_+ \exp(2i\delta_+) \end{bmatrix}, \tag{2}$$

where $\delta_- \equiv \delta_{J-1,J}$, $\delta_+ \equiv \delta_{J+1,J}$, $\eta_+ \equiv \eta_{J+1,J}$, $\eta_- \equiv \eta_{J-1,J}$ and ρ_J is the mixing parameter. The phase shift and mixing parameters are parametrized by polynomial expansion series of the kinetic energy in the laboratory system T_L as,

$$\delta_{\ell,J} = \sum_{i=0}^{n} a^i_{\ell,J}\left(\frac{T_L}{m_p}\right)^i, \tag{3}$$

Table 1. Coefficients a^i in polynomial expansions for each partial wave in Eq. (3).

	a^0	a^1	a^2	a^3	a^4	a^5	a^6
$^1S_0$	-3.73×10^2	8.30×10^2	-1.05×10^3	4.98×10^2			
$^3P_0$	0	4.29×10^2	-4.47×10^3	1.71×10^4	-3.25×10^4	3.00×10^4	-1.08×10^4
$^3P_1$	-1.19×10^2	9.90×10	-3.54×10				
$^3P_2$	1.03×10^2	-1.71×10^2	7.99×10				

Table 2. Coefficients b^i in polynomial expansions for each partial wave in Eq. (4).

	b^0	b^1	b^2	b^3
$^1S_0$	4.97×10^{-2}	-2.20×10^{-1}		
$^3P_0$	-1.48×10^{-1}	2.83×10^{-3}	-2.58×10^{-3}	
$^3P_1$	-1.24×10^{-1}	-1.28×10^{-4}	2.78	-2.54
$^3P_2$	7.97×10^{-1}	$-1.70\times$		

and reflection parameters are parametrized by

$$\eta_{\ell,J} = \exp\left(-\alpha^2\right), \alpha = \sum_{i=0}^{n} b^i_{\ell,J}\left(\frac{T_L}{m_p}\right)^i, \tag{4}$$

where m_p, $a^i_{\ell,J}$ and $b^i_{\ell,J}$ are proton mass and the free parameters. $\eta_{\ell,J}$ are limited to be $\eta_{\ell,J} \leq 1$ for unitarity.

In the energy region from 0.025 to 2.7 GeV, we have not yet obtained final solutions for energy-dependent PSA. In Fig. 1 an example of possible solutions are shown for partial waves of $^1S_0$, $^3P_0$, $^3P_1$ and $^3P_2$, and values of coefficients a^i and b^i are given in Table 1. The solutions obtained by single-enerngy PSA are used for evaluating starting values in energy-dependent PSA. In addition to present energy region, extrapolations of energy-dependent PSA to much higher energy are examined by using our solutions obtained at $T_L - 3.6$, 5 and 11 GeV. For instance, taking $b^0 - 0.157$, $b^1 = 0.221$, and $b^2 = -0.013$, solutions of the reflection parameters (η) from 1 to 11 GeV are reproduced very well as shown in Fig. 2.

In summary, we have developed programs for the energy-dependent PSA for elastic pp scattering at T_L = 0.025 - 1 GeV. Preliminary results are shown using polynomical expansion reflecting the energy dependence of phase shifts, reflection parameters and mixing parameters. Extention to higher energy is examined using same parametrization as ones at intermediate energies.

This work was supported in part by a Grant-in-Aid for Scientific Research(C) from JSPS (No. 17540272).

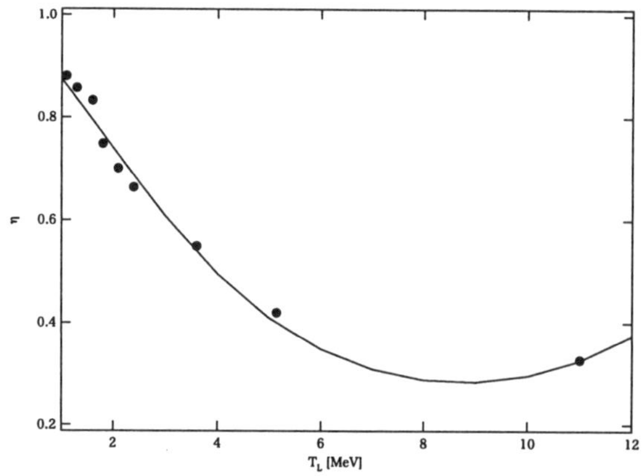

Fig. 2. The reflection parameter of ${}^1G_4$ wave obtained by our single-energy PSA at T_L = 1 - 11 GeV. Solid line is given by taking $b^0 = 0.157$, $b^1 = 0.221$, and $b^2 = -0.013$ in Eq. (4).

References

1. S. Wirth *et al.*, πN Newslett. **16**, 373(2002);P. Kowina *et al.*, Eur. Phys. J. **A18**, 351 (2003).
2. H. Shimizu *et al.*, Phys. Rev. **C42**, 483(1990);Y. Kobayashi *et al.*, Nucl. Phys. **A569**, 791 (1994).
3. For instance, C. D. Lac *et al.*, Nucl. Phys. **B297**653 (1988).
4. C. E. Allgower *et al.*, Eur. Phys. J. **C1**, 131 (1998);J. Arvieux *et al.*, Z. Phys. **C76**, 465 (1997);C. E. Allgower *et al.*, Phys. Rev. **C62**, 064001(2000); Phys. Rev. **C64**, 034003 (2001).
5. D. Adams *et al.*, Acta Polytechnica **36**, 11 (1996).
6. B. P. Adiasevich *et al.*, Z. Phys. **C71**, 65 (1996).
7. D. Albers *et al.*, Phys. Rev. Lett. **78**, 1652 (1997).
8. M. Altmeier *et al.*, Phys. Rev. Lett. **85**, 1819 (2000).
9. F. Bauer *et al.*, Phys. Rev. Lett. **90**, 142301 (2003).
10. R. A. Arndt, I. I. Strakovsky, R. L. Workman, F. Dohrmann, Phys. Rev. **C56**, 3005 (1997); nucl-th/9706003.
11. R. A. Arndt, I. I. Strakovsky, R. L. Workman, Phys. Rev. **C62**, 34005 (2000).
12. J. Nagata, H. Yoshino and M. Matsuda, πN Newslett. **16**, 376 (2002).
13. J. Nagata, H. Yoshino and M. Matsuda, Mod. Phys. Lett. **A18**, 4448(2003).
14. D. Albers *et al.*, Eur. Phys. J. **A22**, 125 (2004).
15. M. Altmeier *et al.*, Eur. Phys. J. **A23**, 351 (2005).
16. F. Bauer *et al.*, nucl-ex/0412014 (2004).

Studying $\gamma^* N \to \Delta \to \pi N$ with Cross Sections and Polarization Observables*

Shalev Gilad

Laboratory for Nuclear Science,
Massachusetts Institute of Technology,
Cambridge, MA 02139, USA
E-mail: sgilad@mitlns.mit.edu

Results are presented of the reaction $\gamma^* N \to \Delta \to \pi N$ that was studied at the Jefferson Laboratory for transferred four momentum of 1.0 $(GeV/c)^2$. Angular distributions of the cross sections and polarizations were measured over much of the reaction phase space. Out of the possible eighteen response functions, fourteen separate responses and two additional response combinations were extracted. All resonance and background s- and p-waves as well as several of the d-wave multipole amplitudes were decomposed. The responses and multipoles were compared to model calculations. Significant differences were observed between the experimental and model values for several responses and multipoles. The EMR and SMR extracted from the fitted multipoles are different from previously reported values.

1. Introduction

In recent years, interest has been renewed in the $\gamma^* N \to \Delta \to \pi N$ reaction as a means to study the deformation of the nucleon. In this context, the signature of deformed nucleon or Δ would be manifested in a non-zero electric or longitudinal quadrupole $N \to \Delta$ transition amplitude in addition to the dominant magnetic dipole one. Significant theoretical and experimental work has been done leading to both predictions and measurements of these ratios, $EMR = \mathbf{Re}(E_{1+}^{3/2}/M_{1+}^{3/2})$ and $SMR = \mathbf{Re}(S_{1+}^{3/2}/M_{1+}^{3/2})$. In most theoretical models, the intrinsic nucleon and Δ quadrupole moments are small and the observed EMR and SMR are dominated by hadronic degrees of freedom, in particular the pion cloud. Experimentally, pion electro-production has been recently measured at the Jefferson Laboratory (JLAB), Bates, and MAMI, while pion photo-production (sensitive to EMR only) has been measured at LEGS and MAMI. In general, the analyses have been performed by extracting from the cross-sections angular distributions coefficients of a few Legendre polynomials (Legendre analysis) that contain the E_{1+} and S_{1+} amplitudes. Usually, the expansion was truncated to s and p partial waves and neglected all

*This work was carried out by members of JLAB E91-011 and Hall A collaborations and supported by the U.S. DOE and NSF, the Canadian NSERC, the Italian INFN, the French CNRS and CEA, and the Swedish VR.

amplitudes that do not contain the dominant M_{1+} multipole (M1 dominance). It has been known that such Legendre analyses yield only approximate values of EMR and SMR. Much of the recent theoretical and experimental developments can be viewed in reference 1 and 2 and references therein.

We present 14 response functions and 2 response combinations (out of the 18 available in this reaction) extracted by a multipole analysis from angular distributions of cross-section and recoil-polarizations measured for the reaction $p(\vec{e}, e'\vec{p})\pi^0$. This analysis provided fits of all the real and imaginary 0+, 1- (except $\mathbf{Im}M_{1-}$), 1+, and **Re**2- multipole amplitudes.

2. The Experiment

The experiment was performed in Hall A of JLAB using the standard Hall A instrumentation that included the focal-plane hadron polarimeter [3]. The electron kinematics was fixed with a nominal four-momentum transfer, $Q^2 = 1.0\ (GeV/c)^2$, at a transverse virtual-photon polarization, $\epsilon = 0.95$ ($E_{beam} = 4.535\ GeV$; $\theta_e = 14.1^o$; $\mathbf{p}_e = 3.066\ GeV/c$; $\theta_{\vec{q}} = 42.0^o$). Because of the Lorentz boost, the protons were ejected in the laboratory frame within a cone of $\theta_{pq} = 17^o$ with respect to the 3-momentum transfer vector, $\vec{q}$, providing a complete center-of-mass (CM) θ_{pq} coverage and a significant out-of-plane (ϕ) coverage of the cross sections and measured recoil-proton polarizations. This enabled the extraction of the response functions and multipole amplitudes in 10 bins of W, the invariant mass of the proton-π^0 system, ranging from 1.17 to 1.35 GeV. More experimental details can be found in references 2.

3. Results and Discussion

We describe here results based on multipole analysis, although we extracted the response functions using the traditional Legendre analysis as well. The multipole analysis is much less model dependent and does not suffer from truncation-related errors that plague the Legendre analysis [2,4]. In the multipole analysis, all the data for each W bin are used to fit simultaneously the real and imaginary parts of the $M_{l\pm}, E_{l\pm}, S_{l\pm}$ amplitudes by varying an appropriate set of lower partial waves while higher multipoles are taken from a baseline model. To minimize model dependence, we used Born amplitudes as a baseline, although we verified that our results show little dependence on the baseline model used.

Figure 1 displays the 14 extracted response functions and 2 Rosenbluth response combinations vs. $x_{cm} = cos(\theta_\pi)$, the center-of-mass angle between the pion and $\vec{q}$. These responses were measured at $Q^2 = 1.0\ (GeV/c)^2$ and $W = 1.23\ GeV$, on top of the $\Delta(1232)$ resonance. The left 2 panels are real-type responses, while the right 2 panels are imaginary-type responses. Only 4 responses, L+T(0), LT(0), TT(0), and LTh(0), have been previously measured. Although on top of the $\Delta(1232)$ resonance, none of the available models - MAID2003 [5], DMT (Dubna-Mainz-Taipei) [6], SAID [7], and SL (Sato-Lee) [8] describes all response functions adequately [2] and disagreements between the data and the models increase with increasing $|W - M_\Delta|$. Also shown in figure 1 are multipole fits of these responses, with the different curves indicating a different subset of amplitudes varying in the fit (see figure caption for details). The red curve, chosen by maximizing the number of fitted amplitudes while maintaining fitting stability, is our final fit where all s-, p- (except $\mathbf{Im}M_{1-}$), and the real part of the 2- amplitude are varied in the fit. Similar data and fits are available for 10 bins of W between 1.17 and 1.35 GeV.

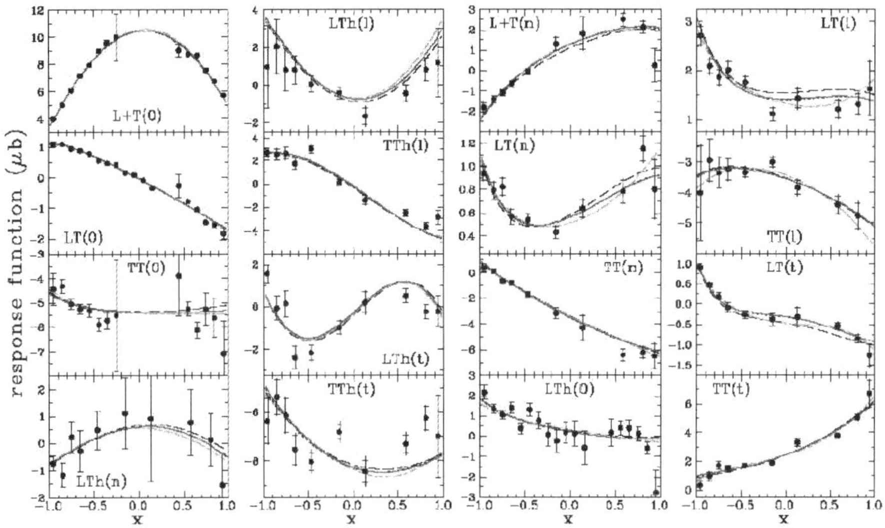

Fig. 1. Data and multipole fits of response functions for $W = 1.23$ GeV using a Born baseline model. Black dashed curves fit all s- and p-wave amplitudes, blue dotted curves also fit real parts of 2- multipoles, and green dashed-dotted curves fit all s-, p-, and d-wave amplitudes. The red solid curves, our final fit, are similar to the blue dotted curves except that $\mathbf{Im}M_{1-}$ is absent.

The multipole amplitudes, extracted from the fits described above, are displayed in figure 2 vs. the invariant mass W. They provide the most detailed information from our measurement. Also displayed are curves generated by several models. As expected, the models reproduce most of the resonant 1+ amplitudes (SAID is doing noticeably worse than others), although none reproduce the $\mathbf{Im}E_{1+}$ amplitude and their slope is too small for $\mathbf{Re}S_{1+}$ at large W. In contrast, large model variations are observed for the non-resonant amplitudes, and significant differences are observed between models and data such that no model describes all the non-resonant amplitudes well. Oddly, SAID reproduces better the 0+ amplitudes, but does significantly worse than the other models in reproducing most other amplitudes. It is interesting that the data does not display any significant deviation from the Born baseline, that omits the Roper resonance, for the $\mathbf{Re}M_{1-}$ amplitude at high W, suggesting that the transverse ${}_pA_{1/2}$ is small. On the other hand, the fitted $\mathbf{Re}S_{1-}$ differs from the Born baseline at high W, suggesting a non-negligible longitudinal contribution consistent with a radial excitation of the Roper.

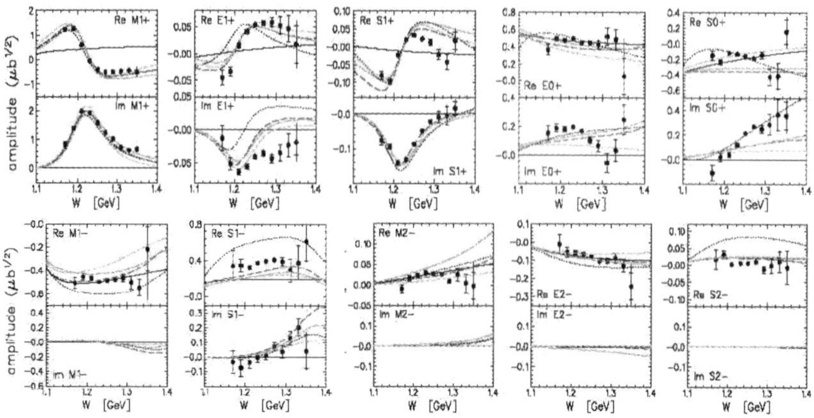

Fig. 2. Fitted multipole amplitudes for $Q^2 = 1.0\ (GeV/c)^2$ using a Born baseline and varying the s-, p-wave (except $\mathbf{Im}M_{1-}$) and $\mathbf{Re}$(2-) amplitudes. Inner error bars with endcaps are statistical, while outer error bars without endcaps include systematic uncertainties. The Born baseline amplitudes are shown in black curves. Also displayed are curves generated by MAID2003 (red dashed), DMT (green dot-dashed), SAID (blue dotted), and SL (cyan short-dashed).

The EMR and SMR were extracted from the measured amplitudes using

$$EMR = \frac{\mathbf{Re}E^*_{1+}M_{1+}}{|M_{1+}|^2} = \frac{\mathbf{Re}E_{1+}\mathbf{Re}M_{1+} + \mathbf{Im}E_{1+}\mathbf{Im}M_{1+}}{\mathbf{Re}M_{1+}\mathbf{Re}M_{1+} + \mathbf{Im}M_{1+}\mathbf{Im}M_{1+}} \tag{1}$$

$$SMR = \frac{\mathbf{Re}S^*_{1+}M_{1+}}{|M_{1+}|^2} = \frac{\mathbf{Re}S_{1+}\mathbf{Re}M_{1+} + \mathbf{Im}S_{1+}\mathbf{Im}M_{1+}}{\mathbf{Re}M_{1+}\mathbf{Re}M_{1+} + \mathbf{Im}M_{1+}\mathbf{Im}M_{1+}}. \tag{2}$$

They are displayed in figure 3, and are compared to MAID2003, DMT, SAID, and SL. The vertical line shows the physical mass of $W = M_\Delta$. As can be deduced from figure 2, all models except SAID reasonably reproduce the values of the quadrupole ratios at $W = M_\Delta$, but differ significantly from each other and from the data for W below and especially above the peak of the resonance. Clearly, our data provide significant input to all models regarding how background multipoles affect the determination of resonance amplitudes.

The experimental values extracted from the multipole analysis, $SMR = -6.84 \pm 0.15\%$ and $EMR = 2.94 \pm 0.19\%$ are significantly different from values obtained by a traditional Legendre analysis, $SMR = 6.11 \pm 0.11\%$ and $EMR = 1.92 \pm 0.14\%$. This difference is traced [4] mainly to the M1 dominance and s- and p-wave truncation approximations employed in the latter, both of which are inadequate and introduce up to 20% error in the Legendre analysis. It also poses a question as to how meaningful are previously reported values of the SMR and EMR that were determined from Legendre analysis of cross-section angular distributions. We suggest that values extracted from Legendre analyses should be accompanied by estimated uncertainties associated with this method.

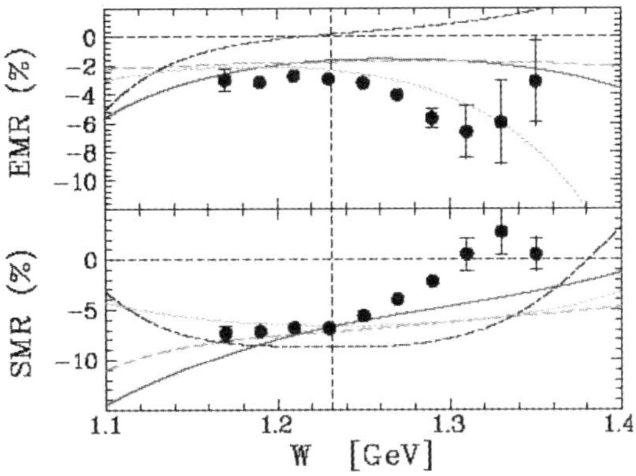

Fig. 3. *EMR* and *SMR* extracted from multipole analysis at $Q^2 = 1.0\ (GeV/c)^2$ are compared with MAID2003 (red solid), DMT (green dashed), SAID (blue dot-dash), and SL (cyan dotted). The vertical line denotes $W = M_\Delta$. Inner error bars with endcaps are statistical, while outer error bars without endcaps include systematic uncertainties.

4. Conclusions

The angular distributions for 16 (14 separate and 2 Rosenbluth combinations) out of a possible 18 response functions in the reaction $\gamma^* N \rightarrow \Delta \rightarrow \pi N$ were extracted at $Q^2 = 1.0\ (GeV/c)^2$ for 10 bins of $1.17 \leq W \leq 1.35\ GeV$, spanning the $\Delta(1232)$ resonance. This was made possible by simultaneously measuring the angular distributions of both cross sections and polarization observables. The response functions were extracted by employing a nearly model independent multipole analysis of the data. The fitted 0+, 1- (except $\mathbf{Im}M_{1-}$), 1+, and **Re**(2-) multipole amplitudes were compared to model calculations. All models (except SAID) predict the resonance amplitudes at $W = M_\Delta$ but fail to predict the background amplitudes. Increasing differences are observed between the experimental multipole amplitudes and those predicted by the models, as well as between the various models, with increasing $|W - M_\Delta|$. The values of EMR and SMR were determined from the fitted multipoles and are significantly different from those determined by a traditional Legendre analysis. This raises a question as to the validity of previously reported values that were determined by Legendre analysis of cross section angular distributions, and suggests that approximation uncertainties should be assigned to those values.

References

1. http://pierre.mit.edu/deformation/
2. J. J. Kelly *et al.*, arXiv:nucl-ex/0509004
3. J. Alcorn *et al.*, *Nucl. Instr. Meth.* **A522**, 294 (2004).
4. J. J. Kelly, arXiv:nucl-ex/0507010
5. L. Tiator *et al.*, *Eur. Phys. Jour.* **A19**, 55 (2004).
6. S. S. Kamalov *et al.*, *Phys. Rev.* **C64**, 032201(R) (2001).
7. R. A. Arndt, I. I. Strkovsky and R. L. Workman, *Phys. Rev.* **C66**, 055213 (2002).
8. T. Sato and T.-S. H. Lee, *Phys. Rev.* **C63**, 055201 (2001).

Parity Violating Electron Scattering at the MAMI Facility in Mainz*

F. E. Maas† for the A4-Collaboration

Johannes Gutenberg Universität Mainz,
Institut für Kernphysik, J.J.Becherweg 45, 55299 Mainz,
E-mail: maas@kph.uni-mainz.de

We report here on a measurement of the parity violating (PV) asymmetry in the scattering of polarized electrons on unpolarized protons performed with the setup of the A4 experiment at the MAMI accelerator facility in Mainz. The physics goal is the extraction of the strangeness contribution G_E^s and G_M^s to the vector form factors of the nucleon. This experiment is the first to use counting techniques in a parity violation experiment. After discussing the experimental context of the experiments, the setup at MAMI and the results are presented.

1. Strangeness in the Nucleon

The understanding of quantum chromodynamics (QCD) in the nonperturbative regime of low energy scales is crucial for understanding the structure of hadronic matter like protons and neutrons (nucleons). The successful description of a wide variety of observables by the concept of effective, heavy ($\approx$ 350 MeV) constituent quarks, which are not the current quarks of QCD, is still a puzzle. There are other equivalent descriptions of hadronic matter at low energy scales in terms of effective fields like chiral perturbation theory (χPT) or Skyrme-type soliton models. The effective fields in these models arise dynamically from a sea of virtual gluons and quark-antiquark pairs. In this context the contribution of strange quarks plays a special role since the nucleon has no net strangeness, and any contribution of strange quarks to the nucleon structure observables is a pure sea-quark effect. Due to the heavier current quark mass of strangeness (m_s) as compared to up (m_u) and down (m_d) with $m_s \approx 140$ MeV $\gg m_u, m_d \approx$ 5-10 MeV, one expects a suppression of strangeness effects in the creation of quark-antiquark pairs. On the other hand the strange quark mass is in the range of the mass scale of QCD ($m_s \approx \Lambda_{QCD}$) so that the dynamic creation of strange sea quark pairs could still be substantial. A determination of the weak vector form factors of the proton ($\tilde{G}_E^p$ and $\tilde{G}_M^p$) by measuring the PV asymmetry in the scattering of longitudinally polarized electron off unpolarized protons allows the determination of the strangeness contribution to the electromagnetic form factors G_E^s and G_M^s [1]. A direct separation of electric (G_E^s)

---

*This work is supported by the DFG in the frame work of the SFB 443
†on leave to the Institut de Physique Nuclaire Orsay

and magnetic (G_M^s) contribution at forward angle is for the first time possible by combining all the world data at 0.1 $(\text{GeV}/c)^2$ [2].

2. The A4 Experimental Setup and Analysis

The PV asymmetry was measured at the MAMI accelerator facility in Mainz [3] using the setup of the A4 experiment [4–8], which is the first parity violation experiment to use counting techniques. The polarized 570.4 and 854.3 MeV electrons were produced using a strained layer GaAs crystal that is illuminated with circularly polarized laser light [9]. Average beam polarization was about 80 %. A 20 ms time window enabled the histogramming in all detector channels and an integration circuit in the beam monitoring and luminosity monitoring systems. The intensity $I = 20\ \mu$A of the electron current was stabilized to better than $\delta I/I \approx 10^{-3}$. An additional $\lambda/2$-plate in the optical system was used to rotate small remaining linear polarization components and to control the helicity correlated asymmetry in the electron beam current to the level of < 10 ppm in each five minute run.
From the source to the target, the electron beam develops fluctuations in beam parameters such as position, energy and intensity which are partly correlated to the reversal of the helicity from +P to -P. We have used a system of microwave resonators in order to monitor beam current, energy, and position in two sets of monitors separated by a drift space of about 7.21 m in front of the hydrogen target. In addition, we have used a system of 10 feed-back loops in order to stabilize current, energy, position, and angle of the beam. The polarization of the electron beam was measured with an accuracy of 2 % using a Møller polarimeter which is located on a beam line in another experimental hall. Due to the fact that we had to interpolate between the weekly Møller measurements, the uncertainty in the knowledge of the beam polarization increased to 4 %. The 10 cm high power, high flow liquid hydrogen target was optimized to guarantee a high degree of turbulence with a Reynolds-number of $R > 2 \times 10^5$ in the target cell in order to increase the effective heat transfer. For the first time, a fast modulation of the beam position of the intense CW 20 μA beam could be avoided. It allowed us to stabilize the beam position on the target cell without target density fluctuations arising from boiling. The total thickness of the entrance and exit aluminum windows was 250 μm. The luminosity L was monitored for each helicity state (R, L) during the experiment using eight water-Cherenkov detectors (LuMo) that detect scattered particles symmetrically around the electron beam for small scattering angles in the range of $\theta_e = 4.4° - 10°$, where the PV asymmetry is negligible. The photomultiplier tube currents of these luminosity detectors were integrated during the 20 ms measurement period by gated integrators and then digitized by customized 16-bit analogue-to-digital converters (ADC). The same method was used for all the beam parameter signals. A correction was applied for the nonlinearity of the luminosity monitor photomultiplier tubes. From the beam current helicity pair data $I^{R,L}$ and luminosity monitor helicity pair $L^{R,L}$ data we calculated the target density $\rho^{R,L} = L^{R,L}/I^{R,L}$ for the two helicity

states independently.
To detect the scattered electrons we developed a new type of a very fast, homogeneous, total absorption calorimeter consisting of individual lead fluoride (PbF_2) crystals [10]. The material is a pure Cherenkov radiator and has been chosen for its fast timing characteristics and its radiation hardness. This is the first time this material has been used in a large scale calorimeter for a physics experiment. The crystals are dimensioned so that an electron deposits 96 % of its total energy in an electromagnetic shower extending over a matrix of 3×3 crystals. Together with the readout electronics this allows us a measurement of the particle energy with a resolution of $3.9\,\%/\sqrt{E}$ and a total dead time of 20 ns. For the data taken at 854.3 MeV only 511 out of 1022 channels of the detector and the readout electronics were operational, for the 570.4 MeV data all the 1022 channels were installed. The particle rate within the acceptance of this solid angle was $\approx 50 \times 10^6$ s^{-1}. Due to the short dead time, the losses due to double hits in the calorimeter were 1 % at 20 μA. This low dead time is only possible because of the special readout electronics employed. The signals from each cluster of 9 crystals were summed and integrated for 20 ns in an analogue summing and triggering circuit and digitized by a transient 8-bit ADC. There was one summation, triggering, and digitization circuit per crystal. The energy, helicity, and impact information were stored together in a three dimensional histogram.
The number of elastic scattered electrons is determined for each detector channel by integrating the number of events in an interval from 1.6 σ_E above pion production threshold to 2.0 σ_E above the elastic peak in each helicity histogram, where σ_E is the energy resolution for nine crystals. These cuts ensure a clean separation between elastic scattering and pion production or Δ-excitation which has an unknown PV cross section asymmetry. The linearity of the PbF_2 detector system with respect to particle counting rates and possible effects due to dead time were investigated by varying the beam current. We calculate the raw normalized detector asymmetry as $A_{\rm raw} = (N_e^R/\rho^R - N_e^L/\rho^L)/(N_e^R/\rho^R + N_e^L/\rho^L)$. The possible dilution of the measured asymmetry by background originating from the production of π^0's that subsequently decays into two photons where one of the photons carries almost the full energy of an elastic scattered electron was estimated using Monte Carlo simulations to be much less than 1 % and is neglected here. The largest background comes from quasi-elastic scattering at the thin aluminum entrance and exit windows of the target cell. We have measured the aluminum quasi-elastic event rate and calculated in a static approximation a correction factor for the aluminum of 1.030 ± 0.003 giving a smaller value for the corrected asymmetry.
Corrections due to false asymmetries arising from helicity correlated changes of beam parameters were applied on a run by run basis. The analysis was based on the five minute runs for which the counted elastic events in the PbF_2 detector were combined with the correlated beam parameter and luminosity measurements. In the analysis we applied reasonable cuts in order to exclude runs where the accelerator or parts of the PbF_2 detector system were malfunctioning. The analysis is based on

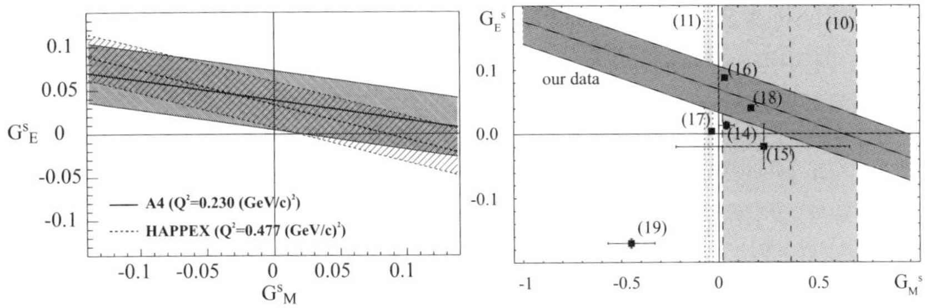

Fig. 1. The left plot shows our result at a Q^2 of 0.230 $(\mathrm{GeV/c})^2$. The solid line represents all possible combinations of $G_E^s+0.225G_M^s$. The densely hatched region represents the 1-σ uncertainty. The recalculated result from the HAPPEX published asymmetry at Q^2 of 0.477 $(\mathrm{GeV/c})^2$ is indicated by the dashed line, the less densely hatched area represents the associated error of the HAPPEX result. The right plot shows our result at $Q^2 = 0.108$ $(\mathrm{GeV/c})^2$. The solid lines represent the result on $G_E^s + 0.106G_M^s$. The hatched region represents in all cases the one-σ-uncertainty with statistical and systematic and theory error added in quadrature. The dashed lines represent the result on G_M^s from the SAMPLE experiment. The dotted lines represent the result of a recent lattice gauge theory calculation for μ_s The boxes represent different model calculations and the numbers denote the references as given in [6].

a total of 7.3×10^6 histograms corresponding to 4.8×10^{12} elastic scattering events for the 854.3 MeV data and $4.8 \cdot 10^6$ histograms corresponding to $2 \cdot 10^{13}$ elastic events for the 570.4 MeV data.

For the correction of helicity correlated beam parameter fluctuations we used multi-dimensional linear regression analysis using the data. The regression parameters have been calculated in addition from the geometry of the precisely surveyed detector geometry. The two different methods agree very well within statistics. The experimental asymmetry is normalized to the electron beam polarization P_e to extract the physics asymmetry, $A_{\mathrm{phys}} = A_{\mathrm{exp}}/P_e$. We have taken half of our data with a second $\lambda/2$-plate inserted between the laser system and the GaAs crystal. This reverses the polarization of the electron beam and allows a stringent test of the understanding of systematic effects. Our measured result for the PV physics asymmetry in the scattering cross section of polarized electrons on unpolarized protons at an average Q^2 value of 0.230 $(\mathrm{GeV/c})^2$ is $A_{LR}(\vec{e}p) = (-5.44 \pm 0.54_{stat} + 0.26_{syst})$ ppm for the 854.3 MeV data and $A_{LR}(\vec{e}p) = (-1.36 \pm 0.29_{stat} + 0.13_{syst})$ ppm for the 570.4 MeV data. The first error represents the statistical accuracy, and the second represents the systematical uncertainties including beam polarization. The absolute accuracy of the experiment represents the most accurate measurement of a PV asymmetry in the elastic scattering of longitudinally polarized electrons on unpolarized protons. From the difference between the measured $A_{LR}(\vec{e}p)$ and the theoretical prediction in the framework of the Standard Model, A_0, we extract a linear combination of the strange electric and magnetic form factors for the 570.4 MeV data at a Q^2 of 0.108 $(\mathrm{GeV/c})^2$ of $G_E^s \pm 0.106\ G_M^s = 0.071 \pm 0.036$. For the data at 854.3 MeV corresponding to a Q^2 value of 0.230 $(\mathrm{GeV/c})^2$ we extract $G_E^s + 0.225\ G_M^s = 0.039$

$\pm$ 0.034. Statistical and systematic error of the measured asymmetry and the error in the theoretical prediction of A_0 been added in quadrature. A recent very accurate determination of the strangeness contribution to the magnetic moment of the proton $\mu_s = G_M^s(Q^2 = 0\ (\mathrm{GeV/c})^2)$ from lattice gauge theory [11] would yield a larger value of $G_E^s = 0.076 \pm 0.036$ if the Q^2 dependence from 0 to 0.108 $(\mathrm{GeV/c})^2$ is neglected. The theoretical expectations for another quenched lattice gauge theory calculation [12], for SU(3) chiral perturbation theory [13], from a chiral soliton model [14], from a quark model [15], from a Skyrme-type soliton model [16] and from an updated vector meson dominance model fit to the form factors [17] are included into Figure 1.
We are preparing a series of measurements of the parity violating asymmetry in the scattering of longitudinally polarized electrons off unpolarized protons and deuterons under backward scattering angles of $140° < \theta_e < 150°$ with the A4 apparatus in order to separate the electric (G_E^s) and magnetic (G_M^s) strangeness contribution to the electromagnetic form factors of the nucleon.

Acknowledgments

This work has been supported by the DFG in the framework of the SFB 201 and SPP 1034. We are indebted to K.H. Kaiser and the whole MAMI crew for their tireless effort to provide us with good electron beam.

References

1. D. B. Kaplan et al., *Nucl. Phys.* **B310**, 527 (1988).
2. See also the contribution of S. Kox in this volume.
3. H. Euteneuer et al., *Proc. of the EPAC 1994* **1**, 506 (1994).
4. F. E. Maas et al., *Phys. Rev. Lett.* **93**, 022002 (2004).
5. F. E. Maas et al., *Phys. Rev. Lett.* **94**, 082001 (2005).
6. F. E. Maas et al., *Phys. Rev. Lett.* **94**, 152001 (2005).
7. F. E. Maas et al., *Eur. Phys. J.* **A 17**, 339 (2003).
8. F. E. Maas et al., *Proc. of the ICATPP-7* World Scientific, 2002, p. 758.
9. K. Aulenbacher et al., *Nucl.Ins.Meth.* **A391**, 498 (1997).
10. P. Achenbach et al., *Nucl. Ins. Meth.* **A465**, 318 (2001).
11. D. B. Leinweber et al., *Phys.Rev.Lett.* **94** , 212001 (2005).
12. R. Lewis et al., *Phys. Rev.* **D67**, 013003 (2003).
13. T. R. Hemmert et al., *Phys. Rev.* **C60**, 045501 (1999).
14. A. Silva et al. *Eur. Phys.J.* **A22**, 481 (2004).
15. V. Lyubovitskij et al., *Phys. Rev.* **C66**, 055204 (2002).
16. H. Weigel et al., *Phys. Lett.* **B353**, 20 (1995).
17. H.-W. Hammer et al., *Phys. Rev.* **C60**, 045204 (1999).

FEW-BODY DYNAMICS IN ATOMS, MOLECULES, BOSE-EINSTEIN CONDENSATES AND QUANTUM DOTS

Four-Body Faddeev-Watson Series Description of Proton-Helium Charge Transfer Process

E. Ghanbari Adivi

Physics Department
Yazd University, Yazd, Iran
E-mail: eghanbari@yazduni.ac.ir

M. A. Bolorizadeh

Physics Department
Shahid Bahonar University, Kerman, Iran.
E-mail: maboliri@mail.uk.ac.ir

The three-body Faddeev formalism is generalized to four-body scattering problems. Six auxiliary transition operators are introduced to split the Lippmann-Schwinger (LS) integral equation governing the whole four-body system into six coupled integral equations. The auxiliary operators are chosen in a manner that all the possible transition matrices corresponding to different initial and final channels, in a four-body scattering system, can be expressed as a linear combination of these auxiliary operators. Solving the coupled integral equations by using the Neumann iteration method leads to some series expansions which called Faddeev-Watson series. The four-body Faddeev-Watson series for proton helium charge transfer process are obtained.

1. Introduction

Considering the scattering of a three-particle system, The LS equation for a typical three-body system is[1]:

$$\mathscr{T} = V + VG_0^+\mathscr{T} \tag{1}$$

in which V is the whole three-body interaction potential, G_0^+ is the free three-body Green function, $\mathscr{T}$ is the whole three-body transition operator. The LS equation, in general, has no unique and definite solution. With the center-of-mass motion being taken into account, the LS equation has ambiguous solutions even in the case of a two-particle system. A similar ambiguity occurs for systems of three and more particles. How could one eliminate the ambiguity?. For two-body systems we choose the center-of-mass frame as the reference frame. For three-body systems, we use a fully quantum mechanical three-body formalism developed by Faddeev [9]. In the Faddeev formalism, the single non-Fredholmian LS integral equation is split into three coupled integral equations. In the Faddeev derived system the two-body transition matrices are the principal dynamical ingredients. The T-matrix operators satisfying the Faddeev equations are not simply related to the scattering and rearrangement T-matrix operators. There is an alternative version of Faddeev coupled equations for the set of the scattering and rearrangement T-matrix operators corresponding to all possible channels; i.e. the Lovelace version of Faddeev equations [3]. In the Lovelace equations, the inhomogeneous term just includes two-body interaction potential, then obtaining the Faddeev-Watson series needs some extra calculations to eliminate the two-body interaction potential. Furthermore, when the Lovelace formalism generalized to four and more interacting particles, the number of coupled equation will be increased very fast when the number of particles increases.

2. Generalization to Four-Body Scattering Process

Here we generalize the Faddeev equations to *four-body* problems. Our method has some advantages: the number of coupled equations is equal to the number of two-body interactions. All scattering and rearrangement T-matrix operators can be written as a linear combination of introduced auxiliary operators, easily. The inhomogeneous term just includes the two-body T-matrices. Obtaining the four-body Faddeev-Watson series from our equations is very simple and straightforward. In a typical four-body collision system, let us assume that the four particles are labeled with 1, 2, 3 and 4, and interact each other just by means of two-body potentials.

The whole potential is:

$$V = \sum_{\substack{i,j \\ i<j}}^{4} V_{ij} = V_k + V^k \qquad ; \quad k = 1, 2, 3 \; or \; 4. \tag{2}$$

For a non-relativistic system, the total Hamiltonian, H, can be written as $H = H_0 + V$ where H_0 is the kinetic energy operator of the whole system with total energy E. The Green's operators corresponding to H and H_0 are denoted by $G^{(\pm)}$ and $G_0^{(\pm)}$ respectively. The Green operators corresponding to interaction potentials V_i, V^i and the two-body interaction potential, V_{ij} are, respectively, given by;

$$\begin{aligned} G_i^{(\pm)}(E) &= \tfrac{1}{E-H_0-V_i \pm i\eta} \\ G^{i(\pm)}(E) &= \tfrac{1}{E-H_0-V^i \pm i\eta} \\ G_{ij}^{(\pm)}(E) &= \tfrac{1}{E-H_0-V_{ij} \pm i\eta}. \end{aligned} \tag{3}$$

Some relations between the introduced Green operators are:

$$\begin{aligned} G &= G_i + G_i V^i G = G^i + G^i V_i G \\ G &= G_i + G V^i G_i = G^i + G V_i G^i \\ G &= G_{ij} + G_{ij}(V - V_{ij})G = G_{ij} + G(V - V_{ij})G_{ij} \\ G_{ij} &= G_0 + G_0 V_{ij} G_{ij} = G_0 + G_{ij} V_{ij} G_0 \end{aligned} \tag{4}$$

The transition operator $\mathcal{T}_{ji}$ is $\mathcal{T}_{ij} = V_i + V_j G V_i$ with $(i, j = 1, 2, 3, 4)$ which satisfies the relations:

$$G^j \mathcal{T}_{ji} = G V_i \; ; \; \mathcal{T}_{ji} = V_i + V_j G^j \mathcal{T}_{ji} \; ; \; G^i \mathcal{T}_{ii} = G V_i \; ; \; \mathcal{T}_{ii} G^i = V_i G \tag{5}$$

The LS equations for direct scattering is;

$$\mathcal{T}_{ii} = V_i + V_i G^i \mathcal{T}_{ii} \quad ; \quad \mathcal{T}_{ii} = V_i + \mathcal{T}_{ii} G^i V_i \tag{6}$$

The scattering matrix $\mathscr{T}(E)$ for the four-particle system is a solution to the whole LS equation;

$$\mathscr{T}(E) = V + VG_0(E)\mathscr{T}(E) \quad \textit{or} \quad \mathscr{T}(E) = V + \mathscr{T}(E)G_0(E)V \tag{7}$$

For a given initial state, suppose that the particle 1 is incident on the bounded subsystem $(2,3,4)$. One can introduce a set of *six auxiliary operators*[4] $\mathfrak{T}_{i1}$ with $i = 1, 2, \cdots, 6$, in terms of the direct (elastic or inelastic(excitation)) scattering operator $\mathcal{T}_{11}$ ($\mathcal{T}_{11} : 1 + (2,3,4) \to 1 + (2,3,4)$) as;

$$\begin{aligned}
\mathfrak{T}_{11} &= V_{23} G^1 \mathcal{T}_{11} \\
\mathfrak{T}_{21} &= V_{24} G^1 \mathcal{T}_{11} \\
\mathfrak{T}_{31} &= V_{34} G^1 \mathcal{T}_{11} \\
\mathfrak{T}_{41} &= V_{12} + V_{12} G^1 \mathcal{T}_{11} \\
\mathfrak{T}_{51} &= V_{13} + V_{13} G^1 \mathcal{T}_{11} \\
\mathfrak{T}_{61} &= V_{14} + V_{14} G^1 \mathcal{T}_{11}
\end{aligned} \tag{8}$$

Faddeev equations governing the dynamical behavior of the set of auxiliary operators can be derived. At first we write the last operators in terms of the fully four-body Green function, G. Then, it is possible to write G, in terms of the two-body Green functions, G_{ij}, corresponding to the two-body interaction potentials, V_{ij}, operating on G from the left side. The final form of the Faddeev equations are obtained by using the two-body LS equations and some rearrangements:

$$\begin{pmatrix} \mathfrak{T}_{11} \\ \mathfrak{T}_{21} \\ \mathfrak{T}_{31} \\ \mathfrak{T}_{14} \\ \mathfrak{T}_{15} \\ \mathfrak{T}_{16} \end{pmatrix} = \begin{pmatrix} 0 \\ 0 \\ 0 \\ T_{12} \\ T_{13} \\ T_{14} \end{pmatrix} + \begin{pmatrix} 0 & T_{23} & T_{23} & T_{23} & T_{23} & T_{23} \\ T_{24} & 0 & T_{24} & T_{24} & T_{24} & T_{24} \\ T_{34} & T_{34} & 0 & T_{34} & T_{34} & T_{34} \\ T_{12} & T_{12} & T_{12} & 0 & T_{12} & T_{12} \\ T_{13} & T_{13} & T_{13} & T_{13} & 0 & T_{13} \\ T_{14} & T_{14} & T_{14} & T_{14} & T_{14} & 0 \end{pmatrix} G_0 \begin{pmatrix} \mathfrak{T}_{11} \\ \mathfrak{T}_{21} \\ \mathfrak{T}_{31} \\ \mathfrak{T}_{14} \\ \mathfrak{T}_{15} \\ \mathfrak{T}_{16} \end{pmatrix} \tag{9}$$

in which $T_{ij}(T_{ij} = V_{ij} + V_{ij} G_{ij} V_{ij})$ are the two-body transition operators. This set of six coupled integral equations, four-body Faddeev equations, contains just the off-shell two-body T-matrices and do not have appearance of the two-body interaction potentials, directly. The potentials appears in these equations, just as in the three-body Faddeev equations, through the two-body off-shell T-matrices. The kernel, $[K] = K_{ij}$ $(i, j = 1, 2, \cdots, 6)$, is a 6×6 matrix operator in which the diagonal elements are zero and remove the self-coupling of the auxiliary T-operators. The system of six coupled integral equations can be solved by Fredholm methods, and in contrast with the LS equation, this system has a single-valued solution. All scattering and rearrangement four-body T-matrices can be expressed as a linear combination of six introduced auxiliary transition operators.

3. Faddeev-Watson Multiple-Scattering Expansions

The first-order solution can be found by using the Neumann iteration method. Inserting the first-order solution in the main equation gives the second-order solution, and so on. This method leads to Faddeev-Watson multiple-scattering expansions[1] for auxiliary operators. Inserting the resultant expansions into the linear expressions for scattering and rearrangement four-body T-matrices results the Faddeev-Watson to describe the desired scattering process. For charge-transfer process in collision of proton with helium[5,6], we have:

$$\begin{aligned} H^+ + He &\to H^* + He^{+*} \\ P + (T + e_1 + e_2) &\to (P + e_1) + (T + e_2) \\ 1 + (2 + 3 + 4) &\to (1 + 3) + (2 + 4) \end{aligned} \tag{10}$$

The Faddeev-Watson series describing this process is:

$$\begin{aligned} \mathcal{T}_R &= V_{13} + \mathfrak{T}_{11} + \mathfrak{T}_{31} + \mathfrak{T}_{41} + \mathfrak{T}_{61}\mathcal{T}_R \\ &= V_{13} + T_{12} + T_{14} + T_{12}G_0T_{13} + T_{12}G_0T_{14} + T_{14}G_0T_{12} \\ &+T_{14}G_0T_{13} + T_{23}G_0T_{12} + T_{23}G_0T_{13} + T_{23}G_0T_{14} \\ &+T_{34}G_0T_{12} + T_{34}G_0T_{13} + T_{34}G_0T_{14} + \cdots \end{aligned} \tag{11}$$

Using the various Jacobi coordinates and two-body Coulomb off-shell transition matrix evaluated by Alston[7] it is possible to calculate the amplitudes in first and/or second-order approximations.

4. Conclusion

Faddeev formalism has been generalized to four-body scattering in an effective manner. A set of six coupled integral equations has been derived to describe all the possible scattering an rearrangement processes in a typical four-body scattering system. Elastic scattering, excitation, ionization, break up and charge transfer processes can be described in terms of the introduced auxiliary matrices. Faddeev-Watson series are derived for single charge transfer process in collision of proton with helium as an example.

References

1. C. J. Joachain, *Quantum Collision Theory*, North-Holland, Amsterdam, 1975.
2. L. D. Faddeev, *Sov. Phys.-JETP* **12**, 1014 (1961).
3. C. Lovelace, *Phys. Rev.* **B135**, 1225 (1964).
4. R. G. Newton, *Scattering Theory of Waves and Particles*, Springer, New York, 1982.
5. E. Ghanbari Adivi and M. A. Bolorizadeh, *J. Phys.* **B37**, 3321 (2004).
6. I. Mančev, V. Mergel and L. Schmidt, *J. Phys.* **B36**, 2733 (2003).
7. S. Alston, *Phys. Rev.* **A54**, 2011 (1996).

Unified Theory of Scattering for Arbitrary Potentials

A. S. Kadyrov[1], I. Bray[1], A. M. Mukhamedzhanov[2] and A. T. Stelbovics[1]

[1] *Division of Science and Engineering,*
Murdoch University,
Perth 6150, Australia

[2] *Cyclotron Institute, Texas A&M University,*
College Station, Texas 77843, USA

Modifications are presented that unify different formulations of potential scattering theory for short and Coulomb long-range interactions.

It is well known that conventional quantum collision theory is formally valid only when the particles interact via short-range potentials. However, in the time-dependent formulation, formal scattering theory can be made to include Coulomb long-range potentials by choosing appropriately modified time evolution operators. This is equivalent to choosing various forms of renormalization methods in the time-independent formulation.

Though the renormalization methods give correct cross sections for the two-body problem, the results from these procedures cannot be regarded as completely satisfactory. For instance, in screening-based renormalization methods different ways of shielding lead to different asymptotic forms for the scattering wave function. Generally, these asymptotic forms differ from the exact one obtained from the solution of the Schrödinger equation. The weakest point about these methods, however, is that they give rise to a scattering amplitude that does not exist on the energy shell. This is because the amplitude obtained in these methods has complex factors which are divergent on the energy shell. These factors must be removed (renormalized) before approaching the on-shell point. Furthermore, the renormalization factors depend on the way the limits are taken when the on-shell point is approached. Thus, the *ad-hoc* renormalization procedure is based on prior knowledge of the exact answer to compare with and has no *ab initio* theoretical justification.

In this work we demonstrate that there is a practical approach to the two-body collision problem with a Coulomb-like potential that does not lead to the formal difficulties described above. We introduce new well-defined forms of the scattering amplitude and wave operators for two-body systems valid for arbitrary interactions. The new definitions of the scattering amplitude not only cover arbitrary potentials but also directly give the physical result. More details can be found in Ref. [1].

We consider scattering in a system of two particles 1 and 2 interacting via an arbitrary spherically-symmetric potential V with a Coulomb long-range tail. Throughout the paper we use such units that $\hbar = 1$. A scattering state of this system is the solution to the Schrödinger equation

$$(E - H)\psi^{\pm}_{\boldsymbol{k}}(\boldsymbol{r}) = 0, \tag{1}$$

where $H = H_0 + V$ is the total two-body Hamiltonian of the system, H_0 is the free Hamiltonian, $E = k^2/2\mu$ is the energy of the system, $\boldsymbol{r}$ is the relative coordinate of the particles 1 and 2 and $\boldsymbol{k}$ is their relative momentum, μ is their reduced mass.

When the potential has the Coulomb tail the scattering wave function $\psi^{\pm}_{\boldsymbol{k}}(\boldsymbol{r})$ asymptotically behaves, in the leading order, like the Coulomb-modified plane wave and a Coulomb-modified outgoing spherical wave

$$\psi^{\pm}_{\boldsymbol{k}}(\boldsymbol{r}) \overset{r\to\infty}{\sim} e^{i\boldsymbol{k}\cdot\boldsymbol{r} \pm i\gamma\ln(kr \mp \boldsymbol{k}\cdot\boldsymbol{r})} + f^{\pm}(\widehat{\boldsymbol{k}}\cdot\widehat{\boldsymbol{r}})\frac{e^{\pm ikr \mp i\gamma\ln(2kr)}}{r}, \tag{2}$$

where $\gamma = z_1 z_2 \mu/k$, z_1 and z_2 are the charges of the particles and $f^{\pm}$ is the scattering amplitude. Note that $\widehat{\boldsymbol{k}}\cdot\widehat{\boldsymbol{r}} \neq \pm 1$.

We can separate $\psi^{\pm}_{\boldsymbol{k}}$ into the so-called 'incident' and the 'scattered' parts as

$$\psi^{\pm}_{\boldsymbol{k}}(\boldsymbol{r}) = \phi^{\pm}_{\boldsymbol{k}}(\boldsymbol{r}) + \chi^{\pm}_{\boldsymbol{k}}(\boldsymbol{r}), \tag{3}$$

where $\phi^{\pm}_{\boldsymbol{k}}$ and $\chi^{\pm}_{\boldsymbol{k}}$ asymptotically behave like the first and the second terms of Eq. (2), respectively. The 'unscattered' wave incident at large distances is given by

$$\phi^{\pm}_{\boldsymbol{k}}(\boldsymbol{r}) = \phi^{(0)\pm}_{\boldsymbol{k}}(\boldsymbol{r}) \sum_{n=0}^{\infty} [(\mp i\gamma)_n]^2 \, (\mp ikr + i\boldsymbol{k}\cdot\boldsymbol{r})^{-n}/n!, \tag{4}$$

$$\phi^{(0)\pm}_{\boldsymbol{k}}(\boldsymbol{r}) = e^{i\boldsymbol{k}\cdot\boldsymbol{r} \pm i\gamma\ln(kr \mp \boldsymbol{k}\cdot\boldsymbol{r})}, \tag{5}$$

where $(z)_n = z(z+1)\cdots(z+n-1)$. Then Eq. (1) can be written in the form

$$(E - H)\left[\psi^{\pm}_{\boldsymbol{k}}(\boldsymbol{r}) - \phi^{\pm}_{\boldsymbol{k}}(\boldsymbol{r})\right] = (H - E)\phi^{\pm}_{\boldsymbol{k}}(\boldsymbol{r}). \tag{6}$$

If we introduce the Green's function G and apply it onto both sides of Eq. (6) we have

$$\psi^{\pm}_{\boldsymbol{k}}(\boldsymbol{r}) = \phi^{\pm}_{\boldsymbol{k}}(\boldsymbol{r}) + \int d\boldsymbol{r}' G(\boldsymbol{r}, \boldsymbol{r}'; E \pm i\epsilon)(\overrightarrow{H} - E)\phi^{\pm}_{\boldsymbol{k}}(\boldsymbol{r}'). \tag{7}$$

where ϵ is a small positive parameter and limit as $\epsilon \to 0$ is assumed. It is easily verified by substitution that this is a formal solution to Eq. (1). We used an arrow on the differential Hamiltonian operator to show the direction in which it acts. It is important to emphasize here a subtle point that the operator $G(\boldsymbol{r}, \boldsymbol{r}', E)(E - \overrightarrow{H})$ in Eq. (7) does not act like a δ-function. In other words, though

$$G(\boldsymbol{r}, \boldsymbol{r}', E)(E - \overleftarrow{H}) = (E - \overrightarrow{H})G(\boldsymbol{r}, \boldsymbol{r}', E) \equiv \delta(\boldsymbol{r} - \boldsymbol{r}'), \tag{8}$$

however, in general

$$G(\boldsymbol{r}, \boldsymbol{r}', E)(E - \overrightarrow{H}) \neq \delta(\boldsymbol{r} - \boldsymbol{r}'). \tag{9}$$

The reason is that the operator $G(\boldsymbol{r},\boldsymbol{r}',E)(E-\overrightarrow{H})$ produces an integral that has a surface-integral component. In order for the action of $G(\boldsymbol{r},\boldsymbol{r}',E)(E-\overrightarrow{H})$ to be equal to that of $G(\boldsymbol{r},\boldsymbol{r}',E)(E-\overleftarrow{H})$ this surface-integral component should be zero.

Using a spectral decomposition for the Green's function we can write Eq. (7) as

$$\psi_{\boldsymbol{k}}^{+}(\boldsymbol{r}) = \phi_{\boldsymbol{k}}^{+}(\boldsymbol{r}) + \int \frac{d\boldsymbol{k}'}{(2\pi)^3} \frac{T(\boldsymbol{k}',\boldsymbol{k})\,\psi_{\boldsymbol{k}'}^{-}(\boldsymbol{r})}{E-k'^2/2\mu+i\epsilon} + \cdots, \tag{10}$$

where dots indicate only the contribution from the bound states. In the above equation we have introduced the notation

$$T(\boldsymbol{k}',\boldsymbol{k}) = \langle \psi_{\boldsymbol{k}'}^{-} | \overrightarrow{H} - E | \phi_{\boldsymbol{k}}^{(0)+} \rangle, \tag{11}$$

Evaluating the integral for $r \to \infty$ we have

$$\psi_{\boldsymbol{k}}^{+}(\boldsymbol{r}) \overset{r\to\infty}{\sim} e^{i\boldsymbol{k}\cdot\boldsymbol{r}+i\gamma\ln(kr-\boldsymbol{k}\cdot\boldsymbol{r})} - \frac{\mu}{2\pi} T(k\widehat{\boldsymbol{r}},\boldsymbol{k}) \frac{e^{ikr}}{r} e^{-i\gamma\ln(2kr)}. \tag{12}$$

By comparing Eqs. (12) and (2) we conclude that $T(\boldsymbol{k}',\boldsymbol{k})$ introduced in Eq. (11) is the transition matrix (T-matrix) which defines the amplitude of scattering of the particles with initial relative momentum $\boldsymbol{k}$ in the direction of $\boldsymbol{r}$:

$$f^{+}(\widehat{\boldsymbol{k}}\cdot\widehat{\boldsymbol{r}}) = -\frac{\mu}{2\pi} T(\boldsymbol{k}',\boldsymbol{k}), \tag{13}$$

where we used a notation $\boldsymbol{k}' = k\widehat{\boldsymbol{r}}$. In analogy with the conventional theory we call the new form for $T(\boldsymbol{k}',\boldsymbol{k})$ as defined in Eq. (11) the *prior* form of the T-matrix.

Repeating the procedure outlined above for $\psi_{\boldsymbol{k}}^{-}(\boldsymbol{r})$ we get

$$T^{post}(\boldsymbol{k}',\boldsymbol{k}) = \langle \phi_{\boldsymbol{k}'}^{(0)-} | \overleftarrow{H} - E | \psi_{\boldsymbol{k}}^{+} \rangle. \tag{14}$$

Again, in analogy this new form is called the *post* form of the T-matrix.

From Eq. (7) we can extract new generalized wave operators

$$\Omega^{\pm} = [1 + G(E \pm i\epsilon)(\overrightarrow{H} - E)], \tag{15}$$

where $G(E\pm i\epsilon) = (E \pm i\epsilon - H)^{-1}$ is the Green's operator associated with the Green's function $G(\boldsymbol{r},\boldsymbol{r}';E\pm i\epsilon)$.

We now show that the results given above are consistent with standard scattering theory. The existing formulation of scattering theory relies on the condition that interaction $V(r)$ decreases faster than the Coulomb interaction when $r \to \infty$ [so that $\phi_{\boldsymbol{k}}^{(0)\pm}(\boldsymbol{r}) \to \phi_{\boldsymbol{k}}^{(0)}(\boldsymbol{r}) = e^{i\boldsymbol{k}\cdot\boldsymbol{r}}$]. Then the initial unscattered wave function satisfies

$$(H-E)\phi_{\boldsymbol{k}}^{(0)}(\boldsymbol{r}) = V\phi_{\boldsymbol{k}}^{(0)}(\boldsymbol{r}). \tag{16}$$

Therefore, Eqs. (11) and (14) reduce to

$$T^{prior}(\boldsymbol{k}',\boldsymbol{k}) = \langle \psi_{\boldsymbol{k}'}^{-} | V | \phi_{\boldsymbol{k}}^{(0)} \rangle, \tag{17}$$

$$T^{post}(\boldsymbol{k}',\boldsymbol{k}) = \langle \phi_{\boldsymbol{k}'}^{(0)} | V | \psi_{\boldsymbol{k}}^{+} \rangle, \tag{18}$$

in agreement with the standard definitions of the T-matrix. Moreover, the generalized wave operators $\Omega^{\pm}$ introduced above reduce to the usual Möller (M) ones:

$$\Omega_M^{\pm} = [1 + G(E\pm i\epsilon)V]. \tag{19}$$

On the other hand, when the interaction is purely Coulomb we have

$$(H-E)\phi_{\boldsymbol{k}}^{(0)\pm}(\boldsymbol{r}) = \frac{\gamma^2 k}{\mu r(kr \mp \boldsymbol{k}\cdot\boldsymbol{r})}\phi_{\boldsymbol{k}}^{(0)\pm}(\boldsymbol{r}). \tag{20}$$

Therefore, Eqs. (11) and (14) transform to

$$T^{prior}(\boldsymbol{k}',\boldsymbol{k}) = \langle\psi_{\boldsymbol{k}'}^{-}|\frac{\gamma^2 k}{\mu r(kr - \boldsymbol{k}\cdot\boldsymbol{r})}|\phi_{\boldsymbol{k}}^{(0)+}\rangle, \tag{21}$$

$$T^{post}(\boldsymbol{k}',\boldsymbol{k}) = \langle\phi_{\boldsymbol{k}'}^{(0)-}|\frac{\gamma^2 k}{\mu r(kr + \boldsymbol{k}\cdot\boldsymbol{r})}|\psi_{\boldsymbol{k}}^{+}\rangle. \tag{22}$$

Here $\psi_{\boldsymbol{k}}^{\pm}$ are the known Coulomb functions. Evaluating the matrix elements we get

$$T^{prior}(\boldsymbol{k}',\boldsymbol{k}) \equiv T^{post}(\boldsymbol{k}',\boldsymbol{k}) = \frac{4\pi z_1 z_2}{|\boldsymbol{k}'-\boldsymbol{k}|^2}\frac{\Gamma(1+i\gamma)}{\Gamma(1-i\gamma)}\left[\frac{4k^2}{|\boldsymbol{k}'-\boldsymbol{k}|^2}\right]^{i\gamma}, \tag{23}$$

which is the well-known full on-shell Coulomb T-matrix.

We have presented a unified theory of potential scattering which is valid for arbitrary interactions including potentials with the long-range Coulomb tail. We introduced new general definitions for the scattering amplitude and wave operators. When the interaction potential is short-ranged the generalized definitions of the scattering amplitude and wave operators transform to the conventional ones used in standard scattering theory. An important feature of the new forms for the scattering amplitude is that they do not contain divergent factors and directly give the physical on-shell scattering amplitude even when the interaction potential is long-ranged. Moreover, the generalized wave operators are also the same for arbitrary potentials including the long-range interactions. Therefore, renormalization methods are nolonger required.

The situation with three charged particles is even more complicated. There is neither theoretical proof nor practical evidence that renormalization approach can be applied to the Faddeev equations. We believe that the method proposed here for solving the two-body problem can also be applied to the three-body problem [2].

The work was supported by the Australian Research Council, U.S. DOE Grant DE-FG03-93ER40773 and NSF Grant PHY-0140343.

References

1. A. S. Kadyrov, I. Bray, A. M. Mukhamedzhanov and A. T. Stelbovics, *Phys. Rev. A* **72**, 032712 (2005).
2. A. S. Kadyrov, A. M. Mukhamedzhanov, A. T. Stelbovics and I. Bray, *Phys. Rev. A* **70**, 062703 (2004).

Application of Few-Body Technique for Spinor Bose-Einstein Condensates

C. G. Bao and Z. B. Li*

The State Key Laboratory of Optoelectronic Materials and Technologies
Department of Physics, Zhongshan University, Guangzhou, 510275, P.R. China

The fractional parentage coefficients are introduced and their exact analytical expressions are given. As an example of the application of these coefficients in many-body systems, we study the spinor Bose-Einstein condensates of Sodium atoms. A set of generalized Gross-Pitaeveskii equations is derived for the eigen wave functions with the total spin S and its Z-component S_z being conserved. The ground band and the first excited band of the condensates are studied numerically.

The technique associated with the fractional parentage coefficients (FPCs) is a powerful tool for few-body systems, thereby related matrix elements can be derived. Usually they are obtained one-by-one through the recurrence relations. These relations will become extremely complicated if N, the number of particles, is larger (say, $N > 10$). Therefore, traditionally, the FPCs are useless for many-body systems. However, we have succeeded to obtain the analytical expressions of the FPCs for the spin-1 N-boson systems, thus they can be used not matter how large N is, they can even be used in the case that N tends to infinity. With them the symmetry involved in spinor condensates can be treated rigorously. In particular in this paper, the FPCs are used to build up the total spin-states for excited states, these spin-states have not only the total spin S and S_z being conserved, but also possess specific permutation symmetry. Then the FPCs are used again to derive all the associated matrix elements. Thereby a set of equations for the low-lying excited states of the condensates has been set up, and the case of ^{23}Na has been studied numerically.

1. The Fractional Parentage Coefficients

For a spin-1 N-body system, let the number of totally symmetric spin states having a specified positive S_Z be denoted as $M_Z(S_Z)$, those having both S_Z and S specified be denoted as $M(S)$. Evidently, if a spin-state has $S_Z + i$ bosons spin-up, i bosons spin-down, while the spin-components of the rest $N - S_Z - 2i$ bosons are zero, then the state is one in $M_Z(S_Z)$. From the requirement that $N - S_Z - 2i \geq 0$, we have

*Supported by NSFC under the grants 10174098, 90103028 and 90306016

$0 \le i \le (N - S_Z)/2$. Thus, we have

$$M_Z(S_Z) = \frac{1}{2}\left[N - S_Z - \frac{1-(-1)^{N-S_Z}}{2}\right] + 1 \tag{1}$$

Denote the normalized totally symmetric spin state of N spin-1 spinors as $\vartheta_S^{\{N\}}$. From eq.(1), the number of $\vartheta_S^{[N]}$, namely $M(S)$, can be derived as

$$M(S) = M_Z(S) - M_Z(S+1) = \left[1 + (-1)^{N-S}\right]/2 \tag{2}$$

From (2) we know that the number of $\vartheta_S^{[N]}$ is *one* if $N - S$ is even, or is *zero* if $N - S$ is odd. Thus the total number of $\vartheta_S^{[N]}$ with S ranging from $N, N-2, \cdots$, to 0 (or 1) is nearly equal to N/2.

The expressions of each $\vartheta_S^{[N]}$ can be deduced based on an iteration procedure. Let us start from a 2-boson system, we have

$$\vartheta_S^{[2]}(12) = (\chi(1)\chi(2))_S \tag{3}$$

where χ is the spin-state of a single boson, the two spins are coupled to $S = 0$ or 2. For $N = 3$, there are two $\vartheta_S^{[N]}$ with $S = 3$ and 1 [1].

$$\vartheta_3^{[3]}(123) = \{(\chi(1)\chi(2))_2\chi(3)\}_3 = \{\vartheta_2^{[2]}\chi(3)\}_3 \tag{4}$$

$$\vartheta_1^{[3]}(123) = \frac{2}{3}\{\vartheta_2^{[2]}\chi(3)\}_1 + \frac{\sqrt{5}}{3}\{\vartheta_0^{[2]}\chi(3)\}_1 \tag{5}$$

Starting from (3), (4), and (5), let us assume that all the $\vartheta_S^{[N-2]}$ have been known, then $\vartheta_{S'}^{[N-1]}$ with $N - 1 - S'$ even can be uniquely expanded as

$$\vartheta_{S'}^{[N-1]} = a_{S'}^{[N-1]}\{\vartheta_{S'+1}^{[N-2]}\chi(N-1)\}_{S'} + b_{S'}^{[N-1]}\{\vartheta_{S'-1}^{[N-2]}\chi(N-1)\}_{S'} \tag{6}$$

where $a_{S'}^{[N-1]}$ and $b_{S'}^{[N-1]}$ are two coefficients to be determined, and $(a_{S'}^{[N-1]})^2 + (b_{S'}^{[N-1]})^2 = 1$ is assumed. This expansion holds because (i) based on the rule of the outer product of permutation groups, only the totally symmetric states of the $(N-2)$-subsystem are involved, (ii) since $N - 1 - S'$ is even, $\vartheta_{S'}^{[N-2]}$ is zero and therefore does not appear in (6).

From (6), a similar formula can be written for $\vartheta_S^{[N]}$ as

$$\vartheta_S^{[N]} = a_S^{[N]}\{\vartheta_{S+1}^{[N-1]}\chi(N)\}_S + b_S^{[N]}\{\vartheta_{S-1}^{[N-1]}\chi(N)\}_S \tag{7}$$

Inserting (6) into (7), we have

$$\begin{aligned}\vartheta_S^{[N]} = {} & a_S^{[N]}a_{S+1}^{[N-1]}\{[\vartheta_{S+2}^{[N-2]}\chi(N-1)]_{S+1}\chi(N)\}_S + a_S^{[N]}b_{S+1}^{[N-1]}\{[\vartheta_S^{[N-2]}\chi(N-1)]_{S+1}\chi(N)\}_S \\ & + b_S^{[N]}a_{S-1}^{[N-1]}\{[\vartheta_S^{[N-2]}\chi(N-1)]_{S-1}\chi(N)\}_S + b_S^{[N]}b_{S-1}^{[N-1]}\{[\vartheta_{S-2}^{[N-2]}\chi(N-1)]_{S-1}\chi(N)\}_S\end{aligned} \tag{8}$$

On the other hand, since the $\vartheta_S^{[N]}$ written in (8) is required to be totally symmetric, it must be invariant against any interchange. Thus we have $\vartheta_S^{[N]} = P_{N,N-1}\vartheta_S^{[N]}$, where $P_{N,N-1}$ denotes the interchange of $(N-1)$ and N. Inserting (8) into both sides of (7), making use of the theory for the recoupling of spins, we arrive at a set of three homogeneous equations as

$$
\begin{aligned}
[\hat{W}_S\,(S+1,S+1)-1]a_S^{[N]}b_{S+1}^{[N-1]}+\hat{W}_S\,(S-1,S+1)b_S^{[N]}a_{S-1}^{[N-1]}&=0\\
\hat{W}_S\,(S+1,S-1)a_S^{[N]}b_{S+1}^{[N-1]}+[\hat{W}_S\,(S-1,S-1)-1]b_S^{[N]}a_{S-1}^{[N-1]}&=0\\
\hat{W}_S\,(S+1,S)a_S^{[N]}b_{S+1}^{[N-1]}+\hat{W}_S\,(S-1,S)b_S^{[N]}a_{S-1}^{[N-1]}&=0
\end{aligned}
\tag{9}
$$

where

$$\hat{W}_S\,(T,T')=\sqrt{(2T+1)(2T'+1)}W(1SS1;TT') \tag{10}$$

and $W(1SS1;TT')$ is the well known Racah coefficients. It turns out that the set of equations (9) are linearly dependent. Taking into account the expression of the Racah coefficients, we obtain

$$a_S^{[N]}=\gamma\frac{\sqrt{2S-1}}{S\sqrt{2S+1}}/b_{S+1}^{[N-1]}\quad ;\quad b_S^{[N]}=\gamma\frac{\sqrt{2S+3}}{(S+1)\sqrt{2S+1}}/a_{S-1}^{[N-1]} \tag{11}$$

if $N-S$ is even, or $a_S^{[N]}=b_S^{[N]}=0$ if $N-S$ is odd, where the constant γ is introduced to assure $(a_S^{[N]})^2+(b_S^{[N]})^2=1$ so that $\vartheta_S^{[N]}$ is normalized. In the case of $S=N$, instead of (11), we have $a_N^{[N]}=0$ and $b_N^{[N]}=1$. In the case of $S=0$, we have $a_0^{[N]}=(1+(-1)^N)/2$ and $b_0^{[N]}=0$. With the above results and equations (9) and (10) in hand, one can deduce all the $\vartheta_S^{[N]}$ of the N-boson system via the iteration procedure. Starting from the case of $N=3$ (cf. (4) and (5)) with $a_3^{[3]}=0$, $b_3^{[3]}=1$, $a_1^{[3]}=\frac{2}{3}$, and $b_1^{[3]}=\frac{\sqrt{5}}{3}$, all the $\vartheta_S^{[N]}$ can be identified.

On the other hand, one can prove directly that eqs.(11) together with the condition of normalization have a unique solution, it reads

$$a_S^{\{N\}}=[(1+(-1)^{N-S})(N-S)(S+1)/(2N(2S+1))]^{1/2} \tag{12}$$

$$b_S^{\{N\}}=[(1+(-1)^{N-S})\;S\;(N+S+1)/(2N(2S+1))]^{1/2} \tag{13}$$

and

$$\gamma_S^{\{N\}}=S(S+1)[\frac{(N-S)(N+S+1)}{N(N-1)(2S+3)(2S-1)}]^{1/2} \tag{14}$$

thus we obtain the analytical form of the fractional parentage coefficients (FPCs).

Now, starting from the total the total symmetric $\vartheta_{S'}^{\{N-1\}}$, let us study the set of total spin-states with the $\{N-1,1\}$ symmetry. Making use of the FPCs, let us define a total spin-state as

$$\Theta_S^i=b_S^{\{N\}}\{\vartheta_{S+1}^{\{N-1\}}(\overset{\times}{i})\chi(i)\}_S-a_S^{\{N\}}\{\vartheta_{S-1}^{\{N-1\}}(\overset{\times}{i})\chi(i)\}_S \tag{15}$$

if $N-S$ is even, or

$$\Theta_S^i=\{\vartheta_S^{\{N-1\}}(\overset{\times}{i})\chi(i)\}_S \tag{16}$$

if $N-S$ is odd, where the notation $\overset{\times}{i}$ represents all the bosons except the $i-th$. From the rule of outer product, (15) and (16) will contain only the symmetries $\{N\}$ and $\{N-1,1\}$. However, when $N-S$ is even, (15) is orthogonal to the unique totally symmetric spin-state $\vartheta_S^{\{N\}}$. Hence, it must have the pure $\{N-1,1\}$ symmetry. When $N-S$ is odd, it has been stated above that $\vartheta_S^{\{N\}}=0$, therefore (16) has also the pure $\{N-1,1\}$ symmetry. It is obvious from (12) and (13) that Θ_S^i does not exist if $S=0$ or N. Thus the S of Θ_S^i is ranged from 1 to $N-1$, while the index i of Θ_S^i is ranged from 1 to N. This set of N states are not mutually orthogonal but linearly dependent. They satisfy $\sum_i \Theta_S^i = 0$. This arises because the summation over i leads to a symmetrization, a state with $\lambda \neq \{N\}$ will not survive after the symmetrization. We have obtained a set of Θ_S^i possessing the $\{N-1,1\}$ symmetry, this is sufficient because it has been proved that, for the $\{N-1,1\}$ representation, the multiplicity of the total spin is one[13].

2. Application of FPC Technique for Spinor Bose-Einstein Condensates

In recent years, accompanying the experimental realization of the spinor Bose-Einstein condensation (SBEC) [2,3], corresponding theories based on the mean field theory have been developed [4–11]. For the eigenstates of the condensates, due to the feature of the interaction, both S_Z and S are conserved. Therefore, when the eigenstates are concerned, a theory with both S_Z and S conserved might be more appropriate. Refs. [12,13] have introduced the total spin-states of the ground band and the first excited band of the SBEC. Applying the FPC technique, the Gross-Pitaevskii equation have been generalized to spinor systems. Here we will review some of the results.

The Hamiltonian of the N spinors confined in a three-dimentional harmonic trap reads as

$$H = \frac{1}{2}\sum_i(-\nabla_i^2 + r_i^2) + \sum_{i<j} U_{ij} \tag{17}$$

with the pair interaction $U_{ij} = \delta(\mathbf{r}_i - \mathbf{r}_j)O_{ij}$, $O_{ij} = c_0 + c_2\mathbf{F}_i \cdot \mathbf{F}_j$. Where $\mathbf{F}_i$ is the spin operator of particle i. The Schrödinger equation is

$$H\Psi_S = E_S\Psi_S \tag{18}$$

Since the interaction is spin-dependent, the adjustment of spins (i.e., a change of S) may cause a change in energy. Thus, the states belonging to the same harmonic oscillator (h.o.) shell may be different in energy if they have different S, and the degenerate levels in the h.o. shell spread into a band. This is the origin of the ground band and the excited bands.

For each member of the ground band, since no spatial excitation is involved, the total spatial wave function can be written as $\Pi_{i=1}^N \phi_S^{(g)}(\mathbf{r}_i)$. Here, a subscript S has been introduced to respond to the possibility that the single-particle state may depend on S. Evidently, in this expression the spatial part is totally symmetric with respect to permutation. Then the total wave function of the ground band has the form

$$\Psi_S^{(g)} = \Pi_{i=1}^N \phi_S^{(g)}(\mathbf{r}_i)\vartheta_S^{[N]} \tag{19}$$

Incidentally, two $\phi_S^{(g)}(\mathbf{r})$ with distinct S are in general not orthogonal but may overlap with each other seriously, the orthogonality is realized by the spin-states.

With the spin wave functions Θ_S^i given in (15) and (16), we can construct the wave functions of the excited bands. Let us confine ourself in the case that only one particle has been excited. The spatial wave functions read

$$F_{S,i} = \varphi_S^b(i)\Pi_{j(\neq i)}\varphi_S^a(j) \tag{20}$$

where φ_S^a and φ_S^b denote the lowest and the second lowest normalized single-particle states, respectively, with the given S . From the rule of outer product, $F_{S,i}$ contains both the $\{N\}$ and $\{N-1,1\}$ symmetries. Therefore, starting from (20), two kinds of total spin eigenstates can be composed as

$$\Psi_S^{(e)\{N\}} = \sum_i F_{S,i}\, \vartheta_S^{\{N\}} \tag{21}$$

and

$$\Psi_S^{(e)\{N-1,1\}} = \sum_i F_{S,i}\, \Theta_S^i \tag{22}$$

where the summation over i leads to a symmetrization as required, and the $\{N-1,1\}$-representation has been introduced. It turns out that (21) is exactly a state of center of mass excitation from the ground band. Therefore, we shall omit the discussion of them and concentrate on the band composed of (22).

Recalling the symmetry of the states and applying the FPCs, the relevant matrix elements of the two-body interaction are obtained as[12,13]

$$g_S =< \vartheta_S^{[N]}|O_{ij}|\vartheta_S^{[N]} > c_0 + \frac{S(S+1)-2N}{N(N+1)}c_2 \tag{23}$$

$$g_S^b = \langle\Theta_S^i|O_{ij}\ |\Theta_S^i + \Theta_S^j\rangle = \frac{N-2}{N-1}(c_0+c_2) - [1+(-1)^{N-S}]\,\frac{S(S+1)}{N\,(N-1)}c_2 \tag{24}$$

$$g_S^a = \langle\Theta_S^i|O_{jk}\ |\Theta_S^i\rangle = c_0 + c_2\frac{1}{(N-1)(N-2)}[\frac{N+1+(-1)^{N-S}}{N}S(S+1)-2(N-1)] \tag{25}$$

With these matrix elements, one can derive the generalized Gross-Pitaevskii (GGP) equations for the spatial wavefunctions of the ground band and the first excited band by the standard variational method.

The GGP equation for the spatial wave function of the ground band is [12]

$$[h_0 + (N-1)g_S\frac{|u_S^g(r)|^2}{4\pi r^2}]\ u_S^g(r) = \varepsilon_S^g\ u_S^g(r) \tag{26}$$

where $u_S^g(r) = \sqrt{4\pi}\, r\phi_S^g$, and

$$h_0 = \frac{1}{2}[-\frac{d^2}{dr^2} + r^2] \tag{27}$$

The eigenenergies of the ground band $E_S^{(g)}$ ($N - S$ must be even) are

$$E_S^{(g)} = N\, \varepsilon_S^{(g)} - \frac{N\,(N-1)}{2} g_S\, \bar{\rho} \tag{28}$$

where

$$\bar{\rho} = \int \mathbf{dr} \phi_S(r)^* \rho \phi_S(r) \tag{29}$$

For the excited band, by rewriting the spatial wave functions as $\varphi_S^a = \frac{1}{\sqrt{4\pi}\, r} u_S^a(r)$ and $\varphi_S^b = \frac{1}{r} u_S^b(r) Y_{lm}(\theta, \phi)$ with $l = 1$, one has [13]

$$[h_0 + g_S^b \frac{|u_S^b|^2}{4\pi r^2} + (N-2) g_S^a \frac{|u_S^a|^2}{4\pi r^2}]\, u_S^a = \varepsilon_a u_S^a \tag{30}$$

$$[h_1 + (N-1) g_S^b \frac{|u_S^a|^2}{4\pi r^2}]\, u_S^b = \varepsilon_b u_S^b \tag{31}$$

where

$$h_1 = \frac{1}{2}[-\frac{d^2}{dr^2} + \frac{2}{r^2} + r^2] \tag{32}$$

The energies of the first excited band are given by

$$E_S^{(e)} = \varepsilon_b + (N-1)\varepsilon_a - \frac{(N-1) g_S^b}{4\pi} \int_0^\infty \frac{dr}{r^2} |u_S^a|^2 |u_S^b|^2 - \frac{(N-1)(N-2) g_S^a}{8\pi} \int_0^\infty \frac{dr}{r^2} |u_S^a|^4 \tag{33}$$

where S can be even or odd, and is ranged from 1 to $N - 1$.

As an instance, we calculated the band structures of ^{23}Na atoms trapped in a harmonic well with $\omega = 100 Hz$. The energies of the members of the first excited band relative to the ground state energy $E_{S_{min}}^{(g)}$ (where $S_{min} = 0$ if N is even, or $= 1$ if odd) are plotted in Fig. 1 . More numerical results can be found in [12,13].

3. Conclusion

The analytical expression of the fractional parentage coefficients (FPCs) is obtained. The successful application of FPCs to the spinor Bose-Einstein condensates has demonstrates that the method based on FPC is a powerful technique that enable us to construct the total spin states of arbitrary number of spinors. It is also useful in the calculation of matrix elements of spinor operators. The analytical form of the FPCs is not only for spinor condensates, but as a general tool, it can also be widely used for various spinor many-body systems. They can even be used for the spatial degrees of freedom when a group of particles each has one unit of orbital angular momentum.

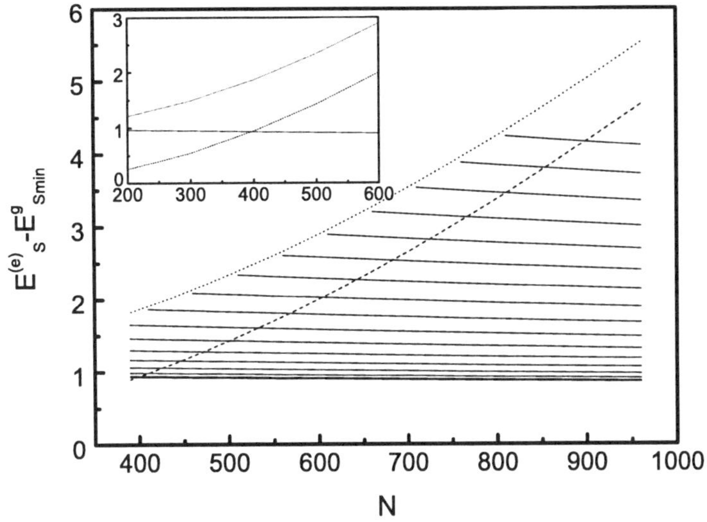

Fig. 1. The levels of the first excited band of the ^{23}Na Bose-Einstein condensates. The energies relative to the energy of the ground state, $E^{(g)}_{S_{min}}$, are plotted versus the atom number. The dashed curve is for the upper bound of the ground band, namely for $E^{(g)}_N$. The solid curves are the levels of selected members of the excited band with various total spins S starts at 1 and each step increases by $\Delta S = 50$ from the lower towards the higher. The dotted curve is for the upper bound of the first excited band, namely for $E^{(e)}_{N-1}$.

References

1. C. G. Bao, T. Y. Shi, Phys. Rev. A **68**, 032509 (2003)
2. C.C. Bradley, C.A. Sackett, J.J. Tollett, and R.G. Hulet, Phys. Rev. Lett. 75, 1687 (1995); 79, 1170 (1997)
3. D.M. Stamper-Kurn, M.R. Andrews, A.P. Chikkatur, S. Inouye, H.-J. Miesner, J. Stenger, and W. Ketterle, Phys. Rev. Lett. 80, 2027 (1998)
4. L.-M. Duan, J.I. Cirac, and P. Zoller, Phys. Rev. A 65, 033619 (2002)
5. C.K. Law, H. Pu, and N.P. Bigelow, Phys. Rev. Lett. 81, 5257 (1998)
6. E. Goldstein and P. Meystre, Phys. Rev. A 59, 3896 (1998)
7. T.-L. Ho and S.K. Yip, Phys. Rev. Lett. 84, 4031 (2000)
8. M. Koashi, and M. Ueda, Phys. Rev. Lett. 84, 1066 (2000)
9. S. Yi, Ö. E. Müstecaphoğlu, C.P. Sun, and L. You, Phys. Rev. A 66, 011601R (2002)
10. S. Ashhab and A.J. Leggett, Phys. Rev. A 68, 063612 (2003)
11. A.A. Svidzinsky and S.T. Chui, Phys. Rev. A 68, 043612 (2003)
12. C.G. Bao and Z.B. Li, Phys. Rev. A. 70, 043620 (2004)
13. C.G. Bao and Z.B. Li, to appear in Phys. Rev. A.

FEW-BODY APPROACHES TO UNSTABLE NUCLEI, NUCLEAR ASTROPHYSICS AND NUCLEAR CLUSTERING ASPECTS

Structure of Oxygen Isotopes Studied with Antisymmetrized Molecular Dynamics

N. Furutachi and S. Oryu

Department of Physics, Faculty of Science and Technology,
Frontier Research Center for Computational Science,
Tokyo University of Science, Noda, Chiba 278-8510, Japan.
E-mail: oryu@ph.noda.tus.ac.jp

Low-lying states of ^{16,18}O have been studied using the framework of antisymmetrized molecular dynamics plus the generator coordinate method. We adopted the MV1 force and G3RS force for the NN interaction. The force parameters are reasonably determined to reproduce the ground state energies of ^{4}He, ^{12}C, ^{16}O and the low-lying energy spectra of ^{12}C. The 0_2^+ state and K^π=$0^\pm$ parity doublet band of ^{16}O are investigated, and the calculations represent the experimental values qualitatively. In the above spectra, the ^{12}C+α cluster-like structure of ^{16}O is also illustrated in our model calculation.

1. Introduction

In light nuclei up to the beginning of the sd-shell, the clustering aspect is an essential feature of nuclei as well as the shell-model-like mean-field structure. Even in the unstable region of the sd-shell nuclei, the cluster structure is significant [1]. ^{16}O is a typical example which indicates a clusterization in sd-shell nuclei, together with ^{20}Ne. Both nuclei have similar K^π=$0^\pm$ bands which were extensively investigated by the variety of cluster models. Nevertheless, understanding of the low-lying excitation energies such as 0_2^+ state of ^{16}O is still incomplete. We study this low-lying states structures of the ^{16}O using the AMD+GCM (Antisymmetrized Molecular Dynamics plus Generator Coordinate Method), as the first step to investigate the cluster aspect of the unstable nuclei. Effectiveness of the AND+GCM method have been proved in the study of ^{20}Ne and Ne isotopes [1,2]. In our approach, the ^{12}C+α cluster structure of ^{16}O nucleus is reasonably represented. We are able to reasonably adjust the interaction parameters, which describe the binding energies of ^{4}He, ^{12}C, ^{16}O and the low lying spectra of ^{12}C. Finally, we examine the possibility of cluster structures in ^{18}O as a next step.

2. Framework

In the AMD framework, the total wave function of an A-nucleon system is described by the parity-projected Slater determinant,

$$|\Phi^\pm(\mathbf{Z})\rangle = (1 \pm P)(1/\sqrt{A!})det[\varphi_i(j)], \qquad \varphi_i = \phi_{\mathbf{Z}_i}\chi_{\xi_i}\tau_i \tag{1}$$

where φ_i is a single particle wave function which is composed of a spatial part $\phi_{\mathbf{Z}_i}$, a spin part $\chi_{\xi i}$, and an isospin part τ_i. That is, we have

$$\langle \mathbf{r}|\phi_{\mathbf{Z}_i}\rangle = (2\nu/\pi)^{3/4} exp\left[-\nu\left(\mathbf{r} - \mathbf{Z}_i/\sqrt{\nu}\right)^2 + \mathbf{Z}_i^2/2\right] \quad (2)$$

$$\chi_{\xi i} = (\frac{1}{2} + \xi_i)\chi\uparrow + (\frac{1}{2} - \xi_i)\chi\downarrow . \quad (3)$$

Here, the i-th Gaussian wave packet center $\mathbf{Z}_i$ and the parameter for the direction of intrinsic spin ξ_i are the variational parameters, and they are optimized by the "frictional cooling method " under a GCM constraint. In this study, we chose the proton quadrupole deformation parameter β for the constraint parameters. The width parameter ν is common to all nucleons, and we set it to 0.17[fm^{-1}] for all nuclei in these calculations. Finally, the wave function is obtained from a superposition of the "angular momentum projected wave functions" with respect to GCM,

$$|\Psi^{J\pm}\rangle = \sum_i c_i P^J_{MK}|\Phi_\pm(\beta_i)\rangle. \quad (4)$$

For this study, we utilized the MV1 force [3] for the central force and the G3RS force [4] for the spin-orbit force. The Coulomb force is approximated by the sum of seven Gaussian-type functions. As mentioned above, we adjust the MV1 and G3RS force parameters to reproduce the ground state energies of ^{4}He, ^{12}C, ^{16}O and the low lying spectra of ^{12}C reasonably. We adjusted only the Majorana parameter m of the MV1 force (the Bartlett and Heisenberg terms are fixed at b=h=0) and the strength of the spin-orbit force u of the G3RS force. It is too difficult to reproduce all of the energies, so we introduced two sets of parameters. The parameters are set to m=0.61, u=3000 MeV for force P1, and to m=0.62, u=3700 for force P2, respectively.

Table 1. Ground state energies of ^{16}O, ^{4}He, ^{12}C and excitation energies of ^{12}C calculated with AMD+GCM using the force P1 and P2.

		B.E[MeV](0^+)	Ex.(2_1^+)[MeV]	Ex.(4_1^+)[MeV]
^{12}C	P1	-88.3	5.6	14.5
	P2	-92.2	9.1	19.3
	Exp.	-92.16	4.44	14.1

		B.E[MeV]			B.E[MeV]
^{16}O	P1	-126.6	^{4}He	P1	26.68
	P2	-124.1		P2	26.68
	Exp.	-127.62		Exp.	28.29

3. Cluster Structures of ^{16,18}O

After the variation with respect to the β constraints and the angular momentum projection, we obtain the energy curves as functions of the proton quadrupole de-

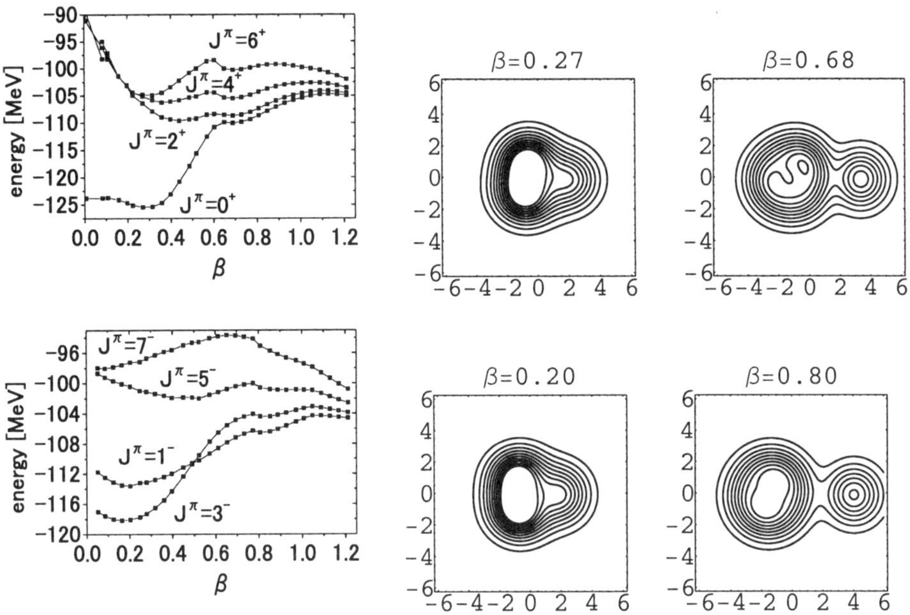

Fig. 1. β-constrained energy curves for the positive and negative parity states of ^{16}O. The density distributions are obtained from the 0^+ or 1^- curve minimum point wave functions, β=0.27 and β=0.68 for positive parity state, β=0.20 and β=0.80 for negative parity state.

formation parameter β, see Figure 1. We can see that the wave function around the minimum at β=0.68 exhibits a ^{12}C+α-like deformed state from the density distribution picture. From the diagonalization of the Hamiltonian matrix, we confirmed that this ^{12}C+α-like state almost corresponds to the 0_2^+ state. The 1^- curve also has two minima and the density distributions have shapes very similar to the positive parity state.

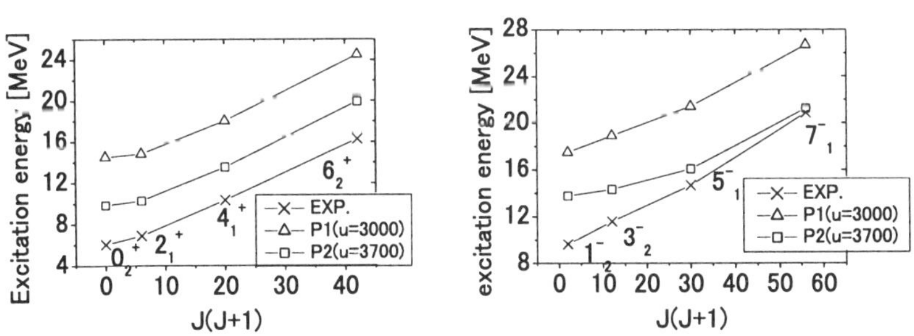

Fig. 2. Excitation energy as function of J(J+1). Only levels which are classified experimentally as $K^\pi=0^\pm$ band are shown.

The energy spectrum (see Figure 2) of ^{16}O is obtained after diagonalizing the Hamiltonian matrix, about ten Slater-determinants are superposed. Level spacing of the $K^\pi=0^\pm$ band is well reproduced with the force P1; however, the excitation energy of the 0_2^+ state is quite large, about 14.5 MeV. The force P2 which has a stronger spin-orbit force, can decrease it to about 10 MeV, but the space of the $K^\pi=0^-$ band can not be reproduced.

Next, the alteration of the cluster structure appearing in the excited state of ^{16}O is investigated. Figure 3 shows the energy curves of ^{18}O as function of β. The energy curve of ^{18}O has a similar shape to that of ^{16}O, while the 0^+ curve has a second minimum around β=0.52, and the density distribution seems to correspond to a ^{14}C+α-like deformed state.

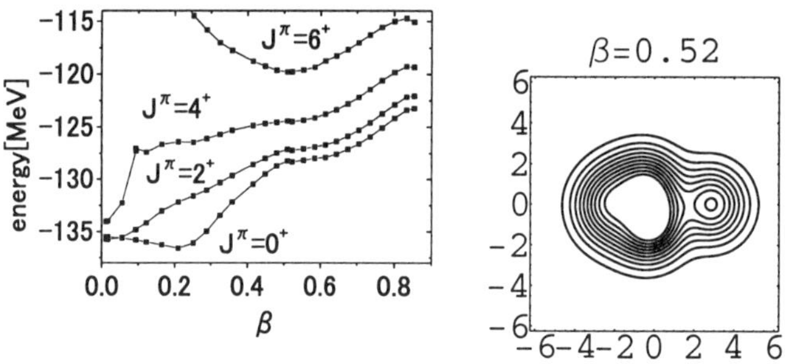

Fig. 3. β-constrained energy curves of the ^{18}O. Density distributions are obtained from the wave functions of β=0.52.

4. Summary

We have studied the cluster structures of ^{16,18}O using the AMD+GCM method. We obtained a ^{12}C+α-like deformed state as the 0_2^+ state of ^{16}O, and the level space of the $K^\pi=0^\pm$ is nicely reproduced. We confirmed that the P1 force could describe the cluster nature of ^{16}O. Furthermore, we investigated the possibility of clustering in ^{18}O. It is expected that the mean field property is enhanced with increasing neutron number, but still ^{18}O seems to exhibit a ^{14}C+α-like cluster structure.

References

1. M. Kimura and H. Horiuchi, *Prog. Theor. Phys.* **111**, 841 (2004)
2. M. Kimura, *Phys. Rev.* **C69**, 044319 (2004).
3. T. Ando, K. Ikeda, and A. Tohsaki, *Prog. Theor. Phys.* **64**, 1608 (1980).
4. N. Yamaguchi, T. Kasahara, S. Nagata, and Y. Akaishi, *Prog. Theor. Phys.* **62**, 1018 (1979); R. Tamagaki, *Prog. Theor. Phys.* **39**, 91 (1968).

Faddeev Three-Cluster Calculation of the Helium Isotopes Including with Core Excitation

M. Yamaguchi and Y. Koike

Science Research Center, Hosei University
17-1 Fujimi, 2-chome, Chiyoda-ku, Tokyo 102-8160, Japan
E-mail: yamagu@rcnp.osaka-u.ac.jp , koike@i.hosei.ac.jp

H. Kamada

Department of Physics, Faculty of Engineering, Kyushu Institute of Technology
1-1 Sensui-cho, Tobata-ku, Kitakyushu-shi, Fukuoka 804-8550, Japan
E-mail: kamada@mns.kyutech.ac.jp

E. Uzu

Department of Physics, Faculty of Science and Technology, Tokyo University of Science
2641 Yamazaki, Noda-shi, Chiba 278-8510, Japan
E-mail: j-uzu@ed.noda.tus.ac.jp

We approach the helium isotopes by using the three-cluster model from the viewpoint that these nuclei make a Borromean-nuclei series. To determine the parameters for two-body subsystem, we make use of the experimental data of the narrow decay width for resonance state. In this report, we treat ^{8}He-n-n (^{10}He) as a example. Our calculation shows that the excited state of inner ^{8}He-cluster is needed to reproduce the very narrow decay width of ^{10}He. Our theoretical prediction identify the J^π of the ground state and excited state for^{9}He to $(1/2^-, 1/2^+)$ respectively, because these state are important for 3-body system.

1. Introduction

The ground states of He-isotopes have been investigated in many experiments.[1] Table 1. shows whether the ground state for ^{4}He - ^{10}He is bound or resonance state from experiments.

Table 1. Experimental information for the ground states of He-isotopes.

^{4}He (α)	^{5}He	^{6}He	^{7}He	^{8}He	^{9}He	^{10}He
bound state.	resonance state. α-n can not be bound.	bound state. α-n-n system is Borromean state.	resonance state. ^{6}He-n can not be bound.	bound state. ^{6}He-n-n system is Borromean state.	resonance state. ^{8}He-n can not be bound.	resonance state. **^{8}He-n-n system is <u>not</u> Borromean state.**

Note: n means neutron

A 3-body bound state, in which any 2-body subsystem would not be bound, is called Borromean nucleus. All He-isotopes of even mass number makes a series of Borromean nuclei (Borromean series) except for ^{10}He. From the point of view of Borromean series, it is natural to expect that the ground state of ^{10}He would be a bound state. However, this theoretical expectation for ^{10}He is not satisfied in the experiments. Thus the theoretical study of ^{10}He is very interesting.

Furthermore, the decay width of ^{9}He and ^{10}He was measured and very narrow width was reported[1] (see Figure 1). Various calculations was done for ^{10}He[3,4] until now, but the narrowness of decay width have not been obtained satisfactorily. From the theoretical pointview, this very narrow decay width may be due to the core excitation effect, like a ^{11}Li.[2] or ^{3}H. In the case of ^{3}H, the lack of the binding energy obtained from 2-body force was compensated for 3-body force. The essence of this mechanism is the introduction of the degrees of freedom of the excited state of nucleon (Δ-isober). In the ^{8}He-n-n system, the effect of the excitation of ^{8}He make the system more stable and make the width of the resonance state of ^{10}He very narrow.

In this paper we calculate the resonance energy E_r and the decay width Γ for ^{10}He by the ^{8}He-n-n 3-body cluster model including with the excited state of ^{8}He.

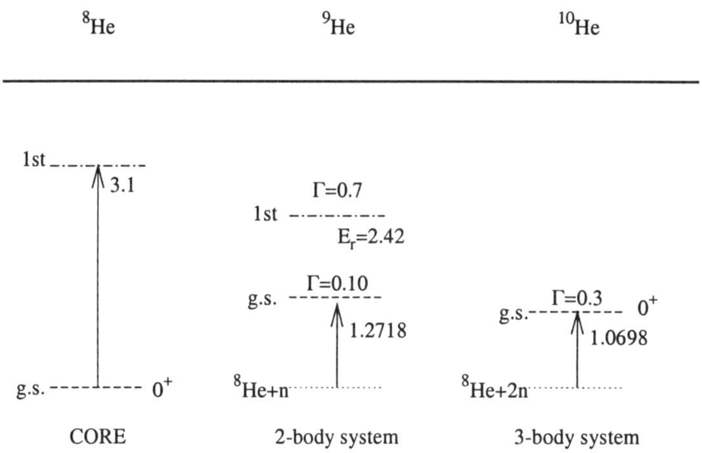

Fig. 1. The energy levels for ^{8}He, ^{9}He and ^{10}He given from the experiments.

2. Formulation

We use a rank-1 separable form potential ,

$$V_l = |g_l\rangle\lambda\langle g_l|, \tag{1}$$

for 2-body (^{8}He - n, n - n) interaction. Form factor $|g_l\rangle$ is represented in momentum space as

$$\langle p|g_l\rangle = N\frac{p^l}{[p^2+\beta^2]^{(l+1)}}, \tag{2}$$

which is depend to angular momentum l for 2-body subsystem. N is a normalization factor.

When the core excitation is taken into account, we extend the form factor to the combination of 2-channel,

$$|g_l\rangle = \gamma_1|A_1\rangle|g_{l_1}\rangle + \gamma_2|A_2\rangle|g_{l_2}\rangle, \tag{3}$$

where subscript 1 and 2 represent the ground state and excited state for ^{8}He, respectively. $|A_i\rangle$ represent the state ket for core and the parameter γ_i is a strength factor for 2-channel state.

The parameters(β and γ) of ^{8}He - n potential were determined to reproduce the resonance energy and decay width for ^{9}He. For example, Table 2 shows these parameters determined in $(1/2^-, 1/2^+)$ for the ground and excited state respectively. The excited state of ^{8}He is dominant in ^{9}He, because γ_2^2 is greater than γ_1^2.

Table 2. The comparison of the strength-factors of interaction when $\beta = 1.5166(fm^{-1})$.

	γ^2	γ_1^2	γ_2^2
^{9}He (ground state)	0.9458	0.0439	0.9019
^{9}He (1st excited state)	0.8851	0.1489	0.7362

Note: $\gamma^2 = \gamma_1^2 + \gamma_2^2$.

We take only $^1S_0$ partial wave of n-n system using parameters in[5].

The following Faddeev equation for ^{8}He-n-n system[5] is solved,

$$\eta(E)f_{C_\alpha}(q_\alpha;E) = \sum_{\gamma\neq\alpha}\sum_{C_\gamma}\int_0^\infty q_\gamma^2 dq_\gamma Z_{C_\alpha,C_\gamma}(q_\alpha,q_\gamma;E)\tau_{j_\gamma}(E-\frac{q_\gamma^2}{2\mu^\gamma})f_{C_\gamma}(q_\gamma;E). \tag{4}$$

This is an eigenvalue equation for the Faddeev component. τ means a T-matrix for 2-body subsystem, Z means 3-body propagator and $\eta(E)$ is a eigenvalue for given energy E. C_α denote the particle channel and partial wave. We search the complex energy E under the condition the eigenvalue $\eta(E) = 1$. The eigenvalue $\eta(0)$ is useful because we can judge whether the system become the bound state ($\eta(0) > 1$) or resonance state ($\eta(0) < 1$). These technique was already applied to the α-α-Λ model of $^9_\Lambda$Be hypernucleus to find the resonance state.[6]

3. Results

Although in section 2 we assume the ground state and the excited state of ^{9}He as $J^\pi = 1/2^-$ and $1/2^+$, unfortunately, J^π of these state have not been determined by

Table 3. Calculation results for ^{10}He ground state 0^+.

J^π of ground ^{9}He	J^π of excited ^{9}He	$\eta(0)$	$E(MeV)$	$\Gamma(MeV)$
$3/2^-$	—	1.015	−0.1872	—
$1/2^-$	—	0.3363		
$1/2^+$	—	0.8922	0.5905	0.1306
$1/2^-$	$3/2^-$	1.034	−0.5340	—
$1/2^+$	$3/2^-$	1.167	−1.862	—
$1/2^+$	$1/2^-$	0.9522	0.3064	0.03436
$1/2^-$	$1/2^+$	0.8904	0.7820	0.6566
experimental value			1.07	0.3

the experiments.[1] We vary J^π of ^{9}He and the theoretical predictions of the energy and the decay width of ^{10}He are shown in Table 3.

Apparently, $3/2^-$ is not suitable for J^π of ground or excited state for ^{9}He because ^{10}He ground state become a bound state for that case, and a bound state have not been observed in experiments $E_r = 1.07(MeV)$ and $\Gamma = 0.3(MeV)$ was indicated from experiments. Comparing the calculation results with the experimental data, $(1/2^-\ ,\ 1/2^+)$ is most suitable for J^π of (ground , excited) state for ^{9}He.

References

1. For the review, D. R. Tilley *et al.*, *Nucl. Phys.* **A708**, 3 (2002). ; D. R. Tilley *et al.*, *ibid.* **A745**, 155 (2004).
2. K. Kato and K. Ikeda, *Prog. Theor. Phys.* **89**, 623 (1993). ; S. Aoyama, K. Kato, T. Myo and K. Ikeda, *Prog. Theor. Phys.* **107**, 543 (2002).
3. S. Aoyama and K. Kato, *Phys. Rev.* **C55**, 2379 (1997).
4. A. A. Korsheninnikov *et al.*, *Nucl. Phys.* **A559**, 208 (1993). ; M. V. Zhukov *et al.*, *Phys. Rev.* **C50**, R1 (1994).
5. A. Eskandarian and I. R. Afnan, *Phys. Rev.* **C46**, 2344 (1992).
6. E. Cravo, A. C. Fonseca and Y. Koike, *Phys. Rev.* **C66**, 014001 (2002).

Isotopic Composition as a Signature for Different Processes Leading to Fragment Production in Midperipheral Ni+Al, Ni, Ag Collisions at 30 MeV/Nucleon

P. M. Milazzo, G. V. Margagliotti, R. Rui

Dipartimento di Fisica and INFN, Trieste, Via Valerio 2, I-34127 Italy

G. Vannini

Dipartimento di Fisica and INFN, Bologna, Via Irnerio 46, I-40126 Italy

N. Colonna

INFN, Bari, Via Amendola 173, I-70126 Italy

F. Gramegna, P. F. Mastinu

INFN, Laboratori Nazionali di Legnaro, Viale dell'Università 2, I-35020 Italy

C. Agodi, R. Alba, G. Bellia, R. Coniglione, A. Del Zoppo, P. Finocchiaro, C. Maiolino, E. Migneco, P. Piattelli, D. Santonocito, P. Sapienza

INFN, Laboratori Nazionali del Sud, Via Santa Sofia 62, I-95123 Catania, Italy

I. Iori, A. Moroni

Dipartimento di Fisica and INFN, Milano, Via Celoria 16, I-20133 Italy

The results of experiments performed to investigate the Ni+Al, Ni+Ni, Ni+Ag reactions at 30 MeV/nucleon are presented. From the study of dissipative midperipheral collisions, it has been possible to detect events in which Intermediate Mass Fragments (IMF) production takes place. The decay of a quasi-projectile has been identified; its excitation energy leads to a multifragmentation totally described in terms of a statistical disassembly of a thermalized system (T$\simeq$4 MeV, E* $\simeq$4 MeV/nucleon). Moreover, for the systems Ni+Ni, Ni+Ag, in the same nuclear reaction, a source with velocity intermediate between that of the quasi-projectile and that of the quasi-target, emitting IMF, is observed. The fragments produced by this source are more neutron rich than the average matter of the overall system, and have a charge distribution different, with respect to those statistically emitted from the quasi-projectile. The above features suggest that this production comes from material which is "surface-like" (since it originates from the overlap of the two nuclei). To investigate the neutron content in both processes the ^{4}He and ^{6}He experimental yields were studied.

1. Introduction

The production of intermediate mass fragments (IMF, Z≥3) is one of the main features of the nuclear reactions at bombarding energies of 30-50 MeV/nucleon[1]. At these energies in peripheral and midperipheral collisions, the quasi-projectile (QP) and the quasi-target (QT) can de-excite giving rise to the production of IMF, following a statistical pattern, in which low density nuclear matter is subject to a liquid-gas phase transition[2]. This is supported by several experimental results[3].

Many experiments have also shown that at midrapidity dynamical mechanisms lead to the production of IMF; this effect is due to the rupture of a neck-like structure joining QP and QT[4,5]. Transport calculations predict that dynamical fluctuations dominate the neck instability allowing the production of IMF[6].

It is then possible to observe the competition between statistical and dynamical mechanisms leading to the production of IMF[5]. To investigate this phenomenon we studied the Ni+Al, Ni+Ni, Ni+Ag midperipheral collisions at 30 MeV/nucleon.

2. Experimental Set-Up

The experiment was performed at the INFN Laboratori Nazionali del Sud, where the superconducting cyclotron delivered a beam of ^{58}Ni at 30 MeV/nucleon, using the MEDEA[7] and MULTICS[8] experimental apparata as detectors.

Since IMF can be produced both in central and midperipheral collisions, it is mandatory to distinguish collisions occurred at different impact parameters. We selected peripheral and midperipheral events when the heaviest fragment (produced by the disassembly of a QP emitting source) in the laboratory frame travels at velocities higher than 80% of that of the projectile. Only "complete" events are analyzed, i.e. when at least 3 IMF are produced (with the heaviest fragment having Z≥9) and more than 80% of the total linear momentum is detected.

3. Dynamical and Statistical IMF Production

The results presented hereafter will refer only to midperipheral collision events, observing the IMF emitted from the QP and from the midvelocity neck, and studying the competitive IMF production mechanisms. In Fig.1 the yields of carbon and oxygen fragments (for the three considered reactions) are plotted as a function of the component of the velocity parallel to the beam axis (v_{par}). The centre of mass velocities for the three systems are 5.18 (Al), 3.80 (Ni) and 2.65 (Ag) cm/ns.

In the upper panels of Fig.1 (Ni+Al), at the centre of mass velocity there is a minimum in the production of Z=6-8 fragments; this fact suggests a negligible formation of a neck-like structure for this light system. On the contrary, in Ni+Ni we notice that at midvelocity a large contribution of IMF is present[5]. The lower panels (Ni+Ag) show evidence of a larger contribution of midvelocity IMF (even if there is a clear efficiency cut). Thus, for mid-peripheral collisions, while the disassembly of a QP is present in all the three considered systems, the production of IMF at midvelocity depends on the size of the target nucleus.

The study of angular, energy and charge distributions of the IMF emitted by the QP shows indications that this system has reached a thermodynamical equilibrium before its disassembly. The QP decay is well described within a statistical framework, and its properties are: a) a size close to that of the incident Ni nucleus, b) an excitation energy around 4 MeV/nucleon, c) a temperature around 4 MeV. Properties b) and c) place the system well inside the plateau of the caloric curve, where statistical multifragmentation is the main decay pattern[9].

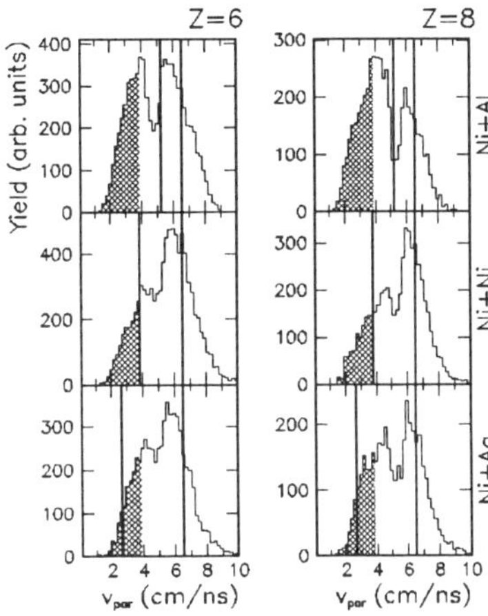

Fig. 1. Experimental v_{par} distributions for Z=6 (left panels) and Z=8 (right panels), for the three studied reactions; vertical lines refer to the centre of mass and QP velocities. Experimental efficiency cut occurs in the shadowed area.

In coincidence with the statistical multifragmentation of the QP, we observed the emission of IMF from a midvelocity source. The yield (Y_{NECK}) from this source have been evaluated by means of fit procedures[5,9] that avoid from possible distortions due to efficiency effects and contaminations from QT at low velocities. In order to compare the statistical and dynamical IMF production in Fig.2a the ratio between the relative yields (Y_{NECK}/Y_{QP}) is presented as a function of the atomic number Z. We observe a bell-like shape, peaked around Z=9. BNV calculations predict an evident massive neck formation (on average a Z=8 fragment in the neck zone), and show that this fragment production comes from material that originates from the overlap of the surfaces of the two nuclei and which could be neutron rich[5,6]. To investigate the neutron content of the neck matter we observed the ^6He experimental yield. Its emission is negligible from a statistical decay, both because the binding energy favours, for Z=2, α particle emission and because the N/Z value of ^6He is quite different from the corresponding ratio of the system (N/Z=2 for ^6He, N/Z=30/28$\simeq$1 for the system). If the surfaces of the interacting nuclei are more neutron rich than the bulk matter or the dynamical process leading to the formation of a neck structure has a particular isospin dependence, the ^6He production should be more abundant in the neck zone with respect to the QP zone. In Fig. 2b the experimental ^4He and ^6He yields are plotted versus v_{par}: while the ^4He distribution is centered around the QP velocity, the ^6He yield is very scarse in the QP zone and starts to increase going towards the midvelocity region. One should take into account that the energy thresholds for mass identification in the He case produce cuts for velocities lower than 4 cm/ns. To better see the isospin effect in Fig.2b the Y(^6He)/Y(^4He) yield ratio is also plotted. The behaviour of this ratio shows the increase of neutron content in the midvelocity zone.

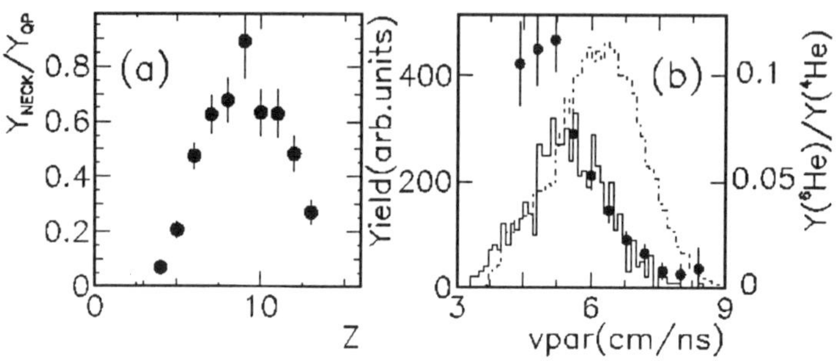

Fig. 2. (a) Ratio of measured yields for neck fragmentation and QP emission; (b) Yields of ^{4}He (dot-dashed line) and ^{6}He (full line, multiplied by a factor 10); Y(^6He)/Y(^4He) yield ratios as a function of v_{par}.

4. Conclusions

In the study of the Ni+Al, Ni, Ag 30 MeV/nucleon dissipative midperipheral collisions it has been possible to reveal events in which IMF are emitted with different mechanisms. We are in presence of a QP, with an excitation energy which leads to the multifragmentation regime; its decay can be fully explained in terms of a statistical disassembly of a thermalized system (T=3.9±0.2 MeV, E* $\simeq$4 MeV/nucleon). Contemporary a midvelocity source is formed, emitting IMF. These clusters are more neutron rich than the average matter of the overall system.

References

1. W. Reisdorf et al., *Phys. Lett.* **B595**, 118 (2004).
2. A. Bonasera, M. Bruno, C.O. Dorso and P.F. Mastinu, *La Rivista del Nuovo Cimento* **23**, 1 (2000); P. Chomaz, *Nucl. Phys.* **A685**, 274 (2001).
3. J. Pochodzalla et al., *Phys. Rev. Lett.* **75**, 1040 (1995); P. F. Mastinu et al., *Phys. Rev. Lett.* **76**, 2646 (1996); J.B. Natowitz et al., *Phys. Rev.* **C65**, 34618 (2002); M. D'Agostino et al., *Nucl. Phys.* **A724**, 455 (2003).
4. C.P. Montoya et al., *Phys. Rev. Lett.* **73**, 3070 (1994); J. Toke et al., *Phys. Rev. Lett.* **75**, 2920 (1995); P. Pawlowski et al., *Phys. Rev.* **C57**, 1711 (1998); Y. Larochelle et al., *Phys. Rev.* **C62**, 051602(R) (2000); J. Lukasik et al., *Phys. Lett.* **B566**, 76 (2003).

5. P.M.Milazzo et al., *Phys. Lett.* **B509**, 204 (2001).
6. V. Baran et al., *Nucl. Phys.* **A703**, 603 (2002).
7. E. Migneco et al., *Nucl. Instr. and Meth. Phys. Res.* **A314**, 31 (1992).
8. I. Iori et al., *Nucl. Instr. and Meth. Phys. Res.* **A325**, 458 (1993).
9. P.M. Milazzo et al., *Nucl. Phys.* **A756**, 39 (2005).

Quasibound States of Eta-Mesic Deuteron and 3He

N. G. Kelkar

Departamento de Fisica, Universidad de los Andes
Cra 1E, 18A-10,
Bogota A. A. 4976, Colombia
E-mail: nkelkar@uniandes.edu.co

K. P. Khemchandani and B. K. Jain

Department of Physics, University of Mumbai,
Kalina, Santacruz (E)
Mumbai 400098, India

The concept of collision time in scattering is extended to evaluate the delay time in off-shell elastic scattering reactions at negative energies and is used to search for the quasibound states of eta mesons and light nuclei. The η-nucleus transition matrix is evaluated by solving few body equations within a finite rank approximation. The elementary interaction between the η meson and a nucleon is described in terms of a coupled channel transition matrix which gives rise to an ηN scattering length of $0.28 + i0.19$ fm. The η^3He state predicted within this approach is in good agreement with the recent claim of an eta-mesic state confirmed from investigations of the photoproduction reaction $\gamma^3\text{He} \rightarrow \eta X$.

1. Searches of Eta-Mesic Nuclei

The first prediction of an attractive ηN interaction[1] (arising basically due to the proximity of the ηN threshold to the $N^*(1535)$ resonance) was made back in 1985. This was followed by the prediction for the existence of new exotic states of eta mesons and nuclei[2]. Several experiments using pion beams[3] were performed to search for eta-mesic nuclei. Though these experiments did not lead to a definite confirmation of such states, they did not exclude the possibility of their existence either. Sokol *et al.*[4] claimed to have observed the eta-mesic nucleus, ${}^{11}_{\eta}C$, however, their claim has been doubted on the basis of poor statistics. Indirect evidence in the form of the strong η-nucleus attraction was also obtained from data on η production in the $pd \rightarrow {}^3\text{He}\,\eta$ [5] and $np \rightarrow d\eta$ [6] reactions which display large enhancements in the cross sections near threshold. Finally, the photoproduction of η-mesic ^{3}He investigated recently by the TAPS collaboration[7] proved to be a step forward in this direction. More information is expected to be soon available from the ongoing program of the COSY-GEM collaboration[8].

Following the early predictions of Haider and Liu[2], there have been a lot of theoretical searches for eta-mesic nuclei using different approaches for the treatment of the eta-nucleus interaction as well as using different techniques to locate such states. One can find a detailed account of the existing literature in a recent work[9], where a search of η-nucleus bound states ranging from $A = 3$ to 208 is made. Here, we shall briefly survey the literature on light nuclei, since we restrict ourselves to the study of ηd and η^3He elastic scattering in the present work. One of the early calculations [10] using the 3-body equation, predicted a quasi-bound state with a mass of 2430 MeV and a width of 10-20 MeV in the πNN-ηNN coupled system. This state led to a remarkable enhancement of the ηd elastic cross section. In yet another Fadeev-type calculation [11,12] of ηd scattering, the existence of a quasi-bound state or resonance at low energies was predicted. The exact values of the predicted poles of course varied with the value of the η-nucleon scattering length, $a_{\eta N}$, which in turn depends on the model of the η-nucleon interaction. The existence of resonances in the above works was inferred from Argand diagrams. Some other recent (also Fadeev-type) calculations, however, ruled out the possibility of a quasi-bound state in the ηd system [13,14]. Predictions of resonances in the light η-nucleus systems, namely, d, ^{3}He and ^{4}He, from the positions and movements of the poles of the amplitude, can be found in [15].

In this work, we shall infer the existence of resonances in the ηd and η^3He systems, from positive peaks in the time delay distributions for eta-nucleus elastic scattering. In the next section we shall briefly discuss the concept of time delay and the transition matrix for eta-nucleus elastic scattering required to evaluate it.

2. Delay Times in Eta-Nucleus Scattering

It was first noticed by Eisenbud and Wigner that a large positive time delay in elastic scattering can be associated with the existence of unstable states[16]. Eisenbud proposed a delay time matrix which was later modified by Smith[17] to define a lifetime matrix for unstable states. An element of this matrix, Δt_{ij}, which is the time delay in the emergence of a particle in the jth channel after being injected in the ith channel is given by,

$$\Delta t_{ij} = \Re e\left[-i\hbar (S_{ij})^{-1}\frac{dS_{ij}}{dE}\right], \tag{1}$$

where S_{ij} is an element of the corresponding S-matrix. The time delay concept which has been very useful in characterizing resonances[18], can also be extended to the search of bound as well as quasibound states[19]. The latter signify states with a negative binding energy but having a finite lifetime. We shall evaluate the time delay in η-d and $\eta-^3$He elastic scattering using the relation

$$S = 1 - \frac{\mu\, i\, k\, t_{\eta A}}{\pi} \tag{2}$$

where $t_{\eta A}$ is the η-nucleus transition matrix.

The ηA transition matrix is evaluated using few body equations solved within a Finite Rank Approximation (FRA) approach, which means that the nucleus in

η-nucleus elastic scattering remains in its ground state, in the intermediate state. Since the η-mesic bound states and resonances are basically low energy phenomena, it seems justified to use the FRA for calculations of the present work. Besides, we present here results using a small ηN scattering length ($a_{\eta N} = (0.28, 0.19)$ fm) since in a comparison of the Alt-Grassberger-Sandhas (AGS) equations (which include the target excitations) with FRA, the authors in Ref. 12 found that the FRA works reasonably well for real parts of the ηN scattering lengths less than 0.5 fm. The η-nucleus t-matrix in this approach is written as [20],

$$t_{\eta A}(\vec{k}', \vec{k}; z) = <\vec{k}'; \psi_0 \,|\, t^0(z) \,|\, \vec{k}; \psi_0> + \varepsilon \int \frac{d\vec{k}''}{(2\pi)^3} \frac{<\vec{k}'; \psi_0 \,|\, t^0(z) \,|\, \vec{k}''; \psi_0>}{(z - \frac{k''^2}{2\mu})(z - \varepsilon - \frac{k''^2}{2\mu})} t_{\eta A}(\vec{k}'', \vec{k}; z), \tag{3}$$

where $z = E - |\varepsilon| + i0$. E is the energy associated with ηA relative motion, ε is the binding energy of the nucleus, ψ_0 is the nuclear wave function and μ is the reduced mass of the ηA system. We use a Gaussian wave function for ^{3}He and a Paris potential parametrization for the deuteron ground state wave function. The operator t^0 describes the scattering of the η meson from nucleons fixed in their space position within the nucleus. It, however, differs from the usual fixed center t-matrices. Here, t^0 is taken off the energy shell and involves the motion of the η meson with respect to the center of mass of the target. The present scheme should not be confused with a conventional optical potential approach which involves the impulse approximation and omits the rescattering of the η meson from the nucleons. The details of the formalism for the ηA t-matrix can be found in Ref. 21.

3. Results and Discussions

In Fig. 1 we show the time delay distribution in η-d and η-^{3}He elastic scattering as a function of the excitation energy, $E = E_{\eta A} - m_\eta - m_A$, where $E_{\eta A}$ is the energy available in the η-nucleus centre of mass system. We find broad bumps centered around $E = -5$ and $E = -12$ MeV in the η^3He and ηd cases respectively. The sharp rise at threshold is probably a mixed effect of the tail of the unstable state and a threshold singularity present in the time delay plots for s-wave scattering. One can imagine the time delay at negative energies as a delay in an off-shell scattering which can be calculated theoretically. Though it cannot be deduced directly from the "measured" phase shifts in an elastic scattering experiment, it would manifest in processes where the particles are produced below their elastic threshold and interact only through off-shell scattering. The production of η-mesic nuclei in nuclear reactions (such as the recently performed photoproduction reaction[7]), indeed corresponds to such a situation. The findings of the present work are in agreement with this photoproduction experiment where a binding energy of -4.4 ± 4.2 MeV for the η-mesic ^{3}He was found.

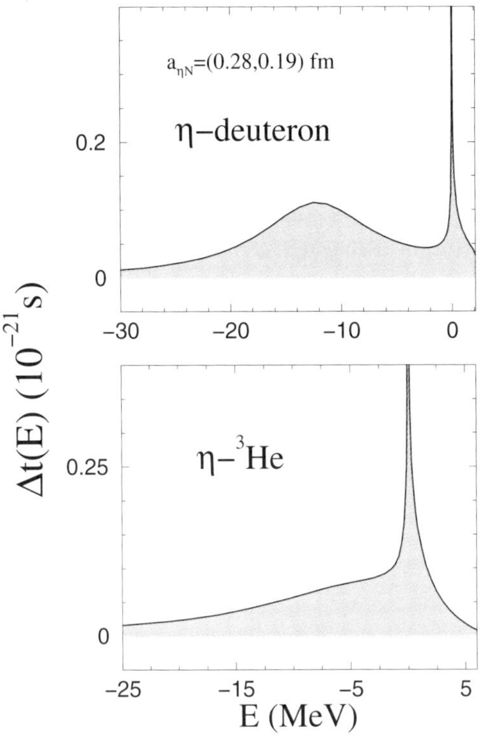

Fig. 1. Time delay in η-deuteron and η^3He elastic scattering as a function of energy, $E = E_{\eta A} - m_\eta - m_A$, where $E_{\eta A}$ is the energy available in the η-nucleus centre of mass system.

References

1. R. S. Bhalerao, L. C. Liu, *Phys. Rev. Lett.* **54**, 865 (1985).
2. Q. Haider, L. Liu, *Phys. Lett.* **B172**, 257 (1986); *ibid* **174**, 465(E) (1986).
3. R. E. Chrien et. al, *Phys. Rev. Lett.* **60**, 2595 (1988); B. J. Lieb and L. C. Liu, Progress at LAMPF, Report No. LA-11670-PR, 1988; J. D. Johnson et al., *Phys. Rev.* **C47**, 2571 (1993).
4. G. A. Sokol *et al.*, *Part. Nucl. Lett.* **102**, 71 (2000); G. A. Sokol and L. N. Pavlyuchenko, arXiv:nucl-ex/0111020.
5. B. Mayer et al., *Phys. Rev.* **C53**, 2068 (1996); J. Berger et al., *Phys. Rcv. Lett.* **61**, 919 (1988).
6. H. Calen *et al.*, *Phys. Rev. Lett.* **80**, 2069 (1998); H. Calen *et al.*, *Phys. Rev. Lett.* **79**, 2642 (1997).
7. M. Pfeiffer *et al.*, *Phys. Rev. Lett.* **92**, 252001 (2004).
8. M. Ulicny [GEM Collaboration], "Measurement of the $p + {}^6Li \rightarrow {}^7Be + \eta$ reaction near threshold energies", *AIP Conf. Proc.* **603**, 543 (2001).
9. Q. Haider and L. C. Liu, *Phys. Rev.* **C66**, 045208 (2002).
10. T. Ueda, *Phys. Rev. Lett.* **66**, 297 (1991).
11. S. A. Rakityansky, S. A. Sofianos, N. V. Shevchenko, V. B. Belyaev, W. Sandhas, *Nucl. Phys.* **A684**, 383 (2001); N. V. Shevchenko, V. B. Belyaev, S. A. Rakityansky, S. A. Sofianos, W. Sandhas, *Eur. Phys. J.* **A9**, 143 (2000), arXiv:nucl-th/9908035.

12. N. V. Shevchenko, S. A. Rakityansky, S. A. Sofianos, V. B. Belyaev, W. Sandhas, *Phys. Rev.* **C58**, R3055 (1998), arXiv:nucl-th/9808009.
13. H. Garcilazo and M. T. Peña, *Phys. Rev.* **C63**, 021001 (2001).
14. A. Deloff, *Phys. Rev.* **C61**, 024004 (2000); A. Fix and H. Arenhövel, *Eur. Phys. J* **A9**, 119 (2000).
15. S. A. Rakityansky, S. A. Sofianos, M. Braun, V. B. Belyaev and W. Sandhas, *Phys. Rev.* **C53**, R2043 (1996); S. A. Rakityansky, S. A. Sofianos, V. B. Belyaev, W. Sandhas, *Few-Body Systems Suppl.*, **9**, 227 (1995).
16. L. Eisenbud, dissertation, Princeton, June 1948 (unpublished); E. P Wigner, *Phys. Rev.* **98**, 145 (1955); E. P. Wigner and L. Eisenbud, *Phys. Rev.* **72**, 29 (1947).
17. F. T. Smith, *Phys. Rev.* **118**, 349 (1960).
18. N. G. Kelkar, M. Nowakowski, K. P. Khemchandani and S. R. Jain, *Nucl. Phys.* **A730**, 121 (2004), hep-ph/0208197; N. G. Kelkar, M. Nowakowski and K. P. Khemchandani, *Nucl. Phys.* **A724**, 357 (2003); N. G. Kelkar, M. Nowakowski and K. P. Khemchandani, *J. Phys. G: Nucl. Part. Phys.* **29**, 1001 (2003), hep-ph/0307134; N.G. Kelkar, M. Nowakowski, K.P. Khemchandani, *Phys. Rev.* **C 70**, 024601 (2004); *ibid, Mod. Phys. Lett.* **A19**, 2001 (2004), nucl-th/0405008.
19. J. Fraxedas and J. Sesma, *Phys. Rev.* **C37**, 2016 (1988).
20. V. B. Belyaev, *Lectures on the theory of FEW BODY SYSTEMS*, Springer series in Nuclear and Particle Physics, 1990.
21. K. P. Khemchandani, N. G. Kelkar and B. K. Jain, *Nucl. Phys.* **A708**, 312 (2002), nucl-th/0112065.

The Resonances of n, α, α - System at Astrophysical Energy Region

N. Takibayev

Kazakh National Pedagogic University, Almaty, Kazakhstan
E-mail: teta@nursat.kz

The series of narrow resonances in n, α, α scattering amplitude at very low energies has been found. The lifetimes of these three body resonances are close to the lifetime of unstable nucleus ^8Be. Consideration of three body scattering has been carried out on base of well-known Faddeev's equations with simple separable forms of two body repulsive potentials. The three body effective potential has been obtained in analytical form in this case. The physical model and explanation of the resonance phenomena in systems consisted of one neutron and two or more α-particles is given to prove appearance of the resonance set. The existence of narrow resonances can stimulate a few-body fusion in systems that give result in many astrophysical phenomena.

1. Introduction

We consider the few-body resonance phenomena at astrophysical energy region, in particular, in system of one neutron and two α - particles [1]. The existence of narrow resonance states in systems gives rise to new type of fusion reactions, which can take place in stellar matter [2]. So, the reactions are resulted from the few body resonance physics [3].

Note that neutron and α - particle can not form a bound state. It is a well-known problem of nuclei with atomic mass $A = 5$. Neutron of very low energy can not be captured by α - particle. Their effective potential is characterized as a repulsive force in this energy region, i.e. in S-wave component of interaction [4].

Two α - particles can not be bound in a stable state. It is the well-known problem, too, concerning nuclei with atomic mass $A = 8$. The scattering amplitude of two α - particles has a very narrow resonance at low energy. The energy of resonance $\simeq 92\ keV$ and the width $\simeq 6.8\ eV$. Moreover, the system of three α-particles has the resonance state of long lifetime, too. The resonance energy and width are $\simeq 0.38\ MeV$ and $\simeq 8.5\ eV$, respectively [4].

Analysis and calculations lead to result that three body system of two α-particles and one neutron has a set of narrow resonances in region below the resonance energy of α, α subsystem.

In the system of three α-particles and one neutron the number of similar resonances may be even more. So, it is proposed that a free neutron can gather few α-particles to form quasi-molecular resonance states.

Therefore, owing to neutron the low energy fusion becomes possible in system containing few α-particles. They can product the different nuclei including nuclei with mass several times superior than α - particle's mass. The neutron can get free and then stimulate resonance reactions again.

Mechanism of neutron catalysis of resonance fusion reminds well known μ - catalysis of d, t nuclear fusion by means of formation of μ, d, t - molecules [5].

It is assumed that inside stars there is presence of area with the high density of helium - "helium core", where can occur described processes.

2. Two Body Potentials

It should be noted that the very simple model has been taken under consideration in order to investigate the resonant three body dynamics. First of all it is a very low energy region - close to the temperature at the center of the Sun ($\approx 1 \div 10\ keV$). So, relativistic effects, contributions of inelastic nuclear channels, etc are not taken into account. Neutron and α-particle are considered as elementary.

Two particle interactions are chosen in the simplest form - in form of separable potentials:

$$\{\lambda V\}^J_{L,S} = |\nu^J_{L,S} > \lambda^J_{L,S} < \nu^J_{L,S}|. \tag{1}$$

In this expression only a partial component of potential is written out. They satisfactory describe resonance behavior in pair subsystems. Hereinafter the angular functions, recoupling coefficients, etc are omitted for the simplification.

Using the separable potential gives an exact analytical solution of two-body problem. It is their obvious quality. The amplitude operator can be written as:

$$\hat{f} = -|\nu > \eta(Z) < \nu|, \quad \eta^{-1} = \lambda^{-1} - A(Z), \quad A =< \nu|G_0(Z)|\nu >, \tag{2}$$

and for simplicity indexes of states are omitted too. The amplitude $f(k, k'; Z)$ comes to the physical one $f(E)$ in limit $Z \to E = k_0^2/2\mu$, where E - the real quantity and $|\vec{k}| = |\vec{k'}| \to k_0$.

In case of potential (1) is represented as a sum of separable ones the expressions of (2) have to be understood as matrix values in relation to corresponding index.

The partial amplitude can get a pole in point of bound state $Z = E_B$, when $\lambda^{-1} = A(E_B)$ and $E_B \leq 0$. But a pole may be in point of resonance state $Z = E_{Res}$, when $\lambda^{-1} = A(E_{Res})$ and $E_{Res} = (k_R + ik_I)^2/2\mu$ is a complex quantity [6].

It is convenient to normalize form-factors with condition $A(0) = -1$ appointing λ as dimensionless constant.

S-wave n, α-potential at low energy region can be chosen as ($x = k/\beta$):

$$< \nu_{n,\alpha}|k >= N_{n,\alpha}/(1 + x^2), \tag{3}$$

where the constant $N_{n,\alpha} = \sqrt{8\pi/(2\mu\beta)}$. So, the parameters fixed by n, α experimental scattering data are equal: $\beta = 0.751 fm^{-1}$ and $\lambda = 14.94$ (see [7] and [8]). Other components of effective n, α-interaction correspond to the attractive forces and the most important among them are P-wave interactions. However, they become essential at higher energy region ($\sim 1\ MeV$).

Obviously, **two α-particle effective potential** has to be a result of several complex interactions and involved forces. It is clear that a determination of exact potential is a very difficult problem of few-body physics. Meanwhile, it is possible to describe at very low energies this resonance within the framework of simple potential model (1). Using S-wave potential form-factors for (α, α)-system in form:

$$< \nu_{\alpha,\alpha}|k >= N_{\alpha,\alpha}/(1+x^2)^3 \ , \tag{4}$$

we can get exact solutions (2) and then determine corresponding parameters which have to be in accordance with experimental data $\lambda = 1.75 \cdot 10^2$, $\beta_{\alpha,\alpha} = 2.12 \cdot 10^{-2}\ fm^{-1}$, $x_R = 4.46$, $x_I = -0.827 \cdot 10^{-4}$ (see [2]).

3. Three-Body Scattering

The quantum theory of three body scattering is excellently presented in classical monographs (see, for instance [9] and [10]).

Here it's enough to refer to general formulas only for understanding of calculation methods in principle. Introducing $Q_{i,j} - \tau_i\delta_{i,j} = |\nu_i > \eta_i P_{i,j}\lambda_j < \nu_j|\varphi_j >$, where $Q_{i,j} = V_i|\Psi_j >$ we get Faddeev's equations in form:

$$P_{i,j} = \Lambda_{i,j} + \sum_{l=1}^{3} \Lambda_{i,l}\eta_l P_{l,j} \ , \tag{5}$$

where $\tau_i = V_i|\varphi_i >$, t_i - the pair transition operator obeys $t_i = V_i + V_i G_0 t_i$, $\overline{\delta}_{i,j} = I - \delta_{i,j}$ and

$$\Lambda_{i,j} =< \nu_i|G_0(Z)|\nu_j > \ , \quad with \quad i \neq j. \tag{6}$$

The sum on intermediate states is intended, of course. Here Z corresponds to the total energy of three body system, $Z_l = Z - k_l^2/2\mu_l$ - subsystem energy of intermediate φ_l state and

$$\eta_l^{-1} = \lambda_l^{-1} - A_l(Z_l), \quad A_l =< \nu_l|G_0(Z_l)|\nu_l > . \tag{7}$$

The amplitudes $P_{i,j}$ have been calculated on base of equations (5) with different point sets, using various numerical schemes. The basic features of n, α, α-system are almost no changed. The imaginary part of the elastic amplitude as function of energy is shown in Figure 1.

Our calculations carried out with Coulomb force between two α-particles have confirmed the fact of existence of resonance set, too.

4. The Effective Potential

In order to understand the reason of resonance phenomenon in n, α, α-system the set of equations (5) can be rewritten for elastic $n, (\alpha, \alpha)$ amplitude, i.e. all transitions

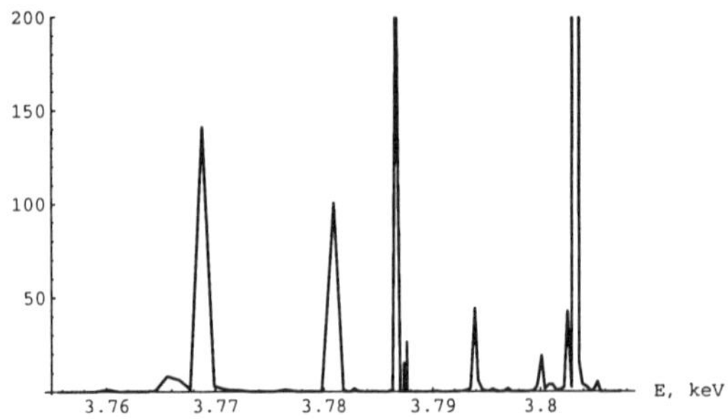

Fig. 1. Imaginary part of elastic $n, 2\alpha$ scattering amplitude in relative units.

between $n, (\alpha, \alpha)$ states marked here as "i" have been collected in special equation:

$$P_{i,i}(k,k') = \Lambda_{i,i}^{eff}(k,k') + \sum_{k_s} \Lambda_{i,i}^{eff}(k,k_s)\eta_i(Z_s)P_{i,i}(k_s,k') \ . \tag{8}$$

Here k, k_s and k' are impulses of neutron in initial, intermediate and final states, respectively. Other transitions named here as "non-elastic" have been collected in separate set of equations and define $\Lambda_{i,i}^{eff}$ - the effective potential of elastic scattering:

$$\Lambda_{i,i}^{eff}(k,k') = \sum_{j,\ l\neq i}^{3} \sum_{k_j,\ k_l} \Lambda_{i,j}(k,k_j)\widetilde{\eta}_{j,l}(Z;k_j,k_l)\Lambda_{l,i}(k_l,k') \ , \tag{9}$$

$$\widetilde{\eta}_{j,l}(Z;k_j,k_l) = \eta_j(Z_j)\delta_{j,l} + \sum_{m\neq i}\sum_{k_m} \eta_j(Z_j)\Lambda_{j,m}(k_j,k_m)\widetilde{\eta}_{m,l}(Z;k_m,k_l) \ . \tag{10}$$

The features of $\Lambda_{i,i}^{eff}$ are mainly formed by α, α resonance interaction - the ratio $|k_I/k_R| \simeq 0.185 \cdot 10^{-4}$ will predestine a behaviour of this potential.

1. For example, the ratios between radii of n,α- and α,α-potentials prove to be very small $\beta_{n,\alpha}^{J,L}/\beta_{\alpha,\alpha} \lesssim 10^{-2}$. These small parameters will provoke the decreasing of contribution of "non-elastic" transitions at very low energies, i.e. rescattering intermediate processes like $\alpha + (n,\alpha') \rightarrow \alpha' + (n,\alpha)$. It means that the energy dependence of n,α potentials becomes inessential, unlike α,α potential.

Under this condition we can find easily the effective potential - Λ_0^{eff} in analytical form (see corresponding expressions in [2]).

2. Note that the effective potential is vanishing almost everywhere excluding narrow strips close to values $x, x' = 2/3\sqrt{1 + E_0/\beta_{\alpha,\alpha}^2}$, where the effective potential can achieve large values [2].

The first point leads to the fact that n,α interaction turns out as the contact one. And the second point signifies that α,α-system acts like the resonator. Similar effects are well-known in quantum physics.

Let's address to the task on scattering of particle by two heavy centers, when repulsive contact interactions act between the particle and every of both centers.

Note, that in this case two-body subsystem consisted of particle and one heavy center can not has any bound state. Nevertheless, three body system consisted of one particle and two heavy centers has a bound state [11]. Similar systems containing three and more heavy centers have more complicated spectrum.

In our problem we consider a resonator instead of heavy centers, the simplest resonator being a resonant state of two α-particles. The resonance state of three α-particles can be assumed in this model as more complex resonator.

Calculations on base of $\Lambda_0^{eff}(k,k')$ give the following results. In case of $E > 0$ calculations confirm the existence of narrow resonances at very low energies. In case of $Z < 0$ the system has the bound state with very small energy $E_B \simeq -4.62keV$. Remarkably, it is in accordance with the above sample of scattering particle and two heavy centers.

5. Summary Conclusions

So, we can say about the new type of fusion - resonance fusion. The neutron can play a role of catalyst to stimulate fusion of few α-particles. By realizing possibility of light nuclei generation in resonance fusion, we can give new way of light nuclei origin.

Resonant fusion in stellar matter can give explanation of periodic activity of stars, in particular, periodic local explosions from the Sun.

It is clear that a lot of interesting effects and phenomena arise from the resonance fusion. Note, that the resonance fusion can be easily put into practice to generate energy.

References

1. N. Takibayev, *The XIX Europ. Conf. on FBP in Physics, Groningen, The Netherlands, 23-27 August 2004, AIP Conference Proceedings* **768**, 349 (2005)
2. N. Takibayev, **Astro-ph/0506734** (2005)
3. N. Takibayev, F. M. Pen'kov, *Yad. Fiz.* **50**, 373 (1989); F. M. Pen'kov, N. Takibayev, *Phys. of Atomic Nuclei* **57**, 1232 (1994); N. Takibayev, *Phys. of Atomic Nuclei* **63**, 574 (2000)
4. F. Ajzenberg-Selone, *Nucl. Phys.* **A490**, 1 (1988); **A506**, 1 (1990)
5. S. S. Gernshtein et al, *Uspekhi Fizicheskikh Nauk* **160**, 158 (1990)
6. N. Takibayev, *Phys. of Atomic Nuclei* **68**, 1147 (2005)
7. R. A. Arndt, D. D. Long, L. D. Roper, *Nucl. Phys.* **A209**, 429 (1973)
8. L. D. Blokhintsev, V. L. Kukulin, D. A. Savin, *Yad. Fiz.* **51**, 1286 (1990)
9. L. D. Faddeev, *Mathematical Aspects of the Three Body Problem in Quantum Scattering Theory. New York* 1965
10. W. Glockle, *The Quantum Mechanical Few-Body Problem. Springer-Verlag* 1983.
11. A. I. Baz', Ya. B. Zel'dovich, A. M. Perelomov, *Scattering, reactions, and decays in nonrelativistic quantum mechanics. (Israel Program of Sci. Transl.) Jerusalem* 1966

HYPERNUCLEAR PHYSICS: HADRON-HYPERON, HYPERON-HYPERON INTERACTIONS

Full Coupling Dynamics of Doubly Strange Hypernuclei

H. Nemura
Advanced Meson Science Laboratory, DRI, RIKEN,
Wako, Saitama,
351-0198, Japan
E-mail: nemura@riken.jp

We describe *ab initio* calculations of doubly strange s-shell hypernuclei (${}^{4}_{\Lambda\Lambda}$H, ${}^{5}_{\Lambda\Lambda}$H and ${}^{6}_{\Lambda\Lambda}$He) by using a set of fully coupled channel potentials. The wave function includes $\Lambda\Lambda$, $\Lambda\Sigma$, $N\Xi$ and $\Sigma\Sigma$ channels. Minnesota NN, D2′ YN, and a simulated YY potential based on the Nijmegen hard-core model, are used. Bound-state solutions of these systems are obtained. We calculate the probabilities of the exotic components such as $N\Xi$, $\Lambda\Sigma$ and $\Sigma\Sigma$. This is a first attempt to explore the few-body problem of the *full*-coupled channel scheme for these systems.

1. Introduction

Few-body problems of strange nuclear systems are important to study and clarify the baryon-baryon interactions including the hyperon-nucleon (YN) and hyperon-hyperon (YY) potentials. There is a long standing problem of few-body systems for s-shell Λ hypernuclei, which is known as anomalously small binding of ${}^{5}_{\Lambda}$He.[1] If one constructs a phenomenological central ΛN potential, which reproduces the experimental $B_\Lambda({}^{3}_{\Lambda}\text{H})$, $B_\Lambda({}^{4}_{\Lambda}\text{H})$, $B_\Lambda({}^{4}_{\Lambda}\text{He})$, $B_\Lambda({}^{4}_{\Lambda}\text{H}^*)$ and $B_\Lambda({}^{4}_{\Lambda}\text{He}^*)$ values as well as the Λp total cross section, this kind of potential would overestimate the $B_\Lambda({}^{5}_{\Lambda}\text{He})$ value[2,3]. According to the recent study by Akaishi *et al.*[4], $\Lambda N - \Sigma N$ coupling has to be explicitly taken into account so as to resolve the ${}^{5}_{\Lambda}$He anomaly.

A $\Lambda\Lambda$ two-body system couples to the $N\Xi$ and $\Sigma\Sigma$ states in free space. Table 1 lists the various coupling potentials of baryon-baryon interaction in the strangeness $S = -1$ (YN) and -2 (YY) sectors. For doubly strange ($S = -2$) hypernuclei, the *core nucleus* + $\Lambda\Lambda$ system couples to the $\Lambda\Sigma$ components as well as the $N\Xi$ and $\Sigma\Sigma$ components. This implies that the $\Lambda N - \Sigma N$ coupling also plays an important role in the binding mechanism of the doubly strange hypernuclei since the $\Lambda N - \Sigma N$ coupling connects the $\Lambda\Lambda$ and $\Lambda\Sigma$ components. Therefore, few-body calculations of the doubly strange hypernuclei should be made in a fully coupled channel formulation with a set of interactions among the octet baryons. The purpose of this study is to describe a study for the s-shell hypernuclei with $S = -2$ in a framework of a fully coupled channel formulation.

Table 1. Baryon-baryon interaction in each isospin (I) and spin channel for strangeness $S=-1$ (YN) and $S=-2$ (YY) sectors.

	YN	
	$^1S_0$	$^3S_1$
$I=\frac{1}{2}$	$\Lambda N-\Sigma N$	$\Lambda N-\Sigma N$
$I=\frac{3}{2}$	ΣN	ΣN

	YY	
	$^1S_0$	$^3S_1$
$I=0$	$\Lambda\Lambda-N\Xi-\Sigma\Sigma$	$N\Xi$
$I=1$	$N\Xi-\Lambda\Sigma$	$N\Xi-\Lambda\Sigma-\Sigma\Sigma$
$I=2$	$\Sigma\Sigma$	

2. Interactions and Method

We assume that the wave function of a system with $S=-2$, comprising $A(=4-6)$ octet baryons, has four isospin-basis components. For example, ${}_{\Lambda\Lambda}^{\;\;6}\mathrm{He}$ has four components as $ppnn\Lambda\Lambda$, $NNNNN\Xi$, $NNNN\Lambda\Sigma$ and $NNNN\Sigma\Sigma$. We abbreviate these components as $\Lambda\Lambda$, $N\Xi$, $\Lambda\Sigma$ and $\Sigma\Sigma$, referring to the last two baryons. The Hamiltonian of the system is hence given by 4×4 components as

$$H=\begin{pmatrix} H_{\Lambda\Lambda} & V_{N\Xi-\Lambda\Lambda} & V_{\Lambda\Sigma-\Lambda\Lambda} & V_{\Sigma\Sigma-\Lambda\Lambda} \\ V_{\Lambda\Lambda-N\Xi} & H_{N\Xi} & V_{\Lambda\Sigma-N\Xi} & V_{\Sigma\Sigma-N\Xi} \\ V_{\Lambda\Lambda-\Lambda\Sigma} & V_{N\Xi-\Lambda\Sigma} & H_{\Lambda\Sigma} & V_{\Sigma\Sigma-\Lambda\Sigma} \\ V_{\Lambda\Lambda-\Sigma\Sigma} & V_{N\Xi-\Sigma\Sigma} & V_{\Lambda\Sigma-\Sigma\Sigma} & H_{\Sigma\Sigma} \end{pmatrix}, \tag{1}$$

where $H_{B_1B_2}$ operates on the B_1B_2 component, and $V_{B_1B_2-B_1'B_2'}$ is the sum of all possible two-body transition potential connecting the B_1B_2 and $B_1'B_2'$ components.

In the present calculations, we use the Minnesota potential[5] for the NN interaction and D2′ for the YN interaction. The Minnesota potential reproduces reasonably well both the binding energies and sizes of few-nucleon systems. The D2′ potential is a modified potential from the original D2 potential[4,6]. For the YY interaction, we use mND$_S$ potential[6], which includes all of the coupling potentials given in Table 1.

The binding energies of various systems are calculated in a complete A-body treatment. Since all of the exotic components ($\Lambda\Lambda$, $N\Xi$, $\Lambda\Sigma$ and $\Sigma\Sigma$) are taken into account, the variational trial function must be flexible enough to search the accurate eigenenergies. The trial function is given by a combination of basis functions:

$$\Psi=\sum_{k=1}^{K}c_k\varphi_k, \qquad \text{with} \qquad \varphi_k=A\{G(\mathbf{x};A_k)\chi_{kJM}\eta_{kIM_I}\}. \tag{2}$$

Here A is an antisymmetrizer acting on the identical particles. For the spin χ_k and the isospin η_k functions, all possible configurations are taken into account. The abbreviation $\mathbf{x}=(\mathbf{x}_1,\cdots,\mathbf{x}_{A-1})$ is a set of relative coordinates. For the spatial part, the basis function is constructed by the correlated Gaussian (CG), $G(\mathbf{x};A_k)$, which is given by

$$G(\mathbf{x};A_k)=\exp\left\{-\frac{1}{2}\sum_{i<j}^{A}\alpha_{kij}(\mathbf{r}_i-\mathbf{r}_j)^2\right\}=\exp\left\{-\frac{1}{2}\sum_{i,j=1}^{A-1}A_{kij}\mathbf{x}_i\cdot\mathbf{x}_j\right\}. \tag{3}$$

The stochastic variational method[7,8] with the above CG basis produces accurate solutions.

3. Results and Discussion

Table 2 lists the $B_{\Lambda\Lambda}$ values for $S = -2$ hypernuclei. Using the mND$_S$ YY potential, we have

$$\Delta B_{\Lambda\Lambda}^{(\text{calc})}({}_{\Lambda\Lambda}^{6}\text{He}) = B_{\Lambda\Lambda}^{(\text{calc})}({}_{\Lambda\Lambda}^{6}\text{He}) - 2B_{\Lambda}^{(\text{calc})}({}_{\Lambda}^{5}\text{He}) = 1.18\text{MeV}, \quad (4)$$

which is in good agreement with the experimental value (Nagara event[10]; $\Delta B_{\Lambda\Lambda}^{(\text{exp})}({}_{\Lambda\Lambda}^{6}\text{He}) = 1.01 \pm 0.20^{+0.18}_{-0.11}$ MeV). Table 3 shows the probabilities of the $N\Xi$, $\Lambda\Sigma$ and $\Sigma\Sigma$ components for the $S = -2$ hypernuclei. The probability of $P_{N\Xi}({}_{\Lambda\Lambda}^{5}\text{H})$ has a surprisingly large value (4.56%).

The normalized energy expectation values of the Hamiltonian (1) for ${}_{\Lambda\Lambda}^{5}\text{H}$ are (given in units of MeV),

$$h = \begin{pmatrix} \frac{\langle H_{\Lambda\Lambda}\rangle}{P_{\Lambda\Lambda}} & \frac{\langle V_{N\Xi-\Lambda\Lambda}\rangle}{\sqrt{P_{\Lambda\Lambda}P_{N\Xi}}} & \frac{\langle V_{\Lambda\Sigma-\Lambda\Lambda}\rangle}{\sqrt{P_{\Lambda\Lambda}P_{\Lambda\Sigma}}} \\ \frac{\langle V_{\Lambda\Lambda-N\Xi}\rangle}{\sqrt{P_{\Lambda\Lambda}P_{N\Xi}}} & \frac{\langle H_{N\Xi}\rangle}{P_{N\Xi}} & \frac{\langle V_{\Lambda\Sigma-N\Xi}\rangle}{\sqrt{P_{N\Xi}P_{\Lambda\Sigma}}} \\ \frac{\langle V_{\Lambda\Lambda-\Lambda\Sigma}\rangle}{\sqrt{P_{\Lambda\Lambda}P_{\Lambda\Sigma}}} & \frac{\langle V_{N\Xi-\Lambda\Sigma}\rangle}{\sqrt{P_{N\Xi}P_{\Lambda\Sigma}}} & \frac{\langle H_{\Lambda\Sigma}\rangle}{P_{\Lambda\Sigma}} \end{pmatrix} = \begin{pmatrix} -9.12 & -1.82 & -14.52 \\ -1.82 & 5.01 & -10.39 \\ -14.52 & -10.39 & 92.41 \end{pmatrix}. \quad (5)$$

Here, we display only the 3×3 components of the Hamiltonian (1), comprising $\Lambda\Lambda$, $N\Xi$, and $\Lambda\Sigma$, since the contributions from the $\Sigma\Sigma$ component are not large. If we solve the eigenvalue problem, $\det(h - \lambda I) = 0$, we obtain the ground state energy, $E = -11.82$ MeV, and the probability, $P_{N\Xi} = 3.99\%$. On the other hand, if we ignore the last row and the last column, the eigenenergy of only the first 2×2 subspace becomes $E = -9.35$ MeV, and the probability, $P_{N\Xi} = 1.57\%$. This clearly means that the couplings between the ($\Lambda\Lambda$, $N\Xi$) and $\Lambda\Sigma$ components play crucial roles to make ${}_{\Lambda\Lambda}^{5}\text{H}$ (and ${}_{\Lambda\Lambda}^{5}\text{He}$) bound. The large coupling potentials, $\langle V_{\Lambda\Lambda-\Lambda\Sigma}\rangle$ and $\langle V_{N\Xi-\Lambda\Sigma}\rangle$, also enhance the $P_{N\Xi}$ probability.

4. Summary

We have performed full-coupled channel *ab initio* calculations for the $A = 4 - 6$ doubly strange s-shell hypernuclei. The mND$_S$ YY potential reproduces the $\Delta B_{\Lambda\Lambda}({}_{\Lambda\Lambda}^{6}\text{He})$ of the Nagara event. We obtained bound-state solutions of ${}_{\Lambda\Lambda}^{4}\text{H}$ and ${}_{\Lambda\Lambda}^{5}\text{H}$ by using the YY potential. We found that the $\Lambda N - \Sigma N$ and $N\Xi - \Lambda\Sigma$ potentials are considerably important to make the ${}_{\Lambda\Lambda}^{5}\text{H}$ bound, and affect the net effect

Table 2. $\Lambda\Lambda$ separation energies, given in units of MeV, of $A = 4 - 6$, $S = -2$ s-shell hypernuclei.

YY	$B_{\Lambda\Lambda}({}_{\Lambda\Lambda}^{4}\text{H})$	$B_{\Lambda\Lambda}({}_{\Lambda\Lambda}^{5}\text{H})$	$B_{\Lambda\Lambda}({}_{\Lambda\Lambda}^{6}\text{He})$
Calc	0.058	3.75	7.54
Exp			$7.25 \pm 0.19^{+0.18}_{-0.11}$

Table 3. Probabilities, given in percentage, of $N\Xi$, $\Lambda\Sigma$ and $\Sigma\Sigma$ components of $A = 4-6$, $S = -2$ s-shell hypernuclei.

	${}_{\Lambda\Lambda}^{4}\mathrm{H}$			${}_{\Lambda\Lambda}^{5}\mathrm{H}$			${}_{\Lambda\Lambda}^{6}\mathrm{He}$		
YY	$P_{N\Xi}$	$P_{\Lambda\Sigma}$	$P_{\Sigma\Sigma}$	$P_{N\Xi}$	$P_{\Lambda\Sigma}$	$P_{\Sigma\Sigma}$	$P_{N\Xi}$	$P_{\Lambda\Sigma}$	$P_{\Sigma\Sigma}$
mND$_S$	0.06	0.25	0.00	4.56	2.49	0.06	0.28	1.17	0.05

of the $\Lambda\Lambda - N\Xi$ coupling potential. The $\Lambda N - \Sigma N$ coupling, which is the key issue to resolve the ${}_{\Lambda}^{5}\mathrm{He}$ anomaly of $S = -1$ s-shell hypernuclei, plays a crucial role even for the $S = -2$ s-shell hypernuclei, particularly for the $A = 5$ systems such as ${}_{\Lambda\Lambda}^{5}\mathrm{H}$.

Acknowledgments

The author is thankful to Prof. Shinmura, Prof. Akaishi and Dr. Khin Swe Myint for the potential parameters and valuable discussions and comments.

References

1. B. F. Gibson and E. V. Hungerford III, *Phys. Rep.* **257**, 349-388 (1995).
2. R. H. Dalitz, R. C. Herndon and Y. C. Tang, *Nucl. Phys.* **B47**, 109-137 (1972).
3. H. Nemura, Y. Suzuki, Y. Fujiwara and C. Nakamoto, *Prog. Theor. Phys.* **103**, 929-958 (2000).
4. Y. Akaishi, *et al., Phys. Rev. Lett.* **84**, 3539-3541 (2000).
5. D. R. Thompson, M. Lemere and Y. C. Tang, *Nucl. Phys. A* **286**, 53-66 (1977).
6. H. Nemura, S. Shinmura, Y. Akaishi and Khin Swe Myint, *Phys. Rev. Lett.* **94**, 202502 (2005).
7. V. I. Kukulin and V. M. Krasnopol'sky, *J. Phys. G*:Nucl. Part. Phys. **3**, 795-811 (1977).
8. K. Varga and Y. Suzuki, *Phys. Rev. C* **52**, 2885-2905 (1995).
9. H. Nemura, Y. Akaishi and Y. Suzuki, *Phys. Rev. Lett.* **89**, 142504 (2002).
10. H. Takahashi, *et al., Phys. Rev. Lett.* **87**, 212502 (2001).

Isoscalar and Isovector Parts of the Σ Single Particle Potential from SCDW Analysis of (π^-, K^+) Inclusive Spectra*

M. Kohno

Physics Division, Kyushu Dental College, Kitakyushu 803-8580, Japan

Analyzing (π^-, K^+) Σ^- formation inclusive spectra on medium-to heavy nuclear targets measured at KEK by using the semiclassical distorted wave model, we obtain the information on the strength of the isovector part of the Σ single-particle potential in nuclei. Together with the repulsive strength of 30∼50 MeV of the isoscalar part obtained from the data on light nuclei, ^{12}C and ^{28}Si, the strength of the isovector part further elucidates the isospin structure of the underlying two-body Σ-N interaction.

1. Introduction

Theoretical studies for the two-body Σ-N force have much ambiguities. In a standard OBEP picture, most models[1,2] of the Nijmegen group predict an attractive Σ s.p. potential in the lowest order G-matrix calculations in symmetric nuclear matter. On the other hand, the recent SU_6 quark model[3] developed by Kyoto-Niigata group predicts that the Σ s.p. potential is repulsive of the order of 10∼ 30 MeV. The repulsion in the $T = 1/2$ $^3S_1$ channel, which originates from quark Pauli effects, overwhelms an attractive contribution in the $T = 1/2$ $^3S_1$ channel. The attractive result was regarded as a nice feature before, because early experimental data[4] was interpreted to show an attractive Σ potential. Recent KEK experiments[5,6] of (π^-, K^+) inclusive spectra on various target nuclei with better precision have indicated, however, that the Σ s.p. potential is strongly repulsive.

Since the determination of the Σ-N interaction on the basis of experimental Σ formation processes is important for our understanding of the physics of octet baryons, we have developed a semi-classical distorted wave (SCDW) method[7,8] to analyze these inclusive spectra. This model accounts well for the experimental spectra without any adjustable parameters nor normalization factors. We obtained[7,8] that the repulsive strength of the Σ s.p. potential in ^{28}Si and ^{12}C is 30 ∼ 50 MeV.

In this report, we extend the SCDW analysis to the data[6] on medium-to heavy nuclear targets: ^{58}Ni, ^{115}In and ^{209}Bi. Since the neutron number is larger than the proton number in these nuclei, we obtain the information on the charge dependence of the Σ s.p. potential, which is connected with the isospin dependence of the

*This work is supported by the Grants-in-Aid for Scientific Research (C) from the Japan Society for the Promotion of Science (Grant No. 17540263).

underlying Σ-N interaction.

2. Semiclassical Distorted Wave Model for Inclusive Spectrum

The starting formula for the double differential cross section in a standard distorted wave model description of the (π^-, K^+) inclusive reaction is expressed as

$$\frac{d^2\sigma}{dWd\Omega} = \frac{\omega_i\omega_f}{(2\pi)^2}\frac{p_f}{p_i}\int\int d\boldsymbol{r}d\boldsymbol{r}'\sum_{p,h}\frac{1}{4\omega_i\omega_f}\chi_f^{(-)*}(\boldsymbol{r})v_{f,p,i,h}\chi_i^{(+)}(\boldsymbol{r})\chi_f^{(-)}(\boldsymbol{r}')v^*_{f,p,i,h}$$
$$\times\chi_i^{(+)*}(\boldsymbol{r}')\phi_p^*(\boldsymbol{r})\phi_h(\boldsymbol{r})\phi_p(\boldsymbol{r}')\phi_h^*(\boldsymbol{r}')\delta(W-\epsilon_p+\epsilon_h)\theta(\epsilon_F-\epsilon_h), \tag{1}$$

where $\chi_i^{(+)}$ and $\chi_f^{(-)}$ represent the incident pion and final kaon wave functions with energies ω_i and ω_f, respectively, and $W = \omega_i - \omega_f$ is the energy transfer. The occupied proton single-particle state is denoted by h, and the unobserved outgoing Σ^- state by p. The elementary amplitude of the process $\pi^- + p \to K^+ + \Sigma^-$ is denoted by $v_{f,p,i,h}$, which depends on the energies and momenta of the particles in the reaction. In order to treat such dependence, it is necessary to introduce momentum space integration. It is desirable, in practice, to develop a tractable and trustful approximation method, avoiding the factorization approximation used frequently.

The semiclassical approximation employs the following idea for the propagation of the wave function. Denoting the midpoint and the relative coordinates of $\boldsymbol{r}$ and $\boldsymbol{r}'$ by $\boldsymbol{R} = \frac{\boldsymbol{r}+\boldsymbol{r}'}{2}$ and $\boldsymbol{s} = \boldsymbol{r}' - \boldsymbol{r}$, respectively, we assume that the propagation of the distorted waves, χ_i and χ_f, from $\boldsymbol{R}$ to $\boldsymbol{r}$ or $\boldsymbol{r}'$ is described by a plane wave with the local classical momentum $\boldsymbol{k}(\boldsymbol{R})$ at the position $\boldsymbol{R}$:

$$\chi\left(\boldsymbol{R}\pm\frac{1}{2}\boldsymbol{s}\right) \simeq \mathrm{e}^{\pm i\frac{1}{2}\boldsymbol{s}\cdot\boldsymbol{k}(\boldsymbol{R})}\chi(\boldsymbol{R}). \tag{2}$$

The local momentum $\boldsymbol{k}(\boldsymbol{R})$ is defined as follows. The direction is specified by the quantum mechanical momentum density $\boldsymbol{k}_q(\boldsymbol{R})$ calculated by

$$\boldsymbol{k}_q(\boldsymbol{R}) = \frac{\Re\{\chi^*(\boldsymbol{R})(-i)\nabla\chi(\boldsymbol{R})\}}{|\chi(\boldsymbol{R})|^2}, \tag{3}$$

where $\Re$ represents taking the real part, and the magnitude is determined by the energy-momentum relation $\frac{\hbar^2}{2\mu}k^2(\boldsymbol{R}) + U_R(\boldsymbol{R}) = E$ at $\boldsymbol{R}$. Here, $U_R(\boldsymbol{R})$ is the real part of the optical potential for the distorted wave function χ with energy E.

We describe unobserved hyperons by a local optical potential of the standard Woods-Saxon form. The standard way to treat the completeness of the final states described by a complex optical potential is to use the Green function method. At this stage, however, we adopt a simplified prescription, employing a real potential and we convolute the result of the calculated spectrum with a Lorentz-type distribution function with the typical width, simulating the effects of inelastic channels. We set the half width for the smearing of the calculated spectra to be the half of the imaginary potential of 20 MeV which is suggested in nuclear matter calculation[9] with

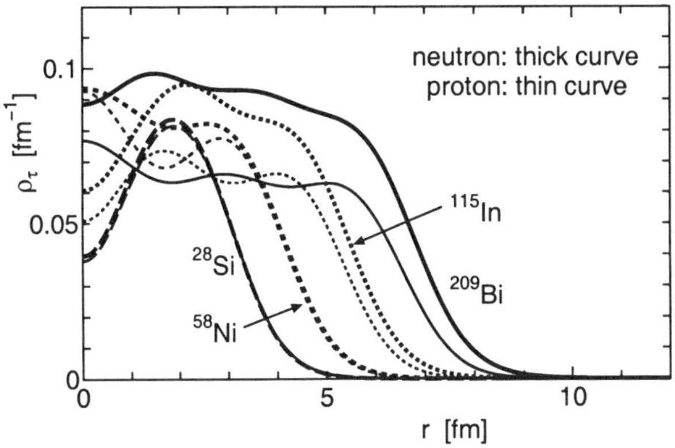

Fig. 1. Point proton and neutron density distributions of ^{28}Si, ^{58}Ni, ^{115}In and ^{209}Bi obtained by DDHF wave functions[10].

the SU_6 quark model interaction, regarding that the hyperon formation processes take place at the surface region. In this prescription, we can use the same SCDW approximation for the Σ^- wave function.

Using these approximations and the Wigner transformation of the nuclear density matrix which is denoted by $\Phi_h^\ell(R, K)$ in the expression below, the double differential cross section in the SCDW model becomes

$$\begin{aligned}\frac{d^2\sigma}{dWd\Omega} = \frac{\omega_i\omega_f}{(2\pi)^2}\frac{p_f}{p_i}\int d\boldsymbol{R}\int d\boldsymbol{K}\,\frac{1}{(2\pi)^3}\int d\boldsymbol{p}\,\frac{1}{4\omega_i\omega_f}|\chi_f^{(-)}(\boldsymbol{R})|^2|\chi_i^{(+)}(\boldsymbol{R})|^2|\phi_p(\boldsymbol{R})|^2 \\ \times(2\pi)^3\sum_h\sum_\ell P_\ell(\cos\hat{\boldsymbol{K}\boldsymbol{R}})\,\Phi_h^\ell(R,K) \\ \times|v_{f,p,i,h}(\boldsymbol{K},\boldsymbol{k}_i)|^2\delta(\boldsymbol{K}+\boldsymbol{k}_i(\boldsymbol{R})-\boldsymbol{k}_f(\boldsymbol{R})-\boldsymbol{k}_p(\boldsymbol{R}))\,\delta(W-\epsilon_p+\epsilon_h). \quad (4)\end{aligned}$$

This final expression admits of the simple interpretation that the reaction in which $\pi^- + p$ yields $K^+ + \Sigma^-$ takes place at the position $\boldsymbol{R}$ and satisfies conservation of the local semiclassical momentum: $\boldsymbol{K} + \boldsymbol{k}_i(\boldsymbol{R}) - \boldsymbol{k}_f(\boldsymbol{R}) + \boldsymbol{k}_p(\boldsymbol{R})$.

3. Description of the Target Nucleus, Σ^-, Pions and Kaons

Single-particle wave functions and energies ϵ_h for ^{28}Si, ^{58}Ni, ^{115}In and ^{209}Bi are given by the density-dependent Hartree-Fock description of Campi-Sprung[10]. The profile of point proton and neutron density distributions of each nucleus is shown in Fig. 1.

The Σ^- hyperon is described by an energy-independent local potential of the Woods-Saxon form. We use the standard geometry: the radius $r_0 = 1.25\times(A-1)^{1/3}$ fm and the surface thickness $a = 0.65$ fm. The Coulomb potential regularized in a nucleus is incorporated.

The incident pions and detected kaons are described by the standard Klein-

Gordon equation with some optical potential model in the form of

$$V(r) = -\frac{k^2}{2E} b_0 \rho(r), \tag{5}$$

where $\rho(r)$ is the one-body nuclear density distribution and the parameters b_0 used in the present calculations are the same as those in the reference[8].

4. Results

In the present calculation, we assume an isotropic angular dependence for the $\pi^- + p \to K^+ + \Sigma^-$ elementary process. The angular dependence in the forward angle region in the laboratory frame is well described by this prescription. The energy dependence of the total cross section is taken from the parameterization by Tsushima et al.[11] Their overall strength was normalized by a factor of 0.82 to match the experimental data taken at KEK[5].

Results for ^{28}Si, ^{58}Ni, ^{115}In and ^{209}Bi are shown in Fig. 2, respectively. It is to be stressed that no renormalization factor is multiplied. Noticing that the experimental data does not show a similarity in shape[6] beyond the energy of a bout 100 MeV in heavier targets and thus still ambiguous in those energy region, the calculated curve is seen to correspond well to the experimental data by choosing a proper potential strength of 30 $\sim$ 70 MeV.

It is instructive to deduce the dependence of the potential strength on the asymmetry parameter $\alpha = (N-Z)/(N+Z)$, even if the strength depends on the geometry assumed and such dependence is not well determined even for the nucleon case. Supposing $U^0_{\Sigma^-} \sim 70$ MeV for ^{209}Bi ($\alpha = 0.21$) and $U^0_{\Sigma^-} \sim 60$ MeV for ^{115}In ($\alpha = 0.16$), we obtain the α-dependence as 200α. If we use $U^0_{\Sigma^-} \sim 50$ MeV for ^{58}Ni ($\alpha = 0.05$) instead of ^{115}In, we obtain 125α. This estimation implies that the strength of the isovector part of the Σ s.p. potential is repulsive and sizable. Note that the corresponding value for the nucleon is $(16 \sim 28)\alpha$.

5. Summary

We have analyzed (π^-, K^+) Σ^- formation inclusive spectra on ^{28}Si, ^{58}Ni, ^{115}In and ^{209}Bi taken at KEK[5,6], using a semiclassical distorted wave model with the Wigner transformation of the nuclear density matrix. This method was shown in previous publications[7,8] to provide a quantitatively reliable model for describing these inclusive spectra without any renormalization factor. Calculations[7,8] for light nuclei, ^{28}Si and ^{12}C, indicated that the Σ^- s.p. potential strength in a standard Woods-Saxon geometry is repulsive of the order of 30-50 MeV.

The s.p. potential needed to reproduce the data on heavier nuclei which having a neutron excess offers important information about the isovector part of the Σ s.p. potential. Since the neutron excess means that the $T = \frac{3}{2}$ contribution becomes larger, the magnitude of the asymmetry dependence, which is found to be sizable, reflects the isospin dependence of the underlying two-body Σ-N interaction.

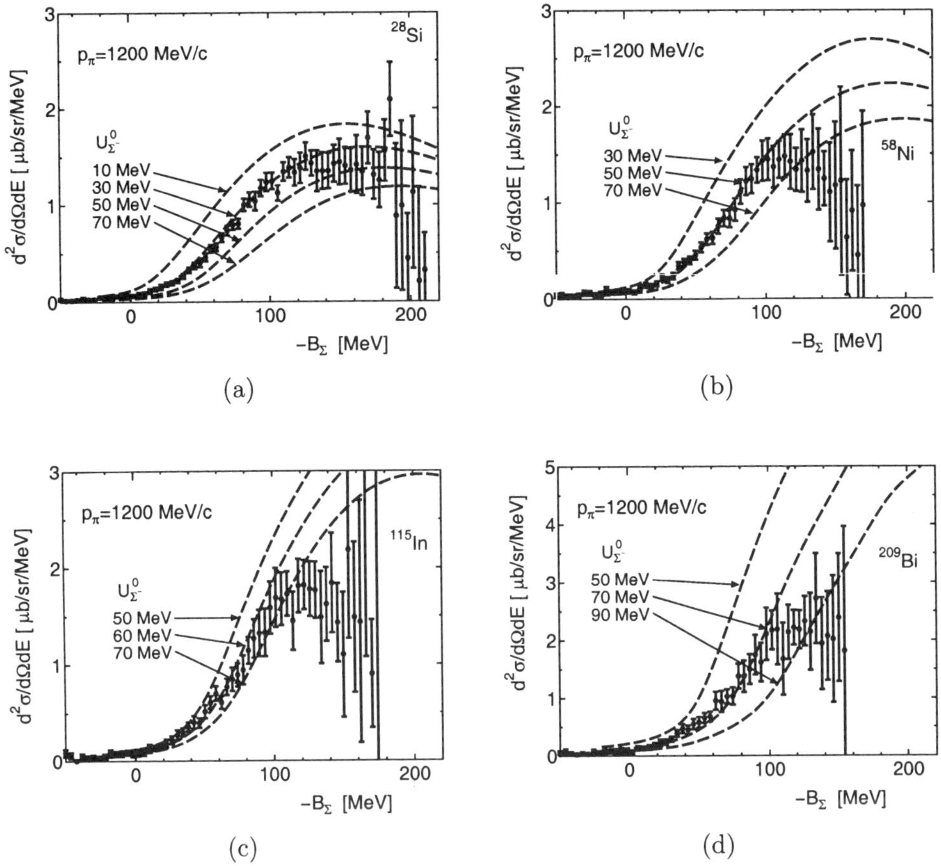

Fig. 2. (π^-, K^+) $(p_\pi = 1200$ MeV/c) Σ^- formation inclusive spectra: (a) ^{28}Si, (b) ^{58}Ni, (c) ^{115}In and (d) ^{209}Bi. Calculated results are shown by dash curves, compared with experimental data from KEK[5,6]. The repulsive strength of the Σ s.p. potential in a Woods-Saxon form used for the calculated curve is shown in the figure.

References

1. M. M. Nagels, T. A. Rijken and J. J. de Swart, Phys. Rev. D **12** (1975), 744; D **15** (1977), 2547; D **20** (1979), 1633.
2. P. M. M. Maessen, T. A. Rijken and J. J. de Swart, Phys. Rev. C **40** (1989), 2226.
3. Y. Fujiwara, C. Nakamoto and Y. Suzuki, Phys. Rev. C **54** (1996), 2180.
4. C. B. Dover, D. J. Millener and A. Gal, Phys. Rep. **184** (1989), 1.
5. H. Noumi *et al.*, Phys. Rev. Lett. **89** (2002), 072301; **90** (2003), 049902(E).
6. P.K. Saha *et al.*, Phys. Rev. **C70** (2004), 044613.
7. M. Kohno, Y. Fujiwara, K. Ogata, Y. Watanabe and M. Kawai, Prog. Thoer. Phys. **112** (2004), 895.
8. M. Kohno, Y. Fujiwara, K. Ogata, Y. Watanabe and M. Kawai, in preparation.
9. M. Kohno *et al.*, Nucl. Phys. A **674** (2000), 229.
10. X. Campi and D. W. Sprung, Nucl. Phys. A **194** (1972), 401.
11. K. Tsushima, S. W. Huang and A. Faessler, Phys. Lett. B **337** (1994), 245.

Signatures of the Three Body Process in the Weak Decay of Λ Hypernuclei

H. Bhang
for KEK-PS E462/E508 collaboration*
School of Physics, Seoul National University
Seoul 151-747, Korea
E-mail: bhang@phya.snu.ac.kr

The emitted pair nucleons in the nonmesonic weak decay (NMWD) of $^5_\Lambda$He and $^{12}_\Lambda$C have been measured for the first time in coincidence method in the KEK-PS E462 and E508 experiments. The angular correlations of np pair from the decay showed the predominant back-to-back kinematic events for both hypernuclei which is the signature of the two-body final state. This is the first experimental verification of the long-believed mechanism of nonmesonic weak decay process. However, those of nn pairs showed significant non-back-to-back kinematic events, in addition to those of back-to-back, which might be considered as a signature of two nucleon induced nonmesonic weak decay, namely the three-body interaction process in the weak decay of Λ hypernuclei.

1. Introduction

Although the importance of NMWD was recognized as early as '60s and many experimental and theoretical efforts have been devoted to the NMWD of Λ hypernuclei, its mechanism is still not clear. Among many important issues of NMWD, especially the Γ_n/Γ_p ratio has been a long standing concern.[1]

Recently there has been a significant developement finding a mistake in the inclusion of K exchange amplitudes whose correction significantly increased the values of Γ_n/Γ_p. Since then the direct quark interaction model calculation produced the ratio for $^5_\Lambda$He up to 0.70.[2] The extensive meson exchange model calculation also produced the increased value for the ratio reaching to 0.53 for $^{12}_\Lambda$C [3] which now agrees well with recent experimental Γ_n/Γ_p ratios.[4,6–8]

Experimental progresses on the NMWD of Λ hypernuclei have been made in a series of experiments at KEK. The accurate determination of proton (KEK-PS E307) and neutron spectra (KEK-PS E369) of NMWD of $^{12}_\Lambda$C made the direct comparison of the spectra possible and showed the ratio significantly smaller than unity, namely (0.45-0.51)±0.15 [4] obtained from the ratio of the normalized neutron to pro-

*Collaborators are S. Ajimura, K. Aoki, A. Banu, H. Bhang, T. Fukuda, O. Hashimoto, J.I. Hwang, S. Kameoka, B.H. Kang, E.H. Kim, J.H. Kim, M.J. Kim, T. Maruta, Y. Miura, Y. Miyake, T. Nagae, M. NAkamura, S.N. Nakamura, H. Noumi, S. Okada, Y. Okayasu, H. Outa, H. Park, P.K. Saha, Y. Sato, M. Sekimoto, T. Takahashi, H. Tamura, K. Tanida, A. Toyoda, K. Tsukada, T. Watanabe and H.J. Yim.

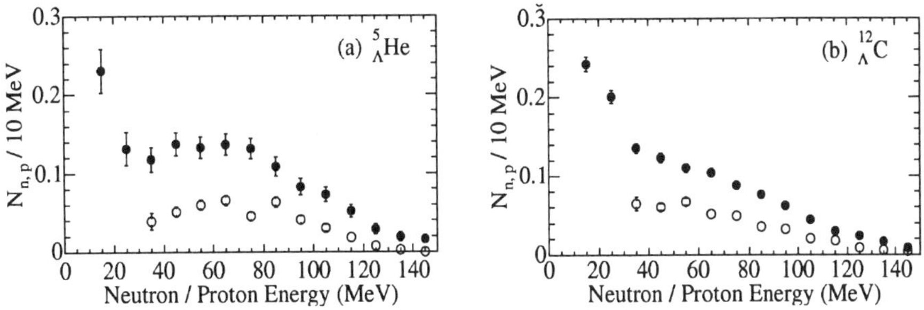

Fig. 1. Proton (open circle) and neutron (dark circle) spectra of the NMWD of $^{5}_{\Lambda}$He (a) and $^{12}_{\Lambda}$C (b) are shown.

ton yields per NMWD, N_n/N_p. It is pointed out that the major part of FSI effects was largely cancelled in the ratio N_n/N_p. However, it still has some uncertainty due to the residual FSI effect and a possibly large one from two nucleon induced (2N) NMWD process, ΛNN→nNN. Though it has not been experimentally identified yet, the strength of the 2N NMWD channel was predicted to be significant.[5] Therefore, it is desirable to separate out its contribution not only for the accurate determination of Γ_n/Γ_p, but also for Γ_n and Γ_p. In this regards, KEK-PS E462 and E508 have been performed for exclusive measurement of the decay of s-shell $^{5}_{\Lambda}$He and p-shell $^{12}_{\Lambda}$C.

2. Experimental results

The experiments were carried out at K6 beam line of KEK 12-GeV PS with 1.05 GeV/c π^+ beam via (π^+,K^+) reactions. Details of the experimental setup has been reported previously.[6]

Figure 1 shows the obtained normalized neutron and proton energy spectra, N_n and N_p, per NMWD in 10 MeV bin for $^{5}_{\Lambda}$He and $^{12}_{\Lambda}$C, respectively.[6] Filled circles are for neutron and open circles for proton. The integrated number ratio, N_n/N_p, for $^{5}_{\Lambda}$He is obtained over the energy region, $60 \leq E \leq 110$ MeV, limiting to 110 MeV in order to avoid the contamination due to the high energy neutron from the $^{5}_{\Lambda}\mathrm{He} \rightarrow \alpha + n$ decay channel, while that of $^{12}_{\Lambda}$C is over the whole energy range above 60 MeV. The obtained number ratios for $^{5}_{\Lambda}$He and $^{12}_{\Lambda}$C are $2.17 \pm 0.15 \pm 0.16$ and $2.00 \pm 0.09 \pm 0.14$, respectively.[6] With these, we get Γ_n/Γ_p ratio 0.59($^{5}_{\Lambda}$He) and 0.50($^{12}_{\Lambda}$C) when FSI and 2N NMWD contribution are not considered in and $0.62 \pm 0.08 \pm 0.09$ ($^{5}_{\Lambda}$He) and $0.57 \pm 0.06 \pm 0.10$ ($^{12}_{\Lambda}$C) when FSI corrections are made applying the same method of E369[4] but assuming no 2N NMWD contribution. These show that the Γ_n/Γ_p($^{5}_{\Lambda}$He) also is significantly smaller than unity and the Γ_n/Γ_p($^{12}_{\Lambda}$C) confirms the proton channel dominance in $^{12}_{\Lambda}$C.[4]

In order to make sure of explicit identification of Λp→np and Λn→nn channels, we measured both nucleons of np and nn pairs in coincidence coming from NMWD of Λ hypernuclei. Since the two nucleons from NMWD are emitted out with back-to-back kinematics, we discriminate the events strongly influenced by FSI and those from 2N processes, 3-body final states, by requiring the back-to-back kinematics condition.

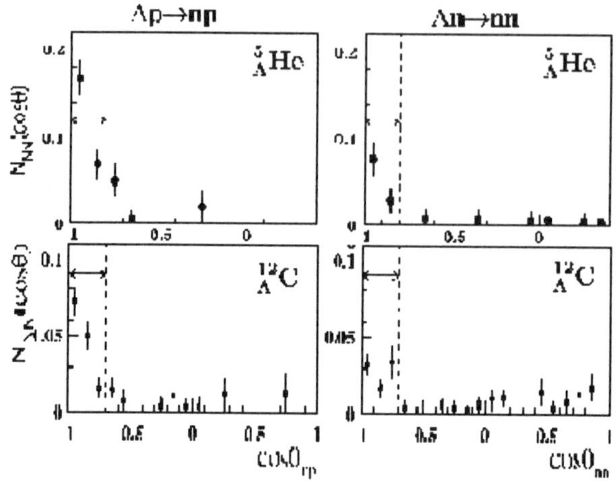

Fig. 2. The pair numbers, N_{NN}, are shown in the opening angle of two nucleons, np in (a) and nn in (b), for ${}^5_\Lambda$He in the upper ones and for ${}^{12}_\Lambda$C in the lower ones.

Figure 2 shows the normalized pair yields of ${}^5_\Lambda$He and ${}^{12}_\Lambda$C per NMWD, $N_{NN}(cos\theta)$, for full solid angle and unit efficiency.[7,8] Specifically, $N_{NN}(cos\theta) = Y_{NN}(cos\theta)/Y_{nm}\epsilon_{NN}(cos\theta)$, where Y_{NN}, Y_{nm} and ϵ_{NN} are the NN coincidence event yield, the total number of NMWD, and a combined efficiency factor for both acceptance and detection efficiency, respectively. The combined efficiency, ϵ_{NN}, were obtained with energy dependent event-by-event Geant simulation.

Back-to-back peaking is clearly observed and dominant in np pair angular correlations in the decay of both ${}^5_\Lambda$He and ${}^{12}_\Lambda$C hypernuclei. In nn pair correlation, it is still the dominant kinematics in ${}^5_\Lambda$He while it is degraded in ${}^{12}_\Lambda$C so that it is no more the dominant one in nn pairs of ${}^{12}_\Lambda$C. Total nn pair number in the non-back-to-back region in ${}^{12}_\Lambda$C is almost comparable to that of back-to-back region. The enhanced nn pair in non-back-to-back region over the np pair show the significant contribution of non-back-to-back kinematics events, namely the 2N NMWD.

3. Discussion

Back-to-back kinematics which we observed in the two nucleon correlation is the signature of two-body final states and can be considered an experimental verification of the 1N-induced NMWD processes, Λp→np and Λn→nn, which have long been considered as the NMWD modes of Λ hypernuclei but without explicit experimental identification. In other words, the angular correlation of non-back-to-back kinematics indicates the contributions other than 1N NMWD, namely, 2N NMWD and FSI effects. Comparision of np pair correlations of ${}^5_\Lambda$He and ${}^{12}_\Lambda$C shows the effect

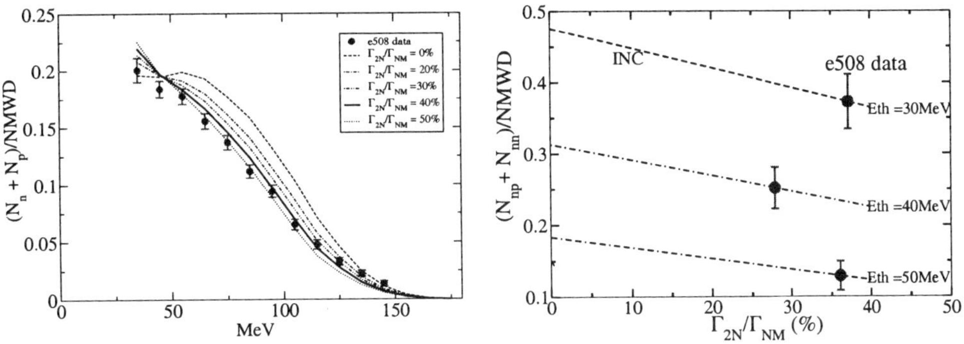

Fig. 3. In the left figure, the sum spectrum of the neutron and proton yields of the decay of ${}^{12}_{\Lambda}$C, $N_n + N_p$, is compared to the INC predictions obtained with various 2N contributions, from 0 to 50 percent of Γ_{1N}. In the right figure, the total pair number sums of ${}^{12}_{\Lambda}$C integrated over all the angular correlation are compared to those of INC calculation as a function of Γ_{2N}.

of FSI and that of np and nn pair correlatins reveals the 2N NMWD contribution.

Γ_n/Γ_p *from pair number ratio:* We consider the integrated pair numbers of np and nn, N_{np} and N_{nn}, in the back-to-back region (cos$\theta \leq$ -0.8 for ${}^{5}_{\Lambda}$He and cos$\theta \leq$ -0.7 for ${}^{12}_{\Lambda}$C) so that the possible FSI effect and 2N NMWD contribution, more or less uniformly distributed in non-back-to-back angle region, are largely rejected. The corresponding uniform-like background components in the back-to-back region also are subtracted from the integrated numbers. Then the ratio N_{nn}/N_{np} can be well approximated to Γ_n/Γ_p. We obtain the Γ_n/Γ_p ratios $0.45 \pm 0.11 \pm 0.03$ and $0.53 \pm 0.12 \pm 0.04$ (preliminary) for ${}^{5}_{\Lambda}$He and ${}^{12}_{\Lambda}$C, respectively. Many uncertainties, like the nonmesonic decay branching ratio, major part of FSI factors and the detection efficiencies, are cancelled out by taking the ratio so that the uncertainty becomes mainly statistical one. Here both Γ_n/Γ_p ratios are somewhat smaller than those derived from the singles nucleon number ratios, N_n/N_p, in the previous section which were derived assuming no 2N NMWD contribution. We pointed out that Γ_n/Γ_p value from N_n/N_p would show a bigger value than the real one if 2N contribution is not properly taken into account in the singles spectrum. The difference manifested here can be considered as another indication of 2N NMWD.

Singles yields sum, $N_n + N_p$: As we explained in the derivation of Γ_n/Γ_p from the experimental N_n/N_p ratio, we pointed out the quenching of singles yields in both proton and neutron spectra. The measured singles total yields, $N_n + N_p$, are compared with the INC yield spectra in the Figure 3.[9] The INC spectra were produced with Γ_{2N} of 20 to 50 percent of Γ_{nm}. Here we assumed the uniform phase space sharing kinematics for the 3 particle produced in the 2N NMWD. This would be somewhat different kinematics from that of the previous calculation [5] in which a low energy nucleon is generated from the weak Λ vertex and two energetic nucleons from the absorption of the virtual pion.[5] Significant quenching of singles yields measured with 30-40 MeV threshold detection energy could not be reproduced by 1N-only INC calculation. In order to reproduce the quenching, a comparable strength of Γ_{2N} to that of Γ_{1N} was required. Please note that since the spectrum is the sum of the proton and neutron, the yield is independent of Γ_n/Γ_p ratio. Therefore the conclusion can be made irrespective of the ratio.

Pair number sum, $N_{nn}+N_{np}$: For the decay of ${}^{12}_{\Lambda}$C, Figure 3 compares the total pair number sum of the decay of ${}^{12}_{\Lambda}$Cintegrated over all the angular correlation to those of INC calculation as a function of Γ_{2N}/Γ_{nm}. Though the exact Γ_{2N} strength would depend on a particular kinematic model of 2N NMWD, here we again find that the pair yield can be explained only with a significant Γ_{2N} strength confirming the observation made with the singles sum yield. Since the sum of both pair nn and np is compared, the conclusion is independent of the particular choice of a combination of initial nucleons involved in the decay process.

In summary, we have measured for the first time the pair nucleons in the NMWD of ${}^{5}_{\Lambda}$He and ${}^{12}_{\Lambda}$C in coincidence and determined the Γ_n/Γ_p of the NMWD for ${}^{5}_{\Lambda}$He and ${}^{12}_{\Lambda}$C accurately from the pair number ratio, N_{nn}/N_{np}, in the back-to-back kinematic region where the effects of FSI and 2N NMWD contribution are minimized. The Γ_n/Γ_p result for ${}^{12}_{\Lambda}$C is preliminary yet. The two nucleon angular correlations of np pair are dominated by the back-to-back kinematic events for both hypernuclei while those of nn pairs show significant non-back-to-back events. This is considered a signature of 2N NMWD along with those observed in the comparison of the singles sum, $N_n + N_p$, and pair number sum, $N_{nn} + N_{np}$, to those of INC calculation.

We are grateful to KEK-PS staff for the support of our experiments and for stable operation of KEK-PS. Author H.B. acknowledges partial supports from KOSEF(R01-2005-000-10050-0) and KRF(2003-070-C00015).

References

1. W. M. Alberico, G. Garbarino, Phys. Rep. 369 (2002) 1 and references therein.
2. K. Sasaki, T. Inoue, and M. Oka, Nuc. Phys. A669 (2000) 331.
3. D. Jido, E. Oset and J. E. Palomar, Nucl. Phys. A694 (2001) 525.
4. J. H. Kim *et al.*, Phys. Rev. C68 (2003) 065201.
5. A. Ramos, E. Oset and L. L. Salcedo, Phys. Rev. C50 (1994) 2314.
6. S. Okada *et al.*, Phys. Lett B597 (2004) 249.
7. B. H. Kang, Ph. D. thesis, Seoul National University, (2004).
8. M. J. Kim *et al.*, Proceedings of the International Conference of DAϕNE 2004: Physics at Meson Factories, Frascati Physics Series XXXVI (2004) 237.
9. M. J. Kim *et al.*, Jour. Kor. Phys. Soc. 46 (2005) 805.

The Physics of the Hypertriton

I. R. Afnan

School of Chemistry, Physics and Earth Sciences
Flinders University
GPO Box 2100, Adelaide 5001, Australia
Iraj.Afnan@Flinders.edu.au

We examine the relative importance of the $N\Lambda - N\Sigma$ coupling, and the coupling due to the tensor force in the $N\Lambda$ channel, in the binding of the hypertriton. We find that the coupling between the $N\Lambda - N\Sigma$ coupling to be more important as it is dominated by one pion exchange, while the tensor force in the $N\Lambda$ channel has no pion exchange and the K and K^* exchange almost cancel. Results are presented for the latest one boson exchange potential in ${}^3_\Lambda$H. To understand these results we examine the relative importance of the two kinds of coupling in a simple model for ${}^3_\Lambda$H.

1. Introduction

The hypertriton (${}^3_\Lambda$H) as the lightest strangeness $S = -1$ nucleus can play as important a role in constraining the $NY, (Y = \Lambda, \Sigma)$ interaction, as the deuteron plays in determining the nucleon-nucleon interaction. Considering the small and limited number of experimental NY data available, the hypertriton may be used to further constrain the NY interaction. This is especially the case within the framework of One Boson Exchange (OBE) potentials where the meson baryon coupling constants satisfy the requirement of $SU(3)$ flavor symmetry. Thus the constraint imposed by ${}^3_\Lambda$H could be used to put limitation on the baryon-baryon interaction in $S = -1, -2, \cdots$. This in turn allows us to predict the properties on heavier hypernuclei and provide the input required for astrophysical model calculations.

The ${}^3_\Lambda$H is a weakly bound state with a separation energy E_{SP} given by

$$E = \epsilon_d - E_{SP} = -2.2246 - (0.130 \pm 0.050) , \tag{1}$$

where ϵ_d is the energy of the deuteron. This weak separation energy allows us to visualize the hypertriton as a weakly bound $d - \Lambda$ system with a large rms radius. In this case, the properties of the ${}^3_\Lambda$H are very sensitive to the long range part of the NY interaction. Within the framework of OBE potentials, this long range interaction should be due to pion exchange. As a result, we expect the coupling between the $N\Lambda - N\Sigma$ to play as important a role in the binding of ${}^3_\Lambda$H as the tensor force in the properties of the deuteron. Furthermore, this $N\Lambda - N\Sigma$ coupling is more important than the tensor force in $N\Lambda$ channel. This is clearly illustrated in Figure 1 for the Nijmegen model F[1] where we observe that the $N\Lambda - N\Sigma$ coupling, present in both the $^1S_0$ and $^3S_1$ channels, is larger than the $^3S_1$-$^3D_1$ coupling in the $N\Lambda$ channel.

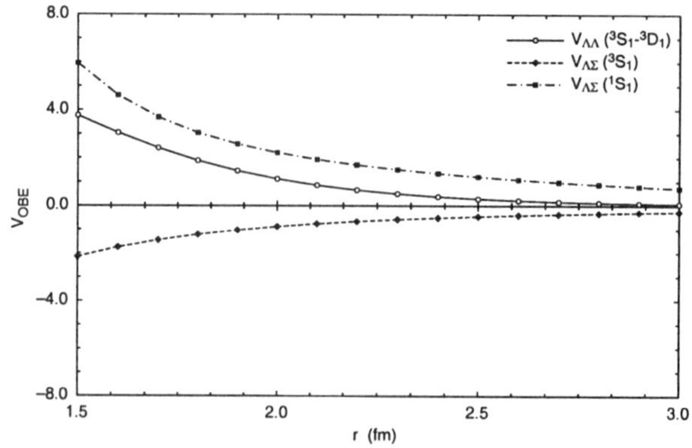

Fig. 1. A comparison of the long range $N\Lambda - N\Sigma$ coupling potential and the tensor force in $N\Lambda$ channel for the Nijmegen model F.

2. $SU(3)_f$ One Boson Exchange Potential and the $^3_\Lambda$H

In $SU(3)_f$ OBE Baryon-Baryon (BB) potentials the masses of the baryon and meson are taken to be the experimental values, while the couplings are related by $SU(3)$ flavor. In the Nijmegen potentials it is common to include a nonets of pseudoscalar, vector, and scalar meson exchanges. As an illustration of how the data determines the coupling constants, let us consider the pseudoscalar meson exchange. In this case we have four parameters for pseudoscalar meson exchange potential that comprise the singlet and octet couplings, the F/D ratio and the mixing angle, *i.e.*

$$g_1 \ , \quad g_8 \ , \quad \alpha = \frac{F}{F+D} \ , \quad \text{and} \quad \theta \ . \tag{2}$$

The NN data can fix three ($g_{\pi NN}$, $g_{\eta NN}$, and $g_{\eta' NN}$) coupling constants, and one needs to extend the data to the $S = -1$ sector to fit all four pseudoscalar parameters. This same arguments applies to the other nonets of mesons exchanged. Thus the $S = -1$ data points play a central role in fixing the BB potential. The only additional parameter one needs in order to determine the potential in $S = -2, -3, \cdots$, is the short range interaction or hard core radius. This could be determined by one data point which could be the ground state energy of the lightest hypernucleus with $S = -2, -3, \cdots$ [2]

To illustrate how one may use the $^3_\Lambda$H as a constraint on the BB potential, let us consider first the results of Nogga *et al.*[3] based on the solution of the Faddeev equations for the hypertriton as presented in Table 1 for several Nijmegen potentials. Here, we first observe that the results are not sensitive to the choice of NN interaction. Second, the results are sensitive to the choice of the NY interaction. The

Table 1. The separation energy of the hypertriton as solution of the Faddeev equation for different BB potentials.

NY potential	NN potential	E_{SP} (keV)
Nijm. SC 89	Nijm 93	143
Nijm. SC 89	Bonn B	155
Nijm. SC 97f	Nijm. 93	80
Nijm. SC 97e	Nijm. 93	23
Exp.		130± 50

fact that the different potentials can either over bind or under bind the hypertriton indicate that we can use the experimental hypertriton to constrain the parameters of the NY potential. This in fact has been attempted by Fujiwara *et al.*[4] within the framework of the $SU(6)$ quark model (see Table 2). Here, they have not only varied the strength of $N\Lambda - N\Sigma$ coupling to reduce the percentage Σ component (P_Σ) in the wave function, but have also changed the mass of the κ-meson (part of the vector meson nonet) to reduce the separation energy of the hypertriton. Here again we see that the hypertriton can be employed as a data point to establish constraints on the BB interaction in the $S = -1$ channel.

Table 2. The separation energy for the hypertriton for BB interaction based on the $SU(6)$ quark model.

potential	E_{SP} (keV)	P_Σ %
FSS	878	1.36
fss2	289	0.8
fss2 with κ	145	0.53
Exp.	130± 50	

3. The $N\Lambda - N\Sigma$ Coupling – A Study

Before one can make use of the hypertriton binding energy to establish constraints on the BB interaction, we need to examine the sensitivity of the hypertriton separation energy to two aspects of the NY interaction that influence this analysis. In the NN channel the long range tensor force is represented by one pion exchange. On the other hand, the long range part of the NY interaction, which dominates the properties of the hypertriton, is also described by one pion exchange, but now in the coupling between the $N\Lambda$ and $N\Sigma$ channel rather than the tensor interaction in the $N\Lambda$ channel (see Figure 1). To illustrate the relative importance of the $N\Lambda - N\Sigma$ coupling and the tensor interaction, we examine a simple model for the hypertriton based on separable potentials.[5] which includes both the tensor interaction and the coupling between the two YN channels.

The introduction of the coupling between the $N\Lambda$ and $N\Sigma$ channels introduces in the ΛNN space an effective three-body force and a dispersive force as illustrated

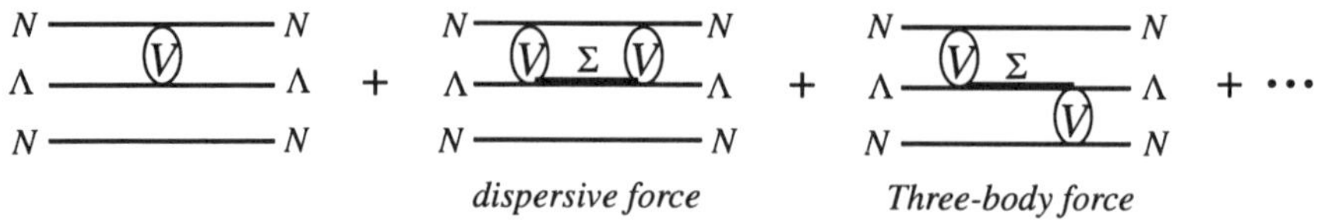

Fig. 2. The contribution of the three-body force and dispersive force to the ΛNN amplitude in perturbation series.

in Figure 2 and given by

$$T_{\Lambda NN} = V_{\Lambda\Lambda} + V_{\Lambda\Sigma}\frac{1}{E^{(3)} - K.E. - \Delta M}V_{\Sigma\Lambda} + \textit{Three-body force} + \cdots . \tag{3}$$

Since $E^{(3)} < E^{(2)} \approx 0$ (no two-body $N\Lambda$ bound state), the dispersive force in the three-body system is less attractive than the corresponding term in the two-body system, and therefore effectively repulsive. On the other hand, the three-body force is attractive. This is illustrated in Figure 3 where the result for V^{eff} is for a potential with no $N\Lambda - N\Sigma$ coupling that gives the same two-body effective range parameters as the potential with the coupling included. Here, we observe that while the three-body force is attractive, the dispersive force, which is not as important as the three-body force but significant, is repulsive.

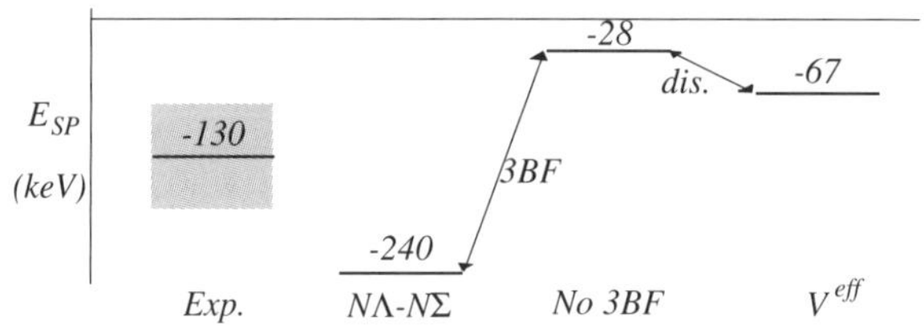

Fig. 3. The separation energy for ${}^3_\Lambda$H including the $N\Lambda - N\Sigma$ coupling, to illustrate the relative contribution of the three-body force ($3BF$) and dispersive (*dis.*) effect.

We now turn to the role of the tensor force in the $N\Lambda$ channel (see Figure 4).Here the additional binding energy due to the tensor force could be a small as 8 keV, *i.e.* an order of magnitude smaller than the three-body force resulting from the coupling between the $N\Lambda$ and $N\Sigma$ channels. This illustrates the importance of the $N\Lambda - N\Sigma$ coupling in $S = -1$ hypernuclei and the need to include this coupling explicitly in any hypernuclear calculation as is the case for the tensor force in ordinary nuclear physics.

4. Conclusions

In the above analysis we have demonstrated that: (i) The latest OBE potential could with some fine tuning reproduce the binding energy of ${}^3_\Lambda$H. (ii) In the hypertriton, the $N\Lambda - N\Sigma$ coupling plays a more important role than the tensor force in $N\Lambda$.

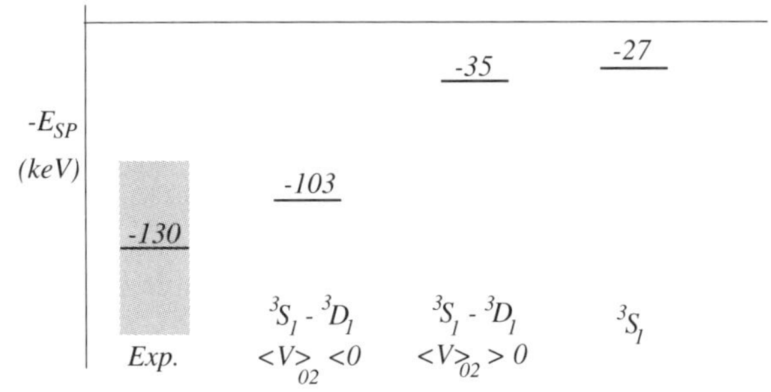

Fig. 4. The role of the tensor force in the $N\Lambda$ channel for the separation energy of $^3_\Lambda$H.

channel. As a result the coupling between the $N\Lambda - N\Sigma$ should be included explicitly in all light hypernuclear calculations, especially since this coupling gives rise to a dispersive force in addition to an effective three-body force. (iii) Light hypernuclei, and especially the $^3_\Lambda$H should be considered an important data point in adjusting the parameters of an $SU(3)_f$ based OBE potential. This opens the way for a better determination of the BB interaction in $S \leq -2$.

Acknowledgments

The author wishes to thank B. F. Gibson for the many enlightening discussions on the subject.

References

1. M. M. Nagels, T. A. Rijken, and J. J. de Swart, Phys. Rev. D **20**, 1633 (1979).
2. S. B. Carr, I. R. Afnan, and B. F. Gibson, Nucl. Phys. **A625**, 143 (1997).
3. A. Nogga, H. Kamada, and W. Glöckle, Phys. Rev. Lett. **88**, 172501 (2002).
4. Y. Fujiwara, K. Miyagawa, M. Kohno, and Y. Suzuki, Phys. Rev. C**70**, 024001 (2004).
5. I. R. Afnan and B. F. Gibson, Phys. Rev. C**41**, 2787 (1990).

Few-Body Aspects of Strangeness Nuclear Physics

E. Hiyama

Department of Physics, Nara Women's University, Nara 630-8506, Japan
E-mail: hiyama@cc.nara-wu.ac.jp

Energy levels of the $A = 7 - 10$ double Λ hypernuclei are predicted on the basis of an $\alpha+x+\Lambda+\Lambda$ four-body model, where $x = n, p, d, t, ^3$He and α, respectively. Furthermore, the binding energy of $NN N\Xi$ is also discussed. Scattering problem of the $uudd\bar{s}$ system, in the standard non-relativistic quark model is solved for the first time, by treating the large five-body modelspace including the NK scattering channel accurately with the Gaussian expansion method and the Kohn-type coupled-channel variational method.

1. Introduction

Recently, strangeness nuclear physics has been getting excited by some epoch-making experimental data on double-Λ hypernuclei at KEK-E373 [1] and observation of pentaquark system called Θ^+ [2].

Furthermore, in the future, at J-PARC, they are planning to produce many double Λ hypernuclei and many Ξ hypernuclei and exotic quark systems such as pentaquarks. Experimental data from J-PARC facility will continuously provide few-body community with encouraging study subjects since there are a lot of interesting and important few-body problems in strangeness nuclear physics.

Here, the following subjects are discussed: First one is to predict energy spectra of $A = 7 \sim 10$ double Λ hypernuclei. Second one is to predict structure of $NNN\Xi$ hypernucleus. Third one is to discuss structure of pentaquark system, Θ^+.

2. Method

In order to solve three- and four-body problem accurately, we have employed Gaussian Expansion Method (GEM) [3]. This method has been successfully applied to the bound states of various three- and four-body systems [4–7]. The basis functions work excellently in describing both the short-range correlations and the long-range tail behavior. If Ref.[3], we proposed infinitesimally-shifted-Gaussian-lobe basis functions, which are mathematically equivalent to the previous ones [4,5] but are different in expression, so that the calculation of the interaction matrix elements can be performed much more easily. Owing to the use of these new basis functions, applicability of our calculational method is much extended even to four-body systems with complicated interactions. As an example, we are making a benchmark test of calculation of four-nucleon bound state (^{4}He) with the use of an AV8 NN potential [8]. The results in the binding energy, r.m.s. radius and two-body correlation are in good agreement with those among seven different calculational methods.

Recently our method has been applied to hypernuclear structure and pentaquark system.

3. Double Λ hypernuclei

Recently, two epoch-making experimental data have been reported by the KEK-E373 experiment.

First one is observation of ${}^{6}_{\Lambda\Lambda}$He, which is called the NAGARA event [1]. The precise experimental value of the $\Lambda\Lambda$ binding energy, $B_{\Lambda\Lambda} = 7.25 \pm 0.19^{+0.18}_{-0.11}$ MeV has been obtained. Second one is the observation of ${}^{10}_{\Lambda\Lambda}$Be, which is called the *Demachi-Yanagi* event [9,10]. The binding energy, $B^{\rm exp}_{\Lambda\Lambda}$, obtained is $12.33^{+0.35}_{-0.21}$ MeV. It has not been, however, determined whether this event was the ground state or the excited state. Analysis of KEK-E373 experiment is still in progress. Furthermore, it is planned to produce many double Λ hypernuclei at J-PARC, GSI and BNL.

The calculated values of $B_{\Lambda\Lambda}$ can be compared with some experimental data, though the data are quite limited at present. The most recent and precise data of the NAGARA event are used as a basic input of our model, so that our $\Lambda\Lambda$ interaction is adjusted to reproduce the experimental value $B^{\rm exp}_{\Lambda\Lambda}({}^{6}_{\Lambda\Lambda}{\rm He} = 7.25 \pm 0.19 \pm^{0.18}_{0.11}$ MeV[1]. It is of particular interest to compare the present result with another datum, *Demachi-Yanagi* event [9,10], which is not used in the fitting procedure. The calculated $B_{\Lambda\Lambda}$ for the 2^+ excited state in ${}^{10}_{\Lambda\Lambda}$Be, 12.28 MeV, agree with the experimental value $B^{\rm exp}_{\Lambda\Lambda} = 12.33^{+0.35}_{-0.21}$ MeV in the *Demachi-Yanagi* event. We therefore interpret this event as the observation of the 2^+ excited state of ${}^{10}_{\Lambda\Lambda}$Be, which is consistent with NAGARA event of ${}^{6}_{\Lambda\Lambda}$He.

Then since our four-body calculations are predictive, hoping to have the data of many double Λ hypernuclei in the future experiments at J-PARC and GSI, we have predicted level structure for double Λ hypernuclei in Fig. 1 with $A = 7 \sim 10$ within the framework of an $\alpha + x + \Lambda + \Lambda$ four-body model, where $x = n, p, d, t, {}^3$He and α. It is our intention that these extensive four-body cluster model calculations should contribute to stimulating comprehensive spectroscopic studies of double Λ hypernuclei.

4. NNNΞ hypernuclei

The reason why to be studied $NNN\Xi$ hypernucleus is as follows:

(1)I can perform 4-body calculation using realistic YY interaction directly.

(2)Although we cannot predict the level structure of $NNN\Xi$ hypernucleus due to the large degree of ambiguity of ΞN interaction, we can investigate what part of ΞN interaction contribute to the binding energy of this hypernucleus.

(3) If this hypernucleus will be observed in the future at J-PARC or GSI, we can suggest how to improve the realistic force directly from the structure study side.

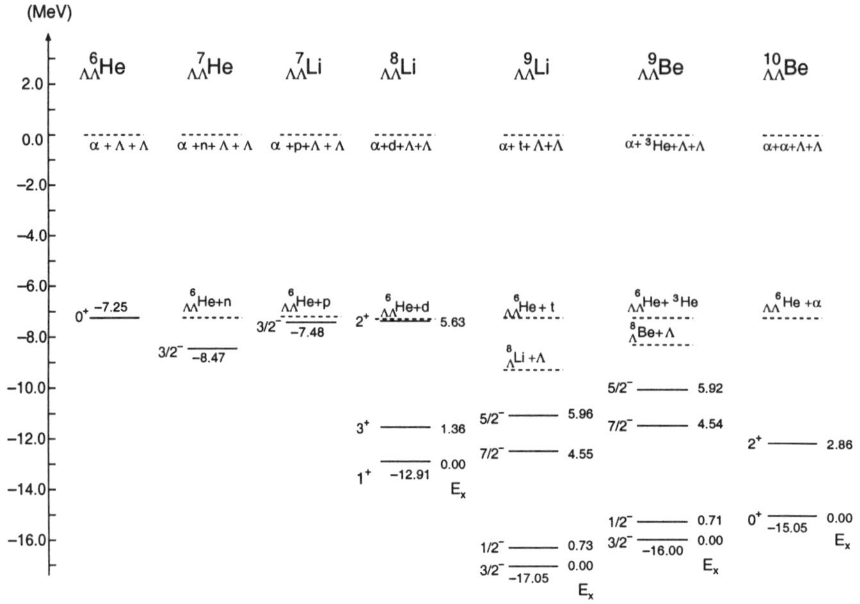

Fig. 1. Calculated energy levels of $A = 7 \sim 10$ double Λ hypernuclei.

For this study, we performed coupled 4-body calculation of $NNN\Xi$ and $NN\Lambda\Lambda$ channel. As for the NN interaction, Minnesota potential was employed. As for the YY interaction, we employed Nijmegen soft core '97 e, f and extended soft core potential.

The calculated binding energies of 0^+ and 1^+ using ESC02 potential are -7.41 MeV and -5.79 MeV with respect to $(NN\Xi) + N$ threshold, respectively.

Why we have bound $NNN\Xi$ state? In ESC02, interaction of isospin 0 and spin 1 part is strongly attractive. The contribution of this part in both states is so large, then we have bound state in this hypernucleus.

Very recently, observation of ${}^{4}_{\Lambda\Lambda}$H was carried out as BNL-E906 and they observed the sequential mesonic weak decay from ${}^{4}_{\Lambda\Lambda}$H. However, they could not determined whether the ground state of ${}^{4}_{\Lambda\Lambda}$H was bound or not. Using this potential, when the four-body calculation taking $NNN\Xi$ and $NN\Lambda\Lambda$ channel was performed, we obtained the ground state of ${}^{4}_{\Lambda\Lambda}$ as a bound state. (The calculated binding energy is 0.29 MeV from the ${}^{3}_{\Lambda}\mathrm{H} + \Lambda$ threshold.)

In the case of NSC97f and e, we have no bound states in $NNN\Xi$ system.

If the $NNN\Xi$ state is observed as a bound state in the future, we can extract information on ΞN interaction. Therefore, I hope to perform search experiment of this hypernucleus.

5. Pentaquark System

The observation of a signal of a narrow resonance at $\sim$ 1540 MeV with $S = +1$ by the LEPS group [2], now called $\Theta^+(1540)$, triggered a lot of theoretical works on multi-quark systems [11], although further experimental reexamination is still needed to establish the state. The question whether the multi-quark baryon $\Theta^+(1540)$ really exists or not is then one of current issues in hadron physics. In order to answer the question theoretically, one has to nonperturbatively evaluate the mass and the decay width of $\Theta^+(1540)$, namely of the $uudd\bar{s}$ resonance. All nonperturbative analyses made so far, however, did not impose any proper boundary condition to the NK scattering component of the pentaquark state. At the present stage, the non-relativistic quark model provides a nonperturbative framework which makes it possible to impose a proper boundary condition to the NK scattering component. Here,, we take the standard quark model of Isgur-Karl [12,13]. The Hamiltonian consists of the confining potential of harmonic oscillator type and the color-magnetic spin-spin interaction. The Hamiltonian satisfactorily reproduces observed properties of ordinary baryons and mesons. The same Hamiltonian is applied to the pentaquark with no adjustable parameter.

Resonant and non-resonant states of the $uudd\bar{s}$ system are nonperturbatively derived with the Kohn-type coupled-channel variational method [14] in which a proper boundary condition is imposed to the NK scattering channel and the total antisymmetrization between quarks is explicitly taken into account. Reliability of the method for scattering problem between composite particles was shown by one of the authors (M.K.) [14], and actually it was already applied to $qqq - qqq$ scattering [15]. As shown in Fig.2, we take five rearrangement channels.

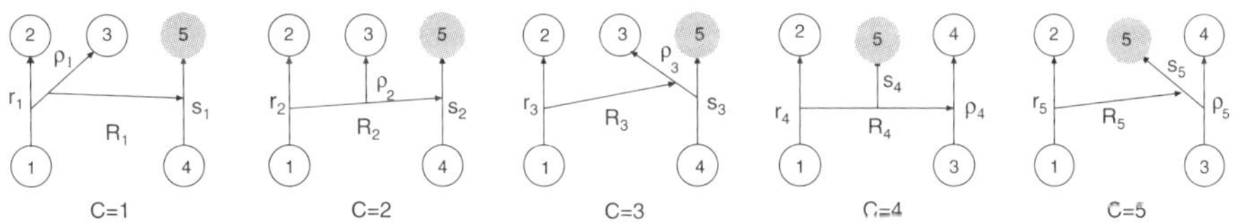

Fig. 2. Five sets of Jacobi coordinates among five quarks. Four u, d quarks, labeled by particle $1-4$, are to be antisymmetrized, while particle 5 stands for $\bar{s}$ quark. Sets $c = 4, 5$ contain two qq correlations, while sets $c = 1-3$ do both $q\bar{q}$ and qq correlations. The NK scattering channel is treated with $c = 1$.

In Fig.3, Calculated phase shifts are shown. Any resonance is not seen at $0-500$ MeV above the NK threshold ($1.4-1.9$ GeV in mass), that is, in the reported energy region of $\Theta^+(1540)$. One does see two resonance around 530 MeV; one is a sharp $J^\pi = \frac{1}{2}^-$ resonance with a width of $\Gamma = 0.12$ MeV located at $E - E_{\rm th} = 540$ MeV and the other is a broad $\frac{1}{2}^+$ resonance at $\sim$ 520 MeV with $\Gamma \sim 110$ MeV.

In this way, we succeeded to perform five-body calculation of bound states and resonance states.

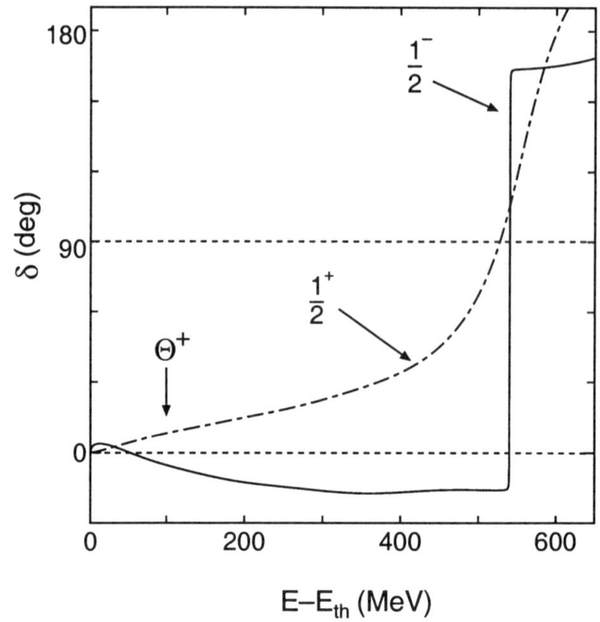

Fig. 3. Calculated phase shifts for $1/2^-$ (solid line) and $1/2^+$ (dashed-dotted line) states. Energies are measured form the The arrow indicates the energy of $\Theta^+(1540)$ in $E-E_{\rm th}$. $E_{\rm th}$ is the calculated NK threshold energy (1422 MeV).

References

1. H. Takahashi *et al.*, Phys. Rev. Lett. **87** (2001), 212502.
2. LEPS Collaboration, T. Nakano *et al.*, Phys. Rev. Lett. **91**, 012002 (2003).
3. E. Hiyama, Y. Kino and M. Kamimura, Prog. Part. Nucl. Phys. **51** (2003), 223.
4. M. Kamimura, Phys. Rev. **A38** (1988) , 621.
5. H. Kameyama, M. Kamimura and Y.Fukushima, Phys. Rev. C**40** (1989), 974.
6. M. Kamimura, Nucl. Phys. A**588** (1990) 17c.
7. E. Hiyama, M. Kamimura, T. Motoba, T. Yamada and Y. Yamamoto, Phys. Rev. C**64** (2002), 011301(R).
8. H. Kamada *et al.*, Phys. Rev. C **64** (2001), 044001.
9. K. Ahn *et al.*, AIP Conference Proceedings of the International Symposium on Hadoron and Nuclei, Seoul, 2001 , p.180.
10. A. Ichikawa, Ph.D. thesis, Kyoto University (2001), unpublished.
11. For a review, for instance, M. Oka, Prog. Theor. Phys. **112**, 1 (2004) and references therein.
12. N. Isgur and G. Karl, Phys. Rev. D **20**, 1191 (1971).
13. J. Weistein and N. Isgur, Phys. Rev. D **27**, 588 (1983).
14. M. Kamimura, Prog. Theor. Phys. Suppl. **62**, 236 (1977).
15. M. Oka and K. Yazaki, Prog. Theor. Phys. **66**, 556, 572 (1981).

OTHERS

Form Factors in Relativistic Quantum Mechanics Approaches and Space-Time Translation Invariance

B. Desplanques
Laboratoire de Physique Subatomique et de Cosmologie (UMR CNRS/IN2P3-UJF-INPG)
F-38026 Grenoble Cedex, France
E-mail: desplanq@lpsc.in2p3.fr

Invariance of form factors under Lorentz boosts is a criterion often advocated to determine whether their estimate in a RQM framework is reliable. It is shown that verifying relations stemming from covariance properties under space-time translations could be a more important criterion. Form factors calculated in different approaches for a simple system are discussed with these various respects. An approximate method is shown to remove the main discrepancies related to a violation of the above relations.

1. Introduction

There are many ways to implement relativity in the description of a few-body system and its properties, such as form factors. Ultimately, their predictions should converge but in an approximate calculation, some approaches may be more efficient than other ones. Relativistic quantum mechanics (RQM) has the advantage over field theory that it is dealing with a fixed number of (effective) degrees of freedom. As a counterpart, relativistic covariance properties are not trivially fulfilled. Following Dirac's work[1], several approaches depending on the symmetry properties of the hyper surface on which physics is formulated have been proposed: they are the instant, front, and point forms. These ones have been used for calculating form factors of various systems, evidencing large discrepancies in the approximation of a one-body current (see Ref. 2 and references therein). This raises the question of determining criteria allowing one to discriminate between the various approaches. Invariance under Lorentz boosts, which can be easily checked, is obviously one of them but relativity also implies other transformations such as space-time translations. The invariance under these ones provides the well-known conservation of energy and momentum, which, evidently, holds globally for the system. However, in an incomplete RQM calculation, this property is not necessarily fulfilled at the interaction vertex of its constituents with an external probe.

In the present contribution, we examined form factors of a simple system with respect to the above transformations. After showing the results obtained in different approaches (Sec. 2), we proceed to their discussion in view of relations which stem from the transformation of currents under space-time translations (Sec. 3).

2. Form Factors in Different Forms: Single-Particle Approximation

The system whose form factors are studied here is a theoretical one. It corresponds to the ground state of the Wick-Cutkoski model (scalar particles exchanging a massless scalar meson). Due to a hidden symmetry, the Bethe-Salpeter equation for this system can be easily solved and the charge and scalar form factor can be calculated exactly. In some sense, this provides our "experiment". It can be compared to form factors calculated in the single-particle current approximation for different RQM approaches and using the same solution of a mass operator. The two types of contributions are shown in Fig. 1, l.h.s. and r.h.s. respectively.

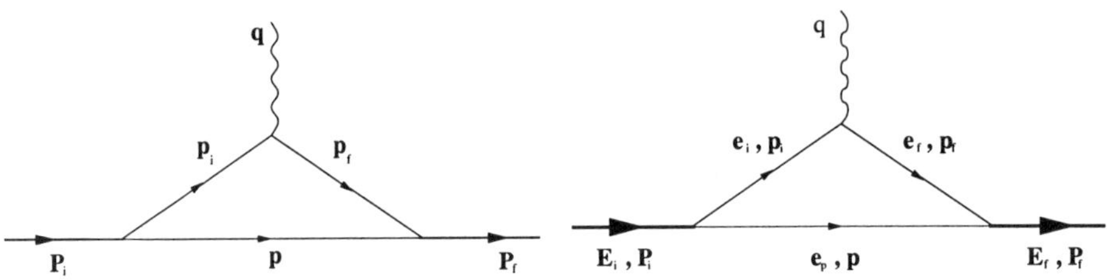

Fig. 1. Single-particle current contribution in field-theory (l.h.s.) and RQM (r.h.s.) approaches. Intermediate particles are respectively off-mass shell and on-mass shell.

In the instant form, where results are not invariant under a Lorentz transformation (boost), form factors can be considered for the Breit frame (I.F. (Breit frame)) and for a frame where the initial- and final-state momenta are parallel while their sum goes to infinity (I.F. (parallel)). In the front form, where results are not invariant under a Lorentz transformation (rotation), they can be considered for the configuration $q^+ = 0$ (F.F. (perp.)) and for a configuration where the initial- and final-state momenta are parallel to the front orientation (F.F. (parallel)). Not surprisingly, the last results are identical to those in the instant form with an infinite average momentum. In the point-form, results turn out to be Lorentz invariant. Different versions may be nevertheless considered. The first one[3,4], extensively used in many works, is an instant form displaying the same symmetry as the point form ("P.F."). The second one[5] is more in the spirit of the Dirac point form, where physics is formulated on an hyperboloid surface (D.P.F.).

Results are presented in Fig. 2 for the charge form factor, $F_1(Q^2)$, at both low and high Q^2. These ranges are aimed to point to the charge radius, $< r^2 >$, and to the asymptotic behavior (expected to be Q^{-4}). In comparison to "experiment" (small diamonds in the figure), it is noticed that the instant-form (Breit-frame) and front-form ($q^+ = 0$) cases do rather well. Lorentz invariance, which underlies point-form results, does not imply good results. Apart from the above instant- and front-form cases, the charge radius scales like the inverse of the mass of the system, leading to the paradox that the former quantity goes to infinity when the latter goes to zero. This suggests the violation of some symmetry. Which one however?

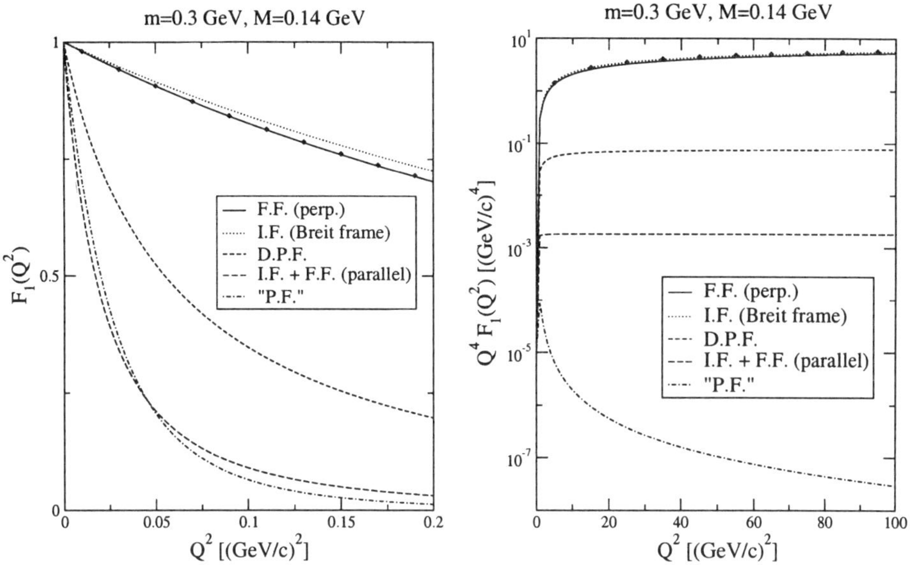

Fig. 2. Form factors at low and large Q^2 (l.h.s. and r.h.s. respectively). Parameters are appropriate to a pion-like system. See text for the definition of abbreviations.

3. Role of Space-Time Translation Invariance

Space-time translation invariance implies the 4-momentum conservation. This has to hold globally for any physical process, giving the relation ($P_i^\mu + q^\mu = P_f^\mu$ in the present case (Fig. 1)). In field-theory, such a relation also holds at the interaction vertex of the constituents with the external probe ($p_i^\mu + q^\mu = p_f^\mu$). However, the last one is not verified in RQM approaches, where constituents are on-mass shell. To quantitatively discuss the consequences of this feature, it is appropriate to consider relations that stem from the transformation of a current under space-time translations[6], which are generated by the operators of the Poincaré algebra, P^μ. Considering matrix elements, one should verify relations like:

$$\langle \,|\Big[P^\mu \,,\, J^\nu(x)\Big]|\,\rangle = -i\langle\, |\partial^\mu J^\nu(x)|\,\rangle,$$

$$\langle\,|\Big[P_\mu \,,\Big[P^\mu \,,\, J^\nu(x)\Big]\Big]|\,\rangle = -\langle\, |\partial_\mu\partial^\mu J^\nu(x)|\,\rangle\,,$$

$$\text{or, here,}\quad \langle\, |q^2\, J^\nu(x)|\,\rangle = \langle\, |(p_i - p_f)^2\, J^\nu(x)|\,\rangle\,, \tag{1}$$

which are automatically fulfilled in field-theory but suppose the existence of many-body currents in RQM approaches. Assuming a single-particle current, it can be verified that the last equation is fulfilled exactly in the front-form ($q^+ = 0$) case or approximately in the instant-form (Breit frame) one. In all other cases, it is violated by large factors, from 30 up to 35000 ("P.F.") here.

Examination of these cases indicates that the factor multiplying Q^2 in calculations misses interaction effects. Correcting this factor such as to remove the above violation factors therefore provides an approximate way to account for the missing

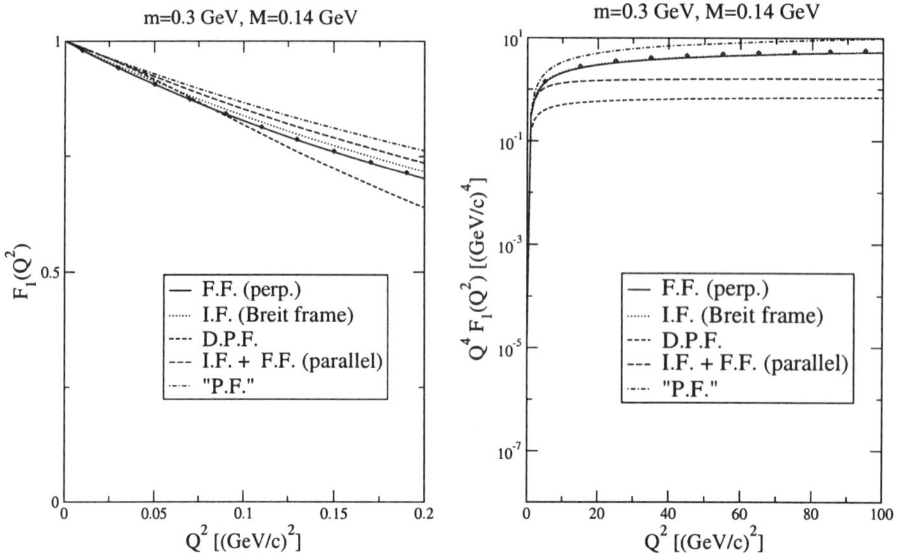

Fig. 3. Form factors with corrections aimed to fulfill relations stemming from transformations of currents under space-time translations, at low and large Q^2 (l.h.s. and r.h.s. respectively).

many-body currents required at all orders in the interaction to fulfill Eqs. 1. This approach has been applied to the form factors shown in Fig. 2. The resulting ones are presented in Fig. 3. It is observed that the largest discrepancies have vanished at low and high Q^2. In particular, the peculiar behavior of form factors in the limit of a zero-mass system is completely removed.

4. Conclusion

The present work shows that constraints from space-time translation transformations could be more important than those from (homogeneous) Lorentz transformations. Invariance under these last ones turns out to be a disadvantage as there is no frame where the violation of the other constraints can be minimized. To a large extent, the present work confirms the standard front-form approach ($q^+ = 0$) and instant-form one (Breit-frame or $E_i = E_f$ case more generally) as more reliable frameworks in the single-particle current approximation.

References

1. P.A.M. Dirac, *Rev. Mod. Phys.* **21**, 392 (1949).
2. A. Amghar, B. Desplanques, L. Theußl, *Nucl. Phys.* **A714**, 213 (2003).
3. B. Bakamjian, *Phys. Rev.* **121**, 1849 (1961).
4. S.N. Sokolov, *Theor. Math. Phys.* **62**, 140 (1985).
5. B. Desplanques, *Nucl. Phys.* **A748**, 139 (2005).
6. F.M. Lev, *Rivista del Nuovo Cimento* **16**, 1 (1993).

Dalitz-esque Treatment of New Heavy Particle Pair Production at the LHC*

Mike Bisset

Center for High Energy Physics and Department of Physics,
Tsinghua University, Beijing, 100084 P.R. China
E-mail: bisset@mail.tsinghua.edu.cn

Processes of the form $pp \to anything \to X_i X_j \to x\bar{x} + y\bar{y}$ $(+\not{E})$ are studied via a technique that may be viewed as an adaptation of time-honored Dalitz plot analyses. X_i and X_j are new heavy states (with $i, j = 1...n$), which may be identical or distinct; and $x\bar{x}$ and $y\bar{y}$ are necessarily distinct Standard Model (SM) fermion pairs whose invariant masses can be measured. A Dalitz-like plot of said invariant masses, $M(x\bar{x})$ *vs.* $M(y\bar{y})$, exhibits a topology connected to the masses and specific decay chains of X_i and X_j. Aside from relatively minor details, observed patterns consist of a collection of box and wedge shapes. This collection is model-dependent: comparison of the observed pattern to the possibilities for a specific model yields information on which new particle pair combinations are actually being produced, information beyond that extractable from conventional one-dimensional invariant mass distributions. The technique is illustrated via application to the Minimal Supersymmetric Standard Model (MSSM) process $pp \to \tilde{g}\tilde{g}, \tilde{g}\tilde{q}, \tilde{q}\tilde{q} \to \tilde{\chi}_i^0 \tilde{\chi}_j^0 \to e^+e^- + \mu^+\mu^-$ $(+\not{E})$. Here the heavy states are neutralinos $\tilde{\chi}_i^0$ $(i = 2, 3, 4)$ — note $\tilde{\chi}_1^0$ is excluded — which are produced in gluino/squark $(\tilde{g}/\tilde{q})$ cascade decay chains. Even with fairly modest expectations for the LHC performance during the first few years, this method still provides substantial insight into the neutralino mass spectrum and couplings if gluino/squark masses are relatively low ($\simeq$ 400 GeV).

Most theoretical extensions of the SM predict new heavy degrees of freedom at or near the TeV-scale. The soon-to-be-completed 14 TeV LHC should readily produce such heavy particles if they couple significantly to SM ones. Introduction of new heavy particle states often comes with the introduction of a new conserved quantum number (or numbers) associated with a new discrete Z_n symmetry (or symmetries) — for example, R-parity[1] conservation in SUSY[a] scenarios[b] — typically dictating the pair production of new particle states as well as the stability of the lightest new particle (in SUSY the "LSP") which is "odd" under the new symmetry.

*Based on the paper "Pair-produced heavy particle topologies: MSSM neutralino properties at the LHC from gluino/squark cascade decays" by M. Bisset, N. Kersting, J. Li, F. Moortgat, S. Moretti and Q.L. Xie. This work and travel to this conference were supported by Tsinghua University. Local hospitality was provided by Suranaree University of Technology.
[a]Henceforth the acronym 'SUSY' will be used for both 'supersymmetry' and 'supersymmetric'.
[b]Other examples include little Higgs models with 'T-parity'[2] and minimal universal extra dimension models with 'KK-parity'[3].

Consider pair production of new *neutral* heavy particles, $X_i X_j$, with $i, j = 1...n$ — the exact value of n being model dependent, but for the present work it may be any integer greater than 1 (thus X_i and X_j may or may not be distinct). These pairs may be produced directly as per $pp \to X_i X_j$ or as a result of cascade decays from the production of other even heavier new particles. The X_i & X_j particles then decay (so neither is an 'LSP') generating processes of the type

$$pp \to X_i X_j \; + \text{ stuff} \to (f_1 \bar{f}_1) \; + \; (f_2 \bar{f}_2) + \begin{array}{c}\text{other}\\ \text{stuff}\end{array} \; . \tag{1}$$

Unlike at an e^+e^- collider, at a hadron collider the center-of-mass energy of the parton-level hard scattering process cannot be controlled, and thus said parton-level center-of-mass energy cannot be incrementally raised to scan through the different $X_i X_j$ thresholds. Rather, all such channels may be produced simultaneously and must subsequently be disentangled to the extent it is possible in the decay analysis.

The new X_i & X_j particles are assumed to decay (possibly indirectly) into pairs of SM fermions $X_i \to f\bar{f}$ (accompanied perhaps by substantial missing energy in the detector). Thus pair production of new heavy particles can result in end-states of the form $f_1\bar{f}_1 + f_2\bar{f}_2$, where $f_1\bar{f}_1$ and $f_2\bar{f}_2$ are *distinct* SM fermion pairs whose invariant masses are measurable with sufficient precision. Mindful of the strong QCD backgrounds expected at the LHC, a particularly attractive choice is same-flavor oppositely-charged lepton pairs: e^+e^- and $\mu^+\mu^-$. A two-dimensional plot of $M(e^+e^-)$ *vs.* $M(\mu^+\mu^-)$ will then be expected to display a distinct topology dependent upon what combinations of new particles are being produced. The basic building blocks of the pattern observed will be boxes from $X_i X_i$ production and wedges from $X_i X_j$ ($i \neq j$) production, as shown below. The use of such two-dimensional plots of two distinct invariant masses is very reminiscent of the Dalitz plots[4] employed for decades in resonance searches.

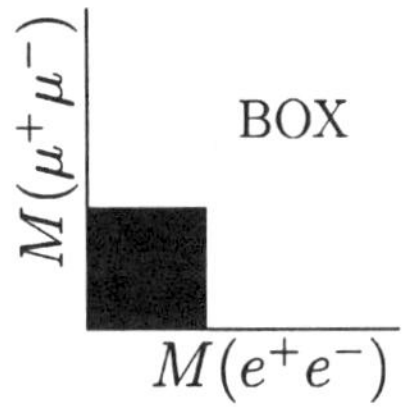

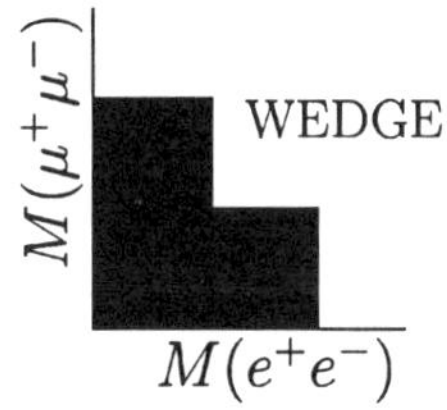

Naturally, other processes besides the sought-after heavy particle decays may also produce an (e^+e^-, $\mu^+\mu^-$)-topology. Thus cuts are needed to purify the event sample, and a partially-contaminated sample may have to suffice. At this point, a specific model of physics beyond the SM should be introduced, and a quite reasonable test subject is the simplest viable SUSY model, the MSSM [1]. At this point I eschew the standard dribble introducing the zoo of new particle states expected in the MSSM, and simply state that the X_i are beasties called "neutralinos," $\widetilde{\chi}_i^0$, or just "–inos" for short, of which there are four (i = 1-4) — excluding the lightest neutralino, $\widetilde{\chi}_1^0$, which happens to be the LSP. Process (1) may now be rewritten as

$$pp \to \widetilde{\chi}_i^0 \widetilde{\chi}_j^0 \; + \text{ stuff} \to (e^+e^-) \; + \; (\mu^+\mu^-) \; + \; \not{E} \; + \begin{array}{c}\text{other}\\ \text{stuff}\end{array} \; , \tag{2}$$

where the missing energy comes mainly from subsequently-produced LSP's.

The $\widetilde{\chi}_i^0\widetilde{\chi}_j^0$ pair may be any of: $\widetilde{\chi}_2^0\widetilde{\chi}_2^0$, $\widetilde{\chi}_2^0\widetilde{\chi}_3^0$, $\widetilde{\chi}_2^0\widetilde{\chi}_4^0$, $\widetilde{\chi}_3^0\widetilde{\chi}_3^0$, $\widetilde{\chi}_3^0\widetilde{\chi}_4^0$ or $\widetilde{\chi}_4^0\widetilde{\chi}_4^0$. The possible Dalitz topologies from –ino decays to lepton pairs in (2) are built from 3 possible boxes and 3 possible wedges giving 63 basic combinations ($\Sigma_{i=1}^{6} C(6,i)$) of boxes and wedges, though of these many are topologically equivalent — the 9 topologically distinct patterns are shown in Fig. 1. These may be augmented by up to three "stripes." Generally expected are the decays $\widetilde{\chi}_i^0 \to \{Z^0, Z^{*0}\} + \widetilde{\chi}_1^0 \to \ell^+\ell^- + \widetilde{\chi}_1^0$ or $\widetilde{\chi}_i^0 \to \ell^\pm + \{\tilde{\ell}^\mp, \tilde{\ell}^{*\mp}\} \to \ell^+\ell^- + \widetilde{\chi}_1^0$, where the two-body decays occur through an on-mass-shell Z^0-boson or slepton and the three-body decays occur when the Z^0-boson or slepton are off-mass-shell. Stripes ensue if cascading decay chains, $\widetilde{\chi}_4^0 \to \widetilde{\chi}_3^0 + \ell^+\ell^-$, $\widetilde{\chi}_4^0 \to \widetilde{\chi}_2^0 + \ell^+\ell^-$ or $\widetilde{\chi}_3^0 \to \widetilde{\chi}_2^0 + \ell^+\ell^-$ are present.

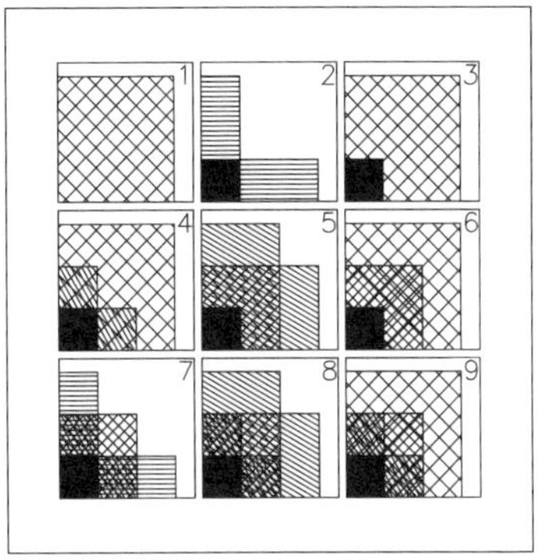

Fig. 1. All possible distinct patterns from production of $\widetilde{\chi}_i^0\widetilde{\chi}_j^0$ pairs ($i,j = 2,3,4$). Densities are shown as uniform for simplicity only.

At the LHC, heavy –ino pair production occurs via virtual SM gauge bosons, via decays of heavy Higgs bosons, and/or via cascade decays of colored squarks and gluinos (see Fig. 2). Due to the strong coupling, production via colored intermediates,

$$pp \to \tilde{g}\tilde{g},\, \tilde{g}\tilde{q},\, \tilde{q}\tilde{q} \to \widetilde{\chi}_i^0\widetilde{\chi}_j^0 \quad (i,j = 2,3,4), \qquad (3)$$

has the potential to yield the largest number of events, if gluinos and squarks are sufficiently light ($\sim 400\,\mathrm{GeV}$). Decaying gluinos/squarks in (3) also typically lead to jet activity in the final state, whereas the other two production mechanisms may be hadronically quiet much or at least some of the time. This may help to distinguish these processes[c], and jets are also useful in background reduction. Since the lightest MSSM Higgs boson (h^0) can only yield LSP-containing –ino pairs, Higgs-mediated production involves heavier MSSM Higgs bosons (H^0 and A^0) with masses in the range to yield sufficient Higgs boson production and yet have open decay modes to heavy –ino pairs. This is documented in another work[5,6]. In this work, we focus on (3), so inputs for gluinos and squarks are set near the lower end of their allowed mass ranges while the input Higgs boson mass is set fairly high ($\sim$700 GeV)[d].

Three points in the MSSM parameter space were chosen for simulation of $pp \to \tilde{g}\tilde{g}, \tilde{g}\tilde{q}, \tilde{q}\tilde{q}, \tilde{g}\widetilde{\chi}, \tilde{q}\widetilde{\chi}$ events utilizing the Monte Carlo package[7] HERWIG 6.5 (with CTEQ6M[8] structure functions) coupled with a detector simulation which assumes a typical LHC experiment:

- Point A: $\tan\beta = 5$, $M_2(M_1) = 200(100)\,\mathrm{GeV}$, $\mu = -150\,\mathrm{GeV}$, $m_{\tilde{\ell}_R} = 150\,\mathrm{GeV}$.
- Point B: $\tan\beta = 20$, $M_2(M_1) = 400(200)\,\mathrm{GeV}$, $\mu = -250\,\mathrm{GeV}$, $m_{\tilde{\ell}_R} = 200\,\mathrm{GeV}$.
- Point C: $\tan\beta = 30$, $M_2(M_1) = 250(125)\,\mathrm{GeV}$, $\mu = 200\,\mathrm{GeV}$, $m_{\tilde{\ell}_R} = 150\,\mathrm{GeV}$.

---

[c] Also note that in (3) the two –inos are produced separately, whereas in the two EW processes there is an –ino–ino$^{(\prime)}$-S vertex (where S is a SM gauge boson or an MSSM heavy Higgs boson). Space prevents further discussion of the consequences of this here.

[d] These restrictions are reversed in the study of heavy MSSM Higgs bosons into –inos in[5,6].

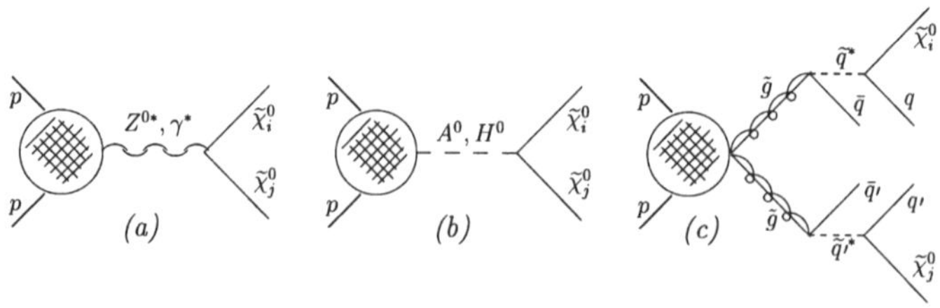

Fig. 2. Feynman diagrams for heavy $(i, j = 2, 3, 4)$ neutralino pair production mechanisms: *(a)* 'direct' production via EW gauge boson; *(b)* Higgs-mediated production; and *(c)* production via cascade decays of gluinos (shown here) or via gluino/squark or squarks — to obtain these diagrams, make one or both of the squarks in (c) on-mass shell and remove the associated gluino(s) and the connected quark(s).

Additional common inputs were $M_{A^0} = 700\,\text{GeV}$, $m_{\tilde{g}} = 400\,\text{GeV}$, and $m_{\tilde{q}} = 500\,\text{GeV}$ (for all soft squark mass inputs). Both $\tilde{g}, \tilde{q}$ signal events and SM & MSSM background events were simulated at each point assuming an integrated luminosity of $30\,\text{fb}^{-1}$ to see how the expected features show up for a realistic sample size. Draconian space limitations preclude showing a table of the sparticle masses.

Combining[e] leptonic BRs of the neutralinos and charginos with their production rates from decays of gluinos and squarks yields the tabular results shown aside the Dalitz-like plots to follow. These numbers may serve as a guideline against which simulation results may be compared, but it should be emphasized that this information will not be accessible with real experiments; *i.e.*, with a simulation, we are of course able to choose what point in the parameter space to simulate.

In event selection and subsequent cuts, stress is put on keeping the cuts reasonably general so that they will hopefully be applicable across a large swath of the allowable parameter space. These cuts can quite probably be further honed once the first evidence(hints) of possible MSSM events is discerned. Events are selected which have exactly four isolated leptons ($\ell = e^{\pm}$ or $\mu^{\pm}$) consisting of exactly one e^+e^- pair and one $\mu^+\mu^-$ pair with $|\eta^{\ell}| < 2.4$ and $E_T^{\ell} > 7, 4\,\text{GeV}$ for $e^{\pm}, \mu^{\pm}$, respectively. The important isolation criterion demands there be no tracks (of charged particles) with $p_T > 1.5\,\text{GeV}$ in a cone of $r = 0.3\,\text{rad}$ around a specific lepton, and also that the energy deposited in the EM calorimeter be $< 3\,\text{GeV}$ for $0.05\,\text{rad} < r < 0.3\,\text{rad}$. The missing transverse energy must also be $\geq 20\,\text{GeV}$, and three or more jets (defined by a cone algorithm with $r = 0.4$ demanding $|\eta^j| < 2.4$ and $E_T^j > 20\,\text{GeV}$) must be present. These cuts effectively eliminate virtually all of the SM backgrounds and, at least for the cases studied herein, almost all of the SUSY "background" processes — primarily $pp \to \tilde{\chi}^{0,\pm}\tilde{\chi}^{*0,*\pm}$ and $pp \to A^0, H^0$.

For the Dalitz-like plot for Point A in Fig. 3, a 'wedge inside of a box' topological structure is apparent, a clear indication that two pairs of –inos, $\widetilde{\chi}_i^0\widetilde{\chi}_j^0$ $(i < j)$ and $\widetilde{\chi}_j^0\widetilde{\chi}_j^0$, are being produced at significant rates. The (3) production mechanism then demands that a $\widetilde{\chi}_i^0\widetilde{\chi}_i^0$ box also be present, the position of which overlaps with that of the low $M(e^+e^-)$, low $M(\mu^+\mu^-)$ corner of the wedge. The shaded bands and dashed

[e]HERWIG lacks the facilities for calculating these cross-sections for each separate $\tilde{g}\tilde{g}$, $\tilde{q}\tilde{q}^{(\prime)}$, $\tilde{g}\chi$ and $\tilde{q}\chi$ process, so these were calculated using ISAJET 7.67 [9] with CTEQ5 [8] parton distributions.

lines included in the plot show the expected locations of hard edges based on the –ino and slepton mass spectrum obtained from ISASUSY for Point A. The 40-45 GeV hard edge corresponds to the 42.8 GeV $\widetilde{\chi}^0_2$-$\widetilde{\chi}^0_1$ mass difference. Here $\widetilde{\chi}^0_2$ is decaying through an off-shell Z^0 or slepton, with $BR(\widetilde{\chi}^0_2 \to \widetilde{\chi}^0_1 \ell^+ \ell^-) = 0.245$ (very unlike the leptonic BR for the Z^0) indicating that the off-shell sleptons are playing large rôles. The other –inos decay mainly through on-mass-shell sleptons. The outer edge at $\sim$140 GeV agrees with the endpoint for the two-body decay chain $\widetilde{\chi}^0_4 \to \widetilde{\ell}\ell \to \widetilde{\chi}^0_1 + \ell^+\ell^-$ So the outer box is from $\widetilde{\chi}^0_4\widetilde{\chi}^0_4$ production and the wedge is from $\widetilde{\chi}^0_2\widetilde{\chi}^0_4$ (including an inner box from $\widetilde{\chi}^0_2\widetilde{\chi}^0_2$ production).

Also discernible inside the wedge are indistinct drops in population densities along both axes at $\sim$85 GeVwhich might be interpreted as evidence for significant

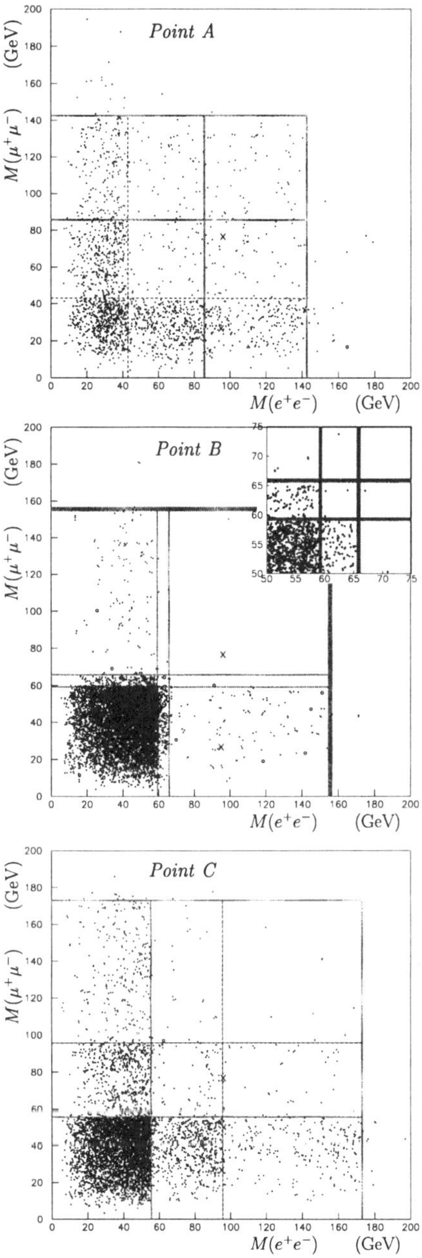

Point A		
$\widetilde{\chi}^0_2\widetilde{\chi}^0_4$	27.45%	(25.2%)
$\widetilde{\chi}^0_2\widetilde{\chi}^0_2$	14.5%	(16.7%)
$\widetilde{\chi}^0_4\widetilde{\chi}^0_4$	11.45%	(8.6%)
$\widetilde{\chi}^0_1\widetilde{\chi}^0_4$	7.8%	(7.6%)
$\widetilde{\chi}^+_2\widetilde{\chi}^-_2$	7.4%	(6.25%)
$\widetilde{\chi}^\pm_1\widetilde{\chi}^0_4$	6.6%	(8.7%)
$\widetilde{\chi}^\pm_2\widetilde{\chi}^\pm_2$	6.5%	(6.25%)
$\widetilde{\chi}^\pm_1\widetilde{\chi}^\mp_2$	4.4%	(5.15%)
$\widetilde{\chi}^\pm_1\widetilde{\chi}^\pm_2$	4.3%	(5.15%)
$\widetilde{\chi}^\pm_2\widetilde{\chi}^0_4$	2.9%	(2.9%)
$\widetilde{\chi}^0_3\widetilde{\chi}^0_4$	2.6%	(2.4%)
$\widetilde{\chi}^\pm_2\widetilde{\chi}^0_2$	1.9%	(2.3%)
$\widetilde{\chi}^0_2\widetilde{\chi}^0_3$	1.8%	(2.1%)
$\widetilde{\chi}^0_1\widetilde{\chi}^0_3$	0.2%	(0.3%)
$\widetilde{\chi}^\pm_1\widetilde{\chi}^0_3$	0.1%	(0.2%)
$\widetilde{\chi}^\pm_2\widetilde{\chi}^0_3$	0.1%	(0.2%)

Point B		
$\widetilde{\chi}^0_2\widetilde{\chi}^0_2$	73.3%	(76.3%)
$\widetilde{\chi}^0_2\widetilde{\chi}^0_3$	20.3%	(22.1%)
$\widetilde{\chi}^0_2\widetilde{\chi}^0_4$	3.2%	(0.0%)
$\widetilde{\chi}^0_3\widetilde{\chi}^0_3$	1.4%	(1.6%)
$\widetilde{\chi}^\pm_2\widetilde{\chi}^0_2$	0.9%	(0.0%)
$\widetilde{\chi}^0_3\widetilde{\chi}^0_4$	0.5%	(0.0%)
$\widetilde{\chi}^\pm_2\widetilde{\chi}^0_3$	0.1%	(0.0%)
$\widetilde{\chi}^\pm_1\widetilde{\chi}^\mp_2$	0.1%	(0.0%)
$\widetilde{\chi}^\pm_1\widetilde{\chi}^\pm_2$	0.08%	(0.0%)

Point C		
$\widetilde{\chi}^0_2\widetilde{\chi}^0_2$	66.6%	(72.7%)
$\widetilde{\chi}^0_2\widetilde{\chi}^0_3$	13.3%	(15.9%)
$\widetilde{\chi}^0_2\widetilde{\chi}^0_4$	12.5%	(6.6%)
$\widetilde{\chi}^\pm_2\widetilde{\chi}^0_2$	2.7%	(1.8%)
$\widetilde{\chi}^0_3\widetilde{\chi}^0_4$	1.3%	(0.7%)
$\widetilde{\chi}^\pm_1\widetilde{\chi}^\mp_2$	0.74%	(0.9%)
$\widetilde{\chi}^0_3\widetilde{\chi}^0_3$	0.68%	(0.9%)
$\widetilde{\chi}^\pm_1\widetilde{\chi}^\pm_2$	0.67%	(0.9%)
$\widetilde{\chi}^0_4\widetilde{\chi}^0_4$	0.5%	(0.15%)
$\widetilde{\chi}^\pm_2\widetilde{\chi}^0_3$	0.3%	(0.2%)
$\widetilde{\chi}^\pm_2\widetilde{\chi}^0_4$	0.3%	(0.1%)
$\widetilde{\chi}^+_2\widetilde{\chi}^-_2$	0.24%	(0.05%)
$\widetilde{\chi}^\pm_2\widetilde{\chi}^\pm_2$	0.15%	(0.05%)

Fig. 3. $M(e^+e^-)$ *vs.* $M(\mu^+\mu^-)$ Dalitz-like plot for Points A, B and C, assuming an integrated luminosity of 30 fb^{-1} (raised to 60 fb^{-1} in the insert for Point B). Signal events from gluino/squark production are denoted by dots, and the few SM (other SUSY processes) background events by X's (open circles). Shaded bands (dashed lines) indicate kinematical endpoints expected from two-body decays (three-body decays) based on ISASUSY. At right are the percentage contributions to *all-inclusive* 4ℓ events from the various neutralino/chargino pair production modes (with gluinos-only contributions in parentheses).

$\tilde{\chi}_2^0\tilde{\chi}_3^0$ production, or as a 'stripe'. In fact, they are due to the latter, associated with the decay chain $\tilde{\chi}_4^0 \to \tilde{\ell}\ell \to \tilde{\chi}_2^0 + \ell^+\ell^-$ which happens 22.8% of the time[f]. *A priori* knowledge is required to designate this feature a stripe, and thus Dalitz-like plots do not always uniquely identify the underlying –ino production/decay modes. These designations agree well with the percentages given to the right of the plot. The events lying outside the outer box in the Dalitz-like plot are due at least in part to production modes including charginos[g].

The plot for Point B in Fig. 3 bespeaks of dominant $\tilde{\chi}_i^0\tilde{\chi}_i^0$ production with weaker contributions from $\tilde{\chi}_i^0\tilde{\chi}_j^0$ and $\tilde{\chi}_i^0\tilde{\chi}_k^0$ $(i < j < k)$ (the latter yielding the outer wedge envelope and the former the short-legged wedge). Note in the MSSM framework i, j and k must be 2, 3 and 4. The short, stubby wedge tells us that two of the heavier –inos, presumably $\tilde{\chi}_2^0$ and $\tilde{\chi}_3^0$, are quite close in mass. This is again in very good agreement with the predictions (also for this point in parameter space, $m_{\tilde{\chi}_3^0}$ is in fact rather close to $m_{\tilde{\chi}_2^0}$). More events from a longer run time would help to clarify the structure in the crucial $(M(e^+e^-), M(\mu^+\mu^-)) =$ (60-65 GeV, 60-65 GeV) region of the Dalitz-like plot (see insert in plot). Events in the legs of the outer wedge are solely from squark production as only the squarks are heavy enough to decay to $\tilde{\chi}_4^0$.

Lastly, the plot for Point C in Fig. 3 indicates fairly dominant $\tilde{\chi}_i^0\tilde{\chi}_i^0$ production, but with significant contributions from $\tilde{\chi}_i^0\tilde{\chi}_j^0$ and $\tilde{\chi}_i^0\tilde{\chi}_k^0$ $(i < j < k)$, and with lesser but still detectable contributions from $\tilde{\chi}_j^0\tilde{\chi}_j^0$ and $\tilde{\chi}_j^0\tilde{\chi}_k^0$. (Again, in the MSSM, i, j and k must be 2, 3 and 4.) Comparing the wedges in the plots for Points B & C, we conclude that for Point C the –ino masses are not so close together. The predicted endpoints from charged-slepton-mediated decays of $\tilde{\chi}_3^0$ and $\tilde{\chi}_4^0$ affirm that the shorter (longer) wedge is from $\tilde{\chi}_2^0\tilde{\chi}_3^0$ ($\tilde{\chi}_2^0\tilde{\chi}_4^0$) production. The more faintly discernible box with edges at $\sim$95 GeV is attributed to $\tilde{\chi}_3^0\tilde{\chi}_3^0$ production, the even more faint wedge of which this box is the corner is from $\tilde{\chi}_3^0\tilde{\chi}_4^0$ production, and the few events in the upper-left corner of the plot are presumably from $\tilde{\chi}_4^0\tilde{\chi}_4^0$ production, all in fairly good agreement with the table given.

In summary, pair production of new heavy particle states at hadron colliders has been studied emphasizing the simple topological forms expected in two-dimensional Dalitz-like plots. A likelihood function indicating how well the a set of data points fits each possibility in a given model may be constructed if visual inspection does not suffice. The 'hard edges' seen in a plot yield information on the –ino mass differences as well as the identities of –inos participating in the decays (though the endpoints need not equal the mass difference of two –inos if on-mass-shell sleptons are involved [10]), while comparing the relative densities of regions populated by

[f]The $\tilde{\chi}_4^0$ decay chain mentioned in the last paragraph occurs 65.8% of the time, and the remaining 10.6% of the $\tilde{\chi}_4^0$ decays are through $\tilde{\chi}_1^{\pm}$.

[g]The inclusive 4ℓ rates say roughly a third of the events for Point A come from production modes including charginos. However, a substantial fraction of these events will not have leptons in same-flavor, opposite-sign pairs. So their effect on this analysis will be diminished. In fact, same-flavor, *like-sign* 4ℓ events (*i.e.*, $e^{\pm}e^{\pm}\mu^{\mp}\mu^{\mp}$ events) could be used to estimate the chargino contribution and then remove it.

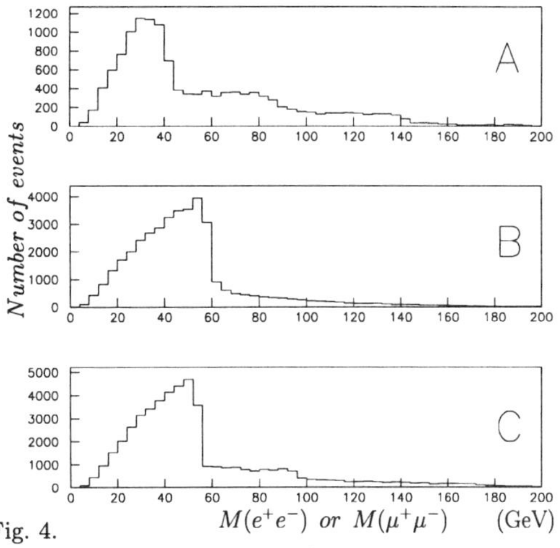

Fig. 4.
One-dimensional projection of points in Fig. 3 assuming an integrated luminosity of 30 fb^{-1}. Information about the decay topology is lost in such projections.

different production channels or combinations of channels has the potential to provide information on the relative production cross-sections times leptonic BRs. Contrast the information apparent in these Dalitz-like plots with that readily obtainable from the more traditional one-dimensional projections shown in Fig. 4.

How much this method can aid in unraveling the –ino mass spectrum will depend on particulars of the point in the MSSM parameter space nature chooses, but clearly very significant information may be extracted. Given that a TeV or sub-TeV scale e^+e^- linear collider is not expected for some time, it is crucial to seek the optimal methods for disentangling the –inos produced at the LHC. Further, information on the heavier –ino states may prove crucial in deciding the reach of a future linear collider to perform the more precise measurements surely required.

References

1. For reviews of the MSSM and SUSY, see: G. Kane, ed., *Perspectives on Supersymmetry* (World Scientific, Singapore, 1998); M. Drees and S.P. Martin, in *Electroweak Symmetry Breaking and New Physics at the TeV Scale* (World Scientific, Singapore, 1996), T.L. Barklow, S. Dawson, H.E. Haber, and J.L. Siegrist (eds.), pages 146–215 (hep-ph/9504324); H. Baer *et al.*, *ibid.*, pages 216–286 (hep-ph/9503479).
2. H.-C. Cheng and I. Low, JHEP **0309** (2003) 051, JHEP **0408** (2004) 061; I. Low, JHEP **0410** (2004) 067; J. Hubisz and P. Meade, Phys. Rev. D **71** (2005) 035016.
3. H.-C. Cheng, K.T. Matchev and M. Schmaltz, Phys. Rev. D **66** (2002) 056006.
4. R.H. Dalitz, Phil. Mag. **44** (1953) 1068; E. Fabri, Nuovo Cimento **1** (1954) 479; M. Ferro Luzzi *et al.*, Nuovo Cimento **36** (1965) 1101.
5. K.A. Assamagan *et al.*, hep-ph/0406152.
6. M. Bisset, N. Kersting, J. Li, F. Moortgat and S. Moretti, *in preparation.*
7. G. Corcella *et al.*, JHEP **0101** (2001) 010, hep-ph/0210213; S. Moretti, K. Odagiri, P. Richardson, M.H. Seymour and B.R. Webber, JHEP **0204** (2002) 028.
8. J. Pumplin *et al.* (CTEQ Collab.) (CTEQ6), JHEP **0207** (2002) 012, *ibid*, **0310** (2003) 046; H.L. Lai *et al.* (CTEQ Collab.) (CTEQ5), Eur. Phys. J. C **12** (2000) 375.
9. H. Baer, F.E. Paige, S.D. Protopopescu and X. Tata, hep-ph/0001086, hep-ph/0312045.
10. F. E. Paige, hep-ph/9609373; I. Hinchliffe, F.E. Paige, M.D. Shapiro, J. Söderqvist and W. Yao, Phys. Rev. D **55** (1997) 5520.

Relativistic Mean Field Models at High Densities

A. Sulaksono, P.T.P. Hutauruk, C.K. Williams, and T. Mart
Departemen Fisika, FMIPA, Universitas Indonesia, Depok 16424, Indonesia

The effect of vector-isovector terms predicted by relativistic mean field models based on effective field theory on the equation of state and the neutrino mean free path in neutron stars has been studied. Using a procedure similar to Ref. [1] in treating the isovector sector, a parameter set (G2*) that predicts a much smaller proton fraction than the standard one (G2) can be obtained. We found that G2* has a softer equation of state compared to G2 at higher densities and the regularity of the neutrino mean free path does not depend on the proton fraction.

1. Introduction

The relativistic mean field (RMF) model allows ground state properties of finite nuclei to be described with relatively high precision (for a review see, e.g., Ref. [2]). If we extrapolate this model to the high density regime, it is known that the RMF model predicts a much stiffer equation of state (EOS) compared to the Bethe Brueckner Goldston (BBG) "data" [2,3], variational calculation of Akmal *et al.* [4], Dirac Brueckner Hartree Fock (DBHF) [5], non relativistic Brueckner Hartree Fock (BHF) with AV14 potential [6], or heavy ion experimental data [7], in the region where baryon density is around $2.0\rho_0 - 4.5\rho_0$, with ρ_0 the nuclear matter saturation density. This model also predicts a too large proton fraction (Y_p) [9] which leads to a too low threshold density to start the direct URCA cooling process in neutron star. Furthermore, the predicted neutrino mean free path (NMFP) shows an irregular behavior. On the other hand, there are parameter sets of RMF model which are specifically parameterized for interstellar purpose (e.g., parameter sets of Refs. [1,10]). This kind of parameter sets predicts a soft EOS at high densities but the predicted finite nuclei ground state properties are quite unsatisfactory. The reason of that is the parameter sets with soft EOS have a relatively large nucleon effective mass (M^*) for finite nuclei density, which has a consequence that their spin-orbit splittings are relatively narrow compared with experimental data [11]. This means that the RMF model is not "effective" enough to cover a wide range of densities and the reason lies in the form of the nonlinear self coupling [3].

In 1996 the chiral effective Lagrangian model was proposed by Furnstahl, Serot and Tang [12], whose mean field treatment is hereafter called the ERMF model. This model has accurate predictions of the ground state properties of nuclei and the extrapolation to high densities yields a soft EOS which is consistent with other calculations [3–6], as well as experimental data of Ref. [7,8]. Nevertheless, it is widely known that this model still predicts a too large Y_p in the neutron star [9]. This is caused by the role of isovector terms, in which the corresponding parameters are poorly constrained by insensitive isovector observables of finite nuclei.

Therefore, in this report, we adjust the isovector-vector channel of the ERMF model in order to have a reasonable neutron star Y_p without changing its ground state predictions in finite nuclei by fine tuning the symmetry energy ($E_{\rm sym}$) of the symmetric nuclear matter (SNM) to achieve a softer $E_{\rm sym}$ at high densities. We follow the prescription of Horowitz and Piekarewicz [1] for the adjustment procedure. Besides that, we also study the regularity of the predicted NMFP in neutron star.

2. Results

In Table 1 we list the parameter sets of the ERMF models (G2 and G2*) along with the parameter set of the standard RMF one. G2 is the standard ERMF parameter set, while G2* is a modified one with adjusted isovector-vector channel couplings.

The effects of the vector-isovector adjustment in $E_{\rm sym}$, Y_p, EOS and M* as a function of baryon density in the range of $(1-5)\rho_0$ are shown in Fig. 1. From the upper-left panel of Fig. 1 we can see that G2 and G2* have a similar value of $E_{\rm sym}$ in the nuclear matter saturation density. Nevertheless, G2* has a softer $E_{\rm sym}$ compared to G2 in higher densities. The agreement of G2* $E_{\rm sym}$ with other calculations [4–6] also appears in this figure.

Since we assume that the neutron star matter consists only of neutrons, protons, electrons, and muons, the relative fraction of each constituent can be determined by the chemical potential equilibrium and the charge neutrality of the neutron star at zero temperature. The results for Y_p of both parameter sets can be seen in the upper-right panel of Fig. 1. Since G2* has a softer $E_{\rm sym}$ than G2, G2* has a higher threshold density ($\rho_B \approx 2.5\rho_0$) for starting direct URCA compared to G2. The

Table 1. Coupling constants of the RMF models.

Parameter	G2	NL-Z	G2*
m_S/M	0.554	0.520	0.554
$g_S/(4\pi)$	0.835	0.801	0.835
$g_V/(4\pi)$	1.016	1.028	1.016
$g_R/(4\pi)$	0.755	0.771	0.938
κ_3	3.247	2.084	3.247
κ_4	0.632	-8.804	0.632
ζ_0	2.642	0	2.642
η_1	0.650	0	0.650
η_2	0.110	0	0.110
η_ρ	0.390	0	4.490

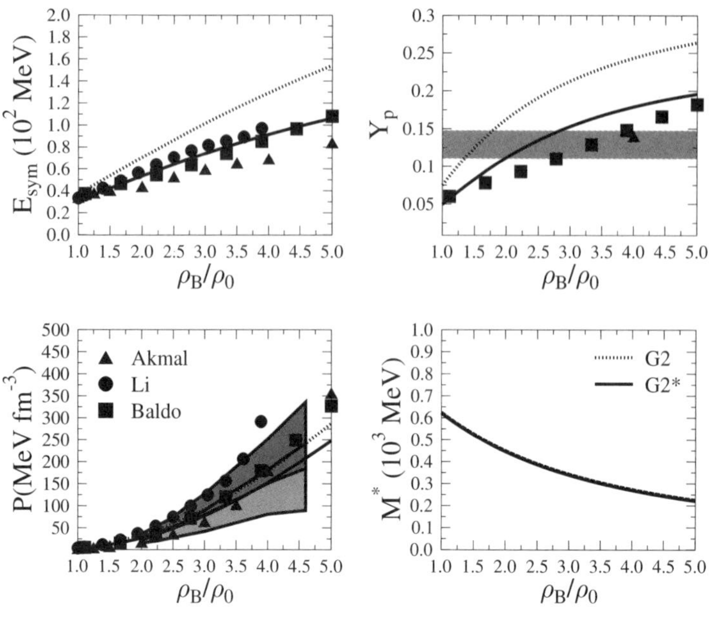

Fig. 1. EOS of the ERMF model at high densities. Shaded region in the upper-right panel corresponds to the proton fraction threshold for the direct URCA process, while shaded region in the lower-left panel corresponds to experimental data. The results obtained by other calculations are also shown.

predicted Y_p of G2* shows a better agreement with calculations of Refs. [4,6] rather than G2.

From the lower-left and lower-right panels of Fig. 1, it is obvious that both parameter sets have a similar trend in the pure neutron matter EOS and a same M^*, i.e., both have a soft EOS and have $M^* \sim 0.6M$ at the nuclear saturation density, but large M^* at high densities. Although less significant, the G2* parameter set has a softer equation of state compared with the G2 one at higher densities. This fact indicates that quantitatively the neutron star properties (masses, radii, etc) predicted by both parameter sets are not too different.

The results for the NMFP of both parameter sets are shown in Fig. 2. Clearly, the NMFP trend of both parameter sets is similar. No irregularity in the NMFP is observed, since both parameter sets have a large M^* at higher densities, only their magnitudes are different. This means that the vector-isovector contributions almost have no influence in controlling the regularity behavior of the NMFP in the neutron star.

3. Conclusion

The standard RMF model needs additional nonlinear terms in order to simultaneously produce accurate ground state predictions of finite nuclei and realistic de-

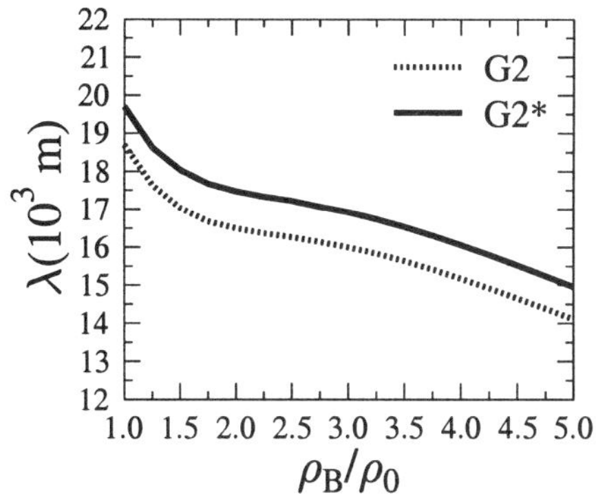

Fig. 2. Neutrino mean free path predicted by ERMF models.

scription in the high densities regime. The ERMF model has a strong theoretical ground to meet this requirement through the language of effective field theory. The reason that the ERMF model can cover a wide range of densities originates from the fact that the model has $M^* \sim 0.6M$ at the nuclear saturation density, but a large M^* and a soft E_{sym} at high densities.

References

1. C. J. Horowitz and J. Piekarewicz, *Phys. Rev. Lett.* **86**, 5647 (2001).
2. P. -G Reinhard, *Rep. Prog. Phys.* **55**, 439 (1989).
3. R. Brockmann and P. -G Reinhard, *Z. Phys.* **A 331**, 367 (1988).
4. A. Akmal, V. R. Pandharipande and D. G Ravenhall, *Phys. Rev.* **C 58**, 1804 (1998).
5. G. Q. Li, R. Machleidt and R. Brockmann, *Phys. Rev.* **C 45**, 2782 (1992).
6. M. Baldo, I. Bombaci, G. F. Burgio, *Astron. Astrophys.* **328**, 774 (1997).
7. P. Danielewicz, R. Lacey and W. G. Lynch, *Science* **298**, 1592 (2002).
8. P. Arumugam, B. K. Sharma, P. K. Sahu, S. K. Patra, T. Sil, M. Centelles, X. Vinãs, *Phys. Lett.* **B 601**, 51 (2004).
9. P.T.P. Hutauruk, C.K. Williams, A. Sulaksono and T. Mart, *Phys. Rev.* **C 70**, 068801 (2004).
10. N. K. Glendenning, *Astrophys. J.* **293**, 470 (1985).
11. A. Sulaksono, T. Mart and C. Bahri, *Phys. Rev.* **C 71**, 034312 (2005).
12. R. J. Furnstahl, B. D. Serot and H. B. Tang, *Nucl. Phys.* **A598**, 539 (1996); *Nucl. Phys.* **A615**, 441 (1997).

Transport Model Analysis of Fuctuations: Baryon-Strangeness Correlations and the Cumulant Method for Elliptic Fow Calculations

S. Haussler[1], X. Zhu[1,2,3], and M. Bleicher[2]

1) Frankfurt Institute for Advanced Studies (FIAS), Johann Wolfgang Goethe Universitt, Max-von-Laue-Str. 1, 60438 Frankfurt am Main, Germany
2) Institut fr Theoretische Physik, Johann Wolfgang Goethe Universitt, Max-von-Laue-Str. 1, 60438 Frankfurt am Main, Germany
3) Physics Department, Tsinghua University, Beijing 100084, China

1. Introduction

An extremely dense and excited phase of deconfined partonic matter (the so called Quark-Gluon-Plasma, QGP) is believed to be created in the course of heavy ion collisions. However, the confinement of the quarks and gluons makes a direct study of the QGP impossible because only hadrons will escape to the detectors after the system might have undergone a phase transition to the QGP. Thus, one has to find observables which will specify the properties of the matter at an early stage of the collision and will not loose this information in the hadronic stage. A number of different signals have been proposed to probe and study the QGP phase. Various results obtained at SPS and RHIC indicate that such a state of deconfined quarks and gluons may have already been obtained in the laboratory.

Among the signals, the ones related to fluctuations might be most adequate because the energy density, temperature and particle density strongly fluctuate on an event-by-event basis [1−3]. E.g., the baryon-strangeness correlations, recently proposed by Koch et al. [4], should allow to get insight about the onset of the deconfinement in temperature and volume.

The measure of the elliptic flow v_2 does also permit to gain information about the properties of the matter at an early stage. More precisely, this observable is sensitive to the early pressure gradient and thus to the equation of state [5]. It is the second Fourier harmonic in the transverse angular distribution of the emitted particles and is defined as $\langle cos2(\phi-\Phi_R)\rangle$. Its experimental determination is difficult, as one has to extract the unknown reaction plane Φ_R out of the data.

In a first part, we will discuss how the baryon-strangeness correlation should allow to gain information about the onset of the QGP phase. It will be studied as a function of the beam energy and centrality. In a second part, the UrQMD model is applied to study the many particle cumulant method used to experimentally determine the elliptic flow. Also for the elliptic flow analysis fluctuations are important and and must not be neglected in a quantitative discussion.

We study the baryon-strangeness coefficient C_{BS} and the elliptic flow v_2 with the help of the Ultra-relativistic Quantum Molecular Dynamics model (UrQMD v2.2). This model describes the full space-time evolution of heavy ions collisions in the energy domain from a few hundreds of MeV to several TeV in the laboratory frame. The interaction of primary and secondary particles is described by the excitation and fragmentation of color strings and via the formation and decay of resonances. This model has been shown to yield a reasonable description of inclusive particle distribution, and event-by-event studies already performed with the UrQMD have given reasonable results [3,6–9]. For a complete review of the model, the reader is referred to [10,11].

2. Baryon-Strangeness Correlations

The baryon-strangeness correlation was proposed [4] to specify the nature of the created matter (ideal QGP or strongly coupled QGP or hadronic matter). The general idea can be understood as follows: if one considers an ideal QGP, strange quarks will transport baryon number in strict proportion to strangeness and thus lead to a strong correlation between strangeness and baryon number. On the contrary, if the degrees of freedoms of the system are of hadronic nature, strange mesons can carry strangeness without baryon number. This in turn makes the correlation between baryon number and strangeness weaker.

To quantify the degree of correlation between baryon number and strangeness, the following coefficient was introduced: $C_{BS} = -3\frac{\langle BS\rangle - \langle B\rangle\langle S\rangle}{\langle S^2\rangle - \langle S\rangle^2}$, where B is the baryon charge and S is the strangeness. In the case of an ideal QGP, the obtained value can be calculated easily and $C_{BS} = 1$. Estimates from lattice QCD studies also give a value close to $C_{BS} \approx 1$. In contrast, for a hadron gas at a temperature $T = 170\ MeV$ and chemical potential $\mu_B = 0$ one finds $C_{BS} \approx 0.6$.

Because of the strong longitudinal flow created in the early stage of the collision, strangeness and baryon number should be frozen in a given reasonably wide rapidity window. The fluctuation might then survive hadronization and the rescattering of the hadronic phase. This leads to the proposal that the measured C_{BS} might reflect the initial stage correlation and therefor allows to extract information on the onset of deconfinement. In the following we use $|y| < 0.5$.

Since the UrQMD model is based on hadronic and string processes, a C_{BS} similar to a hadron gas is expected. Furthermore, the out-of-equilibrium properties of the expanding system and the rescattering among secondaries are taken into account. To be specific, C_{BS} is calculated event-by-event according to:

$$C_{BS} = -3\frac{\frac{1}{N}\sum_n B^{(n)}S^{(n)} - (\frac{1}{N}\sum_n B^{(n)})(\frac{1}{N}\sum_n S^{(n)})}{\frac{1}{N}\sum_n (S^{(n)})^2 - (\frac{1}{N}\sum_n S^{(n)})^2} \qquad (1)$$

$B^{(n)}$ and $S^{(n)}$ stand for the baryon number and strangeness in a given event n.

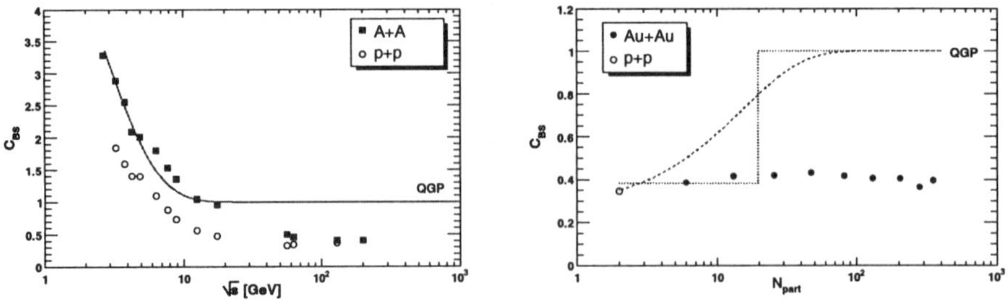

Fig. 1. Left: Correlation coefficient C_{BS} for central Au+Au/Pb+Pb (full symbols) and minimum bias $p+p$ collisions (open symbols) as a function of $\sqrt{s}$. The maximum rapidity accepted is $y_{max} = 0.5$. The solid curve depicts the expectation if a QGP is formed during the reaction. Right: Correlation coefficient for Au+Au collisions at $\sqrt{s} = 200$ AGeV as a function of the number of participants. The maximum rapidity accepted is $y_{max} = 0.5$. Full symbols are the results for Au+Au and the open symbol shows the p+p value. The smooth and step lines depict what is to be expected if the colliding system goes through a QGP phase.

Figure 1 (left) depicts the excitation function of C_{BS} from $E_{\rm lab} = 4$ AGeV to $\sqrt{s} = 200\ AGeV$ for Au+Au/Pb+Pb. As discussed in [4], C_{BS} is found to increase with increasing chemical potential μ_B and therefore with decreasing energy. The solid line illustrates what is to be expected if a QGP is formed in the reaction. A discrepancy with the data should be observed, thus allowing to map out the onset of QGP formation.

The dependence of C_{BS} on the number of participants is plotted in Figure 1 (right). The resulting values of of C_{BS} from the UrQMD model are flat as a function of the centrality. In case of a QGP, C_{BS} should increase form $p+p$ to very central collisions, either smoothly (smooth line) or as a step function (step line). This might allow to gain information on the volume at which the phase transition occurs at RHIC. For details, the reader is referred to [12].

3. Elliptic Flow: Fluctuations and Non-Flow Effects

Let us now turn to the elliptic flow analysis as discussed in detail in [13]. Here, the cumulant method [14] is tested for the calculation of the elliptic flow v_2. experimental problem is that the real reaction plane is not known a priori and is often determined using the flow vector of the event (reaction plane method [15]). It was shown that the reaction plane method is consistent with a 2-particle correlation method [15–17]. However, such a way to determine the reaction plane is influenced by the so called non-flow effects such as resonance decays, momentum conservation and jet production.

The many particle cumulant method was designed to overcome these difficulties. The idea is to extract the flow with many particle correlations, assuming that the non-flow correlations will then be negligible as compared to the 2-particle correlation. However, this method leads to the determination of $\sqrt{\langle v_2^2 \rangle}$ and not $\langle v_2 \rangle$. These two quantities are different if event-by-event fluctuation of v_2 are present.

Even though the v_2 obtained from the UrQMD is only 60% of the experimental value, the model is well suited for a comparison of the different methods. v_2 can be analyzed on an event-by-event basis and non-flow correlation and out-of-equilibrium dynamics of the reaction are naturally present. In addition, the real v_2 can be extracted from the model and can be compared to the different methods.

On Figure 2 (left), v_2 is plotted as a function of centrality. The elliptic flow extracted form $v_2\{2\}$ deviates strongly from the exact v_2. On the contrary, $v_2\{4\}$ and $v_2\{6\}$ show almost no difference and are close to the real v_2 value. Even though the v_2 fluctuations affect the v_2 measurement for the many particle cumulant method, this effect is negligible over a wide range of centralities $(10-50\%)$ and the non-flow correlation are eliminated with the 4 and 6-particles cumulant method.

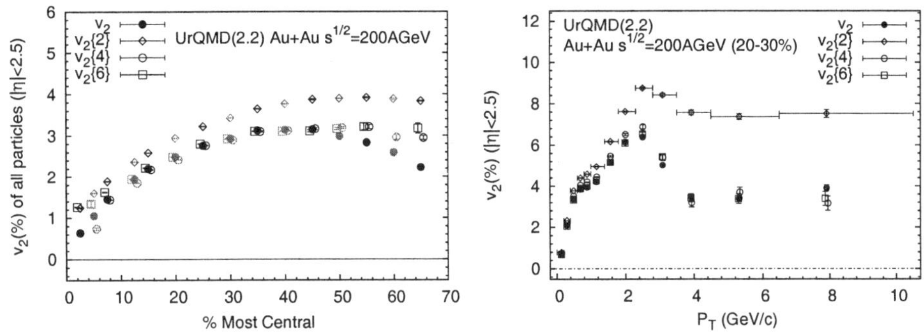

Fig. 2. Left: The integral v_2 results ($v_2\{2\}$, $v_2\{4\}$ and $v_2\{6\}$) from the cumulant method are compared to the exact v_2 in different centrality bins. The points are the corresponding results from the enlarged centrality bins which merge two of the original bins. Right: $v_2(p_T)$ in the semi-central collisions: results from the cumulant method are compared to the exact v_2.

It is also necessary to study the differential v_2 as non-flow effects are expected to give an important contribution at large p_T. Figure 2 (right) depicts the dependence of v_2 on p_T. The exact v_2 goes up with increasing p_T up to a maximum around 2.5 GeV. Then v_2 decreases with a further increase of p_T. $v_2\{2\}$ follows roughly the shape of the exact v_2 at low p_T but then stays roughly constant and higher than the exact v_2. This indicates the importance of non-flow contributions in the two-particles cumulant method at large p_T. This saturation of v_2 is consistent with the STAR measurements [18]. $v_2\{4\}$ follows the exact v_2, even though still slightly overestimating the exact v_2, thus indicating the non-flow contributions are not completely eliminated. The six particle cumulant method $v_2\{6\}$ eliminates the non-flow contribution and gives the same results as the exact v_2 in the whole p_T range.

4. Conclusion

In summary, the baryon-strangeness correlation coefficient C_{BS} was studied from $E_{\rm lab} = 4$ AGeV to $\sqrt{s} = 200$ AGeV with the string-hadronic transport model UrQMD. C_{BS} is shown to decrease from the lowest AGS energy to the top RHIC energies available, reaching about half the value expected in case a QGP existed in the early stage of the collision. Studied as a function of centrality, C_{BS} is roughly constant from $p + p$ to the most central $Au + Au$ collisions at $\sqrt{s} = 200$ $AGeV$. This allows to gain information on the onset of deconfinement in temperature and volume.

From the comparison between the true elliptic flow extracted from the UrQMD calculation with the $v_2\{4\}$ and $v_2\{6\}$ cumulant methods, we conclude that the non-flow contribution are mostly eliminated in $v_2\{n \geq 4\}$. v_2 fluctuations indeed affect the calculations with the cumulant method for central and very peripheral collisions. Nevertheless, the cumulant method gives a good estimate of the elliptic flow over a wide range of centralities ($\sigma/\sigma_{\text{tot}} \sim 10 - 50\%$). Finally, the six particles cumulant method $v_2\{6\}$ eliminates non-flow contributions at large p_T whereas the two particle cumulant $v_2\{2\}$ method is strongly influenced by non-flow effects (e.g. jets).

Acknowledgements

We are grateful to the Center for Scientific Computing (CSC) for providing the computing resources. This work was supported by GSI and BMBF.

References

1. L. Stodolsky, Phys. Rev. Lett. **75** (1995) 1044.
2. E. V. Shuryak, Phys. Lett. B **423** (1998) 9 [arXiv:hep-ph/9704456].
3. M. Bleicher *et al.*, Nucl. Phys. A **638** (1998) 391.
4. V. Koch, A. Majumder and J. Randrup, arXiv:nucl-th/0505052.
5. J.-Y. Ollitrault, Phys. Rev. D **46**, 229 (1992).
6. M. Bleicher *et al.*, Phys. Lett. B **435** (1998) 9 [arXiv:hep-ph/9803345].
7. M. Bleicher, S. Jeon and V. Koch, Phys. Rev. C **62** (2000) 061902 [arXiv:hep-ph/0006201].
8. M. Bleicher, J. Randrup, R. Snellings and X. N. Wang, Phys. Rev. C **62** (2000) 041901 [arXiv:nucl-th/0006047].
9. S. Jeon, L. Shi and M. Bleicher, arXiv:nucl-th/0506025.
10. S. A. Bass *et al.*, Prog. Part. Nucl. Phys. **41** (1998) 225 [arXiv:nucl-th/9803035].
11. M. Bleicher *et al.*, J. Phys. G **25** (1999) 1859 [arXiv:hep-ph/9909407].
12. S. Haussler, H. Stoecker and M. Bleicher, arXiv:hep-ph/0507189.
13. X. Zhu, M. Bleicher and H. Stoecker, arXiv:nucl-th/0509081.
14. N. Borghini, P.M. Dinh and J.-Y. Ollitrault, Phys. Rev. C **63**, 054906 (2001); N. Borghini, P.M. Dinh and J.-Y. Ollitrault, Phys. Rev. C **64**, 054901 (2001); N. Borghini, P.M. Dinh and J.-Y. Ollitrault, nucl-ex/0110016.
15. A.M. Poskanzer and S.A. Voloshin, Phys. Rev. C **58**, 1671 (1998).
16. J. Adams *et al.* (STAR Collaboration), nucl-ex/0409033.
17. C. Adler *et al.* (STAR Collaboration), Phys. Rev. C **66**, 034904 (2002).
18. J. Adams *et al.* (STAR Collaboration), Phys. Rev. Lett. **93**, 252301 (2004)

Some Interesting Features of Particles Produced at High Energy Heavy Ion Collisions

Mustafa Abdusalam Nasr and Selima Aboazoom

High energy physics group
Department of physics
7th October University, Misurata-Libya
Email:bayio@yahoo.com

In order to study some interesting characteristic of particles produced in collisions of silicon ions with nuclear emulsion targets at beam momentum 4.5 GeV/c per nucleon, experimental data have been analysed; average multiplicities of grey, black, shower, compound multiplicity and heavily ionizing tracks obtained in this work are compared with their corresponding values obtained for different projectile nuclei. The variations of $D(N_s)$ and $< N_h >$ with $< N_s >$ and $D(N_c)$ and $< N_h >$ with $< N_c >$ are also investigated. Finally, integral N_h distribution is plotted and fitted with a best- fitting analytical expression.

1. Introduction

Studies of heavy ion collisions at relativistic energies have attracted the attention of particle physicists, owing to the idea[1–6] that these studies might provide a chance to investigate the collective properties of high density and high temperature nuclear matter and invariance of various emission characteristics of particles produced in the collisions of composite systems[7].

The present work deals mainly with the results on the general features of the collisions initiated by silicon nuclei of primary momentum 4.5 AGeV/c with emulsion nuclei. These results are compared with their corresponding values obtained for various other projectiles . Furthermore, multiplicity distributions of different types of secondary particles and their correlation are investigated.

2. Experimental Details

The experimental data has been obtained by investigating nuclear emulsion stacks of NIKFI- BR2 of dimension (16.9 x 9.6 x 0.06) cm^3 irradiated by 4.5 A GeV/c silicon beam at Dubna synchrophasotron, Russia, were used in this study. The stacks were scanned using Nikon Microscopes having traveling stages.

The tracks were picked up at a distance of 3mm from the entrance edge of the stack; the tracks were followed, fast in the forward and slow in the backward directions, until they interacted or left the pellicle. To investigate the emission characteristics of secondary tracks produced in silicon-nucleus collisions, a random sample comprising 727 events with $N_h \geq o$, where N_h denotes the number of tracks produced in an interaction with relative velocities $\beta \leq 0.7$, was collected. Each interaction was looked at under 15X eye- pieces and 95X oil immersion objective. Secondary tracks produced in each interaction were classified as stated below:

i) Shower tracks: Shower tracks have relatively longer ranges in emulsion and have ionization less than 1.4 g_o, where go represents the minimum ionization of a singly charged particle; their number in a collision is designated by N_s.
ii) Grey tracks: Tracks with ionization lying in the range 1.4 g_o to 10 g_o and having ranges in emulsion, $R \geq 3$ mm are called grey tracks. Grey tracks are produced by protons, deuterons, tritons and some slow mesons. The number of grey tracks in an interaction is characterized by N_g.
iii) Black tracks: Tracks with ionization $g \geq 10 g_o$ and having ranges in emulsion, R<3mm are termed as black tracks. Black tracks are due to slow particles and evaporated fragments; their number in an interaction is denoted by N_b. grey and black tracks are jointly referred to as heavy tracks and their number in an interaction is designated by $N_h (= N_g + N_b)$.

3. Results and Discussion

The average values of the multiplicities of grey, black, shower, compound multiplicity and heavily ionizing tracks are listed in Table 1, along with the results reported by other workers[8–9,14–18] for different projectile at various incident energies. It is noticed from the table that the mean multiplicity of black tracks, within the limit of statistical error, is observed to be essentially independent of the mass and energy of the projectile. However, the average values of grey, shower, heavily ionizing tracks and compound multiplicity grow with increasing projectile mass. It may be of interst to note that $< N_s >$ and $< N_g >$ have also been observed by other worker[9] to increase with increasing projectile mass.

The variations of $< N_s >$ with $< N_h >$ and $< N_h >$ with $< N_c >$ for silicon-nucleus collisions are displayed in Fig. 1. Positive correlations are observed to occur in both cases and the data are nicely fitted by a linear relation of the type.

$$< N_i(N_j) >= a < N_j > +b \tag{1}$$

Where $N_i, N_j = N_s, N_h, N_c$ and $i \neq j$. The values of the constants a and b, along with the results reported by other workers[8–9] are presented in Table 2. It may be seen in the table that the values of the slope parameter increase with increasing projectile mass.

The value of dispersion, $D(N_s) = (< N_s^2 > - < N_s >^2)^{1/2}$, which is of much physical significance, is determined at 4.5 GeV/c per nucleon for silicon-nucleus collisions.

Variations of $D(N_s)$ with $< N_s >$ and $D(N_c)$ with $< N_c >$ are shown in Fig. 2. It may be concluded that dispersion grows linearly with the average values of shower and compound multiplicities. This is in fine agreement with the predictions of the KNO scaling. The data are reproduced reasonably well by a linear equation. The values of the constant are listed in Table 2. It is found that positive slopes are observed in the multiplicity correlations of secondary particles produced in silicon-nucleus collisions.

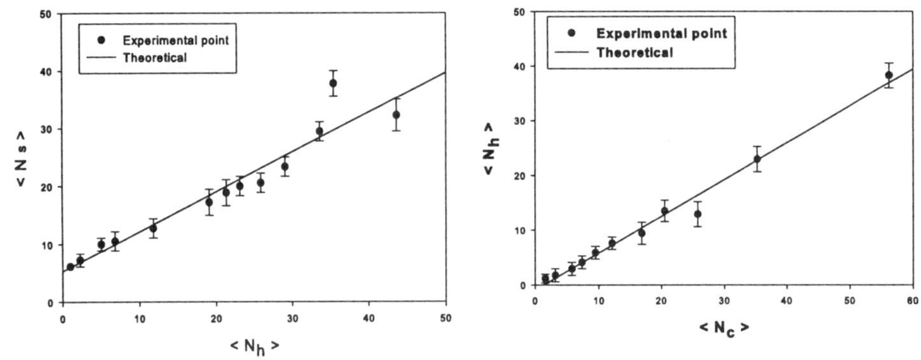

Fig. 1. Left: Multiplicity correction between $\langle N_s \rangle$ and $\langle N_h \rangle$; Right: Multiplicity correction between $\langle N_h \rangle$ and $\langle N_c \rangle$.

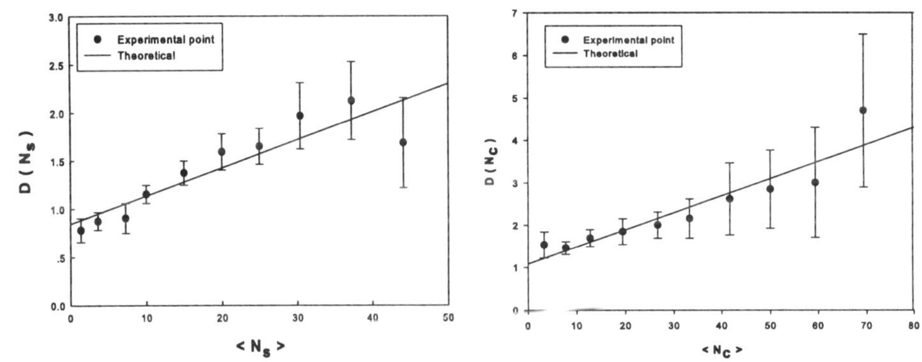

Fig. 2. Left: Variation of $D(N_s)$ with $\langle N_s \rangle$; Right: Variation of $D(N_c)$ with $\langle N_c \rangle$.

Acknowledgments

The authors would like to thank Prof. Mohamed Mahmud Ben Ahmeida Vice-Chancellor of 7th October University for his support and various helps.

Table 1: Mean Multiplicities of Secondary tracks.

Projectile type	Energy GeV/nucleon	$<N_g>$	$<N_b>$	$<N_s>$	$<N_h>$	$<N_c>$	Ref.
p	4.5	2.81±0.06	3.77±0.08	1.63±0.02	6.58±0.11	4.44±0.06	8
d	3.6	2.34±0.07	5.78±0.15	3.08±0.05	8.12±0.17	5.42±0.09	15
α	4.5	4.70±0.20	4.70±0.20	3.90±0.10	9.40±0.28	8.60±0.22	16
C	4.5	5.93±0.34	4.49±0.24	7.67±0.19	10.42±0.42	13.60±0.42	9
N	2.1	5.99±0.31	4.57±0.22	8.85±0.28	9.86±0.38	14.14±0.42	14
O	4.5	7.60±0.60	4.88±0.29	10.50±0.60	12.48±0.67	18.10±0.85	17
Ne	4.1	7.58±0.23	7.63±0.17	8.38±0.20	15.21±0.29	15.96±0.32	18
Si	4.5	8.59±0.26	4.69±0.23	11.26±0.33	13.28±0.41	19.85±0.45	9
Si	**4.5**	**8.60±0.18**	**4.30±0.16**	**11.50±0.20**	**15.69±0.20**	**19.95±0.40**	**This Work**

Table 2: Values of the constants appearing in eq. (1)

Projectile	Correlations	a	b	Ref.
P	$<N_s>$ - $<N_h>$	----	-0.03±0.01	8
C	--do---	----	0.32±0.01	9
Si	--do---	----	0.81±0.02	9
Si	--do---	5.90±1.64	0.70±0.07	This work
Si	$<N_h>$ - $<N_c>$	-0.65±0.66	0.71±0.03	This work
Si	$D(N_s)$ - $<N_s>$	0.83±0.10	0.03±0.004	This work
Si	$D(N_c)$ - $<N_c>$	1.03±0.17	0.04±0.01	This work

References

1. A.M. Baldin, invited talk at VI Int. Conf. on High Energy Physics and Nuclear Structure (Los Alamos and Santa Fe, 1975).
2. G. Byam and S.A. Chin, Phys. Lett. B62, 241 (1976).
3. J. Boguta, Phys. Lett. B109, 251 (1982).
4. M.A. Nasr, T. Ahmad and M. Irfan, J. Phys. Soc. Japan, 63, 2936 (1994).
5. T. Ahmad, M.A Nasr and M. Irfan, Phys. Rev C, 47, 2974 (1993).
6. M.A. Nasr and O. Trifees, XXXI international Symposium on Multiparticle Dynamics, Datong, China, Sept 1-7(2001).
7. J.I. Kapusta, Nucl. Phys. B148, 461 (1979).
8. Z. Phys. A302, 133 (1981).
9. T. Ahmed and M. Irfan, Nuovo Cimento A106, 171(1993).
10. J. Gosset, H.H.Gutbrod, W.G. Meyer, A.M. Poskanzer, A. Sandval, R. Stock and G. D. Westfall, Phys. Rev. C16, 629 (1977).
11. A. Bialas, W. Czyz and W. Furmanski, Acta. Phys. Pol. B8, .585(1977).
12. A. Capella and Y. Tran than Van, Phys. Lett. B93, 146 (1980).
13. K. Kinoshita, A. Minaka and H. symyoshi, Prog. Theor. Phys. 61, 165(1979).
14. G.M. Ghernov, K.G. Gulemov, U.G. Gulyamov, S.Z. Nasyrov and L.N.Svechnikova, Nucl. Phys. A280, 478 (1977).

15. E.S. Basova, G.M. Chernov, K.G. Gulamov, U.G.Gulyamov and B.G. Rakhimbaev, Z.Phys. A297, 383(1978).
16. DGKLMTW Collaboration, JINR Dubna Comm. P. 1-8313(1974).
17. V.A. Antonchik, V.A. Balcaev, A.V. Belousov, S.D. Bogdanov, V.L. Ostroumov and M.V. Pavlovets, Sov. J. Nucl. Phys. 39, 774(1984).
18. B.P. Bannik, A. EL - Naghy, S.A Krasnov, G.S. Shabratova and K. D. Tolstov, Z.Phys. A329, 341(1988).

Concluding Remarks

B. F. Gibson*

Theoretical Division,
Los Alamos National Laboratory,
Los Alamos, New Mexico 87545, USA
E-mail: bfgibson@lanl.gov

These observations represent the author's personal perspective regarding the ideas presented at this third Asia Pacific Conference on Few-Body Problems in Physics. It is not intended as a summary talk in the traditional sense, but is comprised of remarks emphasizing physics *not* speakers. Many of the talks covered results of those other than the speaker, and the ideas discussed will hopefully outlive the rapporteurs bringing these ideas to life at APFB05 held at the Suranaree University of Technology in Nakhon Ratchasima.

1. Introduction

The conferees owe a debt of gratitude to the organizers for the scientific program, one which contained a large number of presentations of significance. Not only were there important plenary talks, but a number of the parallel session papers contained results of interest to numerous participants. Areas of emphasis included astrophysics, meson physics, and quark models in addition to the more traditional nucleon-deuteron scattering/reactions, electromagnetic probes, electroweak physics, and strangeness physics. Non-traditional few-body physics fields which were showcased included relativistic heavy ions, the atomic physics (the Coulomb problem), and charge symmetry breaking. Finally, one should certainly remember the address by C. N. Yang in the formal opening of the conference, in which he explored three themes of 20th century physics: quantum mechanics, symmetry, and phases. Each of Yang's "melodies" plays a significant role in our research efforts in few-body physics.

From the realm of relativistic heavy ion collisions we learned about progress in identifying signals for the quark/gluon plasma. Directed flow plays a key role: 1) the bounce-off motion in the plane of the collision requires additional pressure, beyond that generated by hadron models, to match experiment, and 2) a squeeze out or elliptic flow out of the plane is observed. Jet quenching provides another strong indication: 50% of the jets are suppressed by hadron final-state interactions, but the remainder of the suppression requires absorption by a quark/gluon plasma.

*Supported by the U.S. Department of Energy under contract W-7405-ENG-36

From the atomic physics treatment of the Coulomb three-body problem we learned about the successes of the time-dependent coupled-channel method. Matrix diagonalization is avoided by solving a system of linear algebraic equations. First proposed in the early 1980s, applications in the last decade have provided excellent descriptions of e^- scattering from hydrogen, e^+ scattering from hydrogen, and $\bar{p}$ scattering from hydrogen.

In the study of charge symmetry breaking we heard about the IUCF $d+d \to \alpha + \pi^0$ experiment in which one seeks evidence for the breaking of charge independence and charge symmetry. The difficulty is distinguishing the π^0 which decays into two photons from the $d+d \to \alpha + \gamma + \gamma$ charge-symmetry-allowed reaction. Also, we learned about the TRIUMF measurement for the $n+p \to d+\pi^0$ reaction in which a nonzero fore/aft asymmetry $A_{fb} = [\sigma(\theta) - \sigma(\pi - \theta)]/\ [\sigma(\theta) + \sigma(\pi - \theta)]$ implies charge symmetry breaking. A positive result has been reported:

$$A_{fb} = [17.2 \pm 8.0(stat) \pm 5.5(sys)] \times 10^{-4}\ .$$

2. Astrophysics

In the world of nuclear astrophysics a number of points were emphasized. An effective field theory (EFT) calculation of the $^3\mathrm{He} + \mathrm{n} \to ^4 \mathrm{He} + \gamma$ reaction was found to be in agreement with experiment if the model parameters were fixed by fitting to the magnetic moments of the ^{3}H and ^{3}He. Meson exchange currents play a large role because of the pseudo-orthogonality of the $^3\mathrm{He} + \mathrm{n}$ continuum and the ^{4}He ground state.

Bose-Einstein condensation due to the strongly attractive $I = 0$ s-wave KN interaction and the p-wave πN interaction was suggested to play a significant role in determining the mass/radius relationship in neutron stars. In addition, hyperons produced at the cost of nucleons were suggested to play an essential role. Thus, we are faced with an important question: *What is the role of strangeness in the Equation of State for neutron star matter?*

Finally, the possible role of $n\alpha\alpha$ resonances in nuclear astrophysics reactions was identified. A resonant fusion process in which the neutron acts as a catalyst was proposed.

3. Meson Physics

Assuming that the $\Lambda(1405)$ is a K^-p bound state, one can predict the existence of a $^4\mathrm{He}(\mathrm{K}^-,\mathrm{p})\mathrm{K}^-\mathrm{pnn}$ bound state, the $S^0(3115)$. Evidence for such a state with a binding energy of $\simeq$ 200 MeV has been reported from KEK. A similar prediction can be made for a K^-pp state. Evidence has been reported from DAΦNE for a state with binding energy of 115 MeV. It was suggested that the decay into Λp be sought at COSY, and that observation of $d(e,e'K^+)pnK^-$ might be a possibility at JLab, although the possible effects of the numerous open channels was not addressed.

"Exotic" structures observed in hadron-hadron s-wave scattering (*e.g.*, $\pi\pi$ scattering) appear to result when t-channel exchange plays a role comparable to that of s-channel resonances, which dominate for higher partial waves.

The atomic spectroscopy (energy shifts and decay widths) of the $\bar{p}D$ system is sensitive to the $\bar{N}N$ tensor force, which leads to s-wave and d-wave coupling, and to the finite size of the D core (in p-wave $\bar{p}D$ states).

A meson model description of the three-body decays of the B meson (into $\pi\pi K$ and KKK) under the assumption that $\pi\pi$ and KK pairs interact in isospin zero s-waves, requires the inclusion of charmed penguins (originating from enhanced charmed quark loops) to describe simultaneously the $f_0(600)$ and the $f_0(980)$.

Chiral symmetry (the limit in which $m_\pi \to 0$) plays an important role in NN forces. The contribution to the two-pion-exchange component of the NN potential has been verified; it relates directly to the πN scattering amplitude. We await the further exposition of the role of the three-pion-exchange component as well as the possible contribution to the spin dependence of the three-nucleon force.

4. Quark Models

Several results were reported for chiral quark models. The πN sigma term is estimated to be $\sigma_{\pi n} \simeq 55$ MeV compared with a chiral perturbation theory value of 45 MeV and an experimental estimate of $\simeq 60$ MeV. (The differing results for chiral perturbation theory are likely the result of fitting different experimental data.) An estimate for the value of G_M^S is consistent with zero. For Compton scattering the values of the electric and magnetic polarizability agree qualitatively with experiment (less well than does chiral perturbation theory, but there are fewer parameters). One finds improved KN phase shifts, a ΔK bound state, and a ΛK resonance.

Five-body calculations for the pentaquark system in an Isgur-Karl quark potential model were reported. Although no state in the vicinity of the Θ^+(~100 MeV) was found, there did appear a sharp resonance around 540 MeV with $J^\pi = \frac{1}{2}^-$ and a broad resonance around 520 MeV with $J^\pi = \frac{1}{2}^+$.

In a quark delocalization, color-screening model, predictions for the Θ^+ were in the $E = 1500 - 1700$ MeV range. It was suggested that multi-quark forces were needed to produce such an object. There is no valid calculation yielding the observed narrow width for the Θ^+.

5. Nucleon-Deuteron Scattering and Reactions

The data coming from the laboratories in Japan and The Netherlands have provided opportunities to test our nuclear force models and to push our ability to incorporate the Coulomb force into the two-charged-particle three-body continuum problem.

Nd Radiative capture is an ideal reaction for which to compare theory and experiment. The three-body ground state is of finite size and limits the number of partial waves which contribute significantly. (Higher partial waves should be included, at least in a plane wave approximation.) Unfortunately, the 200 MeV Japanese $p + d \to^3$ He $+ \gamma$ data are in disagreement with the available Faddeev calculation: while A_{yy} shows agreement, A_{xx} ($\simeq A_{yy}$ experimentally) does not.

In contrast, the KVI data at 180 MeV appear to be in agreement with Faddeev calculations.

A total cross section measurement for *nd* scattering from LANL shows a discrepancy between theory and data that becomes noticeable above 100 MeV. The size of the discrepancy increases as the energy of the projectile increases. Detailed calculations were presented which investigated the size of relativistic effects in *nd* scattering. Effects in the cross section were shown to be rather small. Separate calculations regarding the effects of three-body forces in *nd* elastic scattering showed that the model studied failed to explain the polarization data.

From KVI we have *pd* elastic scattering data ranging in energy from 108 MeV to 190 MeV. The data at 135 MeV disagree with similar measurements from Japan. Bochum calculations employing the AV18 + UR9 two-body and three-body force models reproduce the data qualitatively, but the discrepancy increases as the scattering angle increases; the percentage difference approaches 30%-40% at back angles. A Hannover calculation, which uses a Δ isobar rather than a static three-nucleon force, produces similar analyzing powers and a similar differential cross section discrepancy. The inclusion of Coulomb suggests that the discrepancy must be due to another factor, such as relativistic effects or the structure of the three-nucleon force.

Calculations for $\vec{d}p$ breakup at 130 MeV show that Coulomb effects are significant. However, adding a three-body force still leaves a discrepancy with the data from KVI, even though the three-body force effect improves the calculation relative to the data.

At 250 MeV *pd* breakup calculations agree reasonably with *nd* breakup calculations. However, the $d(p,p)pn$ data disagree with the model calculations by 50% for scattering angles larger than 120°.

On the theoretical side there were multiple proposals to implement Coulomb in momentum-space codes. Among the proposals discussed, exponential screening plus renormalization in a two-potential approach appears to provide a sound numerical solution from about 1 MeV to 300 MeV. The Yukawa screening converges much more slowly than the $\exp(\mathrm{r/R})^n$ form demonstrated, where $3 \leq n \leq 6$. Square well screening produces a well known oscillatory behavior. A value of R much less than 100 fm provided good convergence. The calculations presented for elastic scattering and breakup as well as for photonuclear processes were impressive.

The TUNL group reported on a HIGS precision measurement of $^3\mathrm{He}(\gamma,\mathrm{d})\mathrm{p}$ for photon energy in the range of $10-15$ MeV. The high statistics (2%) cross section data agree with older measurements; the best hope for systematic errors is 5%.

6. Electromagnetic Probes

An impressive measurement of the virtual-photon Δ electroproduction from the proton was reported, an almost complete measurement of the response functions. Some 14 of the 18 response functions were determined along with two Rosenbluth cross sections. The goal is the determination of the small quadrupole amplitude and the corresponding deformation of the nucleon.

Radiative pion photoproduction as a tool to determine the magnetic moment of the Δ^+ was reported. Sensitivity to the anomalous magnetic moment was found in the circular polarization of the photon.

The electrodisintegration measurements for few-nucleon systems at JLab were summarized. High momentum structure functions of ^{2}H and ^{4}He have been mapped out using the $(e, e'p)$ reaction. Study of the continuum in ^{3}He(e, e'p)X shows that the three-body breakup channel contains significantly more strength than does the two-body channel.

The neutron structure function investigation at JLab was outlined. Low x data examine the sea quark region while $x > 0.2$ data look at the valence quark region. Results agree with constituent quark model predictions but not with those of pQCD.

The proton form factor measurements at JLab verify that the results from the Rosenbluth cross section separation and the recoil polarization method disagree. Radiative corrections to the former are significant, and heretofore neglected two hard photon contributions affect the form factor observables. (A comparison of e^+p and e^-p provides a direct measurement of two-photon effects.) One can ask whether the $\langle j_\mu|\frac{1}{Q^2}|J^\mu\rangle$ analysis is valid when two-photon processes play such a significant role.

The neutron charge form factor measurements of G_n^E at Bates have been a BLAST success. Using a polarized ^{2}H target and quasielastic scattering of polarized electrons, the collaboration has extracted the ratio G_n^E/G_p^E over a Q^2 range of 0.012-0.80 (GeV/c)^2. The systematics are at the 6% level with half the data analyzed.

7. Electroweak

Even though the nucleon contains no valence quarks of the strange variety, estimates are that as much as 10% of the nucleon mass could arise from its strangeness content. Also, as much as 10% of the proton spin could come from the strange quark. Why the nucleon mass ($\simeq 940MeV$) is so different from the sum of three valence up/down quark masses is a fundamental question. The implication is that the QCD vacuum is not a weakly interacting dilute gas; the density of $q\bar{q}$ pairs in the QCD vacuum must be high. Determining a value for the weak form factors of the proton is the goal of several experiments. One prediction at this meeting by a prominent theorist was, "If the experimental value of G_M^S is positive, then QCD is wrong."

Four parity non conservation (PNC) experiments at MIT-Bates, MAMI, and JLab, using polarized electrons, have been completed or are in progress. In each case one is seeking to measure neutral weak form factors separately from axial current form factors through interference with known electromagnetic currents in the $V - A$ coupling scheme. The SAMPLE measurement at Bates is complete; the resulting errors are large because the measurements were done at back angles, where the cross sections are smaller. PVA4 at MAMI as well as G0 and HAPPEX at JLab have results but are continuing. A4 and SAMPLE are consistent, although the nominal values differ and that from SAMPLE seems high. A4 appears to agree with G0, which are forward angle measurements. A4 and HAPPEX taken together suggest a positive G_M^S. A combined analysis of the currently available data yields the following:

$$G_E^S = -0.013 \pm 0.028 \ ,$$
$$G_M^S = 0.062 \pm 0.31 \pm 0.062(2\sigma) \ ,$$

where the latter uncertainty is an estimate of systematics.

The G0 back angle phase II experiment will take data in 2006. Back angle measurements in A4 are scheduled. HAPPEX ^{4}He represents the future effort at JLab, and should determine G_E^S uniquely. That is needed to define the sign of G_M^S, because all other measurements determine a linear combination, $G_E^S + \eta G_M^S$.

The current situation can be summarized as:

• Be aware that different experiments use different electromagnetic form factors in their analyses.

• G_E^S appears to be unmeasurably small.

• G_M^S appears to be finite, although it is compatible with zero at the 2σ level.

• The lattice extrapolation quoted at this meeting was $G_M^S = -0.046 \pm 0.019$.

8. Strangeness Physics

The strangeness degree of freedom adds a new dimension to our evolving picture of nuclear physics. We should investigate whether the intuition carefully developed in the non strange sector (of conventional nuclei) extrapolates to explain the observations in the strangeness sector of Λ hypernuclei **or** whether these complex models are merely sophisticated interpolation tools valid primarily within the two-dimensional (neutron-proton) isospin space where they were developed.

Strangeness -1 few-body hypernuclei were reviewed comprehensively. These light hypernuclei constrain our models of the baryon-baryon interaction, which play an important role in nuclear astrophysics. They test the assumption of $SU(3)_f$. $^3_\Lambda$H plays the role of ^{2}H in strangeness 0 physics. Its $\Lambda + d$ structure probes the long range properties of the ΛN interaction. Because there is no one-pion-exchange force and the K and K^* exchange cancel, the tensor force plays but a small role in second order. The significant $\Lambda N - \Sigma N$ coupling can be reduced to an effective ΛNN three-body force; there is a dispersive term in which the spectator nucleon generates an energy dependence in the ΛN interaction plus a true three-body-force term. The dispersive term is repulsive, and the true three-body-force term is attractive. Also, in the A=4 isodoublet the true three-body force is attractive. One awaits improved measurements for these interesting systems at J-PARC.

The physics of Λ hypernuclei differs significantly from that of conventional, non strange nuclei. Some of the novel aspects of strangeness -1 flavor nuclear physics include:

• The ΛN interaction has no direct one pion exchange component, but it does involve significant ΣN coupling which plays a prominent role in hypernuclei and results in a significant ΛNN three-body force.

• The hypertriton is barely bound, in contrast to the conventional triton; the $\simeq 130$ MeV Λ separation energy results in the quintessential halo system in which the Λ-^{2}H radius is more than 5 times the radius of the ^{2}H core.

• The $^4_\Lambda$H, $^4_\Lambda$He isodoublet exhibits sizeable charge symmetry breaking and a large spin-flip E_γ between ground and excited states; ΛN-ΣN mixing plays an important role.

The weak mesonic decay of hypernuclei is first cousin to the neutron β decay of conventional nuclei. Decay rates and final-state decay product momentum distributions test our model hypernuclear wave functions. The measured neutron/proton stimulated emission for non-mesonic weak decay now agrees with experiment. Three-body decay processes were suggested to contaminate the singles data in comparison with coincidence measurements.

Double Λ hypernuclei loom large in the future of the new J-PARC facility. On that basis a number of four-body cluster-model $(\alpha + x + \Lambda + \Lambda)$ predictions for the energies of $\Lambda\Lambda$ hypernuclei have been generated using an effective $\Lambda\Lambda$ interaction fitted to the $_{\Lambda\Lambda}^{\ 6}$He binding of the NAGARA event in an $\alpha\Lambda\Lambda$ cluster model. In a separate study, 4-, 5-, and 6-body calculations of s-shell $\Lambda\Lambda$ hypernuclei were generated. A deeply bound $_{\Lambda\Lambda}^{\ 5}$H system was predicted, due to the $\Xi N - \Lambda\Lambda$ conversion that converts the ^{3}H core into a deeply bound ^{4}He core of $^5_\Xi$He..

9. The Pentaquark

The search for the Θ^+, a narrow exotic strangeness +1 system with minimum quark content of $uudd\bar{s}$, has drawn significant attention for the past year. Evidence reported from LEPS and CLAS have stirred active searches of old data by numerous groups. More than 10 experiments have reported evidence for such a state. A number of high energy inclusive experiments have reported negative results in their searches. The primary criticism of the Θ^+ claim has been the low statistics and the variation in the reported value of the mass. The narrow width is not understandable by any simple mechanism.

A new CLAS dedicated, high-statistics experiment has now reported an upper limit of 450 pb for the production cross section. There is no overlap in the kinematics with the LEPS measurement.

Theoretically QCD does not prohibit pentaquark states, but so far they have not been seen. Several quark approaches have predicted pentaquark baryons. However, the predicted masses are generally much higher than the reported values of around 1540 MeV. Also, usually many partner states with different quantum numbers are predicted, states which have not been observed. Moreover, coupling to KN usually leads to the prediction of a broad width, in contradiction to the narrow width claimed for the observed pentaquark structure.

We await the results of the other dedicated experiments before reaching a verdict regarding the existence of the pentaquark.

10. Observations

Phenomenological models have limits. We must understand those limits prior to attempting to drive the models to make predictions beyond the realm which they were designed to describe.

Let us avoid augmenting models in a manner that corresponds to adding epicycles upon epicycles just to fit the data. There should be significant physics underlying a model assumption.

As a community let us address the question: What NN measurements are needed before all of the lower energy accelerators vanish?

Unfortunately it is the case in many areas of nuclear physics research that the overlap between the domain of calculations that are easily performed by theorists and the domain of measurements that are easily made by experimentalists is limited. Nonetheless, nuclear physicists are working diligently to expand that overlap between experiment and theory and our fundamental understanding of the underlying physics. We look forward to the progress to be made in unraveling the mysteries of the nucleus by August 2006 when the 18th International Few-Body Conference is scheduled for Santos, Brazil.

In order to give you something to think about during the intervening months, let me leave you with a question to ponder. You have no doubt heard of the classic atomic physics question: The sky is blue; what is Avagadro's number? Let me pose for you a classic few-body nuclear physics question: Why should Hans Bethe have known in 1932 that $\mathrm{r}^2_{chg}(^3\mathrm{He}) > \mathrm{r}^2_{chg}(^3\mathrm{H})$, even before the measurements were made by Hofstadter in 1964? *Hint*: The answer has nothing to do with the Coulomb repulsion between the protons in ^{3}He.

Acknowledgments

The research of the author is supported by the U.S. Department of Energy under contract W-7405-ENG-36. He thanks the organizers for the invitation to listen so carefully to the presentations at the conference.

LIST OF PARTICIPANTS

Ricardo Alarcon	Department of Physics and Astronomy, Arizona State University, Box 871504, Arizona, U.S.A., alarcon@asu.edu
Shunichi Ando	Theory Group, TRIUMF, 4004 Wesbrook Mall, Vancouver, B.C.V6T 2A3, Canada, sando@triumf.ca
Debades Bandyopadhyay	Saha Institute of Nuclear Physics, 1/AF Bidhannagar, Kolkata 700064, India, debades.bandyopadhyay@saha.ac.in
Hyoung C. Bhang	School of Physics, Seoul National University, Soul 151-747, Korea, bhang@phya.snu.ac.kr
Aleksandra Biegun	Institute of Physics, University of Silesia, PL-40007, Katowice, Poland, abiegun4@nuph.us.edu.pl
Mike Bisset	Department of Physics, Tsinghua University, Beijing 100084, China, bisset@mail.tsinghua.edu.cn
Marcus Bleicher	Institute for Theoretical Physics, J. W. Goethe University, 60064 Frankfurt am Main, Germany, bleicher@th.physik.uni-frankfurt.de
Fu-Guang Cao	Institute of Fundamental Sciences PN461, Massey University, Private Bag 11 222, Palmerston North, New Zealand, F.G.Cao@massey.ac.nz
Wen-Chen Chang	Institute of Physics, Academia Sinica, Taipei 11529, Taiwan, changwc@phys.sinica.edu.tw
Sampart Cheedket	Department of Physics, Faculty of Science, Thaksin University, Song-Khla 90000, Thailand, sampart@hotmail.com
Yu-Chun Chen	Department of Physics, National Taiwan University, Taipei, Taiwan, snyang1@phys.ntu.edu.tw

Achim Czasch	Institut fur Kernphysik, University Frankfurt, D-60438 Frankfurt, Germany, czasch@atom.uni-frankfurt.de
Arnoldas Deltuva	Centro de Fisica Nuclear da Universidade de Lisboa, P-1649-003 Lisboa, Portugal, deltuva@cii.fc.ul.pt
Bertrand Desplanques	Laboratoire de Physique Subatomique et de Cosmologie, F-38026 Grenoble Cedex, France, desplanq@lpsc.in2p3.fr
Yubing Dong	Institute of High Energy Physics, The Chinese Academy of Sciences, Beijing 100049, P. R. China, dongyb@mail.ihep.ac.cn
Wolfgang Eyrich	Physical Institut, University of Erlangen-Nuremberg, Germany, eyrich@physik.uni-erlangen.de
Imam Fachruddin	Departemen Fisika, Universitas Indonesia, Depok 16424, Indonesia, imamf@fisika.ui.ac.id
Amand Faessler	Institute for Theoretical Physics, Tuebingen University, Germany, ptifa01@uni-tuebingen.de
Yoshikazu Fujiwara	Department of Physics, Kyoto University, Kyoto 606-8502, Japan, fujiwara@ruby.scphys.kyoto-u.ac.jp
Tomokazu Fukuda	Osaka Electro-Communication University, Osaka, Japan, fukuda@isc.osakac.ac.jp
Ebrahim Ghanbari Adivi	Physics Department, Yazd University, Yazd, Iran, eghanbari@yazduni.ac.ir
Benjamin Gibson	Los Alamos National Lab, P.O. Box 1663 Los Alamos, NM 87545, USA., bfgibson@lanl.gov
Shalev Gilad	Laboratory of Nuclear Science, Massachusetts Institute of Technology, Massachusetts, USA, sgilad@mitlns.mit.edu
Suresh Chand Goyal	Department of Physics, Agra College, Agra-282002, India, suresh_c_goyal@rediffmail.com
Seema Gupta	Department of Physics, Agra College, Agra-282002, India
Mohammadreza Hadizadeh	Department of Physics, University of Tehran, P.O.Box 14395-547, Tehran, Iran, hadizade@khayam.ut.ac.ir
John Hedditch	Department of Physics, University of Adelaide, Adelaide SA5005, Australia, jhedditc@physics.adelaide.edu.au

Kenneth Hicks	Ohio University, Athens, Ohio 45701, USA., hicks@ohio.edu
Emiko Hiyama	Department of Physics, Nara Women's Univerisity, 630-8506, Japan, hiyama@cc.nara-wu.ac.jp
Nouphy Hompanya	Department of Physics, National University of Laos, DongDok Campus, P.O. Box:7322, Vientiane, Lao PDR., saynamxong.yahoo.com
Atsushi Hosaka	Research Center for Nuclear Physics (RCNP), Osaka University, Japan, hosaka@rcnp.osaka-u.ac.jp
Fei Huang	Institute of High Energy Physics, P.O. Box 918-4, Beijing 100049, PR China, hfbin@mail.ihep.ac.cn
Chang Ho Hyun	Institute of Basic Science, Sungkyunkwan University, Suwon, Korea, hch@color.skku.ac.kr
Souichi Ishikawa	Hosei University, 2-17-1 Fujimi, Chiyoda, Tokyo 102-8160, Japan, ishikawa@i.hosei.ac.jp
Masahiko Iwasaki	Advanced Meson Science Laboratory, DRI, RIKEN, Wako, 351-0198, Japan, advanced_meson@riken.jp
Aksel S. Jensen	Department of Physics and Astronomy, University of Aarhus, DK-8000 Aarhus C, Denmark
Alisher Kadyrov	Division of Science and Engineering, Murdoch University, Perth 6150, Australia
Nasser Kalantar-Nayestanaki	KVI, Groningen, The Netherlands, nasser@kvi.nl
Hiroyuki Kamada	Department of Physics, Faculty of Engineering, Kyushu Institute of Technology, Kitakyushu 804-8550, Japan, kamada@mns.kyutech.ac.jp
Neelima Kelkar	Departamento de Fisica, Universidad de los Andes, Bogota, Colombia, nkelkar@uniandes.edu.co
Michael Kohl	MIT-Bates Linear Accelerator Center Middleton, MA 01949 USA., kohlm@mit.edu
Michio Kohno	Physics Division, Kyushu Dental College, Kitakyushu 803-8580, Japan, kohno@kyudent.ac.jp

Takahisa Koike	Advanced Meson Science Laboratory, RIKEN, Wako-shi, Saitama 351-0198, Japan, tkoike@riken.jp
Serge Kox	LPSC Grenoble, 38026 Grenoble Cedex, France, kox@in2p3.fr
Hironori Kuboki	Department of Physics, University of Tokyo, Tokyo, Japan, kuboki@nucl.phys.s.u-tokyo.ac.jp
Zhibing Li	Department of Physics, Zhongshan University, Guangzhou, 510275, P.R. China, stslzb@zsu.edu.cn
Nilanga Liyanage	University of Virginia, Charlottesville, VA, USA., nilanga@virginia.edu
Benoit Loiseau	Laboratoire de Physique Nucleaire et de Hautes Energies, Groupe Theorie, Univ. P. and M. Curie, 4 Pl. Jussieu, F-75252 Paris, France, loiseau@in2p3.fr
Frank Maas	Institute for Kernphysik, University Mainz, Germany, maas@kph.uni-mainz.de
Victor Mandelzweig	Racah Institute of Physics, Hebrew University, Jerusalem 91904, Israel, victor@phys.huji.ac.il
Terry Mart	Departemen Fisika, FMIPA, Universitas Indonesia, Depok 16424, Indonesia, mart@ui.edu
Paolo Maria Milazzo	Dipartimento di Fisica and INFN, Trieste, Italy, milazzo@ts.infn.it
Khin Swe Myint	Physics Department, Mandalay University, Union of Myanmar, physics@mu-edu.gov.mm
Junichi Nagata	Hiroshima Kokusai Gakuin University, Hiroshima, Japan, j.nagata@hkg.ac.jp
Furutachi Naoya	Department of Physics, Faculty of Science and Technology, Frontier Reserch Center for Computational Science, Tokyo University of Science, Noda, Chiba 278-8510, Japan, j6205702@ed.noda.tus.ac.jp
Mustafa Nasr	Department of Physics, 7th October University, Misurata - Libya, bayio@yahoo.com
Hidekatsu Nemura	Advanced Meson Science Laboratory, DRI, RIKEN, Wako, 351-0198, Japan, nemura@riken.jp
Marek Nowakowski	Departamento de Fisica, Universidad de los Andes, Bogota, Colombia, mnowakos@uniandes.edu.co

Magnus Ogren	Mathematical Physics, Lund Institute of Technology, SE-22100 Lund, Sweden, magnus.ogren@matfys.lth.se
Makoto Oka	Department of Physics, H27, Tokyo Institute of Technology, Meguro, Tokyo 152-8551, Japan, oka@th.phys.titech.ac.jp
Allena Opper	Department of Physics and Astro Edwards Accel Lab, Ohio University Athens, OH 45701 USA., opper@ohiou.edu
Shinsho Oryu	Department of Physics, Tokyo University of Science, 2641 Yamazaki, Noda, Chiba 278-8510, Japan, oryu@ph.noda.tus.ac.jp
Chanxay Outhayavong	Department of Physics, National University of Laos, DongDok Campus, P.O. Box:7322, Vientiane, Lao PDR., chanxay@yahoo.com
Tae-Sun Park	Korea Institute for Advanced Study, 207-43 Cheongnyangni 2-dong, Dongdaemun-gu, Seoul 130-722, Korea, tspark@kias.re.kr
Charles F. Perdrisat	The College of William and Mary, Williamsburg, VA 23187, perdrisa@jlab.org
Lucas Platter	Helmholtz-Institut fur Strahlen- und Kernphysik (Theorie), Universitat Bonn, Nuss allee 14-16, D-53115 Bonn, Germany, platter@itkp.uni-bonn.de
Zhongzhou Ren	Department of Physics, Nanjing University, Nanjing 210008, China, zren@nju.edu.cn
Manoel Roberto Robilotta	Instituto de Fisica, Universidade de Sao Paulo, C.P. 66318, 05315-970, Sao Paulo, SP, Brazil, robilotta@if.usp.br
Kenshi Sagara	Department of Physics, Kyushu University, Hakozaki, Fukuoka 812-8581 Japan, sagara@kutl.kyushu-u.ac.jp
Arunava Saha	Jefferson Lab, 12000 Jefferson Avenue, Newport News, VA 23606, USA., saha@jlab.org
Hideyuki Sakai	Department of Physics, University of Tokyo, Bunkyo, Tokyo 113-0033, Japan, sakai@phys.s.u-tokyo.ac.jp
Peter U. Sauer	Institut fur Theoretische Physik, Universitat Hannover, D-30167 Hannover, Germany, sauer@itp.uni-hannover.de
Kamal Seth	Northwestern University, Evanston, IL 60208, USA, kseth@northwestern.edu
Youhei Shimizu	Research Center for Nuclear Physics (RCNP) Osaka University, Ibaraki, Osaka 567-0047, Japan, yshimizu@rcnp.osaka-u.ac.jp

William Snow	Indiana University and Indiana University Cyclotron Facility, Bloomington, Indiana 47408, USA., snow@iucf.indiana.edu
Horst Stoecker	Institut fuer Theoretische Physik (ITP), J. W. Goethe Universitaet Frankfurt, Germany, stoecker@uni-frankfurt.de
Anto Sulaksono	Departemen Fisika FMIPA Universitas Indonesia, anto@fisika.ui.ac.id
Nurgali Takibayev	Kazakh National Pedagogic University, Almaty, Kazakhstan, teta@nursat.kz
Anthony Thomas	Jefferson Laboratory, 12000 Jefferson Ave, Newport News VA 23606 USA, awthomas@jlab.org
Lauro Tomio	Instituto de Fisica Teorica, Universidade Estadual Paulista, 01405-900 Sao Paulo, Brazil, tomio@ift.unesp.br
Eizo Uzu	Tokyo University of Science, Noda, Chiba, Japan, j-uzu@ed.noda.tus.ac.jp
Thiraphat Vilaithong	Department of Physics, Faculty of Science, Chiang Mai University, Thailand
Guorong Wang	Department of Physics, Nanjing University, China
Takashi Watanabe	Department of Physics, Faculty of Science and Technology, Tokyo University of Science, Noda, Chiba 278-8510, Japan, j6201706@ed.noda.tus.ac.jp
Tornow Werner	Department of Physics, Duke University, Durham, NC 27708 USA., tornow@tunl.duke.edu
Thiengsamone Xouansuandao	Department of Physics, National University of Laos, DongDok Campus, P.O. Box:7322, Vientiane, Lao PDR., thieng_x@yahoo.com
Masahiro Yamaguchi	Science Research Center, Hosei University, 2-17-1, Fujimi, Chiyoda-ku, Tokyo 102, Japan, yamagu@rcnp.osaka-u.ac.jp
Nobuhiro Yamanaka	Department of Physics, University of Tokyo, Tokyo 113-0033, Japan, yam@riken.jp
Chen Ning Yang	Tsinghua University, Beijing 100084, China
Shin Nan Yang	Department of Physics and National Center for Theoretical Sciences at Taipei, National Taiwan University, Taipei 10617, Taiwan, snyang@phys.ntu.edu.tw
Yongle Yu	Mathematical Physics, Lund Institute of Technology, SE-22100 Lund, Sweden, yongle.yu@matfys.lth.se

Bing-song Zou	Institute of High Energy Physics, CAS, P.O.Box 918 (4), Beijing 00049, China, zoubs@ihep.ac.cn
Pornpana Boonma	Department of Physics, Faculty of Science, Thaksin University, Song-Khla 90000, Thailand, pornpanaja@hotmail.com
Chinorat Kobdaj	School of Physics, Suranaree University of Technology, Nakhon Ratchasima 30000, Thailand, kobdaj@sut.ac.th
Prapan Manyum	School of Physics, Suranaree University of Technology, Nakhon Ratchasima 30000, Thailand, pmanyum@sut.ac.th
Prasart Suebka	School of Physics, Suranaree University of Technology, Nakhon Ratchasima 30000, Thailand, prasart@sut.ac.th
Jessada Tanthanuch	School of Mathematics, Suranaree University of Technology, Nakhon Ratchasima 30000, Thailand, jessada@math.sut.ac.th
Worasit Uchai	School of Physics, Suranaree University of Technology, Nakhon Ratchasima 30000, Thailand, uchai@sut.ac.th
Yupeng Yan	School of Physics, Suranaree University of Technology, Nakhon Ratchasima 30000, Thailand, yupeng@sut.ac.th
Moragote Buddhakala	School of Physics, Suranaree University of Technology, Nakhon Ratchasima 30000, Thailand, morakot@rit.ac.th
Sommai Changkian	School of Physics, Suranaree University of Technology, Nakhon Ratchasima 30000, Thailand, csommai@bunga.pn.psu.ac.th
Noppadon Deethae	School of Physics, Suranaree University of Technology, Nakhon Ratchasima 30000, Thailand
Sirichok Jungthawan	School of Physics, Suranaree University of Technology, Nakhon Ratchasima 30000, Thailand, jsirichok@icqmail.com
Kritsada Kittimanapun	School of Physics, Suranaree University of Technology, Nakhon Ratchasima 30000, Thailand, oou_kritsada@hotmail.com
Kanchana Limboonsong	School of Physics, Suranaree University of Technology, Nakhon Ratchasima 30000, Thailand, puna_opal@hotmail.com

Ayut Limphirat	School of Physics, Suranaree University of Technology, Nakhon Ratchasima 30000, Thailand, ayut@physics2.sut.ac.th
Sutassana Naphattaloong	School of Physics, Suranaree University of Technology, Nakhon Ratchasima 30000, Thailand, kkuphysics@yahoo.com
Chakrit Nualchimplee	School of Physics, Suranaree University of Technology, Nakhon Ratchasima 30000, Thailand, nchakrit@hotmail.com
Chalump Oonariya	School of Physics, Suranaree University of Technology, Nakhon Ratchasima 30000, Thailand, chalump@hotmail.com
Wanwisa Pattanasiriwisawa	School of Physics, Suranaree University of Technology, Nakhon Ratchasima 30000, Thailand, w_patta@yahoo.com
Suppiya Siranan	School of Physics, Suranaree University of Technology, Nakhon Ratchasima 30000, Thailand, suppiya007@yahoo.com
Seckson Sukkhasena	School of Physics, Suranaree University of Technology, Nakhon Ratchasima 30000, Thailand, secksons@yahoo.com
Nattapong Yongram	School of Physics, Suranaree University of Technology, Nakhon Ratchasima 30000, Thailand, nattsoccer@yahoo.com

AUTHOR INDEX